PIPELINE DESIGN AND INSTALLATION

Proceedings of the International Conference

Sponsored by the Pipeline Planning Committee of the Pipeline Division of the American Society of Civil Engineers

in cooperation with the:

Nevada Section, ASCE
South Nevada Branch, ASCE

Las Vegas, Nevada
March 25-27, 1990

Edited by Kenneth K. Kienow

Published by the
American Society of Civil Engineers
345 East 47th Street
New York, New York 10017-2398

ABSTRACT

The proceedings of the International Conference on Pipeline Design and Installation, held on March 25-27, 1990 in Las Vegas, Nevada, presents information on all aspects of pipeline design and installation. It covers such broad topics as pipeline materials; product testing, evaluation, and research; design methods; and case histories of pipelines. Specifically, the book discusses such subjects as pipeline rehabilitation, value control, corrosion protection, load testing, and life cycle cost analysis. Therefore, these papers provide the engineer with a valuable resource on this basic topic of pipeline design and installation.

Library of Congress Cataloging-in-Publication Data

Pipeline design and installation: proceedings of the international conference/sponsored by the Pipeline Planning Committee of the Pipeline Division of the American Society of Civil Engineers: in cooperation with the Nevada Section, ASCE [and] South Nevada Branch, ASCE: Las Vegas, Nevada, March 25-27, 1990; edited by Kenneth K. Kienow.

p. cm.

Proceedings of the International Conference on Pipeline Design and Installation.

Includes bibliographical references.

ISBN 0-87262-749-7

1. Pipe lines—Design and construction—Congresses. I. Kienow. Kenneth K. II. American Society of Civil Engineers. Committee on Pipeline Planning. III. American Society of Civil Engineers. Nevada Section. IV. American Society of Civil Engineers. South Nevada Branch. V. International Conference on Pipeline Design and Installation (1990: Las Vegas, Nev.)

TJ930.P553 1990 90-32114

621.8'672—dc20 CIP

Library of Congress Catalog Card No: 90-32114
ISBN 0-87262-749-7
Manufactured in the United States of America.

FOREWORD

The Conference Committee is pleased to present these proceedings on behalf of the Pipeline Division, the Nevada Section, and the Southern Nevada Branch of the American Society of Civil Engineers for the International Conference on Pipeline Design and Installation.

This Conference is the sixth in a series of ASCE Pipeline Division Specialty Conferences begun in 1978. The Conference focus is on pipeline materials; product testing, evaluation, and research; design methods, installation effects; and case histories/performance of pipelines. Specific topics cover a broad range of subjects including performance of ductile iron, steel, plastics, and concrete pipelines, cost effective evaluation techniques, repair, and rehabilitation methods.

Seventy submitted abstracts were carefully screened by the Abstract Review Committee, and each of the papers selected for the Conference has had a minimum of two positive peer reviews. All of the papers are eligible for discussion in the ASCE Journal of Transportation Engineering, and are also eligible for ASCE Awards.

On behalf of the Committee,

Kenneth K. Kienow, P.E.
Editor

Conference Committee:
Kenneth K. Kienow, General Conference Chairman
Kent Olson, Chairman, Local Arrangements Committee
Jey K. Jeyapalan, Chairman, Editorial Committee
Sol Koplowitz, Chairman, Publicity Committee
B. J. Schrock, Chairman, Finance Committee
Malcolm N. Stephens, Chairman, Abstract Review Committee
Bruce A. Bennett
Marshall Marriott, Jr.
Robert L. Meinzer
Ron Metzger
Larry Petroff
Mark B. Pickell
William F. Quinn
Samaan Ladkany
Tony Baez, ASCE Staff
Angela Cappiello, ASCE Staff
Shiela Menaker, ASCE Staff

CONTENTS

ADVANCES IN PIPELINE MATERIALS AND DESIGN IN EUROPE AND NORTH AMERICA

By

Jey K. Jeyapalan, M. ASCE [1]

Abstract

This paper presents an overview of various developments in materials technology and design procedures in the Western Hemisphere over the last 20 years. Some projections for anticipated changes in the industry for the next 10 years are discussed. In addition, worthwhile research directions for the next decade are outlined. Wherever possible, objective comparisons of design and construction practices among technologically competitive countries in Europe and North America are given.

Introduction

Pipelines transport water, storm run-off, process wastes, petroleum, natural gas, slurries, solids, sewage, steam, and other fluids either by gravity or under pressure. Pipelines are also used as utility corridors. The loadings on these pipelines vary considerably depending on the intended use and those usually considered are from soils, differential movement, seismicity, frost, temperature changes, pressure, vaccum, transients, construction, handling and shipping, vibrations, blasting, bends, restraints, weak native soils, wave, self-weight, impact, and live loads. The materials used by engineers and specification writers for pipeline construction could be broadly classified into metals, plastics, concrete, clay, and composites. The last 20 years have seen numerous developments in pipeline materials, design procedures, and standards in Europe and North America, and the primary purpose of this paper is to provide an overview of these significant changes for the benefit of those in teaching, research, and practice of pipeline engineering.

Pipeline Materials

Materials technology has evolved the most among all aspects of pipeline engineering in the last 20 years. Many new materials and products appeared on the market, and an equal number of products and trade names disappeared. In the metals group of materials, structural plate long span culverts were introduced along with evolving shapes and sizes for corrugations. This industry also introduced spiral rib corrugated steel pipe for drainage applications. Numerous coatings from synthetic to natural have also been tried by the metals pipe industry to improve flow characteristics and corrosion potential. For pressure piping which transports gas or petroleum, many alloys and higher grades of steel also have been

1 Vice President, Hardy BBT, Inc., 4172 148th Avenue NE, Redmond, Washington, USA 98052; Ph. 206-869-7070; Fax 206-869-9160

introduced during the last 20 years. The ductile iron pipe industry has developed various wrappings to protect the pipe from corrosion. The developments on both sides of the Atlantic in the metal pipe industry have kept pace with one another.

The concrete pipe industry in North America and Europe has seen numerous developments in manufacturing processes, automation, and use of robotics to some extent. The developments in the area of hardware in Europe have been ahead of North America in some cases. Some Nordic countries and West Germany have also introduced admixtures such as silica fume in concrete pipe construction for better strengths and sulfide corrosion resistance. The use of admixtures in North America is rather minimal compared to European practice, and even in instances where admixtures are used in North America, this is usually fly ash as a cement replacement.

The plastic pipe industry brought numerous developments to the consumer during this period. Resins such as polyvinyl chloride, polybutylene, polyethylene, high density polyethylene, and acrylonitrile butadine styrene entered the industry in smooth wall configurations. Plastics followed the trend of the corrugated metal pipe industry when the metal pipe competed against cast iron, clay, and concrete 30 to 50 years ago. The plastics competed primarily against clay pipe in small sizes initially, and having displaced clay from a good portion of the small size market in North America, began to increase its bore sizes to be able to capture the market shared by concrete, ductile iron, clay, and corrugated metals. The costly smooth-walled construction was in the industry's way while it was trying to accomplish this mission. And then came the advanced extrusion and gluing processes to include cheap fillers, recycled materials, ribs, cores, and profiles of various shapes and sizes to produce pipe rather cheaply in comparison to concrete, clay, or ductile iron. By the mid-eighties, the stiffness of the large diameter plastic pipe dropped to a level requiring either special installation and handling techniques which are rather new to the contractors, or the contractors had trouble installing these low stiffness products. When field performance fell short of the requirements set by the consumer, claims were filed aginst all parties as in any pipe materials. Being new, evolving, and somewhat experimental, bad plastic pipe products settled claims where it was obviously the product which was at fault, and in cases of not-so-obvious causes, the initial blame fell on the contractor or the engineer for either installing the pipe improperly or for not designing the system adequately. In the gas and oil industries, the plastics have penetrated the market rather substantially in the last 20 years and have given good service to the consumer particularly for distribution systems. It is important to recognize that most plastic pipe developments originated in Europe in one form or the other and were tried in the European markets initially. These materials having felt too much resistance in Europe from the consumer and traditional materials such as concrete and clay, eventually found their market niche predominantly in North America. This perhaps explains the use of concrete and clay as primary pipe materials in public works projects in Europe even today. It is very significant to note that even in cases where a European license is used to manufacture a pipe product in North America, the pipe is normally traded under a different name and made with substantially different specifications. Thus, the design engineer needs to do enough research to determine the differences between the parent product in Europe and its spin-off in North America.

The composites have also gained considerable momentum over the years using primarily glass fibers either in the filamant winding process or centrifugally cast process in North America and Europe. The early pipe products made in North America with sand matrix suffered signicantly due to poor quality control, inadequate design procedures and longitudinal strength, and dissappeared from the marketplace in a cloud of litigation, briefly in the mid eighties.

However, improved technology developed in Europe has brought this type of pipe material back to the consumer recently on both sides of the Atlantic.

The clay pipe industry, having been the oldest in perfecting its product, did not offer anything new in its material to the consumer, with the exception of making the pipe for jacking purposes. The joints have also been improved during the last 20 years in North America and Europe.

The cast iron industry transformed into the ductile iron pipe industry by the use of magnesium to have finer microstructure. The industry still provides pipe mostly in the form of pressure piping for water mains.

Pipeline Design

This topic received considerable attention during the last 20 years from all segments of the industry. It is neccessary to group the pipe materials based on their behavior in relation to the surrounding soil; therefore, the remaining discussion in this section will be divided into three catergories. These are rigid, semirigid, and flexible.

Rigid Pipe Design

The rigid pipe is designed by comparing its load carrying capacity in the installed condition to the load applied to ensure that the pipe does not fail in flexural tension . The 20 year period began in North America with the 70 year old Marston load theory, giving the loads and the old bedding factors ranging from 1.1 for Class D to 2.2 for Class A or S. The concrete and clay pipe industries in North America still rely very heavily on these theories. However, much research by these two industries using the finite element method, and field test sections during the last 20 years are beginning to yield good results. A notable achievement is the use of SPIDA by the concrete pipe industry to save cost on the use of steel reinforcement and wall thickness in some cases. The clay pipe industry is contemplating some increases in bedding factors somewhat in line with the British clay pipe industry, which has raised its values rather substantially during the last year. The British clay pipe industry has gone even further to consider discouraging engineers from using the most expensive bedding system of Class S by setting its new bedding factor at 2.5, the same as that for Class B. The Russian rigid pipe industry has used bedding factors as high as 2.9 for Class D and 3.2 for Class S for some time even with loads lower than Marston loads and a factor of safety of only 0.9 in comparison to 1.0 used in West Germany and 1.5 in North America. Needless to say that the rigid pipe materials in Europe use thinner walls, particularly in large diameters, than their counterparts in North America. Furthermore, in Europe these pipes are designed not only for circumferential behavior but also for longitudinal beam behavior. In summary, the rigid pipe industry in North America, in spite of a good supply of new high-tech design tools in its toolbox, lags severely in their introduction into everyday practice, very conservative in the choice of design loads and bedding factors in comparison to most European counterparts.

Semirigid Pipe Design

The semirigid pipe is designed to resist tension caused by internal pressure, and deflection due to pressure and external soil and live loads. The semirigid pipe design is still being done using Spangler's equation, and this industry has made no substantial efforts in North America to change its well-accepted structural design procedures during the last 20 years.

Flexible Pipe Design

The flexible pipe design ensures that the following modes of failure are avoided:

- o material degradation and environmental stress cracking
- o seam separation due to excessive ring compression forces
- o wall crushing due to excessive stresses
- o buckling due to excessive external pressure or internal vacuum
- o excessive deflection leading to leaky joints
- o excessive flexural and compressive or tensile strains leading to yield

The flexible pipe industry composed of corrugated metals, plastics, and some composites still uses Spangler's deflection equation, Watkin's strain equation, and Luscher's buckling equation for designs in North America with the exception of metal culverts. With these culverts, the development of Soil-Culvert Interaction design method by Duncan et al. (1985) is a significant milestone. The Spangler's deflection received the most criticism from numerous researchers and practitioners during the last 20 years in its application to plastic pipe. The reasons for this equation being not reliable for plastic pipe design are as follows:

- o it was developed for corrugated metal pipe
- o it is based on limited lab test data
- o the soil modulus used from Howard's (1977) table is incorrect
- o material behavior is different
- o pipe stiffness is much lower than those of Spangler's test pipes
- o vertical deflection is not equal to horizontal deflection
- o contruction-induced deflection controls

The above and additional reasons make Watkin's strain and Luscher's buckling equations unreliable for plastic pipe design. These reasons were ignored by most plastic pipe manufacturers and designers during the last 20 years. Significant soil box tests, however, were funded by the very same plastic pipe producers to obtain a better solution to the problem, and no changes to their design procedures have been initiated by this industry to date in North America. Some of the significant changes proposed for improving the design tools for plastic pipe are by Jeyapalan et al. (1982-89), Galili (1981), Molin (1985), Howard (1981), Katona (1988), Kienow and Prevost (1983-89), Chambers (1987), Janson (1985), Carlstrom et al. (1985-87) and Leonhardt (1989). In Europe, the ATV A-127 standard has brought numerous changes to plastic pipe design philosophy.

Pipe stiffness property of a plastic pipe became an important issue during the last 10 years despite the claim by plastic pipe suppliers that this property never controlled field perfromance. It is important to note that several new terms to define this property, such as stiffness factor, pipe stiffness factor, pipe class, ring stiffness constant, flexibility factor, and pipe series, were also introduced during the last 20 years, and this added to the confusion among design engineers and consumers. It is necessary to note that the pipe stiffness property is mostly affected by the pipe material behavior, which is controlled by the following parameters:

- o strain
- o temperature
- o stress history
- o creep characteristics
- o manufacturing flaws
- o use of fillers or additives
- o type of resin

The soil modulus used in the plastic pipe design equations comes from those tabulated by Howard. It is important to recognize the following facts regarding the E ' values in this table:

- o An insignificant number of data points used in establishing E' came from plastic pipe installations.
- o The E' value depends only on the type of soil and its compaction density in this table but not on cover depth.
- o The deepest depth of cover of the plastic pipe installations considered was 7 feet.
- o Although a deflection tolerance range is given in the lower half of the table, most plastic pipe producers excluded this information from their design brochures.

Even from basic soil-structure interaction principles an experienced engineer would tend to expect the pipe-soil stiffness ratio to have some effect on the choice of E' value for design. It is rather interesting to note that the range of E' used in much stiffer ductile iron pipe design ranges from 150 to 700 psi, while softer plastic pipe designs are based on E' in the range of 1000 to 3000 psi. Furthermore, even as early as during 1965-70, the steel pipe industry used a maximum E' value of as low as 700 psi in their design. The recent works by Jeyapalan and Jaramillo (1989) and Duncan and Hartley (1987) provide better E' values for plastic pipe design and steel pipe design, respectively.

The choice of appropriate trench width for flexible pipe construction also was a major issue during the last 10 years, and the guidelines in AASHTO, ASTM, Bureau of Reclamation specifications have been somewhat contradictory to date. Particularly in weak native soils, such as swelling clays, silty clays, clayey silts, silts, sensitive clays, loose sand, muck, peat, and highly compressive materials, the choice of trench width becomes a critical question. The factors which should determine the trench width are as follows:

- o stiffness of the pipe
- o stiffness of the native soil
- o stiffness of the trenchfill
- o location of the water table
- o loading
- o depth of cover

Besides the use of appropriate finite element analyses to account for all of these factors in a rational manner, there are no other reliable techniques available at this time in North America. In Europe, engineers are using Leonhardt's factor to adjust the E' value as a function of the trench width and the ratio of E' of the native soil to that of the backfill. Soil migration from native ground to trenchfill also became an issue in flexible pipe installations. In order to ensure that the fines from the native soil do not migrate into the voids of the trenchfill, the design engineer needs to check the filter criteria. It is important to recognize the differences in bedding systems between rigid, semirigid, and flexible pipe materials, and the specifications prepared for construction and cost estimates should reflect these. The parameter pipe-soil stiffness ratio defined by Jeyapalan, Saleira, and Boldon (1987) provides a good yardstick for the contractor and the engineer to know in advance whether the pipe chosen for the bedding condition is stiff enough to withstand construction-induced loads and the associated deflections. Even ASTM standard D-2321 is proposed to have some provisions for construction-induced deflections in its latest set of revisions. It is important to recognize that the engineer has to account for construction-induced deflection in addition to time-dependent and load-induced deflections in the estimate of the field deflection of the flexible pipe. The construction-induced deflection would depend on the following factors:

- o stiffness of the pipe
- o quality control of the pipe material and wall construction
- o stiffness of the soil embedment
- o stiffness of the native soil
- o type of shoring systems used
- o dewatering techniques used
- o skill of the contractor
- o handling of the pipe
- o compaction control
- o level of inspection provided in the field

Some efforts were made by Molin (1985), Howard (1981), and NCHRP (1980) to give a simple relationship for estimating the construction-induced deflection in plastic pipe, but there are no definitive quidelines available to the engineer to date.

Despite these major difficulties with the choice of E' during the last 20 years, soil properties played an important role in pipeline design. The engineers required boring logs, basic soils tests for soil classification, and compaction tests in the laboratory and in the field for ensuring that important parameters were taken into consideration in underground pipeline design. In computer models, those who relied on finite element analyses, used hyperbolic model soil parameters developed by Duncan and Chang (1971) for representing soil behavior in soil-pipe interaction computations. In Europe, the extent to which finite element analyses were used as design tools lagged considerably behind North America during the last 20 years. The state-of-the-art of finite element analyses today provides engineers with tools which could model the following aspects in pipeline design:

- o nonlinear soil behavior
- o layered construction and sequential loading
- o time-dependent long term behavior
- o buckling behavior
- o behavior of concrete, clay, metal, plastic, and composites
- o ring behavior
- o longitudinal beam behavior
- o three-dimensional behavior

The research by Duncan, Heger, Jeyapalan, and Katona has contributed significantly to the above developments in North America. The common misconception among practicing engineers is that the use of finite element analysis method of design would require more data on the project. The type of data neccessary to generate finite element based designs are as follows:

- o geometry of the trench and the pipeline
- o material properties for the pipeline
- o properties of native soil and trench backfill
  - o unit weight
  - o compaction density or blow count or unconfined strength
  - o soil classification
  - o index properties, grain size curve
- o loadings
- o location of the water table

It is interesting to note that these data are neccesary even for conventional design procedures. The types of results obtained from these analyses are quite extensive and these provide a comprehensive picture of the behavior of the pipeline under a variety of loading conditions such as:

- o deflected shape of the pipe along the length of the line
- o deflected shape of the pipe in chosen cross-sections
- o stresses and strains along the length of the pipe wall
- o stresses and strains along the circumferential direction
- o stresses and movements in the soil around the pipe

The primary advantages of the finite element based designs are as follows:

- o it is cost-effective
- o can analyze several cases of loading or trench conditions rather quickly
- o results are very easy to interpret
- o can simulate abnormal construction conditions anticipated
- o can analyze field conditions accurately

In summary, the flexible pipe design practice as disseminated by the plastic pipe industry in North America is unconservative in comparison to that in Europe, and a good example is the recent introduction of the interim AASHTO Bridge Standard-Section 18 on Soil-Thermoplastic Interaction Systems.

Expert Systems In Pipe Design and Selection

It is important to recognize that the soil-pipe system is the outcome of the engineering, and therefore, all significant parameters need to be taken into consideration in the design process. During the last few years, some research efforts have focused on pipe durability evaluation and life cycle cost studies. Significant efforts in North America were by the National Corrugated Steel Pipe Association (NSCPA), U.S. Army Engineer Waterways Experiment Station, and the Ohio Department of Transportation. An interested reader could refer Hurd et al. (1982), Potter (1988), and the reports by NCSPA on corrosion condition surveys for further details. The engineers in Europe and North America also questioned the validity of choosing pipe material based solely on first time material cost. This led to the inclusion of parameters such as the following in the decision making process:

- o Corrosion Potential
- o Abrasion Resistance
- o Expected Service Life
- o Construction Cost
- o Structural Performance
- o Quality Assurance
- o Flow Characteristics
- o Maintenance
- o Future Construction
- o Joint Types
- o Type of Warranty
- o Track Record
- o Characteristics of the Fluid Transported

An intercepter project in the city of Fontana, California was the first to include the above parameters in a systematic fashion into the decision, and the design engineers relied heavily on hand-computations to tabulate the composite scores for various pipe materials which were considered for that project. Some efforts have been made in North America to implement this into an expert system for the selection of sanitary sewer pipe materials, and an example of such an effort is that by Jeyapalan and Kallas (1988). A similar approach was attempted by the Water Research Center (1986) in the U.K. for watermains. Similar expert systems could be developed for pipelines in the following applications to arrive at the most suitable pipe material among those listed below as a function of important selection criteria:

- Potable Water Pipelines
  - ductile iron
  - pvc
  - hdpe
  - steel
  - prestressed concrete
  - rpm
- Gas Pipelines
  - steel
  - cast iron
  - ductile iron
  - hdpe
  - pvc
  - composite
- Power Transmission Pipelines
  - steel
  - prestressed concrete
  - thermosetting resin
  - rpm
  - ductile iron
  - concrete
- Irrigation Pipelines
  - ductile iron
  - pvc
  - hdpe
  - steel
  - rpm
  - concrete
- Storm Drainage Pipelines
  - concrete
  - reinforced concrete
  - cast-in-place concrete
  - corrugated steel/aluminum
  - pvc
  - hdpe
- Oil Pipelines
  - steel
  - hdpe
  - rtr

Standards

The most significant difference between European standards and North American standards is that in Europe the standards are more concept-oriented than pipe product-oriented. The standards in Europe are far more rigid than those in North America. The standard writing bodies receive much higher level of input from the consumers than from the producers of one type of pipe. In North America, ASTM, AWWA, AASHTO, ANSI and CSA standard committees, subcommittees, and task committees and groups were very busy for the last 20 years. The results of their hardwork is listed below with the counterpart German standards:

Asbestos Cement Pipe

ASTM C-296, 428, 508, 644, 663, 668, and AWWA C-400, 402
DIN 19800, 19831, 19841, 19830, 19850

Concrete Pipe

ASTM C-14, 76, 118, 361, 412, 443, 444, 478, 505, 506, 507, 655, 789, 850
DIN 1045, 4032, 4035, 19695

Concrete Pressure Pipe

AWWA C-300, 301, 302, 303

Clay Pipe

ASTM C-700
DIN 1230

Cast Iron Pipe

ANSI A 21.6, 21.4, and AWWA C-115
DIN 19519, 19690, 19500, 19501, 19502, 19503, 19507, 19509, 19511, 19513

Ductile Iron Pipe

ASTM A-716, 746, ANSI A 21.4, and AWWA C-110, 111, 115, 150, 151
DIN 19691, 19692, 28600

Corrugated Steel and Structural Plate Pipe

AASHTO M-36, 245, 36, 167, and ASTM A-760, 762, 761

Asphalt Coated Corrugated Steel Pipe

AASHTO M-190

Invert Paved Steel Sewer Pipe

AASHTO M-190

Fully Lined Steel Sewer Pipe

AASHTO M-190, 36

Welded Steel Pipe

AWWA C-200
DIN 1626, 1629

Aluminum Pressure Pipe

ASTM B-241, 345, 547

ABS Solid Wall Pipe

ASTM D-2751

PB Solid Wall Pipe

ASTM D-809. AWWA C-902

PE Solid Wall Pipe

ASTM D-2239, 3035, F-678, 714, and AWWA C-901
DIN 19537, 19533, 16932, 8075, 8074

PVC Solid Wall Pipe

ASTM D-1785, 2241, 2466, 2467, 3033, 3034, 2672, 2740, F-679, 789, and AWWA C-900
DIN 19534, 8061, 8062, 19532,

PE Corrugated and Profile Wall Pipe

ASTM F-892, 894

PVC Corrugated and Profile Wall Pipe

ASTM F-794, 949

ABS Composite Pipe

ASTM D-2680

Thermosetting Resin Pipe

ASTM D-2310, 2517, 2996, 2997, 3262, 3517, 3754, and AWWA C-950

In addition to the above, there are installation standards for some of the plastic pipe materials in North America. These standards are however minimal in comparison to those available in West Germany. For example, the DIN standards common to all pipe materials are as follows: DIN 4033, 1054, 1986, 4124, 18125, 18127, 18196, 1053, 1055, 1072, 2402, 4022, 4030, and 4051. These standards deal with topics such as construction practices, design loads, design procedures, dewatering requirements, material quality control, site inspection, field monitoring, etc. It is important to recognize that many of these standards are continually undergoing revisions by member ballots.

The most significant standard in Europe has been the West German ATV A-127, which provides a unified approach to the structural design of all pipe materials used for drainage and sewer construction. A detailed evaluation of this standard is given in Jeyapalan and Ben Hamida (1988). It is important to recognize that this standard was developed by a joint effort among representatives from all of the competing pipe materials and the consumers in West Germany. Unlike a design engineer in North America, the West German engineer has more design tools and fewer material standards to deal with in pipeline design. Furthermore, the design procedures used by the engineer are given in ATV A-127 for all pipe materials and not in each pipe material standard or pipe supplier's brochure.

Future of the Industry

It is anticipated that the rigid and the semirigid pipe industries will have a tendency in the future to cut down wall thickness, and lower the steel requirement by updating their design procedures. The design procedures expected to be proposed will either lower design loads or increase the strength capacity of the bedded pipe. These industries also will attempt to bring tools based on the finite element method to offer to the consumer tailormade designs at a significant cost saving. These industries will also continue to update their manufacturing capabilities to include more automation. These materials will also offer new product lines for pipeline rehabilitation applications. It is also expected that the concrete pipe industry is likely to include more and innovative admixtures to cut down the cost and time of production of precast concrete pipe. There is also the potential for the industry to introduce a lighter pipe using materials such as the gasified or autoclaved concrete.

The flexible pipe industry will continue to increase its pipe sizes. It will introduce wider applications of various product lines. The industry will be under intense pressure from various public funding agencies to update its design tools for the consumer to be able to have better products with longer life. Nearly 50 % of the plastics sorted from regular household garbage are recycled in Europe compared to only 10 % in North America. The research programs in progress at the present time in North America for producing various products from recycled plastics will provide better technology for the plastic pipe industry to produce improved pipes from recycled materials. This would assist the industry in its attempts to enter new market niches. Many new composite materials from the spin-offs of aerospace materials research efforts are also likely to have an impact on the pipe industry.

Conclusions

The following conclusions can be drawn based on the observations made during the last 20 years:

1. All pipe materials have their share in the marketplace, and significant progress has taken place in the area of materials technology on both sides of the Atlantic. Although plastics and composites are new on the block, these materials have made substantial gains in the marketplace. Many engineers have chosen these materials because these materials come with the appeal of newness and an evolutionary process in technology. There are, however, some serious concerns about the design technology used by this industry. This industry needs to address these issues.

2. The rigid pipe industry in North America needs to look at the issue of lack of newness rather seriously. The design practice in Europe is not as conservative as it is in North America, and this has resulted in clay and concrete still being the dominant pipe materials for public works projects in Europe.

3, The primary aspects in which the practices in Europe and North America compare well are as follows: types of applications, materials, manufacturing processes, trench design, and loadings in some cases. The differences are in design methods, installation and inspection procedures, writing standards, and handling of disputes.

4. In summary, in Europe the rigid pipe materials are thinner and lighter, and the flexible pipe materials are thicker and stiffer. It is important for the pipe industry and the designers in North America to ask the key questions: WHY and HOW?

Acknowledgements

It is a pleasure to thank my former students Felix Osequeda, Mark Gardner, Bruce Boldon, Jim Hutchison, Bachar Kallas, Hassan Ben Hamida, Kanji Kubo, Peggy Jaramillo, Mike Thiyagaram, Naiyi Jiang, Beckry Abdel-Magid, and Wesley Saleira for the stimulating intellectual partnership over the past decade. The research which led to this paper was partially funded by numerous public agencies, consulting firms, and pipe manufacturers, and their support is acknowledged.

References

1. AASHTO (1988). "Interim Specifications Bridges - Section 18 on Soil Thermoplastic Pipe Interaction Systems," Washington, D.C.

2. ATV (1984). "Specifications A127 for the Structural Design of Wastewater Drains and Sewers," West Germany.

3. Bakht, B. (1985). "Live Load Response of a Soil-Steel Structure with a Relieving Slab," TRB Record 1008.

4. Bennett, B.A. (1988). "Proceedings of the ASCE conference on Pipeline Infrastructure," Boston.

5. Bland, C.E.G. and Shepard, K. J. (1985). "Investigations into the Structural Performance of Clay Pipes, " Proceedings of the ASCE Conference on Advances in Underground Pipeline Engineering in Madison.

6. Carlstrom, B., et.al. (1985). "Successful Use of Centrifugally Cast Glassfiber Reinforced Plastic Pressure Pipes for Public Works in Sweden," Proceedings of the ASCE Conference on Advances in Underground Pipeline Engineering in Madison.

7. Carlstrom, B., Leonhart G., and Schneider, H. (1987). " Static Calculation of Underground Glass Fiber Reinforced Plastic Pipe," Proceedings of the International Conference on Pipeline Construction, Hamburg.

8. Chambers, R.E. (1987). "Corrugated PE Pipe: Structural Design and Installation in Storm Water Applications," 20th Annual Water Resources Conference, University of Minnesota.

9. DIN (1988). "Specifications 4033, 1054, 1986, 4124, 18125, 18127, 18196, 4060, 4062, 19543, 4279, 19800, 19830, 19850, 1045, 4032, 4034, 4035, 19695, 19519, 19690, 19534, and 19537."

10. Duncan, J.M., et.al. (1985). "Design of Corrugated Metal Box Culverts," TRB Record 1008.

11. Fidjestol, P. (1987). "Silica - Concrete in Sewage Installations," Proceedings of the International Conference on Pipeline Construction, Hamburg.

12. Galili, N. and Shmulevich, I. (1981). "A Refined Elastic Model for Soil-Pipe Interaction," Proceedings of the ASCE Conference on Underground Plastic Pipe, New Orleans.

13. Greatorex, C.B. (1981). "The Relationship Between the Stiffness of a GRP Pipe and Its Performance when Installed," Proceedings of the ASCE conference on Underground Plastic Pipe, New Orleans.

14. Hartley, J.D., and Duncan, J.M. (1987). "E' and its Variation with Depth," Journal of Transportation Engineering, ASCE, Vol. 113, No. 5, September.

15. Heger, F.J. (1988). "Earth Load and Pressure Distribution and New Installation Criteria for Buried Concrete Pipe," Proceedings of the ASCE conference on Pipeline Infrastructure, Boston.

16. Howard, A.K. (1977). "Modulus of Soil Reaction Values for Buried Flexible Pipe," Journal of the Geotechnical Engineering Division, ASCE, Vol. 103, No. 1.

17. Howard, A.K. (1981). "The USBR Equation for Predicting Flexible Pipe Deflection," Proceedings of the ASCE Conference on Underground Plastic Pipe, New Orleans.

18. Hurd, J.O., et.al. (1982). "Ohio Culvert Durability Study," Report of ODOT/L&D/82-1.

19. Janson. L. (1985). "The Relative Strain on a Design Criterion for Buried PVC Gravity Sewer Pipes," Proceedings of the ASCE Conference on Advances in Underground Pipeline Engineering, Madison.

20. Jeyapalan, J.K. , and Osequeda, F. (1982). "Finite Element Analyses of Plastic Pipe-Soil Interaction," Research Report from the Texas Transportation Institute.

21. Jeyapalan, J.K. (1982). "Stress Reduction in Flexible Box Culverts Due to Overlays of Geofabric," Proceedings of the Second International Conference on Geotextiles, pp.814-823.

22. Jeyapalan, J.K. (1982). "Soil-Structure Interaction Analyses of Plastic Pipes," Paper presented at the ASCE Fall Convention in New Orleans, Preprint 82-511.

23. Jeyapalan, J.K., and Duncan, J.M. (1982). "Deflection of Flexible Culverts Due to Backfill Compaction," Transportation Research Record No. 878. pp. 10-17.

24. Jeyapalan, J.K., and Gardner, M.P. (1983). "Preliminary Analyses of the Behavior of Reinforced Concrete Box Culverts Under Backfill Loads," Report of the Texas Transportation Institute to SDHPT and FHWA.

25. Jeyapalan, J.K., and Gardner, M.P. (1983). "The Behavior of Reinforced Concrete Box Culverts Under Symmetrical and Unsymmetrical Live Loads," Report to the Texas Highway Department and FHWA.

26. Jeyapalan, J.K. (1983). "Geofabric Stabilization of Soft Backfill Materials for Plastic Sewer Pipe Installations," Proceedings of the ASCE Specialty Conference on Pipelines in Adverse Environments, II , San Diego, California, pp. 188-198.

27. Jeyapalan, J.K., and Abdelmagid, A.M. (1984). Importance of Relative Pipe-Soil Stiffness in Plastic Pipe Design," Special Technical Publication entitled Pipeline Materials and Design of the Journal of the Transportation Engineering of ASCE, pp. 66-92.

28. Jeyapalan, J.K. (1985). "Soil-Structure Interaction Behavior of Reinforced Box Culverts," Preprint of the 64th Transportation Research Board Meeting.

29. Jeyapalan, J.K. (1985). "Advances in Underground Pipeline Engineering," Proceedings of the International ASCE Speciality Conference, Madison Wisconsin, 610 pages.

30. Jeyapalan, J.K. , and Jiang, N. (1986). "Load Reduction Factors for Buried Clay Pipes," Paper in the pipeline division of the Journal of Transportation Engineering of ASCE, pp. 236-249.

31. Jeyapalan, J.K., and Boldon, B.A. (1986). "Effects of Embedment on the Selection and Performance of Rigid and Flexible Pipes," in the Journal of Transportation Engineering of ASCE, Vol. 112. No. 5, pp. 507 to 524.

32. Jeyapalan,J.K., and Abdel-Magid, A.M. (1987). "Considerations of

Longitudinal Stresses and Strains in the Design of Buried Reinforced Plastic Mortar Pipes," Journal of Transportation Engineering of ASCE, Vol. 113, No. 3, pp. 315-331.

33. Jeyapalan, J.K., and Ethiyajeevakaruna, W.S., and Boldon, B.A. (1987). "Behavior and Design of Buried Very Flexible Plastic Pipes," Journal of Transportation Engineering ASCE, Vol. 114, No. 6, pp. 1-16.

34. Jeyapalan, J.K., and Jiang, N. (1987). "New Bedding Factors for Vitrified Clay Pipes," to appear in Transportation Research Record.

35. Jeyapalan, J.K., and Hampton T. (1987). "Design of Siphon Penstocks for Low Head Hydros," Paper in the Proceedings of the ASCE Conference WATERPOWER 87, Portland, Oregon.

36. Jeyapalan, J.K., and Saleira, W.E., and Jiang, N. (1987). "New Design Procedures ffor Vitrified Clay Pipes," Report to the National Clay Pipe Institute, Lake Geneva, Wisconsin.

37. Jeyapalan, J.K., and Schrock, B.J. (1987). "Underground Pipelines Design," Seminar Notes published during April to September.

38. Jeyapalan, J.K., and Saleira, W.E. (1987). "Finite Element Analyses of the Response of HDPE Pipes Under Backfill and Traffic Loads Using Micro-computers," paper in the proceedings of the 10th Fuel Gas Plastic Pipe Symposium, New Orleans.

39. Jeyapalan, J.K.,and Saleira, W.E., and Kienow, K.K. (1987). "Deflections and Strains in HDPE Pipes Under Traffic Loads," Paper in the Proceedings of the International Conference on Pipeline Construction, Hamburg, Germany.

40. Jeyapalan, J.K., and Saleira, W.E. (1988). "Design of Underground Pipelines and Storage Tanks Using Personal Computers," Paper in the Proceedings of the ASCE Specialty conference on Computing in Civil Engineering, Alexandria, Virgina.

42. Jeyapalan, J.K., and Saleira, W.E. (1988). "Soil-Structure Interaction Analyses on Micro-computers," Proceedings of the 6th International Conference on Numerical Methods in Geomechanics, Innsbruck, Austria.

43. Jeyapalan, J.K., and Saleira, W.E., and Abdel Magid, B. (1988). "An Investigation of Field Performance of Reinforced Plastic Mortar Pipes," Paper in the proceedings of the ASCE Speciality Conference on Pipeline Infrastructure, pp. 148-157, Boston.

44. Jeyapalan, J.K., and Kallas, B. (1988). "Priority Integrated Pipe Evaluation," Paper in the proceedings of the ASCE Specialty Conference on Pipeline Infrastructure, pp. 1-9, Boston.

45. Jeyapalan, J.K. (1988). "Pipeline Materials and Design," Notes for the Seminar sponsored by the American Society of Civil Engineers, Boston.

46. Jeyapalan, J.K., and Ben Hamida, H. (1988). "Comparison of Marston to German Design Methods for Clay Pipes," Journal of Transportation Engineering of ASCE, Vol. 114, No. 3. pp. 420-434.

47. Jeyapalan, J.K., and Saleira, W. E. (1988). "New Design Procedures for Very

Flexible Plastic Pipes, " Research Report.

48. Jeyapalan, J.K., and Thiyagaram, M., Saleira, W., and Jaramillo, C. (1988). "Forces on the Jennings Randolph Dam Penstock Inside the Outlet Tunnel and Design Considerations," Proceedings of the International Conference on Small Hydro, Toronto, Canada.

49. Jeyapalan, J.K. , and Thiyagaram, M, Saleira, W.E., and Magid, B.A. (1989). "Analysis and Design of RPM and Other Composite Underground Pipelines," Journal of Transportation Engineering of ASCE, pp. 219-231.

50. Jeyapalan, J.K. , and Jaramillo, C.A., and Saleira, W.E. (1989). "Design Considerations for a Penstock Located in an Outlet Tunnel at the Jennings Randolph Dam in West Virginia," Proceedings of the ASCE Conference WATERPOWER 89, Niagra Falls, Canada.

51. Jeyapalan, J.K., and Jaramillo, C.A. (1989). "Hydraulic Model Tests of Flow Around Penstock Suspended Inside the Tunnel for Design of a Small Hdyro," Proceedings of the ASCE Conference WATERPOWER 89, Niagra Falls, Canada.

52. Jeyapalan, J.K., and Ben Hamida, H. (1987) "Evaluation of Current Design Methods for Rigid Pipes in Trenches," Research Report.

53. Jeyapalan, J.K., and Jaramillo, P.A. (1989) . "Review of Field Deflections of Plastic Pipes Under Soil Loads," Research Report.

54. Katona, M.G. (1987). "Allowable Fill Heights for Corrugated PE Pipe," TRB paper.

55. Kienow, K.K., and Prevost, R.C. (1983). "Stiff Soils - An Adverse Environment for Low Stiffness Pipes," Proceedings of ASCE Conference on Pipelines in Adverse Environments, II, San Diego.

56. Kienow, K.K., and Prevost, R.C. (1989). "Pipe/Soil Stiffness Ratio Effect on Flexible Pipe Buckling Threshold," Journal of Transportation Engineering, Vol. 115, No. 2. March.

57. Kurdziel, J.M. and McGrath, T.J. (1989). "SPIDA Method for Reinforced Concrete Pipe Design," ASCE Session at New Orleans Convention.

58. Lang, D.C. and Howard, A.K. (1985) "Buried Fiberglass Pipe Response to Field Installation Methodss," Proceedings of the ASCE Conference on Advances in Underground Pipeline Engineering, Madison.

59. Leonards, G.A., et al. (1985). "Predicting Performance of Buried Metal Conduits," TRB Record 1008.

60. Leonhardt, G. (1989). "Statische Berechnung von Abwasser-Kanalen und leitungen aus Kunststoffen nach dem ATV-Arbeitsblatt A 127," Proceedings of the SKZ Conference in Munchen.

61. Molin, J. (1985). "Long -Term Deflection of Buried Plastic Sewer Pipes," Proceedings of the ASCE Conference on Advances in Underground Pipeline Engineering, Madison.

62. Moore, I. D. (1985). "The Elastic Stability of Buried Tubes, " Ph.D. Dissertation at the University of Sydney, Australia.

63. NCHRP (1980). "Plastic Pipe for Subsurface Drainage of Transportation Facilities," Research Report, Washington, D.C.

64. Pickell, M. (1983). "Proceedings of the ASCE Conference on Pipelines in Adverse Environments, II," San Diego.

65. Potter, J.C. (1988). "Life Cycle Cost for Drainage Structures," Report GL-88-2, U.S. Army Corps of Engineers, Vicksburg.

66. Prevost, R.C. and Kienow, K.K. (1985). "Design of Non-Pressure Very Flexible Pipe," Proceedings of the ASCE Conference on Advances in Underground Pipeline Engineering, Madison.

67. Prevost, R.C. (1989). "Alternate Flexible Pipe Structural Designs, Part I and II," Pipes & Pipelines International, Vol. 34, Nos. 5 & 6, October and December.

68. Prevost, R.C. (1990). " Flexible Pipe Design Revisited, " Proceedings of the ASCE Conference on Pipeline Design and Installation, Las Vegas, March.

69. Schrock, B. J. (1981). "Proceedings of the ASCE Conference on Underground Plastic Pipe, " New Orleans.

70. Taprogge, R.H. (1981). "Large Diameter Polyethylene Profile Wall Pipes in Sewer Application," Proceedings of the ASCE Conference on Underground Plastic Pipe, New Orleans.

71. Tohda, J., et al. (1985). "Earth Pressure on Underground Rigid Pipe in Centrifuge," Proceedings of the ASCE Conference on Advances in Underground Pipeline Engineering, Madison.

72. WAA Sewers and Water Mains Committee. (1988). "Revised Bedding Factors for Vitrified Clay Drains and Sewers," England.

73. Water Research Center (1985). "A Review of UK and Overseas Research on Bedding Factors and Other Aspects of Rigid Sewer Design," Report, Swindon, England.

74. Water Research Center (1986). "Selection of Pipe Materials for Water Transmission," Report, Swindon, England.

EARTH PRESSURE ACTING ON BURIED FLEXIBLE PIPES IN CENTRIFUGED MODELS

J. Tohda[1], H. Yoshimura[2], T. Morimoto[3] and H. Seki[4]

ABSTRACT

Distributions of both normal and tangential earth pressures acting on three flexible model pipes having different wall thicknesses were successfully measured in centrifuge model tests. The accuracy in the measurement was checked by comparing bending strains measured along the pipe circumference with bending strains calculated under the measured earth pressure conditions, together with by the external force equilibrium. The effects of both density of model sand ground and pipe installation type on the measured earth pressures were also discussed.

INTRODUCTION

Earth pressure acting on buried flexible pipes is indispensable for the interpretation of their actual behavior in the ground and for the development of design. However, reliable data have not been obtained yet, within the authors' knowledge, because of the difficulty of measurement due to the pipe deformation, although a lot of efforts have been made. One of the authors, J. Tohda, has extensively investigated the earth pressure on buried rigid pipes during this decade [Tohda. 1989]. Through this research, he developed a technique for measuring accurately earth pressure on buried rigid pipes in centrifuged models [Tohda et al. 1985, 1986a]. A similar technique was applied to flexible model pipes successfully.

---

1 Lecturer, 2 Post graduate student, Civil Eng. Dept., Osaka City University, 3-3-138, Sugimoto, sumiyoshi-ku, Osaka, 558, Japan.
3 Technical manager, 4 Engineer, Pipe and Industrial Equipment Eng. Center, Sekisui Chemical Co. Ltd., Nojiri Ritto-cho, Kurita-gun, Shiga Prefecture, Japan.

The authors first examined the cause of error in the earth pressure measurement on buried flexible pipes. On the basis of the examination, three model pipes having different wall thicknesses were designed. Centrifuge model tests were carried out using these pipes to measure the followings: 1) circumferential distribution of both normal and tangential earth pressures on the pipe, 2) circumferential distribution of bending strains in the pipe wall, and 3) vertical deflection of the pipe. The accuracy in measurement of these three items was validated by comparing them with the following calculated results: distribution of bending strains under the measured earth pressure conditions, and vertical deflection under the measured bending strain conditions. The measured earth pressures were also supported by the satisfaction of the external force equilibrium.

This paper details the measuring technique of the earth pressures, and presents the measured results for the three pipes of different flexibility.

MODEL PIPES

The three model pipes, made of aluminium alloy, are named as Rigid-pipe, Medium-pipe and Flexible-pipe. These pipes and their dimensions are shown in Fig.1 and Table 1. The outer diameters D of the pipes are 90 mm, except for Medium-pipe, the length being 148 mm. The ring in their middle portion is divided into either 20 or 40 segments fixed on the supporting beams by means of small prismatic bars; these three parts (segment, supporting beam, and prismatic bar) compose a load cell to measure the earth pressures. The pipes are also longitudinally divided into two parts (the pipe-parts); they are fixed to each other by means of the supporting beams. The end of each pipe-part, on which the supporting beams are fixed, is named as the pipe-end (cf. Fig.2). The surface of the pipe-parts is finished smooth. For one of the pipe-parts, strain gauges are pasted at 12 points on the outer and inner surface to measure the bending strains, and a deflection gauge of cantilever type is installed inside.

Fig.2 shows details of the load cells. For each load cell mounted in left-half section of the pipe, wire strain gauges are pasted on both sides of the prismatic bars to measure their axial and bending strains, through which normal and tangential earth pressures acting on the segment surface are obtained. Right-half section of the pipe has load cells measuring only axial strains to obtain the normal earth pressures. The width 2.5 mm and thickness 1 mm of the prismatic bars are determined,

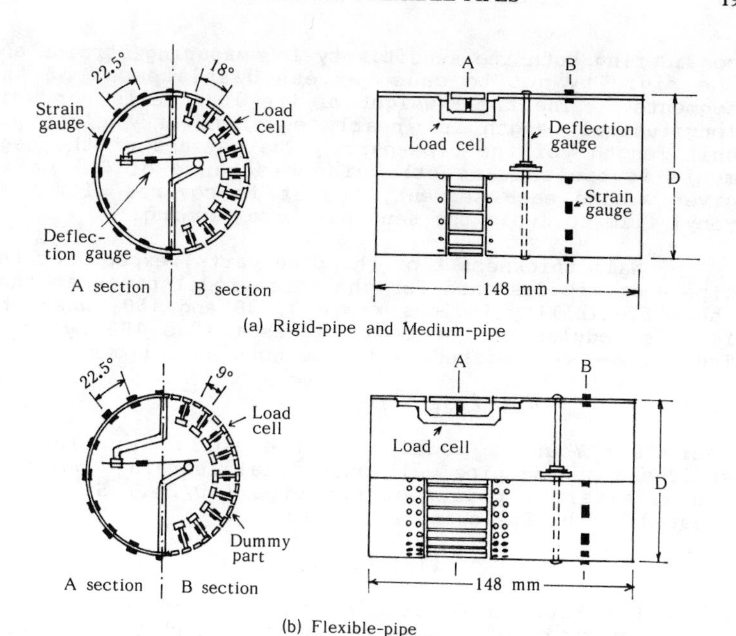

(a) Rigid-pipe and Medium-pipe

(b) Flexible-pipe

Fig. 1 Three Model Pipes

Table 1 Dimensions of Model Pipes

Pipe	Outer diameter D (mm)	Wall thickness t (mm)	E_p (kgf/cm^2)	ν_p	S_f ($kgf \cdot cm^2/cm$)	κ	Weight of pipe-part (gf/cm)
Rigid-pipe	90	3.5			2967	3	35
Medium-pipe	86	1.5	740,000	0.33	234	30	20
Flexible-pipe	90	0.95			59	150	31

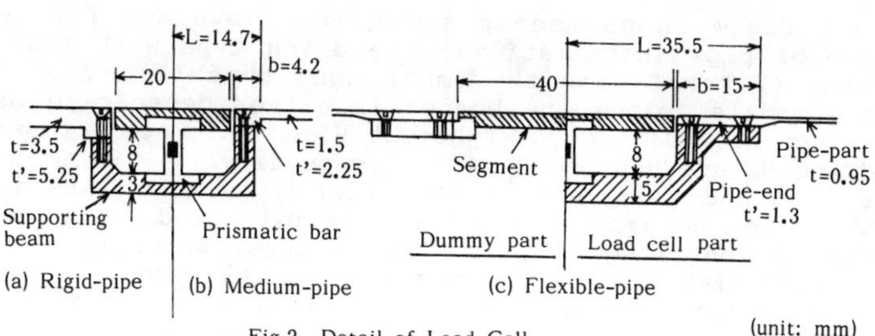

(a) Rigid-pipe (b) Medium-pipe (c) Flexible-pipe

(unit: mm)

Fig.2 Detail of Load Cell

considering both the sensitivity in measuring strains and the rigidity not to cause excess displacement of the segments. The total weight of the load cells per unit longitudinal length is nearly equal to the weight per unit length of the pipe-part. The surface of the segments is finished smooth. The clearance of 0.5 mm is given around each segment, and it is covered with thin vinyl film to avoid the sand particle jamming.

Wall thickness t of the pipe-parts, except for the pipe-ends, is designed for the three model pipes so that their flexibility indices κ are 3, 30 and 150, when the Young's modulus of the soil is assumed as 100 kgf/cm². The index κ was defined by the authors as follows:

$$\kappa = E/(S_f/a^3)$$

where E : Young's modulus of the soil, S_f: flexural stiffness of the pipe wall under plain strain condition, and a: external radius of the pipe (=D/2). S_f is expressed by the following equation:

$$S_f = E_p t^3/\{12(1-\nu_p^2)\}$$

where E_p: Young's modulus of the pipe material, ν_p: Poisson's ratio of the pipe material, and t: pipe wall thickness. Fig.3 shows the range of κ for the three model pipes and different pipe materials. This figure is obtained through the modification of the figure originated by J.E.Gumble [1983], based on the authors' elastic analysis for two types of pipe installations (ditch type with sheet-piling and embankment type) [Tohda et al. 1986b]. The κ value of 3 for Rigid-pipe lies near the border classifying the rigid and the medium pipe behavior. The κ values of 30 and 150 for Medium-pipe and Flexible-pipe lie within the classified range of the medium and flexible pipe behavior, respectively.

MODELS AND TEST EQUIPMENT

Fig.4 shows models and testing systems for two types of pipe installations: the ditch type with sheet-piling (Ditch-S) and the embankment type (Embk.). In these models, pipes are buried in either dense or loose dry-sand ground; the cover height H and thickness of sand bedding H_b are H=D and H_b=4/9D, respectively. In Ditch-S model, a pair of model sheet-piles 5 mm thick are installed in the ground, and they are extracted simultaneously in centrifuge flight through steel wires by means of a pair of hydraulic cylinders. In Embk. model, a pair of rigid steel plates with a smooth surface are installed as the lateral boundaries of the model ground. The

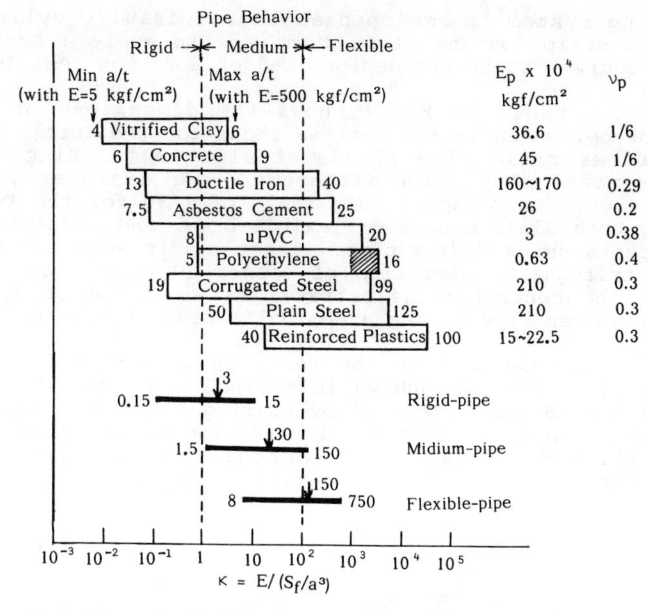

Fig.3 Range of κ for Three Model Pipes and Different Pipe Materials

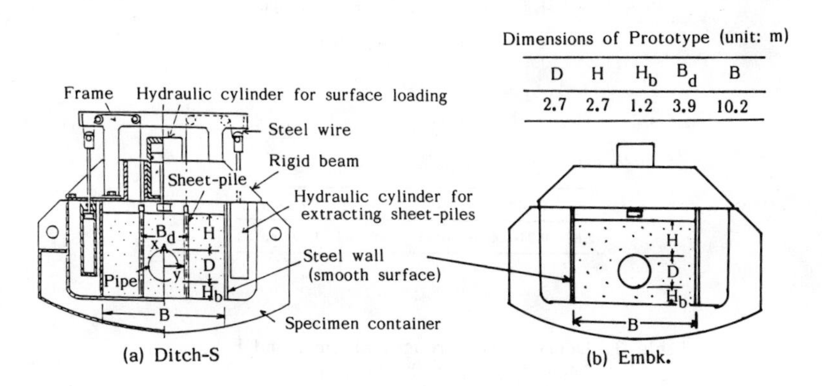

Dimensions of Prototype (unit: m)

D	H	H_b	B_d	B
2.7	2.7	1.2	3.9	10.2

Fig.4 Models for Two Types of Pipe Installations

testing system is equipped another hydraulic cylinder with a strip loading plate 2 cm wide to apply a surface load equivalent to the design load of a 20 ton vehicle.

The table in Fig.4 shows the dimensions of the prototype, which is the same as those for the tests using the other rigid pipe [Tohda et al. 1990]. Since the three model pipes have different outer diameters, the models were constructed in a scale of 1/30 for the tests using both Rigid-pipe and Flexible-pipe, and 1/31.4 for the tests using Medium-pipe. These models were put into a centrifugal acceleration of either 30 g or 31.4 g by using the centrifuge illustrated in Fig.5, which is of swing bucket type having a nominal radius of 2.56 m.

The sand used in the tests is an air-dried sand, whose properties are shown in Table 2 and Table 3. The model ground was prepared by pouring the sand into the container in the longitudinal direction of the pipe from a small hopper through a circular nozzle of 1 cm diameter.

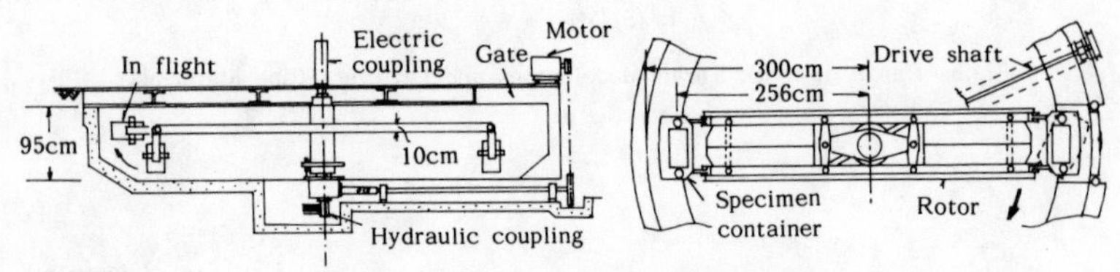

Fig.5 Centrifuge

Table 2 Properties of Sand

G_s	Grain Size	Uc	ρ_{dmax}	ρ_{dmin}
2.65	0.24-1.4mm	1.75	1.58t/m³	1.32t/m³

Table 3 Density and Strength of Sand and Friction Angle ϕ_p against the Pipe Surface

Ground	ρ_d	c_d	ϕ_d	ϕ_p
Dense	1.55 t/m³	0	47°	17°
Loose	1.43	0	36°	16°

EXAMINATION ON EARTH PRESSURE MEASUREMENT

Causes of Error in Measurement

When measuring earth pressure on flexible pipes, various problems occur owing to the pipe deformation. The measuring method employed in this study does not include the difficulties in the traditional method using earth pressure cells, such as pressure concentration on the earth pressure cells. Nevertheless, it involves the following problems:

(a) The earth pressure acting on the load cells are transmitted through the supporting beams to the pipe-parts. This generates non-uniform pipe deformation along the longitudinal axis of the pipe(Fig.6a).

(b) The deformation in the pipe section generates the difference in the curvature between the flexible pipe-parts and rigid segments (Fig.6b). Fig.6b shows that the measured normal earth pressure will be less than the actual one at the pipe top and bottom, and greater at the pipe sides.

(c) The supporting beams are fixed to the pipe-ends. This generates the null slope along the circumferential direction of the pipe-ends at the connection parts of the beam edges; as a result, the sectional deformation of the pipe-ends becomes different from that of the pipe-parts.

Countermeasures for Each Problem

The following countermeasures were employed for each problem:

(a) Non-uniform longitudinal deformation: The thickness of the pipe-ends, t', was increased so that their moment of inertia, I, become L/b times as great as I of the pipe-parts. Since the load intensity acting on the pipe-ends is L/b times as great as that on the pipe-parts, the deformation of the pipe-ends ought to become the same as that of the pipe-parts, when assuming the pipe as an

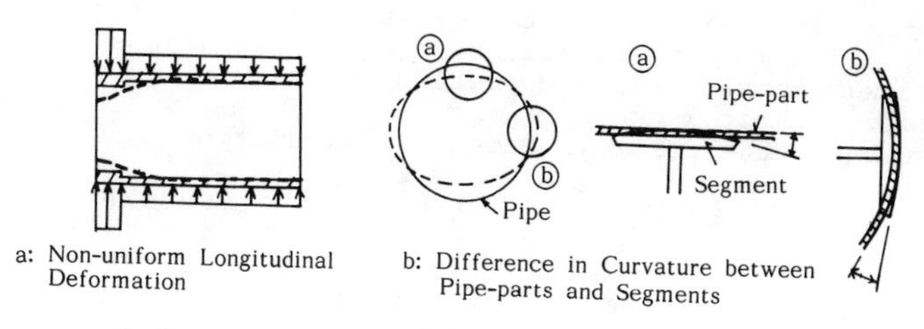

a: Non-uniform Longitudinal Deformation

b: Difference in Curvature between Pipe-parts and Segments

Fig.6 Causes of Error in Earth Pressure Measurement

elastic ring. Actually, the pipe is not an elastic ring and, therefore, non-uniform longitudinal deformation will occur owing to the different wall thicknesses of the pipe-part. However, an elastic analysis revealed that this deformation can be practically ignored.
(b) Difference in the curvature: The countermeasure for this problem is the increase in the division number of the ring to decrease the segment width. The test results showed that both Rigid-pipe and Medium-pipe having 20 segments give satisfactory results because of the small pipe deflection, while the other flexible pipe having the same division pattern does not. Thus, the middle portion of Flexible-pipe was divided into 40 segments, instead of 20.
(c) Sectional deformation at the pipe-ends: The following three countermeasures were employed: 1) the supporting beams were fixed on the pipe-parts by the small screws without bonding, 2) their widths were minimized as far as possible, and 3) their edges were finished aslant.

Confirmation of Accuracy in Measurement

In spite of the above countermeasures, the error in the earth pressure measurement can not still be avoided in the restrict sense. Therefore, the accuracy of the measured data must be confirmed in some way. In this study, the following two methods were employed.

(a) The force equilibrium in the vertical and horizontal directions ($\Sigma V = \Sigma H = 0$) was checked, when the measured earth pressures and the pipe weight were applied as the external forces. In the calculation, distributions of the tangential earth pressures on the right-half of the pipe were assumed to be symmetrically the same as those measured on the left-half of the pipe and, therefore, another force equilibrium of bending moment ($\Sigma M = 0$) is naturally satisfied.
(b) Distribution of bending strains produced on the pipe wall was calculated through the principle of minimum work under the same external load condition; it was compared with the distribution of the measured bending strains.
In these calculations, the distributions of the measured normal and tangential earth pressures were translated into quadratic curves in 10 degree intervals to minimize the error in integral calculations.

In addition, the accuracy in the measurement of both the bending strains and the vertical pipe deflection was confirmed by comparing them with the vertical deflection calculated from the distribution of the measured bending strains through the principle of virtual work.

TEST RESULTS OF DITCH-S MODEL

Fig.7 shows change of the normal earth pressures σ at the pipe top and bottom, measured during Ditch-S tests for the three pipes. This figure refers to the dense ground condition; the loose ground generated a similar earth pressure change. In the sheet-pile extraction process, the following differences in earth pressure among the three pipes are observed:

(a) Rigid-pipe: The earth pressure σ on the pipe top reaches its peak value when the lower ends of the sheet-piles pass through the pipe top (④ in the figure), and then decreases to the value recorded just before the sheet-pile extraction. The pressure σ on the pipe bottom increases markedly when the lower ends of sheet-piles pass through the level of the pipe bottom (② in the figure), and maintains its high value thereafter.

(b) Medium-pipe: The pressure σ on the pipe top once decreases in the early stage of sheet-pile extraction, and then steeply increases to the peak value to decrease afterward. The pressure σ on the pipe bottom at first increases, and then decreases to some extent.

(c) Flexible-pipe: The pressures σ on the pipe top and bottom are always smaller than the corresponding ones of the former two cases, particularly on the pipe bottom.

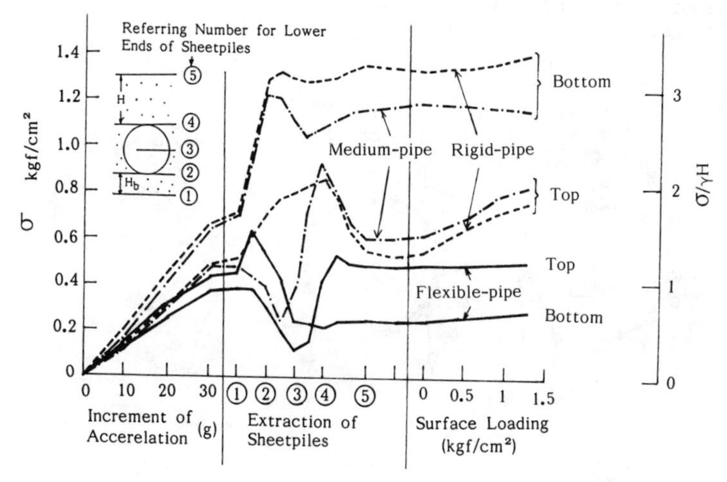

Fig.7 Change of Measured Normal Earth Pressures during Ditch-S Test (Dense Ground)

Fig.8 illustrates distributions of the normal and tangential earth pressures on the three pipes, measured for the dense ground in the three test stages: before, during, and after the sheet-pile extraction. This figure is illustrated in polar coordinates; the compressive normal pressure and downward tangential pressure are taken as positive. The data noted as "During Sheet-pile Extraction" are those when the pressure σ on the top of Rigid-pipe reaches its peak value (④ in Fig.7).

Before the sheet-pile extraction, the greater pipe flexibility generates the more uniform distribution of σ. During and after the sheet-pile extraction, the pressures σ on the pipe top and bottom for Rigid-pipe and Medium-pipe are considerably greater than those on the pipe sides, whereas the pressure σ on Flexible-pipe shows an unique distribution shape utterly different from the former two cases. The tangential pressures τ on the three pipes are always considerably smaller than σ at every measuring points.

Furthermore, all earth pressure distributions shown in Fig.8 have revealed to satisfy the force equilibrium as expected.

Fig.9 shows change of the vertical deflection ΔD of the three pipes, measured during the tests for the dense and loose grounds. The greater the pipe flexibility is, the greater the pipe deflection it generates. The deflections of all pipes remain unchanged after the lower ends of sheet-piles pass through the pipe top level.

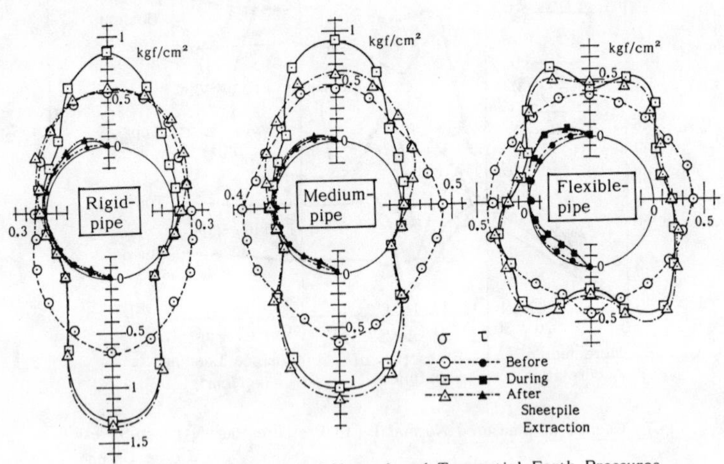

Fig.8 Distribution of Normal and Tangential Earth Pressures on Three Pipes (Ditch-S, Dense Ground)

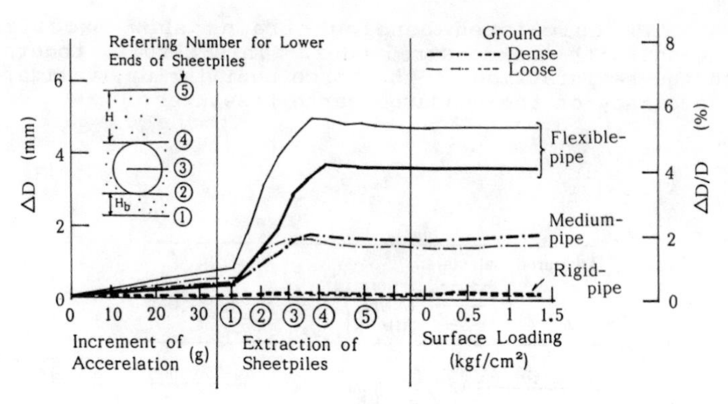

Fig.9 Change of Vertical Pipe Deflection during Ditch-S test

Table 4 shows the measured and calculated vertical deflections of the three pipes during the sheet-pile extraction. The calculated deflections are almost coincident with the measured ones, confirming the accurate measurement of both the bending strains and the pipe deflections.

Fig.10 shows the measured and calculated bending strains of the three pipes for the three test stages; the horizontal axis denotes the degree θ measured from the pipe top. In the figure, the measured results are plotted as the "marks" (⊙, ⊡, △), and the calculated ones are illustrated as the "lines". The measured results show that the greater the pipe flexibility is, the greater the bending strains it generates, whereas the generated bending moments are somewhat less.

Table 4 Measured and Calculated Vertical Deflection of Three Pipes during Sheet-pile Extraction for Ditch-S model (unit: mm)

	Dense Ground			Loose Ground		
	Rigid-pipe	Medium-pipe	Flexible-pipe	Rigid-pipe	Medium-pipe	Flexible-pipe
Measured	0.13	1.74	3.69	0.12	1.59	5.00
Calculated	0.15	1.62	3.96	0.13	1.50	5.45

The calculated bending strains show excellent agreement with the measured ones. This result, together with the satisfaction of the force equilibrium, validates the accuracy of the measured earth pressures.

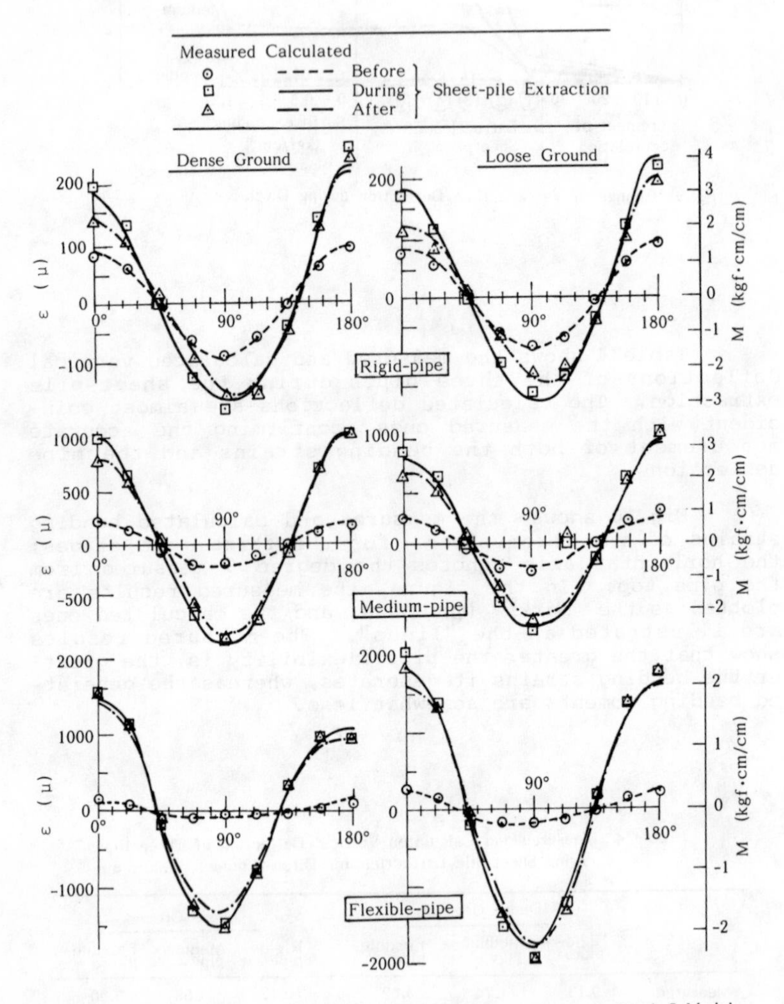

Fig.10 Distribution of Measured and Calculated Bending Strains in Ditch-S Model

TEST RESULTS OF EMBK. MODEL

Fig.11 shows distributions of the earth pressures σ and τ, measured before the surface loading in Embk. model for the dense and loose grounds. With an increase of pipe flexibility, the distribution of σ changes its shape to be uniform. Table 5 and Fig.12 show that the pipe flexibility affects the measured pipe deflection and bending strains in a similar way in Ditch-S model; the calculated results agree well with the measured ones, validating the accuracy of the measurement.

LOAD DISTRIBUTION ON THE THREE PIPES

The measured earth pressures in the whole tests are summarized in Fig.13 in terms of the vertical load p_v, vertical reaction load p_r and horizontal load p_h, which are calculated as follows:

$$p_v \text{ and } p_r = \sigma + \tau \tan\theta, \quad p_h = \sigma - \tau \tan\theta$$

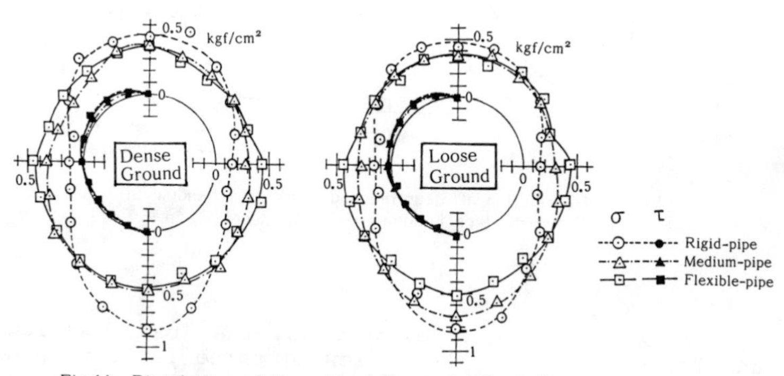

Fig.11 Distribution of Normal and Tangential Earth Pressures on Three Pipes (Embk.)

Table 5 Measured and Calculated Vertical Deflection of Three Pipes before Surface Loading for Embk. Model (unit: mm)

	Dense Ground			Loose Ground		
	Rigid-pipe	Medium-pipe	Flexible-pipe	Rigid-pipe	Medium-pipe	Flexible-pipe
Measured	0.07	0.43	0.74	0.05	0.54	0.82
Calculated	0.09	0.39	0.77	0.07	0.53	0.77

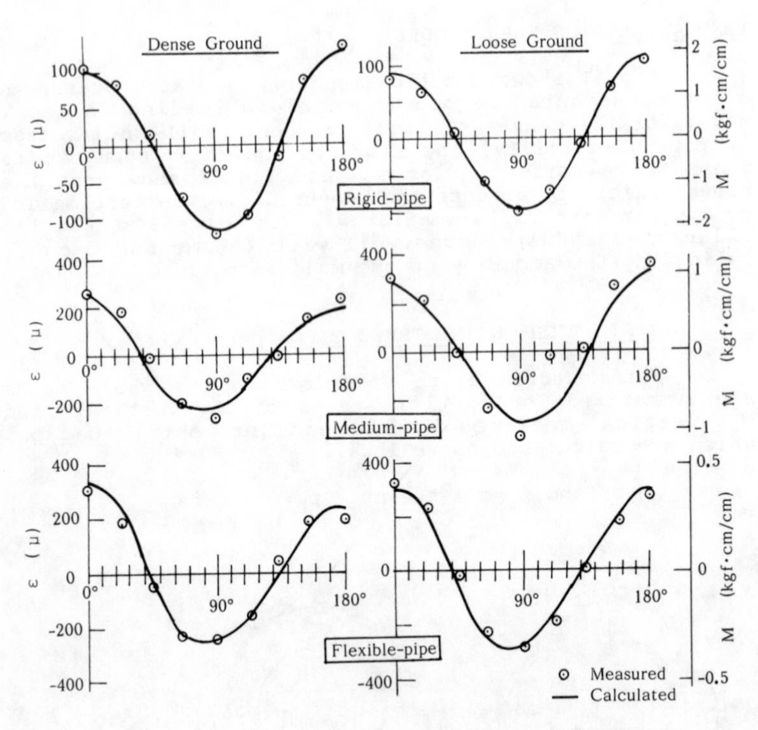

Fig.12 Distribution of Measured and Calculated Bending Strains Before Surface Loading in Embk. Model

This figure shows clearly how the load distribution changes owing to the following three factors: pipe flexibility, type of pipe installation, and density of the sand ground.

CONCLUSIONS

Centrifuge model tests yielded very accurate earth pressures on the three buried pipes of different flexibility. The accuracy of both measured bending strains on the pipe wall and the measured deflection of the pipe was also confirmed. Furthermore, the test results clarified the effects of both pipe installation type and density of sand ground on earth pressure and deformation of buried flexible pipes.

In general, actual behavior of flexible pipes in

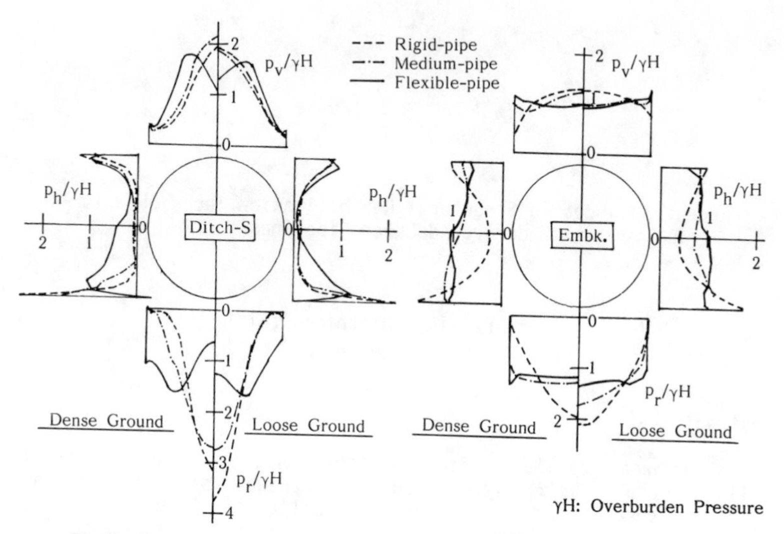

Fig.13 Distribution of Vertical and Horizontal Loads (Ditch-S: During Sheet-pile Extraction, Embk.: Before Surface Loading)

the ground has not been sufficiently clarified, owing to the difficulty of accurate earth pressure measurement. The authors believe that this obstacle will be substantially removed by this study.

REFERENCES

Gumble J.E. 1983. Analysis and design of of buried flexible pipes, PH.D. thesis submitted to Univ. of Surrey.

Tohda J. et al. 1985. Earth pressure on underground rigid pipe in a centrifuge, Proc. of ASCE Int. Conf. on Advances in Underground Pipeline Eng.

Tohda J. et al. 1986a. A study of earth pressure on underground rigid pipes by centrifuged models. Proc. of JSCE. Vol.376/3-6. (in Japanese)

Tohda J. et al. 1986b. A study of earth pressure on underground pipes based on theory of elasticity. Proc. of JSCE. Vol.376/3-6. (in Japanese)

Tohda J. 1989. Earth pressure acting on buried pipes. Proc. of 13th AIT Sympo. on Underground Excavations in Soils and Rocks.

Tohda J. et al. 1990. FE elastic analysis of measured earth pressure on buried rigid pipes in centrifuged models. Proc. of ASCE Int. Conf. on Pipeline Design and Installation.

Recent Advances in Structural Rehabilitation Techniques for Underground Sewer Pipelines

Prof. Dr.-Ing. Dietrich Stein[1]

Abstract

A great challenge for the coming years, if not decades, will be the repair, renovation and renewal of the existing sewerage systems in the Federal Republic of Germany in the interest of ground water protection and their adaptation to the more stringent water protection regulations.

A multitude of methods for rehabilitation are available but also new concepts for the creation of functionally safe sewerage systems which above all are easy to maintain, as described in this article.

1. Introduction

According to estimates, approximately 10 %, perhaps even 20 %, of the public sewers in the Federal Republic of Germany are defective and approx. 30 to 60 billion US $ will be required for repairs by the year 2000 for remedying the damages. These figures do not include house connections i.e. the private areas of sewerage. Their structural condition is virtually unknown. It is assumed today that house connections present an above average defect and risk potential and are the weakest point of the sewage system, requiring considerable investment [3].

[1]Ruhr-Universität Bochum, Universitätsstr. 150, 4630 Bochum, West-Germany

The methods established to correct these faults are [1, 2]:
- repair,
- renovation, and
- renewal.

The essential criteria for their application are the extent of the defects, the necessary drainage capacity, the possibilities for technical modifications, the economic efficiency of such modifications, and finally the decision whether to continue the use of network or network areas (figure 1).

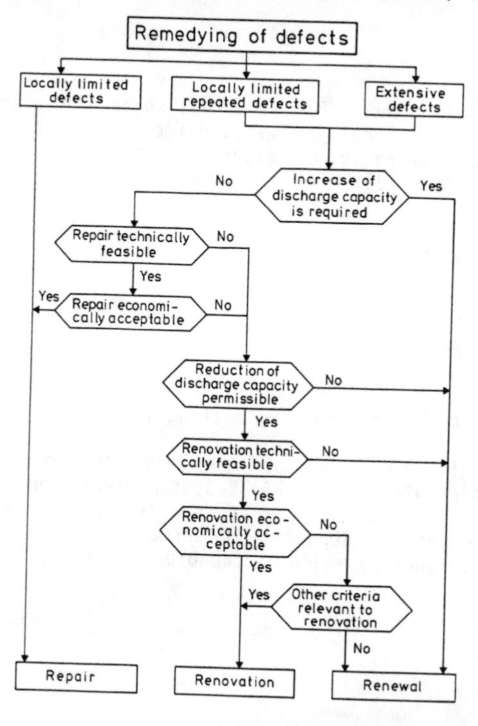

Figure 1.
Decision process for selecting the appropriate solution for remedying defects [1].

Any method for remedying must be aimed at establishing a standard condition which meets the same requirements as those valid for new sewerage systems. This is naturally also applicable to the materials and structural elements used. In order to avoid repetition the causes of any defects should also be eliminated when defects are being remedied.

The following is a summary of the methods themselves and of their fields of application, of relevant new developments in the Federal Republic of Germany, and of the possibilities for developing new sewerage concepts in the course of remedying defects.

2. Repair

The term repair designates methods for re-establishing the standard condition in the case of locally limited defects. In respect of the methods applied internally or externally the following differentiation should be made [2]:

- Repair:
  Elimination of structural defects

- Injection method:
  Sealing and/or solidification of defect areas by forcing in fluid materials under pressure

- Sealing method:
  By-passing of the defect area by applying sealing materials

New developments in the sphere of sewer repairs comprise chiefly robots and manipulators. The KA-TE-System developed in Switzerland has been used for approx. 2 years now in inaccessible sewers > DN 250 in the Federal Republic of Germany. This is a remote-controlled robot (figure 2) which is capable of sealing

Figure 2. Remote-controlled robot KA-TE-System

locally limited defects of various kinds such as cracks and leaking pipe joints by means of crack grouting or injection and of milling off deposits and protruding drainage obstacles.

A robot development of the Hochtief AG, which is at present at the trial stage, is designed for the mechanical and automated clearance and grouting of defective joints in accessible, brick-lined sewers. This development is of particular importance as it enables the repair in a simple and cost saving way of the weak points of corroded brickwork joints in numerous, otherwise proven, brick (-lined, waste water) sewers. Parallel to the machine technology, corrosion-proof gap mortars are being developed and tested by international mortar research teams, in order to obtain solid gap fillings.

The development of the partial hose relining method serves to supporting the efforts aimed at extending the repair methods also in inaccessible sewers. Various versions of this system are now available on the German market.

A hose-shaped fabric, impregnated with epoxy resin is slipped over an inflated packer of an external diameter being about 20 mm smaller than the nominal internal diameter of the defect pipe. The external surface of the pipe is provided with a protecting coat of PE for protection against damages. Equipped like this, the packer is then inflated, pulled to the defect sewer section by using a winch, positioned in this section which is observed by camera, and inflated with hot air. Having terminated the hardening process caused by the hot air, the packer is ventilated and removed.

Thus, a glass fibre-reinforced inliner, flattening out on both sides and effecting an adhesive gluing, remains in the defect area.

The Sanipor method developed in Hungary is a further development in the field of injection methods. It is at present being tested in the Federal Republic of Germany. This method is applied in separate sections. In the initial stage one component of the injection agent is filled into the sewer section to be sealed. Due to the hydrostatic pressure produced it penetrates through the leakages of the sewer into the pores of the surrounding ground and is pumped out after a suitably long retention time. In the second stage, the second component is introduced and reacts with the first component remaining in the cavities to form a watertight gel. After the reaction time, the second component is also pumped out again (figure 3).
In this way the manholes, the sewer section between them and the house connections are sealed at the same time.

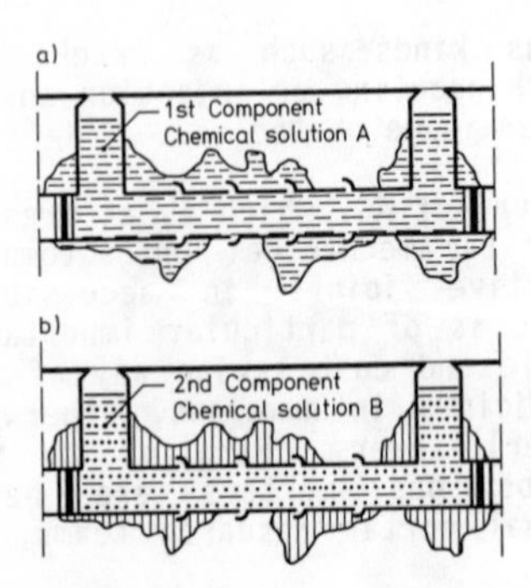

Figure 3.
Sanipor method

replaced by attached drawing

3. Renovation

Renovation includes measures for the restoration of the standard condition of defective sewers through their technical modification and preservation of their substance. These methods, which can only be applied internally and at least by sections (length of a conduit between two manholes and/or special constructions) include [2]:

- Lining method:
  Application of a lining e.g. of cement mortar, reaction resin modified cement mortar, reaction resin mortar on the inner wall. Depending on the method of application, one distinguishes between the pressing out, displacement, spraying on, and centrifugal methods.

- Relining method:
  . Pipeline relining (installation of continuous plastic pipes in the section to be renovated),
  . Long pipe relining (pulling, pushing or drawing-in of individual pipes over excavations),
  . Short pipe relining (pulling, pushing or drawing individual pipes over existing manholes),
  . Winding pipe relining (pushing in of a winding pipe made in the manhole)
  . Hose relining (installation of a felt hose impregnated with hardening materials in the section to be renovated via the manhole and pressing this against the inner wall of the pipe for hardening in situ).

- Assembly method:
  Installation of self-supporting or non-self-supporting lining elements either manually and/or by means of suitable devices on site as a partial or full lining of accessible sewers.

All of these methods result in a greater or lesser reduction of the cross-section and, depending on the circumstances, also in a reduction of the hydraulic capacity.

Whereas in the accessible nominal width range such minor reductions of the cross-section are generally tolerated, they are often not acceptable for hydraulical reasons in inaccessible sewers.

In the Federal Republic of Germany the new developments in the field of renovation methods particularly concern pipeline relining, winding pipe relining and the use of new materials for the assembly methods.

A new development for pipeline relining is the elimination of the normally required pulling-in excavation by the introduction of the pipeline through the manhole into the section to be renovated. This is realized by the compression of the PE-HD pipeline which is heated to approx. 70 °C and the respective 90° turn above and inside of the manhole. Immediately before the introduction of the pipeline into the sewer section, it is reshaped to the original circular cross-section. This method is known under the term "Thermoline" (figure 4).

Figure 4.
Thermoline-System
left: 90° - turn of PE-HD pipeline
right: View on the reshaped liner after 90° - turn inside the manhole

A recent adaptation concerns the spiral liners made from a specially designed PVC-profile by using a winding machine lowered into the manhole.
As usual the liner is formed to a diameter smaller than that of the pipe to be renovated. One through to the next manhole the spiral joint of the tube is made to slip, expanding in size equal to and intimate with the inner diameter of the pipe in need of renewal (figure 5).
This close contact with the existing pipe wall not only enhances the structural strength of the liner, but also ensures that any loss of pipe crossection is kept to an absolute minimum and the need for grouting is elimited.

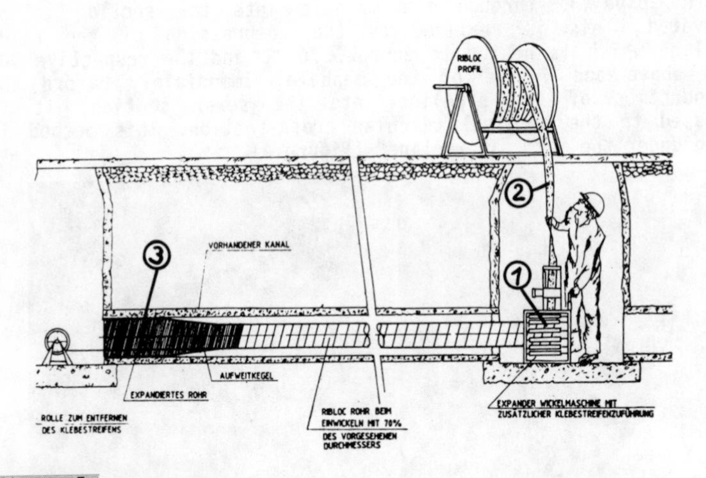

Figure 5.
Expanda Pipe System
(1) Winding machine
(2) PVC-profile
(3) Expanded inliner

In the field of assembly methods, the introduction of glass elements is noteworthy as a partial or full lining in corroded, accessible sewers. The lining consists of a special glass which has been developed for plant construction and which is particularly resistent to shock, impact and bending stresses as well as to thermal strains. The development of this system is not yet completed (figure 6).

Figure 6.
Sewer lining with glass elements

4. Renewal

4.1 General

Sewers which for hydraulical, statical, chemical or also economical reasons can no longer be repaired or renovated must be renewed. Renewal means that new sewers are built to take over the function of the old ones which are placed out of operation. This can be carried out at the same place by exchange (destruction of substance) or at another place (loss of substance).

The following methods can be applied:

- open trench construction method
- trenchless construction method.

The realization of a renewal according to the open trench construction method is basically comparable with the installation of a new sewerage. It is the standard procedure in the Federal Republic of Germany when renewing old sewerage systems. This regularly results in the opening of the roads at intervals and then re-closing them.

When this method is applied in urban areas for the construction of new sewers in the area of the roadways it is increasingly subject to political-ecological constraint, particularly in the case of relatively deep-lying sewers.

The renewal of waste water sewers in urban areas is rendered particularly difficult due to the existing supply network. These often present a considerable obstacle during construction and this is also reflected in the costs. In the case of narrow roads, the obstacle can be such that, for instance, in the course of sewer renewals all other supply lines existing in the cross-section of the road must also be renewed.

4.2 Trenchless Construction Method

When defective pipe systems must be renewed at the same place (in the same line and gradient) according to the closed type of construction the exchange is effected underground, i.e. without excavating an open ditch. The following methods are mainly applied [2, 5]:

- Drifts or tunnels with framework are driven as in the mining industry

- Shield or pipe jacking with accessible cross-section

- Bursting method

- Passing over non-accessible sewers.

New and further developments concern the bursting method as well as passing over non-accessible sewers.

The pipe bursting method is characterized by pulling or pushing a bursting body through the sewer section to be renewed whereby the pipe wall is simultaneously destroyed by static or dynamic forces and the displacement of the fragments into the ground. The new conduit (pipeline or short pipes) is installed immediately behind the bursting body (figure 7).

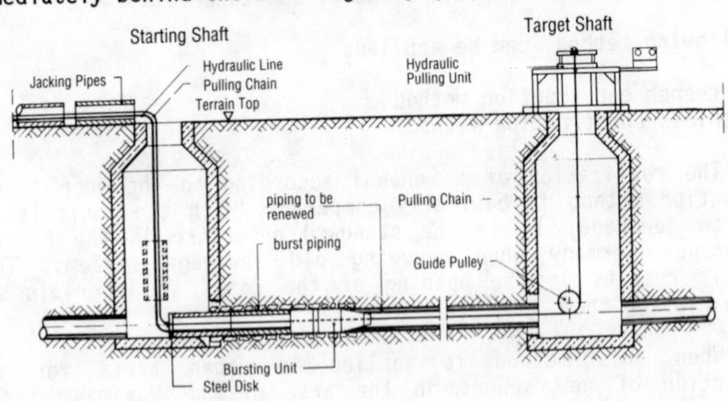

Figure 7. Static Pipe-Bursting System KM-Berstlining combinated with short pipes

New developments for the bursting method mainly concern modifications to the statically operating bursting body, in the use of brittle pipe materials in dynamically operating methods, and in the improvement of the bedding of the renewed sewer.

The modification of the statically operating bursting body mainly concerns the widening mechanism. Modifications to the design prevent the penetration of soil particles and thereby also any blocking of the equipment.

Dampers installed between the dynamically operating bursting body and the product pipe minimize the stress on the pipe during bursting, so that, in contrast with former practice, also brittle materials can be used.

The improvement of the bedding conditions is achieved by carrying along steel bands or a felt hose. Both methods serve to prevent the occurrence of point or line loads via the displaced pipe fragments on the new sewer.

In the case of the latest development, i.e. passing over non-accessible sewers, the sewer to be renewed is re-bored, destroyed and conveyed by means of a unmanned, remote control tube advancing system and at the same time the new sewer is built with the same or, if required, a larger nominal width in the same line (figure 8).

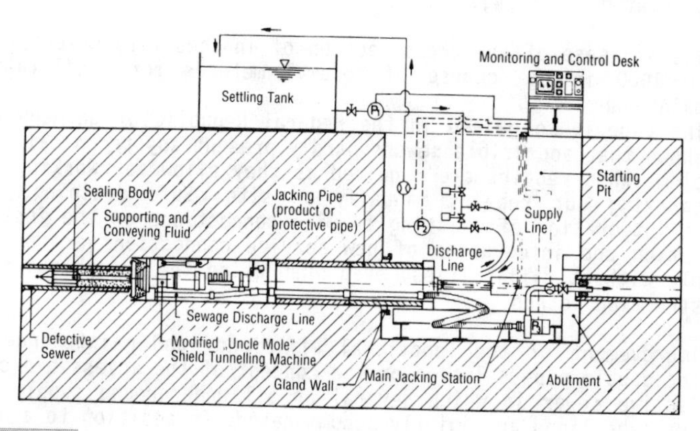

Figure 8.
Renewal of nonman-size sewer ducts by pipe replacing method [5]

The machine systems which are on the market are designed for sections having a length of approx. 50 m to 100 m and a pipe bottom depth of 3 m to 7 m. Depending on the type of machine, the ground water over pipe bottom does not require any special water pumping measures.

This method has already been applied successfully with different advancing systems in the Federal Republic of Germany. Advancing lengths of up to 100 m were reached from one starting pit and the advancing speeds averaged 12 m/8 h [5].

5. Possibilities for Realizing New Sewerage Concepts during Sewer Rehabilitation

The comprehensive sewer rehabilitation program for the sewerage systems in the Federal Republic of Germany will entail in some cases considerable interference with the existing structural substance. This not only offers the opportunity to re-establish the original standard condition but also to eliminate any structural or drainage weaknesses which may be found or to carry out improvements for an adaptation to new or changed requirements to be expected in future.

From the point of view of obtaining a functional, watertight and solid sewerage system which also fully meets the maintenance requirements the following solutions are being discussed at present [4]:

1. The increase of the cross-section of inaccessible sewers to > DN 1000 in the course of renewal methods for facilitating maintenance.
   The experience gained in the Federal Republic of Germany has shown that accessible sewers hardly present any technical problems when repairs are required as they allow for working and carrying out repairs manually on site. Safe draining, the re-integration of existing service connections or the installation and integration of new service connections are possible without difficulties, even when using the trenchless construction method.

2. Improvement of the functional efficiency and maintenance by the double pipe solution or the concept of the accessible collective sewer in which several or all underground supply and drainage lines are jointly accommodated. In addition to a number of other advantages which cannot be outlined within this context, a supplementary protective sheathing is available in both cases to prevent the exfiltration of waste water into the ground. Environmental risks due to defective sewers are therefore excluded.

3. The largest problem area of our sewerage systems are the numerous service connections and their direct integration into the road sewer. This arrangement not only impedes a correct inspection but particularly also any repair methods.

   In the case of many renovation and renewal procedures, before work is carried out in the open trench construction they must be disconnected from the sewer section and subsequently connected again, in order to avoid e.g. uncontrolled pipe breaks or to allow for a correct integration. Difficulties and costs increase with the number of service connections, so that in many cases it is only possible to renew the section in the open trench construction method, the disadvantages of which are known.

   A way out of this situation seems to be a sewerage network in which all service connections open out exclusively into manholes (figure 9).

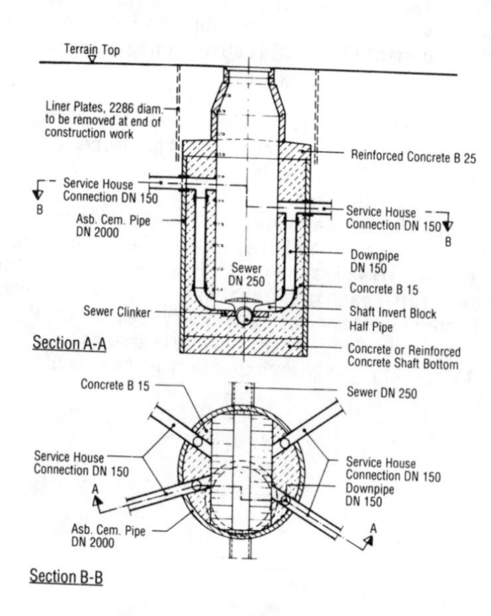

Figure 9. Shaft designed for link up of several service house connections [5]

This indirect integration of the service connections with the road sewer will offer the following advantages:

- Easier maintenance and inspection
- Easier rehabilitation both in the service connections and the road sewer
- Checking of the waste water fed in from previous at any time which appears to the imperative for indirect feeding trading and industrial operations.

This concept has already been realized in individual cases in the Federal Republic of Germany, also with economical success.

6. Summary

In the field of remedying defects, a multitude of methods are available, all of them presenting advantages and disadvantages, as well as restrictions concerning their application. There is no universal method, which is in an appropriate and equal manner technically and economically applicable, under any possible marginal condition.

The smaller the diameter of the sewer and the larger the number of connecting sewers within a section, the more difficulties arise in applying most of the methods.

In the Federal Republic of Germany, new and further process developments, presented in this article, have been pursued in order to eliminate these disadvantages. However, it becomes obvious, that without far-reaching changes in the present practice, planning and construction of sewer systems, there will be no way to meet all present and future requirements for these sewerage systems, particularly with respect to the aspects of environmental protection.

Literature

/1/ Abwassertechnische Vereinigung e.V.: Merkblatt M 143 Teil 1 Inspektion, Instandsetzung, Sanierung und Erneuerung von Entwässerungskanälen und -leitungen - Grundlagen

/2/ Stein, D.; Niederehe, W.: Instandhaltung von Kanalisationen Verlag Ernst & Sohn, Berlin 1986.

/3/ Stein, D.: Undichte Kanalisationen - ein kommunales Problemfeld der Zukunft aus der Sicht des Gewässerschuztes. Zeitschrift für angewandte Umweltforschung 1(1988) H. 1, S. 65 - 74.

/4/ Stein, D.: Schadensbehebung als Chance zur Durchsetzung neuer Kanalisationskonzeptionen. Korrespondenz Abwasser 36(1989) H. 8, S. 842 - 850.

/5/ Stein, D.; Möllers, K.; Bielecki, R.: Microtunnelling - Installation and Renewal of Nonman-Size Supply and Sewage Lines by the Trenchless Construction Method. Verlag Ernst & Sohn, Berlin 1989

The paper that was to have appeared on pages 46 - 64 has been withdrawn by the author.

UPRATING AN IN-SERVICE OFFSHORE PIPELINE

David G Jones[1] and Glenn Nespeca[2]

ABSTRACT

The Ekofisk to Emden 36 inch pipeline, transmits gas 275 miles from the Ekofisk field to Emden, West Germany, and has been in operation since 1977. The operator Phillips Petroleum Company Norway (PPCoN) in cooperation with the owner Norpipe A/S has recently conducted extensive studies to evaluate the feasibility of increasing the capacity of the pipeline.

This paper shows that it would be extremely costly to perform a hydrostatic re-test as the basis for uprating. An alternative fitness-for-purpose assessment, based on the findings of an intelligent pig inspection, is described and is shown to provide a better assurance of the pipeline integrity following uprating. The fitness-for-purpose assessment and inspection were conducted under contract by British Gas.

The results of all the studies have been utilised by PPCoN/Norpipe as the basis for discussions with the Regulatory Authorities to approve uprating of the pipeline without the need to perform a hydrostatic re-test.

INTRODUCTION AND BACKGROUND

The Ekofisk-Emden Gas pipeline extends, as shown in Figure 1, 275 miles from the Ekofisk Field to Emden, West Germany. The pipeline was designed to transmit in excess of 2000 mmscf/d and currently serves as the only transport of Norwegian gas to continental Europe.

---

[1]British Gas plc, Engineering Research Station, PO Box 1LH, Newcastle upon Tyne NE99 1LH, UK.
[2]Phillips Petroleum Company Norway, PO Box 220, N-4056 Tananger, Norway.

The 36 in (914.4 mm) diameter pipeline has a nominal wall thickness (wt) of 0.875 in (22 mm) for the submarine section and 0.984 in (25 mm) for the riser and onshore sections in Germany. The pipeline, was designed in accordance with ANSI/ASME B31.8-68 using steel conforming to API 5LX Grade 60 material. The current maximum allowable operating pressure (MAOP) of 1945 lbf/in² (134 bar) corresponds to a design factor of 0.67xSMYS. The offshore section was successfully hydrotested to 1.25xMAOP and the onshore section to 1.4xMAOP.

Increased Norwegian gas reserves and production potential indicate the capacity to be inadequate in the 1990's. Consequently, the operator Phillips Petroleum Company Norway (PPCoN) in cooperation with the owner, Norpipe A/S, evaluated various alternatives for increasing the gas deliveries. It was concluded that uprating the MAOP to 2100 lbf/in² (145 bar) is the most feasible method of obtaining the required increase in capacity.

The Regulatory Agencies have to date required hydrostatic re-testing to the original levels (1.25xMAOP offshore and 1.4xMAOP onshore) as the basis for uprating.

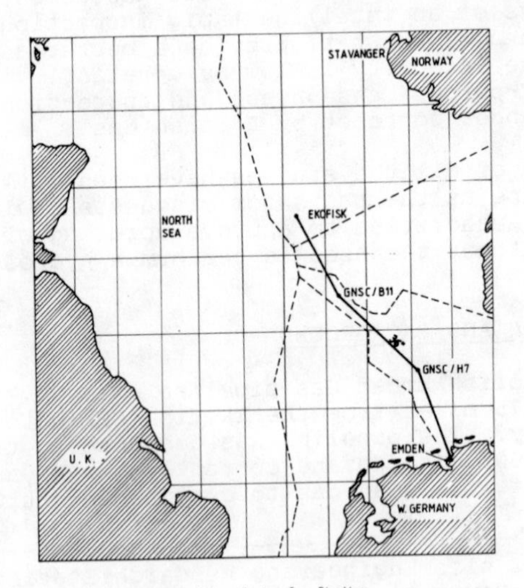

Fig1. Ekofisk - Emden Gas Pipeline.

IMPRACTICALITIES OF HYDROSTATIC RE-TESTING

The feasibility of conducting a hydrostatic re-test on the pipeline was investigated. It was found that there are serious practical difficulties in shutting down, testing, recommissioning and starting up production on an existing in-service offshore pipeline.

The pipeline pressure would be blowndown to 1200 lbf/in² (83 bar), before waterfilling the pipeline from Emden using shore-based equipment. This would permit water filling without the need for large tankers offshore and offshore flaring of the displaced gas at Ekofisk. Inhibited fresh water would be needed from shore during filling at a rate of approximately 260,000 gallons per hour (1000 m³/H).

A waterfilling train, consisting of a series of bi-directional pigs and chemical gels would be recovered at each booster platform. The pigs with fresh chemical gels would then be re-introduced and run into the next section of pipeline.

As the filling progressed, the gas would be displaced around the compressor stations, utilising existing smaller diameter bypass loops on each offshore platform. Once filled, the pipeline would be hydrotested consecutively, one section at a time. Some 28 days without contingency were estimated to be required for blowdown, filling and testing. However, an additional 15 days without contingency were also required to depressure the pipeline in order to remove and replace the three discharge risers as part of the uprating plan requirements. It was also assumed that topside piping would require replacement before the planned hydrotest.

Dewatering and recommissioning of the pipeline would commence in the first section from Ekofisk to B-11. A series of bi-directional pigs separated by slugs of chemical gels and methanol would be introduced using low pressure gas to propel the train to the receiving platform. Due to expected small amounts of gas entering gel slugs, the material would require special processing when received in addition to recovery of the methanol into small tankers stationed adjacent to the platforms. It was also anticipated that a second methanol train would be required to suppress hydrate formation. The test water was planned to be discharged to the sea. After each section was dewatered and pressured to 250 lbf/in² (17 bar), production could be gradually raised and further drying provided by the low dew point gas from Ekofisk. Some 20 days without contingency were estimated to be required to dewater, dry and resume production in the pipeline. Therefore, in total some 63 days plus contingency were considered necessary. Based on current economics, the cost, exclusive of lost sales during shutdown, to perform a hydrostatic re-test is estimated to be approximately ten (10) times the cost required to test a similar new pipeline today.

Thus, this approach is not only costly, but also requires an extended shutdown and poses serious practical difficulties. Consequently, this led to the consideration of the fitness-for-purpose approach as an alternative to hydrostatic re-testing.

FITNESS-FOR-PURPOSE - AN ALTERNATIVE APPROACH

Fitness-for-purpose analyses utilising fracture mechanics methods are accepted worldwide. Many International Standards now permit the use of fitness-for-purpose assessments, including ASME X1 1986, BS 5500 1986, API 1104, 1983, CSA Z184-M1979, and BS 4515, 1984. Formalised procedures (routes) and guidelines on performing fitness-for-purpose assessments are available, particularly BSI PD6493, 1980 and Milne et al, 1986. Additionally, analytical methods developed specifically for line pipe are available. These analytical methods are well established, generally accepted, published and proven (eg Hahn et al 1969, Kiefner and Duffy 1973, Kiefner et al 1973, Shannon 1974, Folias 1965, Kiefner and Maxey 1982).

All defect assessments using fitness-for-purpose require three inputs:

(i) Defect size
(ii) Applied stress
(iii) Fracture toughness

The data required for defect sizes (length and through thickness depth) were available from the intelligent pig inspection performed for PPCON/Norpipe between March and September 1987 (Nespeca, Hveding, 1988). The following is a summary of the reported findings:

EKOFISK 2/4R TO B-11

10 internal metal loss defects
1 external metal loss defect

It was also reported that this section contained internal corrosion of depth less than the reporting level (20% wt).

B-11 TO H-7

1 girth weld anomaly

H-7 TO EMDEN

3 offshore girth weld anomalies
1 onshore girth weld anomaly (removed in 1987)
1 external metal loss defect (burn mark on H-7 riser)

An estimate of all the stresses acting on the pipeline was required for the assessment. The operating pressures were readily available. The stresses due to internal pressure cycles and environmental loads were not readily available. It was necessary to obtain and analyse all the compressor station inlet and outlet pressure fluctuation histories for the period June 1977 to July 1988. Because 99% (approximately) of the pipeline is buried and exposed areas are subject to frequent inspection, it is not anticipated that any significant external loads and resulting axial stresses are likely to occur.

Modern pipeline materials are sufficiently tough that defect failure is controlled by plastic collapse mechanisms and the critical defect size is associated with the tensile properties. Some 1,273 test certificates were analysed to confirm the toughness of the linepipe. The toughness of the girth welds was quantified by CTOD (crack tip opening displacement) testing and four full scale tests.

The following assessment was conducted to investigate the fitness of the pipeline for uprating without hydrostatic re-testing.

VERIFICATION OF THE INTELLIGENT PIG INSPECTION

The fitness-for-purpose assessment depends critically on the acceptance of the inspection findings. Consequently, PPCoN verified that the intelligent pig had reliable detection and sizing capability:-

Statoil experience PPCoN was aware of the results of a verification test performed by Statoil in 1986, whereby, a 36" inspection vehicle was run through a test loop, containing two concrete coated pipes (36" x 22 & 27 mm wt) with 79 metal loss features. The metal loss features were internal and external pitting defects of depths 14 to 61% wt, general corrosion of depths 11 to 52% wt, gouges oriented in both the axial and circumferential direction, and girth and seam weld irregularities.

The test involved one run through the pipes where the defects had been filled smooth with non-metallic material and externally taped to cover the pipe surface. After completion of the initial report of the findings, additional runs were made to assess the repeatability of the tool at velocities from 0.7 to 2.8 m/s. The reported defect dimensions were compared with the actual dimensions. The conclusion of the test was that no spurious defects were reported and the repeatability of the tool was within the specification. As a result of this verification the intelligent pig has regularly been used on the Norwegian Continental Shelf since 1986.

PPCoN experience The inspection vehicle has been used in the Emden gas and Teesside oil pipelines and most of the oil and gas pipelines in the Ekofisk area. PPCoN has made several successful comparisons of the reported findings with diver visual and NDT measurements. The most convincing PPCoN verification relates to the reported girth weld crack in the onshore German section of the Emden pipeline. The presence of a crack was confirmed by ultrasonic and x-ray testing at the location and orientation reported by the intelligent pig. The defective girth weld was removed and subsequent investigations included sectioning which confirmed an 70 mm long x 15 mm deep crack. The defective girth weld was also inspected by manual ultrasonics under laboratory conditions; the sizing was no more accurate than the intelligent pig estimate.

On the basis of the above, PPCoN and the pipeline owners concluded that an intelligent pig inspection provides an accurate determination of the condition of a pipeline, and thus, the basis for the fitness-for-purpose assessment is technically sound.

FITNESS-FOR-PURPOSE ASSESSMENT

Potential Uprated MAOP

The purpose of a pre-service high-level pressure test is to prove the integrity of a pipeline. By deliberately testing to a pressure higher than the MAOP, all defects large enough to cause failure during service are removed. Furthermore, the high-level test removes defects smaller than those which would fail at the MAOP.

Consequently, a safety margin is provided which also allows the pipeline to operate safely even when defect growth occurs during service. For the Ekofisk to Emden pipeline, the following current test/MAOP ratios were adopted to comply with the relevant code requirements:

Pipeline Section	Test Level		MAOP		Test/ MAOP Ratio
	lbf/in^2	bar	lbf/in^2	bar	
Offshore (22.2 mm)	2432	168	1945	134	1.25
Onshore (25 mm)	2718	187	1945	134	1.40

None of the features reported by the intelligent pig inspection would have failed if the pipeline was hydrostatically re-tested to "yield" (100% SMYS). Thus, there is the potential to uprate the MAOP to 2150 lbf/in^2 (148 bar) whilst maintaining the current test/MAOP ratios:-

Pipeline Section	Inspection Equivalent Test Level (%SMYS)	Potential Uprated MAOP			"Inspection" /MAOP Ratio
		lbf/in^2	bar	(%SMYS*)	
Offshore (22.2 mm)	100	2150	148	80	1.25
Onshore (25 mm)	100	2150	148	71	1.40

*Conservatively, based on API min wt (92% nom wt).
SMYS = specified min yield stress.

The integrity of the pipeline was investigated at both the current MAOP and the potential uprated MAOP by quantifying the defect safety margin and the defect growth for the relevant test/MAOP levels. The aim was to determine if there would be any reduction in the pipeline integrity during operation at 2150 lbf/in^2 (148 bar). Note that the 2150 lbf/in^2 (148 bar) pressure is higher than the 2100 lbf/in^2 (145 bar) to obtain the required increase in capacity.

Defect Safety Margin

The relationship between defect size and failure pressure is shown in Figures 2-6. These show that a defect which can remain after the hydrostatic test is much smaller than a defect which would fail in service. These defect sizes and the safety margin between them can be quantified as follows:

(a) The safety margin provided by the pre-service hydrostatic test for operation at the current MAOP (1945 lbf/in² (134 bar)) (Figures 2 and 4).

(b) There is a reduction in the above defect safety margin provided by the pre-service hydrostatic test for operation at 2150 lbf/in² (148 bar) (Figures 2 and 4).

(c) The safety margin provided by a hydrostatic re-test to yield for subsequent operation at 2150 lbf/in² (148 bar) is shown in Figure 3 and 5.

(d) The safety margin based on the maximum defect depth reported by the intelligent pig and that which is necessary for failure at 2150 lbf/in² (148 bar) is shown in Figure 6; it exceeds the safety margin (a) and (c) above.

The conclusion is that operation at 2150 lbf/in² (148 bar) following the intelligent pig inspection gives a greater safety margin (Figure 7) than operation at 1945 lbf/in² (134 bar) following the original pre-service hydrostatic test (Figure 2).

Defect Growth

The defects reported by the intelligent pig could grow by fatigue or corrosion.

Fatigue The reported features were assumed to be subjected to internal pressure fluctuations only. All the compressor station inlet and outlet pressure fluctuation data for the period June 1977 to July 1988 was obtained. Analysis of the data identified the most severe pressure fluctuations which were used in the subsequent fatigue analysis.

A fatigue analysis was conducted conservatively assuming that the features were crack-like defects. The estimated fatigue life of all the reported features exceeds 100 years (at both 1945 and 2150 lbf/in² (134

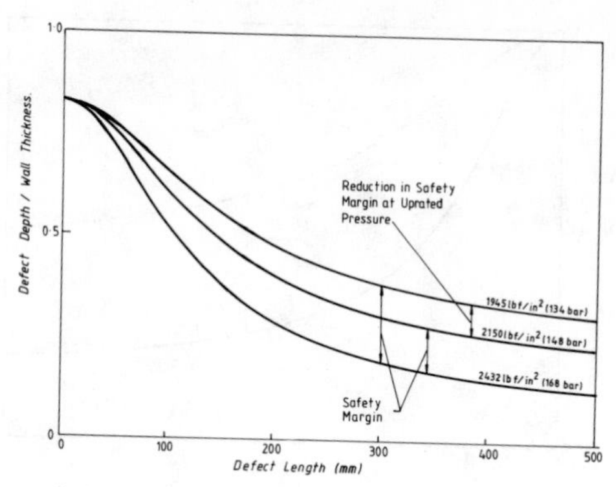

Fig. 2. Safety Margin provided by Pre-Service Pressure Test. (2432 lbf/in²) (22·2 mm wt.)

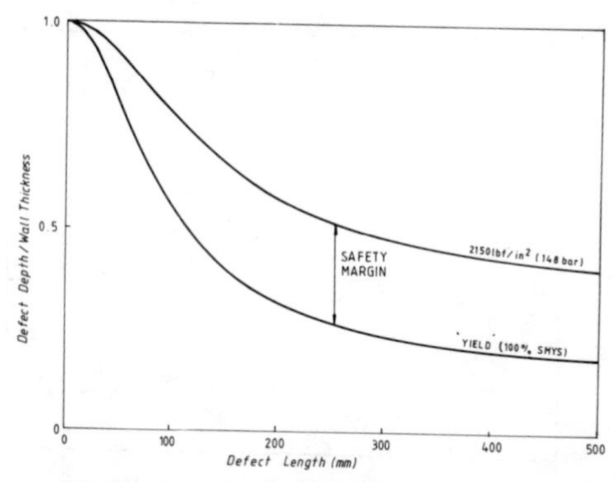

Fig.3. Safety Margin provided by Revalidation Test to Yield. (22·2 mm wt)

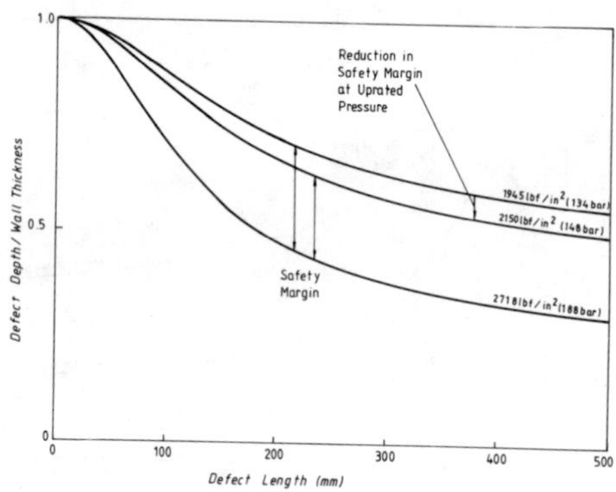

Fig.4. Safety Margin provided by Pre-Service Pressure Test. (2718 lbf/in²) (25 mm. wt)

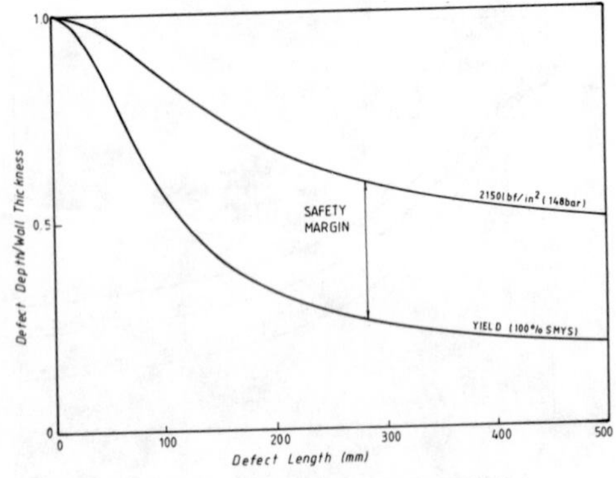

Fig.5. Safety Margin provided by Revalidation Test to Yield. (25 mm. wt.)

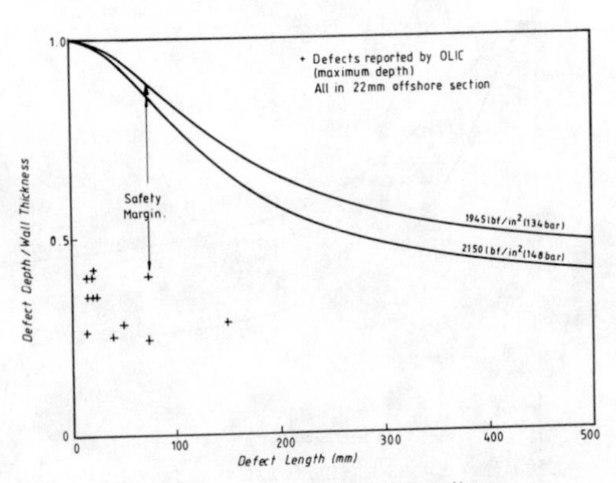

Fig 6. Safety Margin provided by OLIC Inspection.

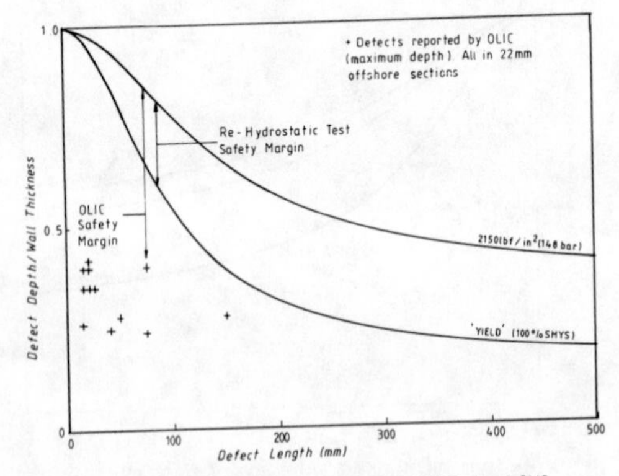

Fig.7. Safety Margins provided by Re Hydrostatic Yield Test and OLIC.

and 148 bar) MAOP's). It is concluded that provided there is no significant change in the pressure fluctuations during the uprated period (circa 10 years) none of the reported features would fail by fatigue.

Corrosion For all the reported corrosion features ("initial" size), we have calculated the ("final") size which is tolerable at 1945 and 2150 lbf/in² (134 and 148 bar). The difference between the "final" and "initial" feature sizes is the margin available for corrosion growth. The smallest margin available for corrosion growth is 4 mm. Consequently, during the anticipated uprated period of 10 years the corrosion growth rate must be restricted in the worst case to less than 0.4 mm annually.

Other Factors

All conceivable cracks which might be in the pipeline and their influence on the integrity of the pipeline were also considered:

Girth Weld Cracks The intelligent pig reported one onshore and four offshore girth weld anomalies. The pipeline was uncovered to locate the onshore weld and investigated to confirm the presence of a crack at the edge of an internal root weld. A section of the pipeline containing the defect was removed. Subsequent extensive investigations including analysis and four full scale tests with the actual and simulated defects, confirmed that the defective girth welds have insignificant effect on the integrity of the pipeline. In particular, the adjacent pipe would yield before defect failure in the girth weld, ie uprating the internal pressure from 1945 to 2150 lbf/in² (134 to 148 bar) has no effect. It was also confirmed that the cracks cannot extend by fatigue in 100 years to a size which would fail at 2150 lbf/in² (148 bar).

Seam Weld Cracks A full fracture and fatigue analysis was conducted to investigate the conditions under which seam weld cracks could fail. Axially orientated surface breaking cracks, the most typical and severe defects were considered.

The only conceivable failure mechanism is that a seam weld defect pre-existing when the pipeline was commissioned in 1977 extends by fatigue under the action of internal pressure cycles to a size at which failure would occur at 2150 lbf/in² (148 bar). However, a pre-existing, infinitely long, surface-breaking seam weld defect of depth 22% wt (offshore) and 24% wt (shore approach and onshore) is necessary for failure in 100 years. Pre-existing seam weld cracks of the above dimensions have not been experienced in "modern" linepipe (Fearnehough, 1986) as utilised in the Ekofisk to Emden pipeline, ie seam weld failure would not occur and no inspection of the seam weld to detect and/or size cracks is necessary.

Environmental Cracking Sulphide stress corrosion (SCC) and hydrogen induced cracking (HIC) are not conceivable under the dry gas conditions specified for the pipeline and the hydrogen sulphide content (1 grain per 100 scf) is too low to qualify the gas as sour. The only other possible environmental cracking mechanism is external stress corrosion cracking (SCC). However, SCC is not conceivable in a coal tar coated, weight coated and cathodically protected offshore pipeline and, indeed, SCC has never been observed in an offshore pipeline.

Cracks Associated with Mechanical Damage Cracks occurring as a result of mechanical damage are associated with gouges and/or dents. No gouges or dents were found during the intelligent pig inspection.

Fracture Propagation A feature which is unique to gas pipeline is that a fracture once initiated can propagate for long distances. However, toughness requirements can be defined to prevent long running fractures. Drop weight tear test (DWTT) requirements can be defined to prevent long running brittle fractures; Charpy upper shelf energy can be defined to prevent long running shear fractures (Vogt et al 1983).

1,273 test certificates were examined to determine the actual pipe toughness properties. The DWTT (API) specification was satisfied and it is concluded that long running brittle fracture would not occur. A Charpy of 54 ft lb is required for shear fracture arrest at 1945 lbf/in² (134 bar) and 60 ft lb at 2150 lbf/in² (148 bar). Statistical analysis of the data confirmed that the actual Charpy toughnesses exceed the minimum specification (45 ft lb) by a sufficient margin to ensure shear fracture arrest at the current and the potential uprated MAOP.

SUMMARY AND CONCLUSIONS

The required increase in the capacity of the Ekofisk to Emden pipeline for the 1990's can be met by uprating the MAOP from 1945 to 2100 lbf/in² (134 to 145 bar). It is impractical to uprate on the basis of a hydrostatic re-test because not only is it costly, it also requires extended shutdown and poses severe practical difficulties. Consequently, an alternative fitness-for-purpose assessment was conducted. Based on the results of the fitness-for-purpose assessment and particularly the intelligent pig inspection, it is concluded that the pipeline could be operated at a higher pressure 2150 lbf/in² (148 bar) without hydrostatic re-testing.

The adoption of a MAOP based on a fitness-for-purpose assessment does not mean that there would be a reduction in the integrity or safety of the pipeline. This is because a thorough knowledge and quantification of the stresses, toughness and defect sizes are required before the fitness-for-purpose assessment can be conducted. This knowledge is not normally required prior to a hydrostatic re-test.

Indeed it is concluded that the intelligent pig inspection and subsequent fitness-for-purpose assessment has provided a better knowledge of the pipeline than would be possible from a hydrostatic re-test. In particular the following information was provided on the integrity of the Ekofisk to Emden pipeline during operation at 2150 lbf/in² (148 bar):

(1) The defect safety margin based on the defects revealed by the intelligent pig inspection exceeds the safety margin which would be provided by a hydrostatic re-test to yield as shown in Figure 7.

(2) Defect corrosion growth is not a problem unless it exceeds 0.4 mm per year during the anticipated 10-year period of uprating.

(3) Internal corrosion was reported in the first section of the pipeline from Ekofisk to B-11. The low level corrosion, attributed to periodic operational upsets, was identified and positive measures implemented to prevent future occurrence. The corrosion would not have been detected by a hydrostatic re-test.

(4) Fatigue failure would not occur unless the magnitude and frequency of internal pressure cycles are significantly increased.

(5) Extensive analytical and experimental studies, including four full scale tests, confirmed that the defective girth welds revealed by the intelligent pig have no influence on the integrity of the pipeline; the adjoining parent pipe material would yield first.

(6) Failure from seam weld defects is not conceivable and any inspection of the seam weld to detect/size defects is unnecessary.

(7) The pipeline toughness is high enough to prevent the occurrence of long running brittle or shear fractures.

(8) No other cracking mechanism or evidence could be found or conceived which would have a detrimental effect on the integrity of the pipeline for the intended uprating.

REFERENCES

API 1104, 1983, Standard for Welding Pipelines and Related Facilities.

ASME Section XI, 1986, Boiler and Pressure Code, ASME

BS 4515, 1984, Process of Welding of Steel Pipelines on Land and Offshore, British Standard Specification.

BS 5500, 1986, Specification for Unfired Fusion Welded Pressure Vessels, British Standard Specification.

BSI PD6493, 1980, Guidance on Some Methods for the Derivation of Acceptance Levels for Defects in Fusion Welded Joints, British Standards Institution.

CSA Z184, 1979, Gas Pipeline Systems, Canadian Standards Association.

Fearnehough G D, Nov 1986, Safe Pipeline Performance, The Welding Institute 3rd Int Conf Welding and Performance of Pipelines, London.

Folias E S, 1965, The Stresses in a Cylindrical Shell Containing an Exial Crack, Int J Fracture Mechanics, pp 104-13.

Hahn G T, Sarate M, Rosenfield A R, 1969, Criteria for Crack Extension in Cylindrical Pressure Vessels, Int J Fracture Mechanics, pp 187-210.

Kiefner J F, Duffy A R, 1973, Criteria for Determining the Strength of Corroded Areas of Gas Transmission Lines, AGA, Operating Section Transmission Conference, Paper No 73-7-S.

Kiefner J F, Maxey W A, 1982, Judging Defect Severity in Offshore Pipelines, Oil and Gas Journal, pp68-74.

Kiefner J F, Maxey W A, Eiber R J, Duffy A R, 1973, Failure Stress Levels of Flaws in Pressurised Cylinders, Progress in Flaw Growth and Fracture Toughness Testing, ASTM STP 536, pp 461-481.

Milne I, Ainsworth R A, Dowling A R, Stewart A J, May 1986, Assessment of the Integrity of Structures Containing Defects, CEGB Report R/HR6 - Rev 3.

Nespeca G A, Hveding K B, 1988, Intelligent Pigging of the Ekofisk-Emden 36 in Gas Pipeline, 63rd Annual Technical SPE Conference, Houston, Texas.

Shannon R W E, 1974, The Failure Behaviour of Pipeline Defects, Int J Press Ves and Piping (2).

Vogt G H, Bramante M, Jones D G, March 1983, EPRG Report on Toughness for Crack Arrest in Gas Transmission Pipelines, 3R International 22, Jahrgang Heft 3.

ACKNOWLEDGEMENTS

The authors wish to thank British Gas plc for permission to publish the paper. The authors also wish to thank the following Phillips Norway Group partners for their approval to publish the paper: Phillips Petroleum Company Norway, Fina Exploration Norway Inc, Norsk A-gip A/S, Norsk Hydro a.s., Elf Aquitaine Norge, Total Marine Norsk A/S, Den Norsk Oljeselskap a.s., Norpipe A/S, Eurafrep, Norge A/S, Coparex Norge A/S, Cofranord Norge A/S.

EVALUATING THERMAL PRESSURE CHANGES IN LIQUID PACKED PIPING

Rodney F. Wedge[1], Member ASCE

Abstract

Theoretical and practical issues involved with the phenomenon of thermally induced pressure change in closed piping systems are investigated. The mathematical derivation of the applicable expressions is used to illustrate points of interest to designers. The language of the American National Standards Institute (ANSI) piping design codes is discussed as it applies to typical situations. Sources of useful physical design data are mentioned. Tables and figures are presented to illustrate how the ANSI piping design codes could be more explicit about this subject.

Introduction

Thermal pressure rise in fully isolated piping is a subject that most piping engineers know something about, but perhaps not enough. Thermal pressure changes are usually considered only incidentally during the design of piping systems. This probably happens because there never seems to be enough time to look at every aspect of a system in detail. As a consequence, many engineers and designers never take the opportunity to acquire reliable criteria and rules of thumb.

One objective of this paper is to separate, to some extent, fact from fiction about thermal pressure rise in blocked-in piping. Another objective of this discussion is to provide the reader with a frame of reference for use of the applicable provisions of the American National Standards Institute (ANSI) piping codes. The codes specifically mentioned are ANSI B31.3, *Chemical Plant and Petroleum Refinery Piping*, and ANSI B31.4, *Liquid Transportation Systems for Hydrocarbons, Liquid Petroleum Gas, Anhydrous Ammonia and Alcohols*.

---

[1]Project Engineer/Project Manager, Willbros Butler Engineers, Inc., P. O. B. 701650, Tulsa, OK 74170.

Chapter II of ANSI B31.3 contains the following terse requirement:

"Piping or piping components, or portions thereof, that can be blocked in or isolated form a pressure relieving device shall be designed for not less than the maximum pressure that can be developed under those conditions, considering short term variations..."

ANSI B31.4 is even more brief in stating only this:

"Provision shall be made in the design either to withstand or to relieve increased pressure caused by the heating of static fluid in a piping component."

Do these code statements imply that it is normally practical to design for containment of thermal pressure changes? If the piping segment to be blocked in is buried, will that assure that liquid thermal expansion relief provisions are unnecessary? If the isolated piping run is very short, can the liquid thermal expansion effects be ignored? How would a test engineer begin to develop criteria for corrections to account for thermal pressure changes in a system to be leak tested with a liquid other than water? When the fluids being transported are hazardous, it can become important for the engineer to supply correct answers to questions like these.

Most of us would be daunted if a quick answer to one of these questions was critical. Many of the readily available handbooks do not provide much help. However, with the aid of some of the published material on related subjects and a look at the fundamentals, reasonable solutions to these and similar problems are available. The following paragraphs will explore the fundamentals of this topic.

The Fundamentals

It is worth the exercise to look at the derivation of thermal pressure rise relationships. This will answer some of the basic questions just raised.

The following notation is used in the expressions below:

A = Original inside cross-sectional area of the pipe.
a = Coefficient of linear thermal expansion of pipe wall material.

B = Coefficient of volume thermal expansion of liquid, $dV/(VdT)$.
D = Original inside diameter of pipe.
d = Denotes unit incremental change; dP, incremental pressure change; dV, incremental volume change; dT, incremental temperature change, etc.
E = Modulus of elasticity of pipe wall material.
e = Unit strain of pipe wall material; e_x, axial; e_y transverse.
k = Bulk modulus of liquid, $-dP/(dV/V)$.
L = Total length of pipe segment.
P = Internal pressure in pipe segment.
S = Unit stress in pipe wall; S_x, axial stress; S_y hoop stress.
T = Temperature of pipe/liquid system.
t = Wall thickness of pipe material.
u = Poisson's ratio of the pipe wall material.
V = Original volume of the pipe segment or contained liquid.
π = Pi, 3.14159...

The General Case

Most liquids expand thermally more per unit volume than does the pipe volume. Therefore, with an increase in temperature the pipe is stretched by the expanding liquid and the liquid is compressed by the confinement of the pipe. This is a phenomenon which has been recognized for a long time, but the methods available to quantify the change in pressure as the temperature rises are not widely appreciated.

For any pipe segment which is fully closed, it is obvious that the volume of the pipe must always be equal to the volume of the contained liquid. Likewise, the incremental volume change of the expanding liquid must be equal to the incremental volume change of the pipe:

$$dV_{pipe} = dV_{liquid} \tag{1}$$

The pipe and the liquid are both affected by temperature and pressure changes. The incremental change in the volume of the liquid due to temperature change is defined by rearranging the expression for the mean coefficient of volume thermal expansion. Similarly, the incremental volume change of the liquid due to pressure change is taken from the definition of bulk modulus. The right side of equation (1) becomes:

$$dV_{liquid} = BVdT - (dPV)/k \tag{2}$$

Evaluation of the left side of equation (1) depends upon the degree of restraint of the pipe segment.

For continuously welded pipelines, there are basically two conditions of interest. These are unrestrained and fully restrained.

Unrestrained Pipe

Unrestrained pipe constructed of elastic material is free to expand axially and radially due to changes in temperature and pressure. The circumferential, or hoop, stress is equal to PD/2t and the axial stress is equal to one half of this. The pipe is free to expand thermally without significant additional stress being induced in the pipe wall by the action of restraints. The case of an unrestrained pipe segment is as modeled in Figure 1.

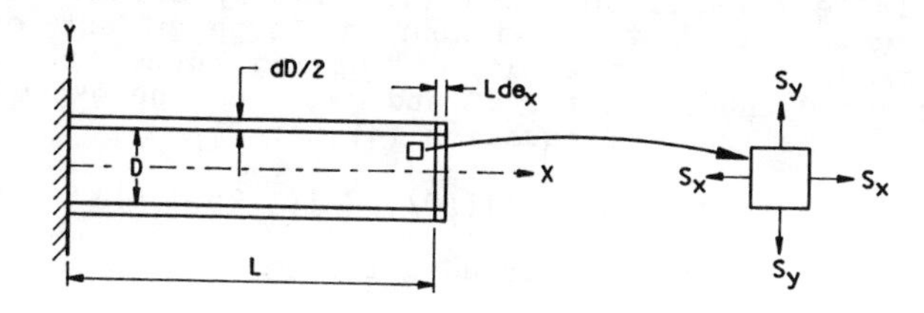

Figure 1
Thermal and Pressure Volume Change, Unrestrained Pipe

In this case, the incremental change in volume of the pipe is equal to the change in volume of the pipe with respect to pressure change plus the change in the volume of the pipe with respect to temperature change. The left side of equation (1) can be written:

$$dV_{pipe} = dV_{pipe,\ p} + dV_{pipe,\ t} \qquad (3)$$

The incremental volume change due to pressure change will be defined first. In doing so, it is necessary to recall from mechanics of materials the expressions for the unit strain of materials under biaxial tensile stress:

$$e_x = (S_x - uS_y)/E \quad \text{and} \quad e_y = (S_y - uS_x)/E$$

By substituting expressions for the hoop and axial pressure stress in the above and writing appropriate terms to represent incremental unit changes we get:

$$de_x = [(dPD)/(2tE)](0.5 - u) \quad \text{and}$$
$$de_y = [(dPD)/(2tE)](1 - 0.5u)$$

Expressions for incremental change in diameter and length are as follows:

$$dL = Lde_x = L[(dPD)/(2Et)](0.5 - u) \quad \text{and}$$
$$dD = Dde_y = [(dPD^2)/(2tE)](1 - 0.5u)$$

The increase in volume per unit of the original length of pipe is equal to the increase in the cross-sectional area of the pipe. The increase in cross-sectional area is very close to: $(\pi DdD)/2$. We can now write the components of volume change of the pipe with respect to pressure change as follows:

$$dV_{pipe,\,p} = (\pi DdDL)/2 + (\pi DdDLde_x)/2 + (\pi D^2Lde_x)/4 \quad (4)$$

At this point, we can save ourselves a lot of trouble by recognizing that the product of dD and de_x in the middle term of equation (4) is very small. So, we can neglect this term without a large effect on the precision of the result. Having done this and substituted the terms for dD and de_x from the expressions above, we can rewrite equation (4) as:

$$dV_{pipe,\,p} = dP[(\pi D^2L)/4]\{[D/(tE)](1.25 - u)\} \quad (5)$$

This can be simplified a little more as follows:

$$dV_{pipe,\,p} = VdP\{[D/(tE)](1.25 - u)\} \quad (6)$$

We now need the expression for the thermal effect on the pipe volume. By again referring to the model of figure (1), we can write terms for the components of the thermal volume change of the pipe in a manner very similar to that for the pressure induced volume change. This time we have the following:

$$dV_{pipe,\,t} = (\pi DdDL)/2 + (\pi DdDdL)/2 + (\pi D^2dL)/4 \quad (7)$$

In which, $dD = DadT$ and $dL = LadT$

Again, since the product of dD and dL in the middle term of equation (7) will be very small compared to the other two terms, the middle term can be neglected. Having done this and having made the appropriate substitutions, equation (7) can be reduced to just:

$$dV_{pipe,\,t} = 3VadT \quad (8)$$

By combining equations (3), (6) and (8), we have defined the left side of equation (1) for the case of unrestrained pipe. We now have:

$$dV_{pipe} = VdP[D/(tE)](1.25 - u) + 3VadT \quad (9)$$

By combining equations (9) and (2) we get:

$$VdP[D/(tE)](1.25 - u) + 3VadT = BVdT - (dPV)/k \quad (10)$$

Note that each term of equation (10) contains V, the original volume of the pipe segment. This term drops out of the result, but the original diameter remains. We can conclude that the length of the pipe segment has nothing at all to do with the magnitude of the possible thermally induced pressure rise if the pipe segment is completely closed.

To obtain the final result, V is divided out of equation (10), and the expression is rearranged to leave:

$$dP = (B - 3a)\{(k)/[1 + (Dk/Et)(1.25 - u)]\}dT \quad (11)$$

Equation (11) gives the pressure change with respect to temperature change for unrestrained pipe.

Restrained Pipe:

Long, buried, welded pipelines are usually fully restrained by soil friction acting along the pipe length against axial expansion. In this case, it can be assumed that the axial strain is zero. Because of the axial restraint, a significant axial stress can be induced from a change in temperature. This is the fundamental difference between restrained piping and unrestrained piping. In the case we are investigating, the axial stress can become compressive with only a modest rise in the temperature. A simple model of this case is as shown in Figure 2.

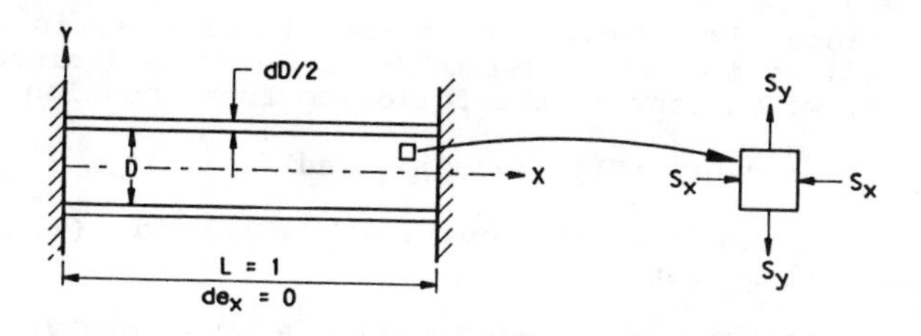

Figure 2
Thermal and Pressure Volume Change, Restrained Pipe

The general case outlined above applies to restrained piping as it does to unrestrained piping. In addition, we know that the length of the restrained pipe segment will not change by the nature of the restraint, so we need to concentrate only on the change in diameter to find an expression for the unit volume change. A compressive axial stress, tensile hoop stress, and thermal expansion all tend to increase the diameter of the restrained pipe segment. We can readily write the expressions for change in diametric strain due to an increase in hoop stress and thermal expansion, but we also need to quantify the axial stress. An expression for axial stress in a restrained pipeline can be derived. However, to save space here, we will borrow it from ANSI B31.4. This expression is as follows:

$$S_x = EadT - uS_y$$

Using this expression, the unit change in diametric strain due to a change in hoop stress, axial stress and thermal expansion for a restrained pipe segment is as follows:

$$de_y = dS_y/E + u(adT - udS_y/E) + adT \quad (12)$$

The change in the pipe diameter would be: $dD = Dde_y$. Since the length of the restrained pipe segment does not change as the temperature rises, the change in volume per unit length is equal to the change in the cross-sectional area. The change in the cross-sectional area is: $dA = dV = \pi DdD/2 = (\pi D/2)Dde_y$. By substituting equation (12) into this expression and simplifying, we can obtain the following:

$$dV_{pipe} = (\pi D^2/2)[(dS_y/E)(1 - u^2) + adT(1 + u)] \quad (13)$$

Which can be rewritten as:

$$dV_{pipe} = (\pi D^2/4)[(2dS_y/E)(1 - u^2) + 2adT(1 + u)] \quad (14)$$

Since the original cross-sectional area is equal to the volume for our model of unit length and since $dS_y = dPD/2t$, we can obtain the following from equation (14):

$$dV_{pipe} = V\{[(dPD)/(tE)](1 - u^2) + 2adT(1 + u)]\} \quad (15)$$

By substituting equation (15) and (2) into equation (1) we get:

$$V\{[(dPD)/(tE)](1-u^2) + 2adT(1+u)]\} = BVdT - (dPV)/k \quad (16)$$

As was the case with unrestrained pipe, the original volume drops out of the expression. The original diameter remains a component of the equation. This again indicates that the length of a blocked-in pipe segment has nothing at all to do with the magnitude of the potential thermal pressure rise. This is an important factor for piping engineers and designers to remember.

Equation (16) can be rearranged in a form similar to equation (11) to give us a general expression for thermal pressure rise in restrained pipe.

$$dP = [B - 2a(1 + u)]\{(k)/[1 + (Dk/Et)(1 - u^2)]\}dT \qquad (17)$$

Equations (11) and (17) are both in the form referred to by Arnold (1981). That is, the thermal pressure change is the product of the apparent volume change (thermal volume change of the liquid less the thermal volume change of the pipe), the effective bulk modulus and the change in temperature. It is worth noting from a comparison of equations (11) and (17) that the thermal pressure rise in a restrained line will always be greater (but only slightly greater) than that in an unrestrained line with all other conditions the same.

For restrained pipelines under hydrostatic test, Gray (1976) has expressed the apparent volume change as (B - 2a), rather than [B - 2a(1 + u)]. This neglects the effect of axial thermal stress on the blocked-in pressure change. Because the volume thermal expansion coefficient of water is generally an order of magnitude greater than the coefficient of thermal expansion of steel, the effect of neglecting the thermal axial stress is such that computed values will be only about five percent (or less) higher than would be obtained by the use of equation (17). Methods generally used in the United States for computing the volume balance corrections necessary to certify pipeline hydrostatic tests do account for the thermal axial stress.

The Real World

We now have handy, direct expressions for the thermal pressure rise in restrained and unrestrained piping. However, we have not taken into account the fact that most of the material properties we have treated as constants in the above derivations are themselves affected by changes in temperature and pressure. The coefficient of volume thermal expansion of most liquids changes significantly with temperature change. This is also true of the thermal expansion coefficient of pipe materials. The bulk modulus of liquids is affected by both temperature and pressure changes. Even the modulus of elasticity of pipe materials changes somewhat with temperature change within the ambient range. Nevertheless, if we keep these facts in mind as we select representative material property values for design or analytical calculations, we can still make reasonably good estimates.

Since the volume thermal expansion coefficient of many liquids is large compared to the thermal expansion coefficient of many common pipe materials, this single variable usually has the most significant impact on the computed result. As we shall see below, this is especially true of cases involving water or petroleum in steel piping. It is worth some extra effort to find liquid thermal expansion data that will fit the case needing analysis. Fortunately, a coefficient of liquid thermal volume change can be easily derived from density data. The value used must be relative to the initial and final temperature as must the coefficient of thermal expansion of the pipe material.

Since the bulk modulus of liquids varies with both temperature and pressure, values must be corrected for both. If the span of temperature and pressure change is not too large, a value computed at the average temperature and the average system pressure will probably give good enough results. Otherwise, the estimated pressure change can be calculated as the sum of incremental changes.

A few good sources of data for these kinds of calculations are as follows:

Variable:	Available Reference:
a	ANSI B31.3, Appendix C
B	Via Density Data from the *Handbook of Chemistry and Physics*, or via volume correction tables such as from Chapter 11, Section 1 of the API *Manual of Petroleum Measurement Standards*
E	ANSI B31.3, Appendix C
k	*Handbook of Chemistry and Physics* or SAE Coordinating Research Council, *Handbook of Aviation Fuel Properties*
u	ANSI B31.3 (data should be consistent with reported value of E.)

Even if we have correctly modeled the mechanics and have gone to the trouble of assembling good data, it is often difficult to predict what will actually take place in the field. Valve seals, stem packing or gaskets often leak slightly thereby relieving pressure. Perhaps the system under scrutiny is never really completely sealed. Perhaps the temperature will not change enough to cause a problem, or will it?

The design must be based upon the worst set of probable circumstances. If the piping system can be blocked in while filled with liquid, sooner or later it will be. So, for design purposes, we only need an idea of how much temperature change the system can stand and compare that to what the system might actually experience.

Table 1, shown below, gives the estimated pressure rise for water and Jet A (43.0 degrees API gravity jet fuel) in twelve inch, standard wall thickness, steel pipe. Figures 3 through 6 illustrate how quickly pressure can rise from expansion of these same liquids when the temperature rises from 60 degrees Fahrenheit, and the system pressure rises from zero gauge. The pressure in water filled, twelve inch steel pipe can rise approximately 200 psi with just a ten degree F rise in temperature. The same pipe filled with Jet A fuel would experience a pressure rise of almost four times that, or approximately 750 psi. The degree of restraint has relatively little impact.

Table 1
Thermal Pressure Rise in Closed Piping

	Restrained Pipe		Unrestrained Pipe	
Filled With:	Water	Jet A	Water	Jet A
Variable:				
T_1 (^oF)	60	60	60	60
T_2 (^oF)	61	61	61	61
B (psiX10^5)	8.94	50.5	8.94	50.5
a (psiX10^6)	6.05	6.05	6.05	6.05
u	0.3	0.3	0.3	0.3
k (psiX10^{-5})	3.10	1.81	3.10	1.81
D (in.)	11.25	11.25	11.25	11.25
t (in.)	0.375	0.375	0.375	0.375
E (psiX10^{-7})	2.79	2.79	2.79	2.79
dP (psi)	17.5	75.2	16.8	74.4

Since the initial system shut-in pressure will normally be well above zero gauge, one can see that very little temperature change can be tolerated before some system limit (flange rating, valve differential rating, equipment maximum, etc.) is reached. Even buried lines can experience several degrees of temperature change if left closed over a period of time. The resulting pressure rise could be unexpectedly dramatic. If the designer's assumptions about potential temperature change are too low, the problem is compounded. In the writer's opinion, one should rarely infer from the piping code language that it is practical to design to withstand thermal pressure change. A means of relieving excessive thermal pressure is too easily provided, and system limits are too easily exceeded, to assume otherwise.

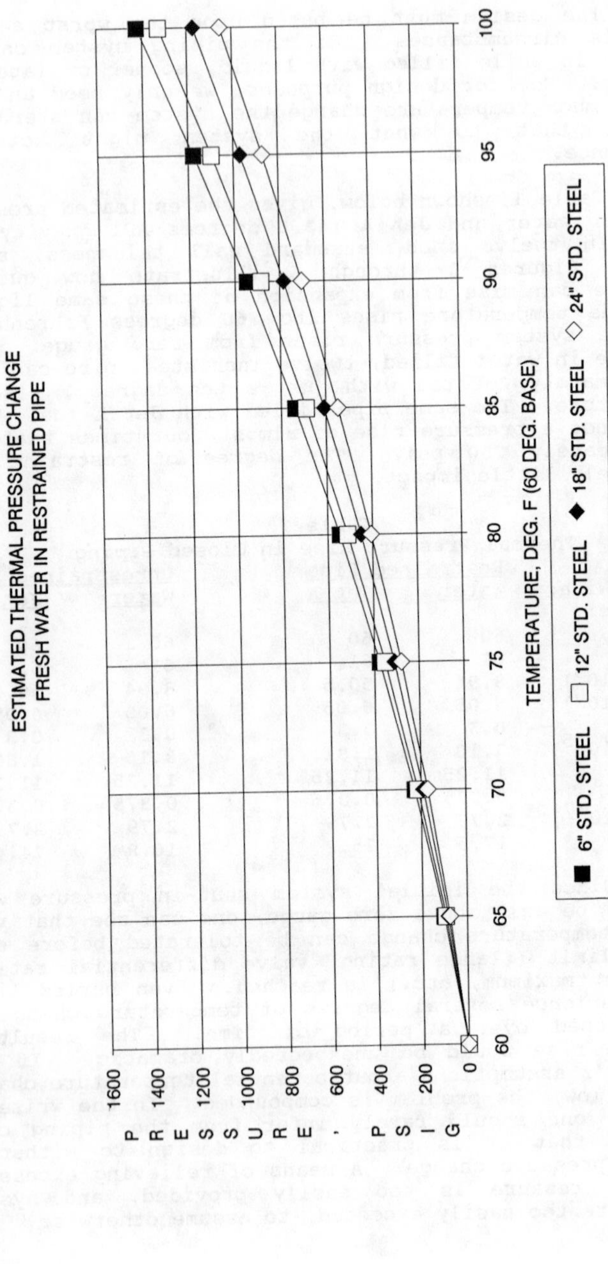

ESTIMATED THERMAL PRESSURE CHANGE
FRESH WATER IN RESTRAINED PIPE

PRESSURE - PSIG

TEMPERATURE, DEG. F (60 DEG. BASE)

■ 6" STD. STEEL □ 12" STD. STEEL ◆ 18" STD. STEEL ◇ 24" STD. STEEL

Figure 3

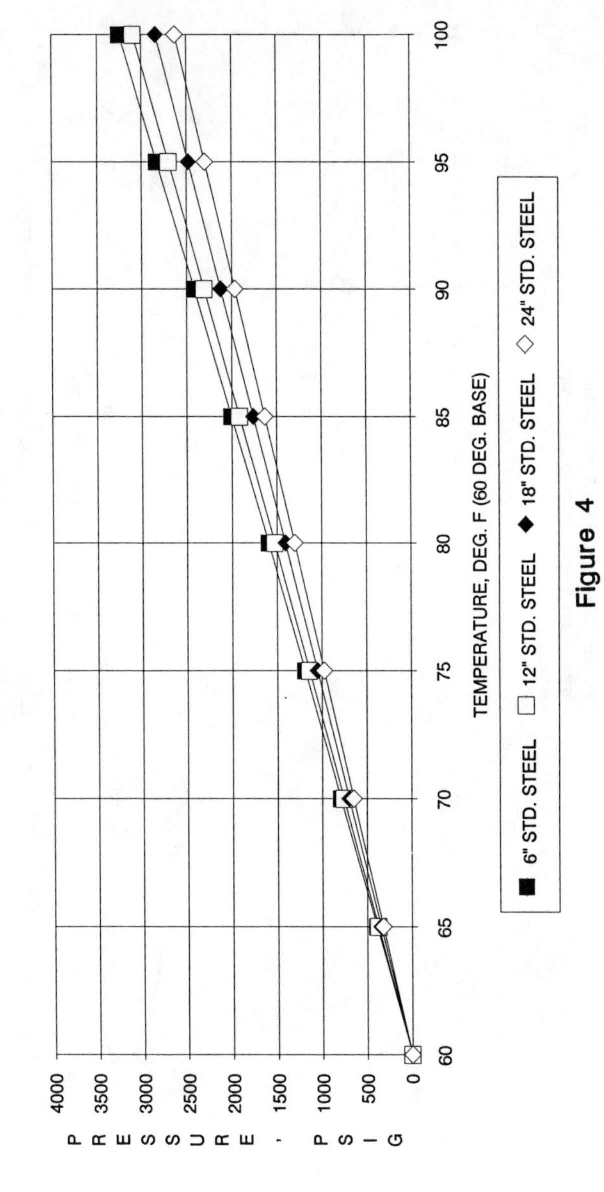
ESTIMATED THERMAL PRESSURE CHANGE
JET A FUEL IN RESTRAINED PIPE
PRESSURE - PSIG
4000
3500
3000
2500
2000
1500
1000
500
0
60
65
70
75
80
85
90
95
100
TEMPERATURE, DEG. F (60 DEG. BASE)
6" STD. STEEL
12" STD. STEEL
18" STD. STEEL
24" STD. STEEL

Figure 4

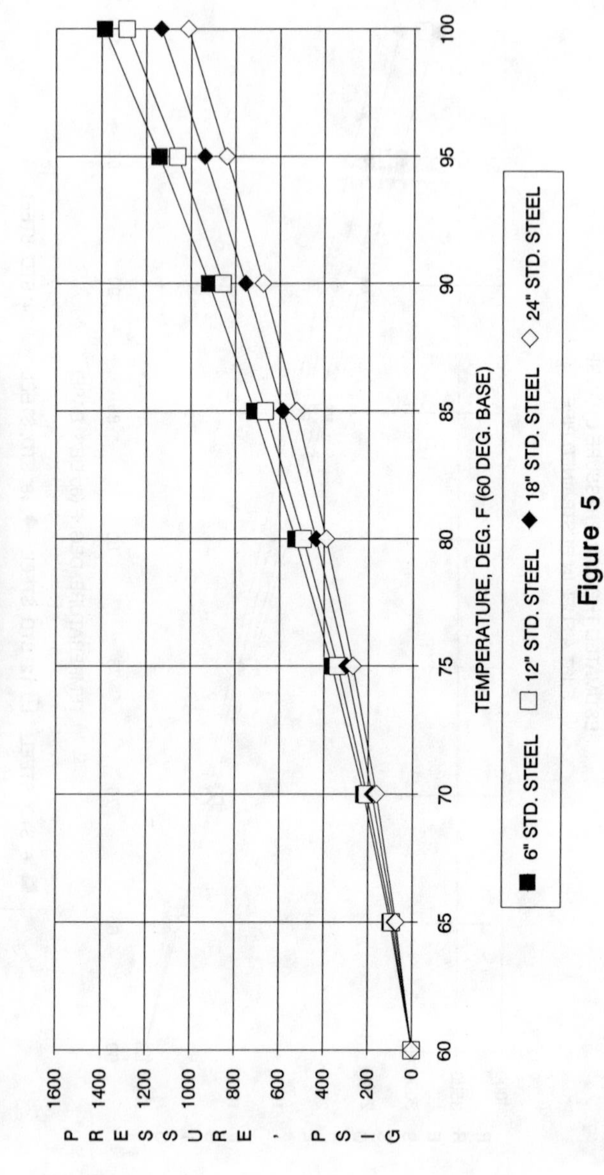
ESTIMATED THERMAL PRESSURE CHANGE
FRESH WATER IN UNRESTRAINED PIPE
PRESSURE - PSIG
1600
1400
1200
1000
800
600
400
200
0
60
65
70
75
80
85
90
95
100
TEMPERATURE, DEG. F (60 DEG. BASE)
6" STD. STEEL
12" STD. STEEL
18" STD. STEEL
24" STD. STEEL

Figure 5

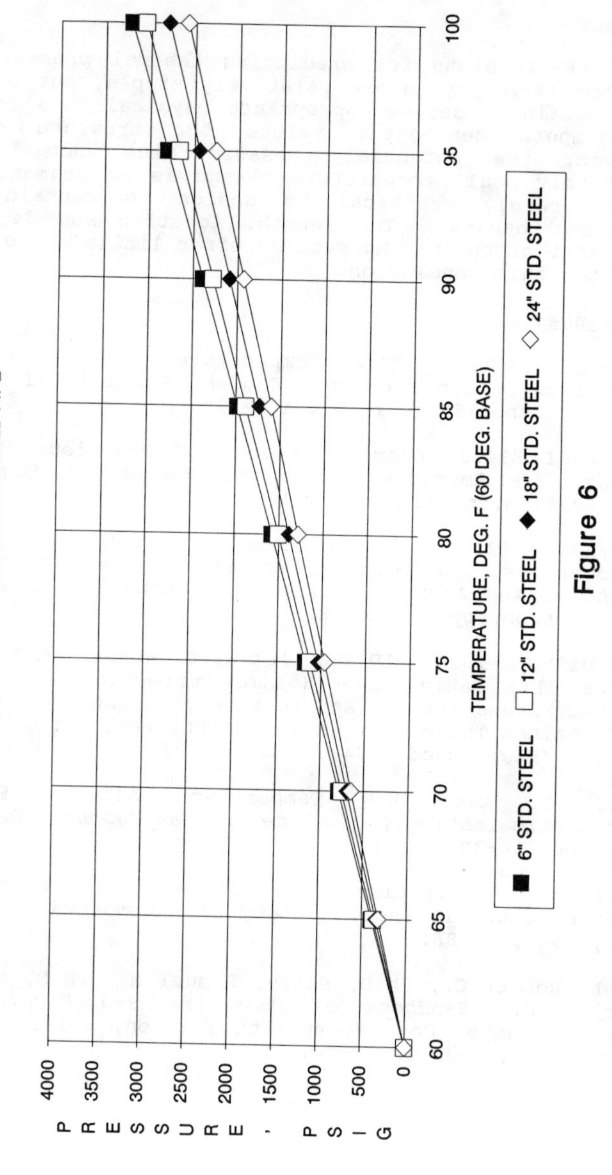
ESTIMATED THERMAL PRESSURE CHANGE
JET A FUEL IN UNRESTRAINED PIPE
PRESSURE - PSIG
4000
3500
3000
2500
2000
1500
1000
500
0
60
65
70
75
80
85
90
95
100
TEMPERATURE, DEG. F (60 DEG. BASE)
6" STD. STEEL
12" STD. STEEL
18" STD. STEEL
24" STD. STEEL

Figure 6

Summary

Expressions for predicting thermal pressure rise in blocked-in piping are relatively simple, but one must also obtain or derive appropriate physical data in order to compute meaningful values for pressure change. However, the potential pressure rise caused by a relatively small temperature change is so dramatic that it is rarely practical to design to contain these pressure changes. The length, location and degree of restraint of the piping segment offer little or no reason to alter this conclusion.

References

Arnold, C. L., "How Temperature Affects Pipeline Hydraulics Design for Heavy Crudes, Resid," *Oil and Gas Journal*, June 15, 1981, pp. 104-120.

ASME/ANSI B31.3, *Chemical Plant and Petroleum Refinery Piping*, The American Society of Mechanical Engineers, 1984 Edition, p. 11 and pp. 189-202.

ASME/ANSI B31.4, *Liquid Transportation Systems for Hydrocarbons, Liquid Petroleum Gas, Anhydrous Ammonia and Alcohols*, The American Society of Mechanical Engineers, 1986 Edition, pp. 7 and 28.

ASTM D1250/API 2540/IP 200, *Petroleum Measurement Tables*, Volume III, Table 6C - Volume Correction Factors for Individual and Special Applications, Volume Correction to 60°F Against Thermal Expansion Coefficients at 60°F, 1980 Edition (Reapproved 1984).

Gray, J. C., "How Temperature Affects Pipeline Hydrostatic Testing," *Pipeline and Gas Journal*, December, 1976, pp. 26-30.

Handbook of Aviation Fuel Properties, Coordinating Research Council, Inc., Society of Automotive Engineers, Inc., 1983, p. 86.

Weast, Robert C., Ph.D, Selby, Samuel M., Ph.D, Hodgman, Charles D., *Handbook of Chemistry and Physics*, The Chemical Rubber Co., Forty-Fith Edition, 1964, pp. F-4, F-9 to F-11, F-33 and F-60.

The NOVA SPAN Bridge System

Keith A. Wilson[1]

Abstract

The NOVA SPAN Bridge concept provides the basis for the design and construction of a new variant in the field of bridges and culverts which, in its performance and operation, incorporates many of the features and characteristics of both culverts and bridges. The system accommodates almost any desired shape, height of cover and design loading. It is most cost-effective for applications with low heights of cover.

Introduction

The concept and principles of flexible metal conduits interacting with their structural backfill envelope are well understood and accepted. Similarly, the composite action between concrete and steel formwork /tensile membranes is widely used for economical and efficient construction.

NOVA SPAN Bridges effectively combine these two principles of modern structural engineering to produce an evolution in Long Span SPCSP technology.

A significant feature of NOVA SPAN Bridges is that not only is the construction period shortened considerably (usually in the order of 20-30%), but the the total installed costs are also significantly reduced (by up to 40%). The system is effective for spans ranging from 6.0 metres (20'-0") to 15.0 metres (50'-0"), and is used for culvert applications and grade separations. Non-standard loads can be easily accommodated in the design phase, while standard road and railway live loads are applied automatically.

----

[1] Chief Engineer, Fletcher Manufacturing Pty. Ltd. 127-141 Bath Road, Sutherland, Sydney, Australia. 2232

Probably the main advantage of this system is its ability to function with a minimal height of cover over the structure. Often SPCSP or Long Span structures are precluded from consideration for certain applications due to minimum height of cover restrictions. With the NOVA SPAN system however, the wearing surface of the road can, if required, be installed directly in contact with the composite concrete deck. For instance, a long-span SPCSP structure (eg "Super-Span") of maximum span 9.5 metres (31'-2") requires about 900mm (3'-0") of cover, plus pavement construction. By contrast, a NOVA SPAN of exactly the same configuration and loading needs only 400mm (<16"), resulting in a saving of 500mm (20") of formation construction.

The System

The NOVA SPAN System is essentially a SPCSP structure constructed with a composite reinforced concrete deck. The deck is attached to the top arc of the structure by means of welded shear studs (high strength bolts have also been used as shear connectors). These shear connectors cause the steel top arc and concrete deck to act compositely in resisting the applied design loads.

Horizontal ellipses are probably the shapes most commonly used with the system. However, high and low profile arches are also quite suitable. Corrugated metal walls always form the walls in order to maintain the advantages of flexible structures. The concrete segments of the structure are usually cast-in-situ (but may also be precast) and are reinforced or prestressed. Cast-in-situ roof slabs often necessitate the use of temporary propping to support the weight of the wet concrete.

The concrete roof section is extended across the structure to both sides, beyond the width of the structure, to form 'roof arms'. These arms bear directly on the surrounding backfill. Sufficient reaction is provided to resist the forces for which the roof and its arms are designed. Vertical reactions are comfortably achievable due to the inherent strength of the compacted structural backfill, while horizontal reactions are generated by a combination of friction on the arms and passive earth pressure on the wall.

The arm lengths and thicknesses are designed to produce requisite factors of safety against failure of the structure. Failure mechanisms to be designed against are:

* failure of the structural backfill - checks should also be made of the strength of the embankment beyond the zone of structural backfill;
* failure of the metal wall - ring compression procedures apply;
* failure of the arms - check for bending, shear and compression buckling using normal concrete design procedures.

A proprietary computer programme has been developed to analyse and design this composite section, and is basically comprised of two sections; Analysis and Design. The analysis section assumes the top arc to be a two-pinned arch, hinged at the extremities of the top arc, divided into ten equal segments. See Figure 1. The programme allows for analysis of:

Dead load of SPCSP shell only
Dead load of SPCSP and wet concrete
Dead load of SPCSP, concrete and soil backfill
Total dead load, plus live load(s).

The live load can be entered as a standard loading (eg HS25), or special loadings can be entered axle by axle (eg, off-highway haul trucks). Railway live loads can also be catered for, either in standard or special configurations.

The structure, viewed as an arch supported at the roof-arm junction by the side-wall plates, acting in axial compression, and the arm effects, is analysed for maximum and minimum moments, shear and axial forces for each loading condition requested. Any load to the arms caused by the backfill acting against them due to outward thrust of the side plates is then considered as causing additional positive bending in the roof: this extra moment is added to that derived in the arch analysis.

The design section takes data derived in the analysis and, given prescribed further data, calculates the top and/or bottom reinforcing steel, the SPCSP thickness, depth of concrete deck, shear connector spacings, etc. A check can also be made to verify structural adequacy if all the variables are known.

The output for both the analysis and design sections is clear and of a form that could be presented for checking by an independent engineer. Analysis results are printed for both Working Stess Design (WSD) and for Limit State Design (LSD). Currently only the WSD results are used in the design segment.

Equilibrium of the forces shown in Figure 1 is the key to the operation of the structure. In the region marked "A", a wedge of backfill (usually well-graded and compacted crushed granular material) provides reactions to support the slab. This reaction is balanced by vectored forces generated by the thrust of the metal wall on the backfill, and the passive-earth resistance of the backfill itself.

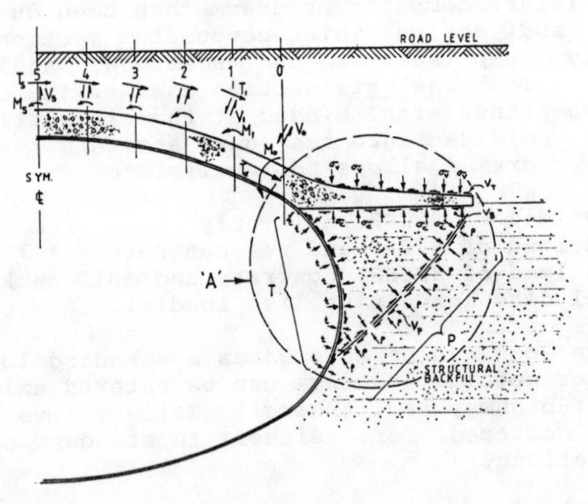

Figure 1

These three forces are shown more clearly in Figure 2, where:

* R = the total reaction of all dead and live loads acting on the structure,
* T = the thrust generated by the flexible metal walls due to ring compression, and
* P = the total passive earth resistance.

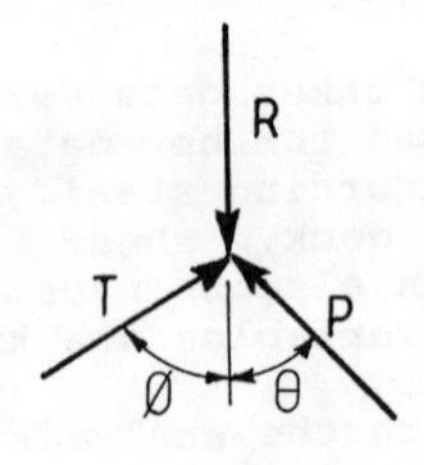

Figure 2

The forces T and P are combined by adjusting the relevant geometrical parameters such that their vertical components are equal to or greater than R. The required factors of safety can be set by adjusting T(max) and P(max) as described.

The Thrust can be increased as desired by reducing the wall radius, Rs, while maintaining a suitable plate thickness. The Passive forces can be increased as desired by using more or better structural backfill materials, or by increasing arm length. Both R(max) and its location can be adjusted by means of altering the arm length to suit requirements.

It should be noted that the granular backfill is quite lightly loaded in relation to its confined strength, and hence its strength will seldom be a limiting factor. Within limits, the further the arm is extended, the thinner can be the SPCSP, since more of the reaction R is directed to be resisted by the passive earth pressure. This must be balanced, of course, by the longer arm needing to be thicker to resist the greater moments caused by a longer lever-arm.

Confinement of the backfill area due to arm extension also increases with longer arms. The geometry of the SPCSP structure and slab dimensions can therefore be balanced for maximum performance and economy.

Should arm settlement occur without corresponding footing or floor settlement, the wall radius, Rs, will tend to become smaller. This effect will automatically maintain the thrust capacity of the SPCSP and its factor of safety.

Hydraulics

The hydraulic capacity of these NOVA SPAN Bridges is similar to any other CSP culvert. Generally, of course, higher velocities can be tolerated through a NOVA SPAN culvert than through a conventional bridge construction. This is due to the presence of invert plates and suitable bed preparation and protection at both inlet and outlet.

One of the main features of NOVA SPAN Bridges is their low profile and low cover requirements. In order to facilitate determination of a particular profile's hydraulic capacity, an estimating chart was devised. This chart plots Design Discharge (in cumecs and c.f.s.) against Roadway Height above Streambed (in

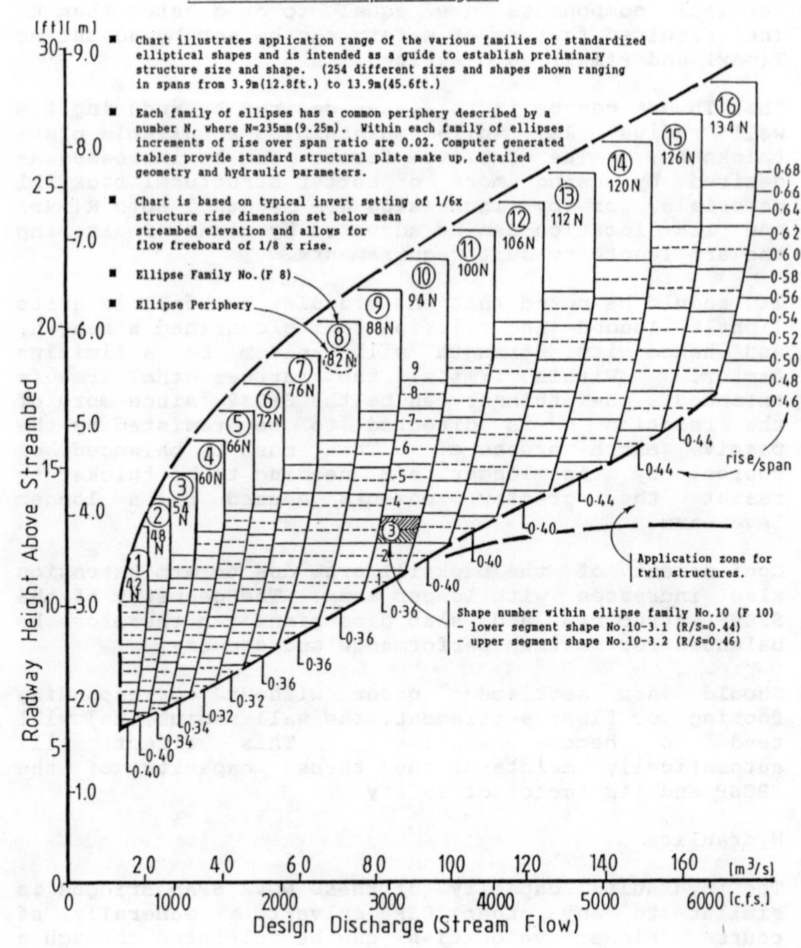
APPLICATION CHART
(for horizontal ellipse designs)
Chart illustrates application range of the various families of standardized elliptical shapes and is intended as a guide to establish preliminary structure size and shape. (254 different sizes and shapes shown ranging in spans from 3.9m(12.8ft.) to 13.9m(45.6ft.)
Each family of ellipses has a common periphery described by a number N, where N=235mm(9.25m). Within each family of ellipses increments of rise over span ratio are 0.02. Computer generated tables provide standard structural plate make up, detailed geometry and hydraulic parameters.
Chart is based on typical invert setting of 1/6x structure rise dimension set below mean streambed elevation and allows for flow freeboard of 1/8 x rise.
Ellipse Family No.(F 8)
Ellipse Periphery
Shape number within ellipse family No.10 (F 10)
- lower segment shape No.10-3.1 (R/S=0.44)
- upper segment shape No.10-3.2 (R/S=0.46)
Application zone for twin structures.
rise/span
1 42 N
2 48 N
3 54 N
4 60N
5 66N
6 72N
7 76N
8 82N
9 88N
10 94N
11 100N
12 106N
13 112N
14 120N
15 126N
16 134N
0·68
0·66
0·64
0·62
0·60
0·58
0·56
0·54
0·52
0·50
0·48
0·46
0·44
0·40
0·36
0·34
0·32
Roadway Height Above Streambed
[ft][m]
Design Discharge (Stream Flow)
[m³/s]
[c.f.s.]

Figure 3

metres and feet). See Figure 3. Over 250 ellipse shapes are shown on the chart with Rise/Span ratios ranging 0.32 to 0.68, thus acommodating a large range of topographies. Spans range 3875mm (12.7 ft) to 13885mm (45.6 ft). These profiles are grouped into 'families', each family sharing a common periphery but varying in their rise/span ratios. This variation allows the most efficient and economical shape to be utilised at each individual location.

A computer-generated hydraulics manual was also developed to be read in conjunction with, and to supplement the chart. The manual provides all data necessary for a complete hydraulic analysis of each culvert selected.

The common practice in Canada is to set the invert of the ellipse one sixth (1/6) of the structure's rise dimension below the streambed elevation. Also a freeboard or aflux of one eighth (1/8) of the rise is allowed for debris and ice-flow.

The designer enters the chart with the design flow on the horizontal axis. The height from roadway to streambed is then projected across from the vertical axis. This height comprises headwater depth, estimated headloss, freeboard, structure depth (plate plus concrete), fill (if any), and pavement depth. At the intersection of these two ordinates, a structure profile is selected. The designer then references the appropriate pages of the hydraulics manual to carry out detailed hydraulic calculations.

Testing

In order to verify that this evolutionary development performed as well in service as in theory, the Alberta Transportation and Utilities Department, in conjunction with the Alberta Research Council, tested a full-size NOVA SPAN Bridge in November, 1983.

The test site was at Blairmore Creek in southern Alberta. The structure, known in Canada as an ABC Structure (Arch-Beam Culvert), was a horizontal ellipse 8.6m span x 4.5m rise (28'-3" x 14'-9"), having an obvert about 28.6m (93'-10") long.

The stiffening cap of reinforced concrete was a nominal 510mm (20") with arms extending about 1.8m (6'-0") horizontally on each side.

Strain gauges were placed at four cross-sections,

spaced at 4.6m (15'-0") centres, in the central section of the pipe. Wires leading from the gauges to a digital strain indicator at one headwall, were housed in two conduits within the concrete cap. Vertical and horizontal deflections were measured by a tape extensometer attached to eye-bolts located at the same cross-sections as the strain gauges. Soil pressure cells were also located at selected locations.

Two static load tests were carried out. The first involved a self-propelled earthmover which imposed gross axle weights of 311 and 440 kN (70,000 and 99,000 pounds). Recordings were made when soil cover was only 100mm (4").

The second test was made when cover had increased to 740mm (29"). In this instance, the load was applied via two 0.61 x 0.91m (2ft x 3ft) H-pile pads, placed 2.1m (7ft) apart. These pads, of size to simulate a standard highway vehicle footprint, were then loaded with precast concrete bridge-deck planks. Strain gauge measurements were made for various load stages ranging from 462 to 1,080 kN (104 to 243 kips).

The maximum deflection measured under the heaviest static load (1080 kN) was 1.5mm (0.059"), with corresponding outward deflections at the springline of 0.8mm (0.031"). The recorded tensile strains indicated that no cracking of the concrete occured, despite the loads being some six times the specified highway design load. By correlating the obtained data with shell theory, it appears that the loads are carried predominantly by arching, with some longitudinal load distribution.

In Australia, we are about to build a high profile arch, 13.8m (45'-3") span x 84.0m (275'-6") long. Plans are well underway to instrument this structure during construction in order to fully investigate its structural behaviour under load. This structure will carry a highway interchange feeder ramp over twin railway tracks. The combination of super-elevation from a horizontal curve with the natural grade of the ramp, will result in cover heights ranging from an absolute minimum at one end to about 2.0m (6'-5") at the other. Hence, a comprehensive range of loading conditions will be available for monitoring.

Warrnambool, Victoria

BRIDGE TYPE	SUPER-SPAN	NOVA SPAN BRIDGE	BEBO ARCH	REINFORCED EARTH	CONVENTIONAL
BRIDGE PROFILE	+ 500	0.0	+ 100	+ 100	+ 100
$000's	220	245	320	420	590
CONSTRUCTION WEEKS	6	8	7–8	10–12	15–16

Brighton, South Australia

BRIDGE TYPE		NOVA SPAN BRIDGE		STEEL BEAMS/ CONCRETE DECK	PRESTRESSED CONCRETE BEAMS
BRIDGE PROFILE					
$000's		190		310	390
CONSTRUCTION WEEKS		10–12		16–18	22–24

COMPARISON OF BRIDGE ALTERNATIVES

Table 1

Construction

The construction of a NOVA SPAN Bridge, as for a Long-Span SPCSP Structure, essentially comprises four basic stages, viz:

Site preparation - involves foundation investigation, and if necessary, remedial works; locating exact site; preshaping of bed for ellipses or casting strip footings for arches.

Plate Assembly - involves plate stacking and handling; panel preassembly; bolt tightening; temporary props and tie-rods; shape tolerances.

Backfilling - involves selection suitable material; equipment selection & supervision; adequate placing and compaction; shape control.

Concreting - involves fixing of reinforcing steel; construction of formwork; placement of concrete in the specified sequence; screeding and finishing; stripping of formwork.

Construction of a NOVA SPAN Bridge varies from the well known Long-Span SPCSP Structure in some notable areas, viz:

* As the roof segments (top arcs) are usually of quite large radius (up to 12.0m or 40 ft), the preassembled panels are often lifted using spreader beams to avoid buckling in both the longitudinal and transverse directions. These spreader beams often take the form of a large angles with 'key-hole' slots cut into one leg: these slots slip over the shear connector heads and lock on to the connector shanks.

* Due again to the large top arc radii, the roof sections usually need temporary propping for structures over 7.6m (25 ft) span. The temporary props are positioned accurately such that the true theoretical profile is achieved during assembly, and maintained until the concrete cap has attained adequate strength. During backfilling operations, the roof segment may 'peak', thus lifting off the props. In some cases, no peaking will be experienced. Whichever situation eventuates, it is necessary that there is a 25mm (1") gap between the props and the underside of the SPCSP shell.

* Concreting is poured in a specific sequence which is important for structural reasons. Firstly the arms are poured and extended to about 1.0m (3'-0")

up the arc, and 5 to 6 metres along the structure. Secondly, a central section of the cap - about 3.0m (10'-0") wide - is poured symmetrically about thestructure centre-line, and 3 to 4 metres longitudinally. The remainder of the top arc cap is then poured, as far as practicable maintaining the loads each side of the centre-line balanced.

This sequence is repeated along the structure. Headwalls are poured when the concrete at the cap interface has sufficient set to resist being displaced by the fresh concrete being poured into the headwall forms, but prior to that concrete going 'off' and resulting in a cold joint.

The purpose of the 25mm gap above the props is to allow the shell to settle downwards with the weight of the wet concrete being poured over the top arc. If the shell comes to rest on the props, it cannot snap-through to collapse, the propping having been designed to support the full weight of the steel shell and wet concrete. If, however, the shell settles only part of the 25mm, there is still no cause for alarm: this means that the top arc is supporting the full load by arching across its full width. In either case, whatever settlement has occured will have caused a certain amount of movement outwards at the backfill interface. This movement will have the effect of activating passive pressure in the backfill. This passive pressure is part of the structural mechanism by which the whole system supports the applied loads.

Shape control is considerably easier with NOVA SPANS than other Long Spans because of the temporary propping. Hence, the owner can confidently expect that the profile and clearances ordered will be what is built: this is most important in the case of grade separations.

Cost Comparison

1. AUSTRALIA

The following are two examples of NOVA SPAN Bridges that have been built in Australia recently. The first, for the Warrnambool City Council in Victoria, was built in early 1988. The second was built in Adelaide, South Australia, and completed in May 1989.

1.1 Warrnambool

This structure was originally offered as a Super-Span. However, the Super-Span concept was considered unacceptable due to the minimum height of cover necessary for this type of structure: the height necessitated an increase in road grade line that was precluded by environmental constraints on both approaches. The NOVA SPAN concept allowed the Council to build exactly to the desired grade line. Other structure types investigated by the Council were:

Bebo Arch - a precast, reinforced concrete arch.
Reinforced Earth - precast, prestressed concrete planks supported on precast concrete panels, tied into the fill with galvanised strips.
Conventional Bridge - a three-span, prestressed concrete bridge.

1.2 Adelaide
In this case the Council called tenders for a Design and Construct contract to replace a 40 year old concrete portal bridge. Initially, the contractor's consultant designed a bridge with steel beams and a cast-in-situ, composite concrete deck. An alternate design was also prepared with precast concrete girders and a concrete wearing surface. At that stage we were approached for a NOVA SPAN proposal, which was ultimately accepted.

A summary of the costs and construction periods for each projects is shown in Table 1 below.

2. CANADA

Another example is drawn from the files of the Alberta Transportation Department. The details were:

Design Data:

Roadway over stream	skew 30 degrees
Design Q	57 cumecs (2010 cfs)
Finished roadway width	13.75m (45')
Height, roadway to streambed	4.15m (13.6')
Standard highway loading	MS 225 (HS 25)
Natural streambed width	7.6m (25')
Streambed width under structure	6m (20')
Min. opening width for ice/debris	6m (20')
Min. freeboard for ice/debris	0.8m (2.5')

Alternatives:
Conventional Bridge
2:1 batters need bridge length of 24m (80')
Precast girders on concrete and steel foundation - estimated rates are $800 to $900 / sq m ($74 to $84 / sq ft).

Estimate 13.7m x 24m = 329 sq m
@ $800 / sq m
= 329 x 800
= $263,200

NOVA SPAN Bridge
Horizontal ellipse 7.7m span x 4.0m rise
(26' x 13')
Sized to satisfy all hydraulic, environmental and structural design criteria

Total estimated cost = $160,000

CONCLUSION

The conventional and the NOVA SPAN Bridges satisfy all the structural, hydraulic and environmental criteria

The NOVA SPAN Bridge provides a direct cost saving of $103,200, or approximately 40% over the standard alternative

RECOMMENDATION

Select the NOVA SPAN Bridge alternative!!

3. AUSTRALIA

More recently, the State Electricity Commission of Victoria called tenders for two bridges in their open-cut coal mine at Morwell in eastern Victoria. The bridges were designed to carry off-highway haul trucks with an axle load of 390 kN (44 US tons). One option was a cast-in-situ, prestressed concrete deck, 650mm (25.6") thick, supported by Reinforced Earth walls.

The second option for the contractors to bid on was the NOVA SPAN system with 600mm (23.6") of reinforced concrete over the 7mm (0.276") corrugated steel plates.

The NOVA SPAN alternative saved the Commission approximately $150,000 in a total contract bid price of $1.3 million. Construction will commence in January 1990.

Safety Assessment of Welded Pipelines Undergoing Large Ground Deformation

Nobuhisa Suzuki[1] and Akira Hagio[2]

Abstract

Spatial distribution and modelling of liquefaction induced permanent large ground deformation are presented. The permanent large ground displacement patterns required for the safety assessment of buried pipelines are idealized on the basis of case history information of the 1964 Niigata Earthquake and the 1983 Nihonkai Chubu Earthquake. The idealized patterns are simply presented as function of spatial extent of longitudinal transient zone, lateral spreading zone and ground settlement and the maximum displacements. The margin of safety of the continuous buried pipelines of 12- and 24-in. outside diameters subject to the permanent ground deformation are demonstrated in comparisons with the local buckling strength of the pipes.

Introduction

The liquefaction-induced permanent large ground movement, which has a close relationship to soil layer profiles of estimated liquefaction zone, is a significant parameter for the safety assessment of the buried pipelines in a liquifiable zone (Refs.1-5).

We have to estimate the spatial distribution of the ground displacement for the seismic design and deformation analysis of the buried pipelines. It may be difficult to estimate the displacement because we have to deal with a lot of information over a wide area in which lifeline networks are extensively constructed.

The objectives in this paper are therefore twofold. The first objective is to perform the idealization or

---

[1]Senior Researcher and [2]Manager of Technical Research Center, NKK Corporation, Kawasaki, 210 Japan.

modelling of the permanent ground deformation required for the safety assessment of the buried pipelines through investigating the permanent ground displacement patterns occurred during the 1964 Niigata Earthquake and the 1983 Nihonkai Chubu Earthquake. The second objective is to investigate the effects of the permanent ground displacement on the continuous buried pipelines and to demonstrate the margin of safety of the pipelines in comparisons with the local buckling strength of thin-wall cylindrical shells.

Measured Permanent Ground Displacement

Figure 1 shows measured displacement vectors in a 1000*800m rectangular area of interest in Niigata City. The Niigata Railway Station is located in the south of the area which covers the center of Niigata City where lifeline network systems for gas and water distribution had been constructed before 1964.

Figure 2 represents the measured permanent ground displacement vectors occurred during the 1983 Nihonkai Chubu Earthquake in two rectangular areas of 800*1600m wide in the northern and southern part of Noshiro City.

The displacement vectors shown in Figs.1 and 2 were measured at a number of specified points on the ground such as manholes, the lower end of poles and corners of buildings. While, we have neglected the data measured at roof edges, fences and guard rails because they are not supposed to have behaved similarly to the ground movement. The areas of interest illustrated in Figs.1 and 2 are discretized into 50*50m regular squares.

Spatial Distribution of Permanent Ground Displacement

As shown in Figs.1 and 2, the measured displacement vectors are distributed over the entire areas of interest. In order to estimate the spatial distribution of the permanent ground displacement we have to evaluate the magnitude of the permanent ground displacement at any point. Figure 3 shows a schematic illustration of interpolation of the displacement vector at a point of intersection (A) using several adjacent displacement vectors. If we express the deformation of the circular region including the point of intersection as

$$u=ax^2+bxy+cy^2+dx+ey+f \qquad (1)$$

$$v=gx^2+hxy+iy^2+jx+ky+l \qquad (2)$$

the displacement at the point can be calculated by knowing the coefficients from (a) through (l) by the

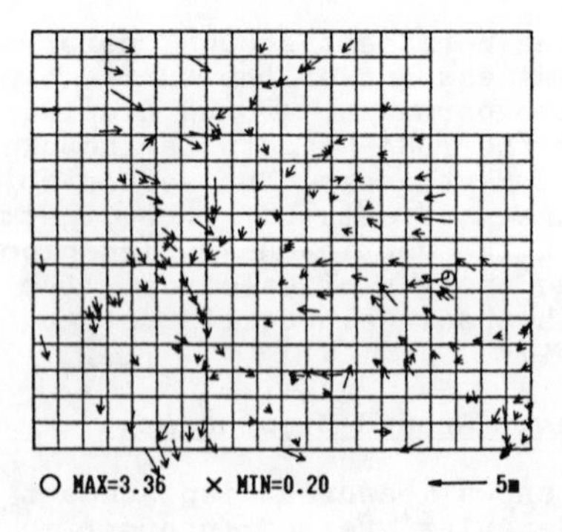

Figure 1. Measured Displacement in Niigata City

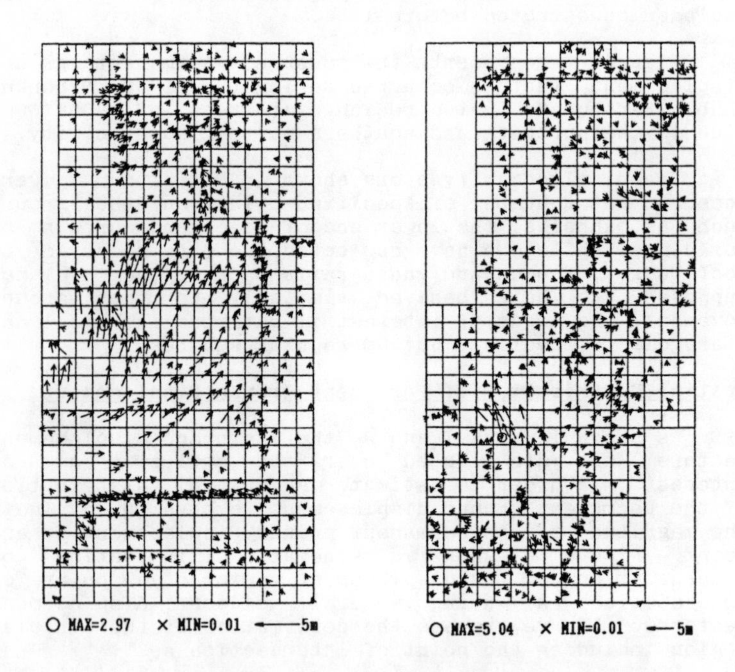

Figure 2. Measured Displacement in Noshiro City
(a)Northern Part, (b)Southern Part

Least Square Method. Where u and v express the displacement vector in the horizontal and vertical direction, respectively.

Figures 4 and 5 show calculated results of the spatial distribution of the PGD and the deformed configuration of the rectangular area in Niigata City. Furthermore, Figs.6 and 7 are provided for supplementary data for the permanent ground displacement vectors separated along the grid lines and perpendicular to the grid lines. Also the same methodology was applied to the analysis of the spatial distribution of vertical ground displacement and the results are shown in Fig.8.

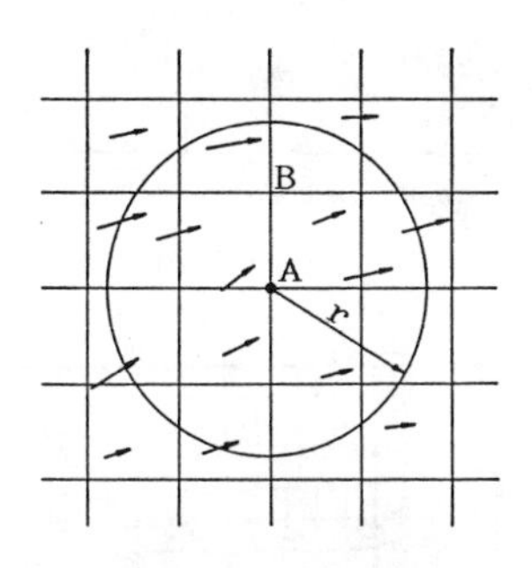

Figure 3. Interporation of Displacement Vector

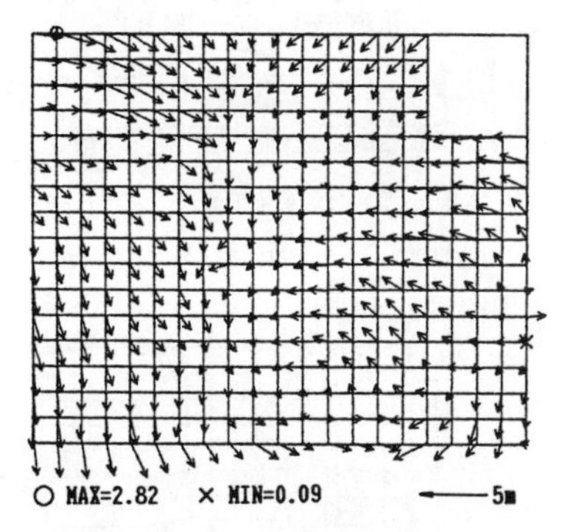

Figure 4. Estimated Vector in Niigata City

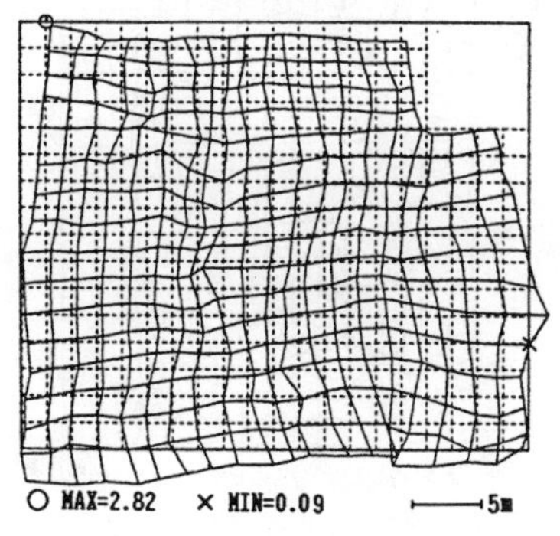

Figure 5. Deformation in Niigata City

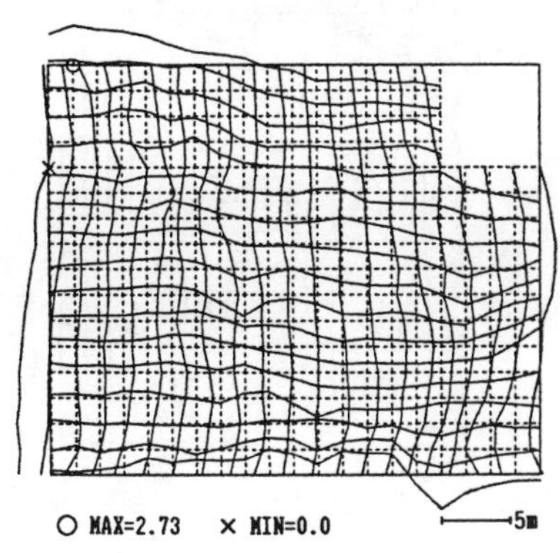

Figure 6. Longitudanal Displacement

The spatial distribution in the northern part and southern part of Noshiro City were estimated in the same manner as in Niigata City. The deformed configurations of the areas are presented in Figs.9 and 10.

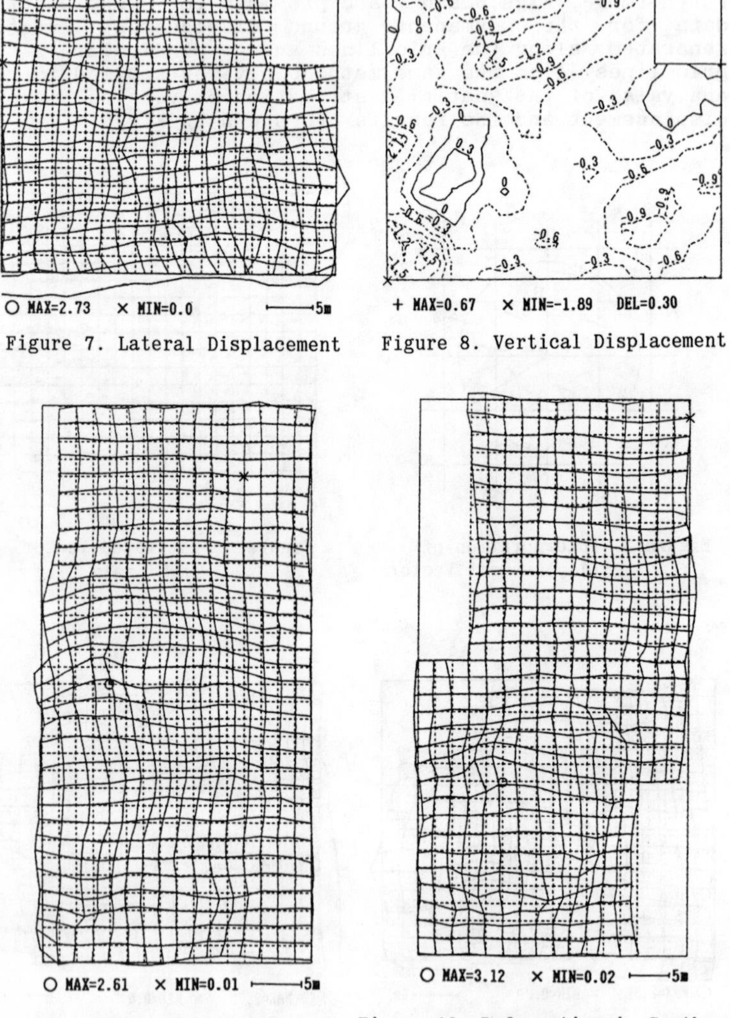

Figure 7. Lateral Displacement

Figure 8. Vertical Displacement

Figure 9. Deformation in Northern Part of Noshiro City

Figure 10. Deformation in Southern Part of Noshiro City

Modelling of Permanent Ground Deformation

It would be actually possible for us to estimate the ground displacement along the buried pipelines having arbitrary shape by the above-mentioned methodology. The ground displacement estimated in this manner, however, is peculiar to the corresponding pipeline so that it can not express a spatial ground distribution. If we regard the grid lines as the straight continuous pipelines, we can express the general spatial distribution by investigating the PGD along the grid lines presented in Figs.1 and 2.

Then the liquefaction-induced permanent ground displacement for the straight pipelines can be idealized as Figs.11, 12 and 13. Figure 11 shows the longitudinal displacement patterns of the PGD, which may result in the axial deformation of the buried pipelines. The axial ground deformation can be defined by the length of the transition zone L and D, the relative ground displacement.

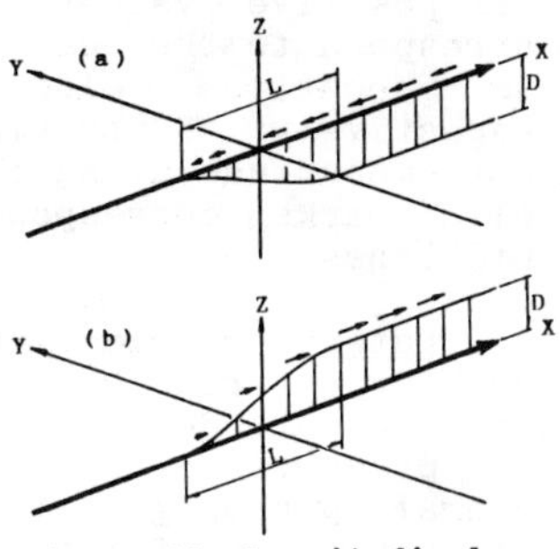

Figure 11. Longitudinal displacement

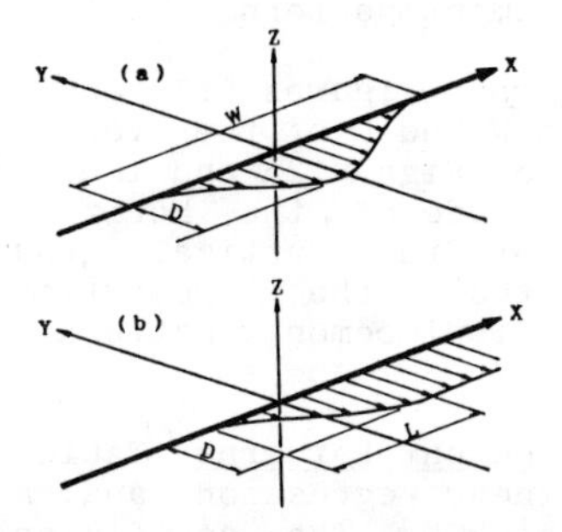

Figure 12. Lateral Displacement

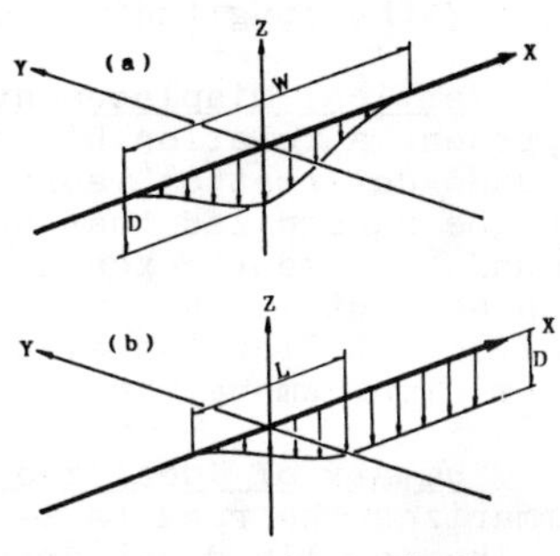

Figure 13. Vertical Displacement

While the buried pipelines may subject to flexural deformation under the displacement patterns shown in Figs.12 and 13. These patterns correspond to the lateral spreading and the ground settlement or rise in which the displacement vectors are perpendicular to the pipeline axis. The pattern of these PGD can be expressed by parameters of the spatial extent of the ground displacement W or L and the maximum ground displacement D. Figures 12a and 13a represent the lateral spreading or the settlement expressed by a sinusoidal function. While, Figs. 12b and 13b represent the shear mode deformation.

Longitudinal Displacement Pattern Figure 14 shows the relationship between L and D of the longitudinal ground displacement along the grid lines. The length of transient zone and the maximum permanent ground displacement are expressed by L and D. The solid lines and the dotted lines represent the results of the linear regression analysis and envelopes of the calculated data. The positive values of the relative displacement, D, correspond to the tensile deformation of the pipeline and the negative values the compressive deformation, respectively. Furthermore, the open marks represent the data measured along the horizontal grid lines and the solid marks correspond to the data along the vertical grid lines.

Lateral Displacement Pattern Figures 15 and 16 show the W or L versus D relationships for the lateral ground displacement defined as Figs.12a and 12b, respectively. The proportional coefficient related to the sinusoidal lateral spreading, illustrated as Fig.12a, is less about 60% than the coefficient for the longitudinal deformation pattern shown in Fig.4. On the other hand, the coefficient for the lateral shear mode spreading, expressed as Fig.12b, is approximately the same value as that of the longitudinal displacement pattern.

Vertical Displacement Pattern Figures 17 and 18 represent a relationship between W and D of the vertical ground deformation represented as Figs. 13a and 13b. It can be recognized that the magnitude of the horizontal ground movement exceed those of the vertical ground displacement from the fact that the proportional coefficient for the sinusoidal displacement pattern is the minimum among the coefficients.

Summary of Idealized Displacement Patterns Table 1 summarizes the results of the linear regression analysis of the idealized patterns of the PGD. The coefficients for the longitudinal ground displacement are almost constant for both of the tensile and compressive deformation and are the largest values in the table. The sinusoidal and the shear mode deformation in horizontal direction are two times as large as the deformation in vertical direction. On the contrary, the shear mode deformation in horizontal and vertical direction are also approximately double the sinusoidal displacement pattern. The envelopes were defined bi-linear curves and drawn considering the coefficient obtained from the regression

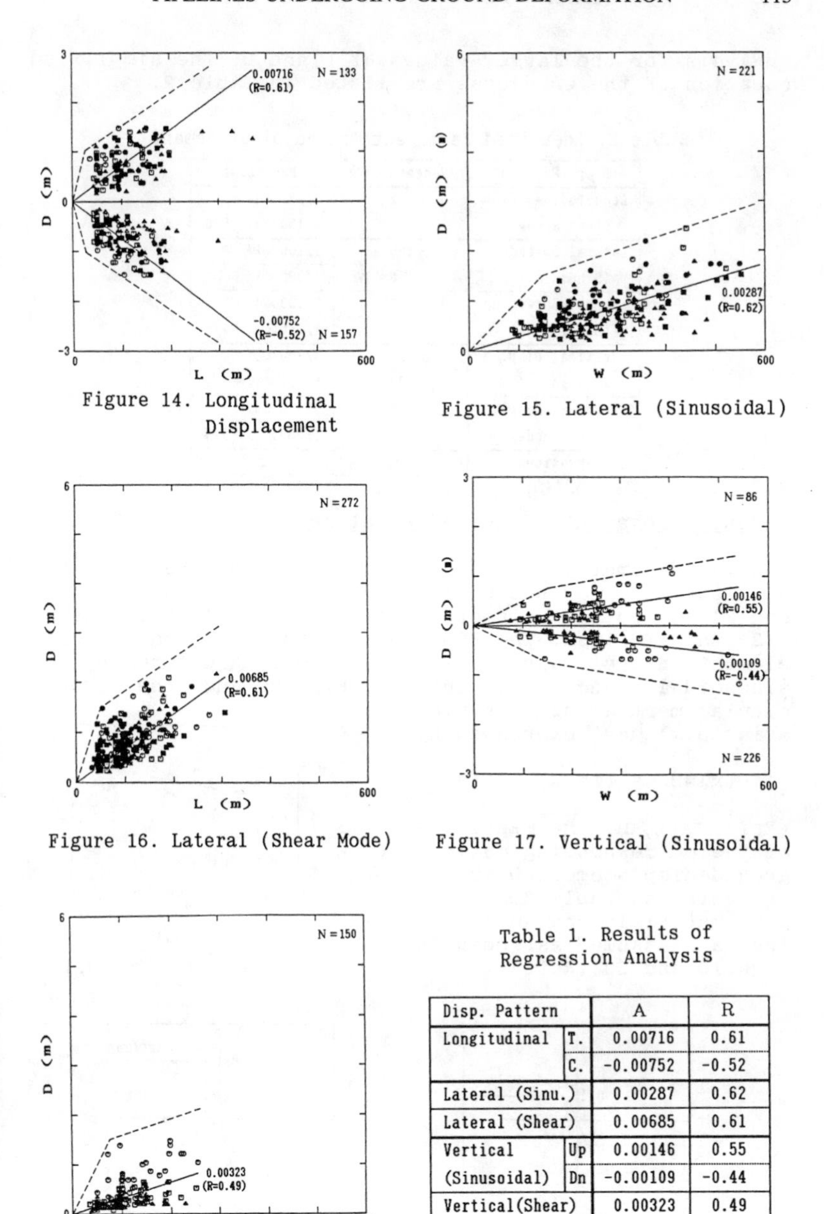

Figure 14. Longitudinal Displacement

Figure 15. Lateral (Sinusoidal)

Figure 16. Lateral (Shear Mode)

Figure 17. Vertical (Sinusoidal)

Figure 18. Vertical (Shear Mode)

Table 1. Results of Regression Analysis

Disp. Pattern		A	R
Longitudinal	T.	0.00716	0.61
	C.	-0.00752	-0.52
Lateral (Sinu.)		0.00287	0.62
Lateral (Shear)		0.00685	0.61
Vertical (Sinusoidal)	Up	0.00146	0.55
	Dn	-0.00109	-0.44
Vertical(Shear)		0.00323	0.49

analysis for the larger values of L and D. The simplified equation of the envelopes are listed in Table 2.

Table 2. Idealized Permanent Ground Displacement

Disp. Pattern	Range(L,W)	Max. Disp.
Longitudinal Disp (Ten. & Com.)	$L \leqq 25$	$1/25 \times L$
	$25 < L$	$1/150 \times L + 5/6$
Lateral Spreading (Sinusoidal)	$W \leqq 150$	$1/100 \times W$
	$150 < W$	$1/300 \times W + 1$
Lataral Spreading (Shear Mode)	$L \leqq 50$	$3/100 \times L$
	$50 < L$	$1/150 \times L + 7/6$
Vertical Disp. (Sinusoidal)	$W \leqq 100$	$1/150 \times W$
	$100 < W$	$1/600 \times W + 1/2$
Vertical Disp. (Shear Mode)	$L \leqq 75$	$1/50 \times L$
	$75 < L$	$1/300 \times L + 5/4$
Dimensions	m	m

Safety Assessment of Buried Pipelines

A parametric study on deformation analysis was carried out for the buried continuous pipelines of 12- and 24-in. outside diameters as well as wall thickness of 1/3 and 1/2 inch. The PGD used for the nonlinear finite element analysis was the lateral spreading defined by the sinusoidal function as shown in Fig.12a and the ground displacement along the pipe axis D(x) was expressed by

$$D(x) = D\cos^2(\pi x/W) \qquad (3)$$

(Refs.2,4) and the magnitude of the corresponding maximum ground displacement D is represented as Table 2. Also, pipe materials and soil springs are assumed as shown in Figs.19 and 20 (Ref.4).

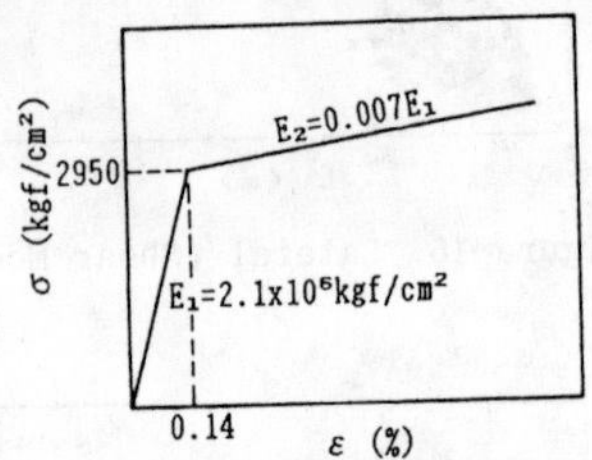

Figure 19. Pipe Material

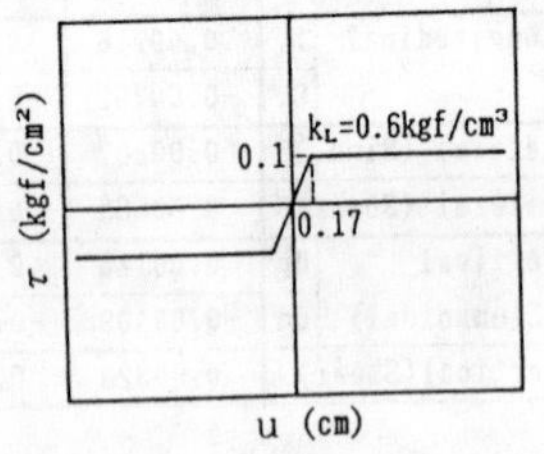

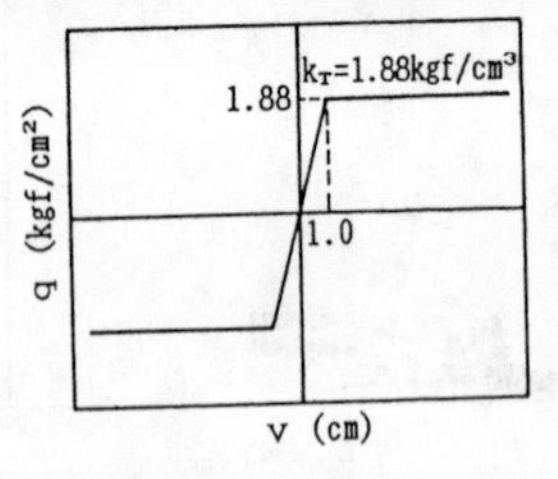

Figure 20. Soil Spring

Maximum strains induced in the axis of symmetry are presented in Fig.21 with respect to various extent of the lateral spreading of the PGD. The circular and triangular marks represent the fiber strains of the 12- and 24-in. pipelines, respectively.

The fiber strains of the pipelines present peak values at W=10 and 20m and decrease rapidly as W increases. Bending strains are predominant when W is less than 200m and the fiber strains decrease gradually and the pipelines present a cable-like behavior at the values of W larger than 300m. However, slight axial tensile strains are induced in the pipelines due to geometric nonlinearlity.

For the safety assessment of the pipelines, the local buckling strains of the pipes due to axial compression are estimated as 0.97% and 0.73% for the 12- and 24-in. pipelines (Ref.6). The margin of safety of the pipelines can be recognized from the fact that the maximum fiber strains illustrated in Fig.21 are less than the buckling strains.

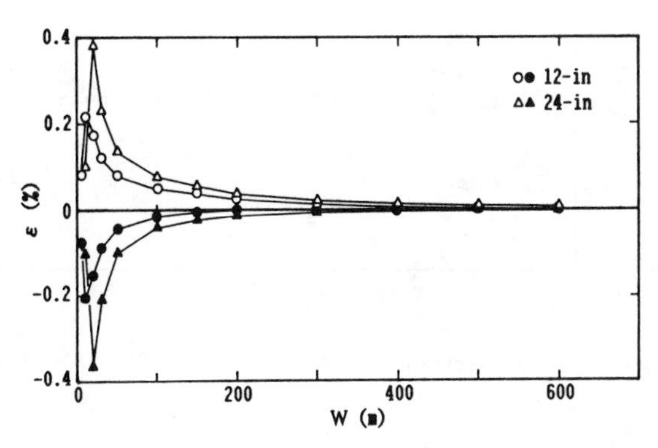

Figure 21. Maximum Fiber Strain due to Lateral Spreading

Figures 22 and 23 show the maximum fiber strains at the axis of symmetry versus the ground displacement relationship for various extent of the PGD. The fiber strains increase rapidly as the ground displacement increases when W=50m, however, the increase of the strains are restricted by the limitation of the permanent ground displacement written as Table 2. On the other hand when W=600m, the fiber strains do not reach the higher values as in the case of W=50m and the fiber strains remain small.

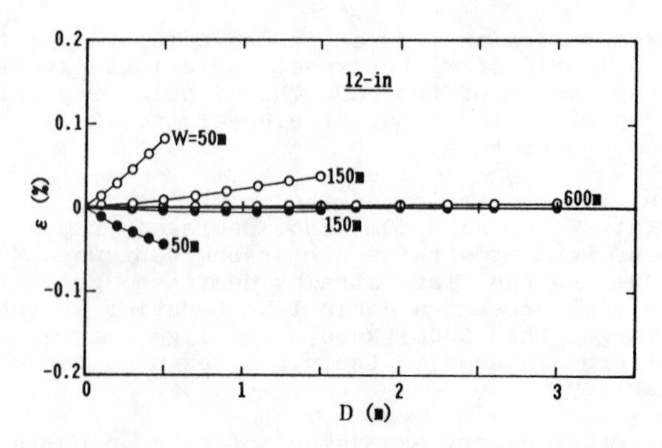

Figure 22. Development of Maximum Fiber Strain (12-in.)

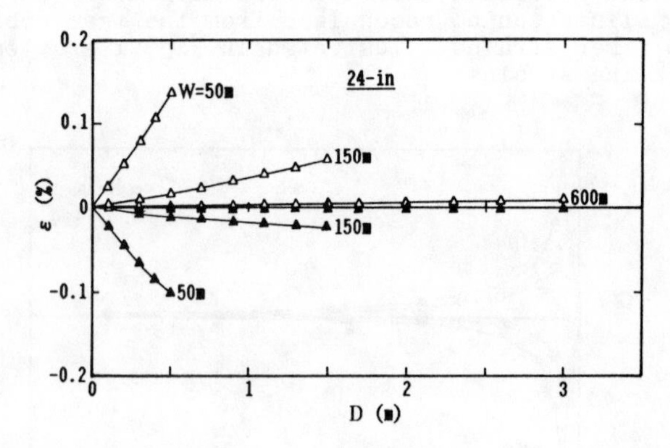

Figure 23. Development of Maximum Fiber Strain (24-in.)

Conclusions

The spatial permanent ground displacement observed in the 1000*800m rectangular area in Niigata City and the 800*1600m rectangular areas in Noshiro City were discussed on the basis of case history information of the 1964 Niigata Earthquake and the 1983 Nihonkai Chubu Earthquake. The permanent displacement patterns of the ground movement along the grid lines were quantitatively investigated, which patterns consist of the longitudinal displacement, the lateral spreading and the vertical displacement. However, the ground displacement pattern was restricted to the lateral sinusoidal spreading, the margin of safety of the continuous buried pipelines are also demonstrated in comparisons with the local buckling strain of cylindrical shells.

We consequently concluded as the following.

(1) The extent of the permanent ground displacement to be considered for the safety assessment can be several hundred meters.

(2) The lower strain can be indeduced in the pipeline with the wider liquifiable zone, and the higher strain can be observed with the narrower liquifiable zone.

(3) We have not focused on the local deformation which width were less than 50 meters. The local ground deformation with narrow with can present a significant effect on the nonlinear deformation of the buried pipelines.

Acknowledgements

The authors also would like to thank Mr.I.Kubo and Mrs.M.Nishigaki of the Japan Industrial Technology Co., Ltd. for their help in the preparation of this paper.

References

1. Hamada, H., Yasuda, S., Isoyama., R. and Emoto, K., "Study on Liquefaction Induced Permanent Ground Displacements," ADEP, 1986, 87p.

2. Kobayashi, T., "Recommended Practice for the Earthquake-Resistant Design of Gas Pipelines, with Liquefaction Considered," Proc. of First Japan-U.S. Workshop, 1988, pp.212-221.

3. O'Rouke, T.D., "Critical Aspects of Soil-Pipeline Interaction for Large Ground Deformation," Proc. of First Japan-U.S. Wokshop, 1988, pp.118-126.

4. Suzuki,N., Arata,O. and Suzuki,I., "Parametric Study on Deformation Analysis of Welded Pipeline Subject to Liquefaction-Induced Permanent Ground Dislacement," Proc. of First Japan-U.S. Workshop, 1988, pp.155-162.

5. O'Rouke M.J, "Approximate Analysis Procedures for Permanent Ground Deformation Effects on Buried Pipelines," Proc. of Second U.S.-Japan Workshop on Liquefaction, 1989.

6. Japan Gas Association, <u>Earthquake Resistant Design Code for Gas Pipeline</u>, 1982.

Prestressed Concrete Pipe Field Test

John R. Thurston,[1] Member ASCE

Abstract

Strain changes in the prestressing wires and deflections of a 2130-mm-diameter precast (embedded cylinder), prestressed concrete pipe were measured during earth backfilling operations. The purpose of the test program was to determine how well strains and deflections that were calculated--using the Bureau of Reclamation method of analysis for calculating earth loads on concrete pipe--compared with measured strains and deflection in a field installation. Measured strains were less than the calculated strains. An alternate method of calculating the earth loads and resulting strains gave a better comparison.

Introduction

The Bureau of Reclamation has calculated the internal forces in concrete pipe using the (Olander 1950) method for many years. Concrete pipe designed by this method has had an excellent performance history.

This article presents the results of a test conducted in 1989 to compare measured concrete pipe strains caused by earth backfill with calculated strains using the Olander method.

Test Program Preparations

The pipeline tested is the intake to the Black Mountain Pumping Plant near Tucson, Arizona. The plant is part of Reclamation's Tucson Aqueduct of the Central Arizona Project. Figure 1 shows the pipeline in the pipe trench.

---

[1]Technical Specialist, Concrete Pipe, Bureau of Reclamation, Denver Office, PO Box 25007, Denver, CO 80225-0007.

The pipeline consists of precast pipe sections having dimensions as shown on figures 2 and 3.

The test program was conducted upon two adjoining pipe sections. The purpose of the test was to measure the changes in strains and temperature on the prestressing wire during earth backfilling. Pipe deflections also were measured.

Figure 3 shows locations of the strain gauges and thermocouples. Eyebolts that were used for measuring the pipe deflections are shown also.

Before mortar coating placement, Styrofoam blocks, 50 mm by 150 mm, were attached to the prestressing wires where the strain gauges and thermocouples were to be attached.

The Styrofoam blocks were removed after the coating was placed, and the strain gauges and thermocouples were attached to the prestressing wires by spot welding. Then, the openings in the coating were filled with hand-placed mortar and painted with coal tar epoxy.

Strain gauge and thermocouple lead wires were led circumferentially around the outside of the pipe to the top of the pipe. Fiberglass bands then were placed over the wires to protect them during the earth and lean concrete backfilling operation.

The preceding preparations were performed at the pipe manufacturing plant in Phoenix, Arizona. The pipe sections were later transported to the pumping plant site.

The trench bottom was prepared by placing a 300-mm thick layer of earth backfill (as shown on Fig. 2) compacted to 95-percent laboratory maximum dry density. After the pipe sections were laid on the earth backfill, a lean concrete bedding was placed.

Earth backfill--compacted as defined above--was placed between the surface of the lean concrete and the top of the pipeline.

The strain gauge and thermocouple wires were led along the tops of the two pipe sections to their common joint. Two polyvinyl chloride (PVC) conduits (150 mm) were placed at right angles to the pipe sections at the joint. The conduits extended to the trench wall, up the trench wall, and through a shallow trench to a nearby building. The strain gauge and thermocouple wires were pulled through the conduits and through a hole in the

Fig. 1. Precast pipe sections in the pipe trench showing the two test sections with fiberglass bands.

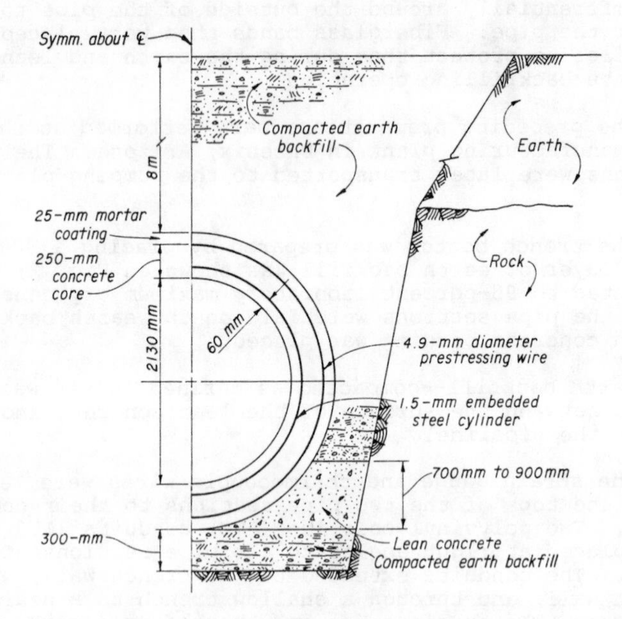

Fig. 2. Pipe dimensions and lean concrete and earth backfill.

wall of the building. The strain gauge and thermocouple wires from each pipe section were attached to separate electronic data recorders located inside the building. Readings were taken of each thermocouple and strain gauge at 30-minute intervals until the test program was completed several weeks later.

Conducting the Test Program

Earth backfilling operations, above the top of the pipe sections, began several days after the test program preparations were completed. Backfill was compacted to 95-percent laboratory maximum dry density.

The backfilling operations required about 40 days. The procedure consisted of placing layers of earth and compacting the layers until the required density was obtained. The thickness placed varied each day; on some days none was placed.

Soil densities and earth backfill elevations were recorded at the end of each working day. The backfill surface was kept level between the trench walls and for 10 meters beyond the end of each test section.

Test Results

Figure 4 shows the average measured strains at backfill completion.

The temperature of the prestressing wire decreased about 1 °C during the backfilling. Additional thermocouples were installed on the inner surfaces of the pipe sections after backfilling began. The inner surface was less than 1 °C cooler than the prestressing wire. Attempt was not made to correct the strains for temperature differential.

The earth backfill was material that had been removed during the trench excavation. It was clayey sand with gravel classified SC. Forty four density tests of the compacted backfill were performed. The average in-place wet unit density was about 2080 kg/m^3.

The deflection measurements were obtained using a dial gauge and a steel tape that was connected between the eyebolts.

Comparison of Test Results With Theory

Strains were calculated, at the strain gauge location, for the typical Olander earth loading shown

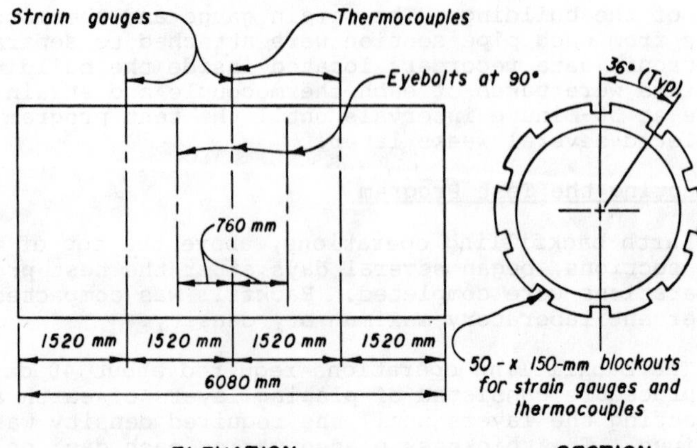

Fig. 3. Locations of strain gauges thermocouples, and eyebolts in each pipe section.

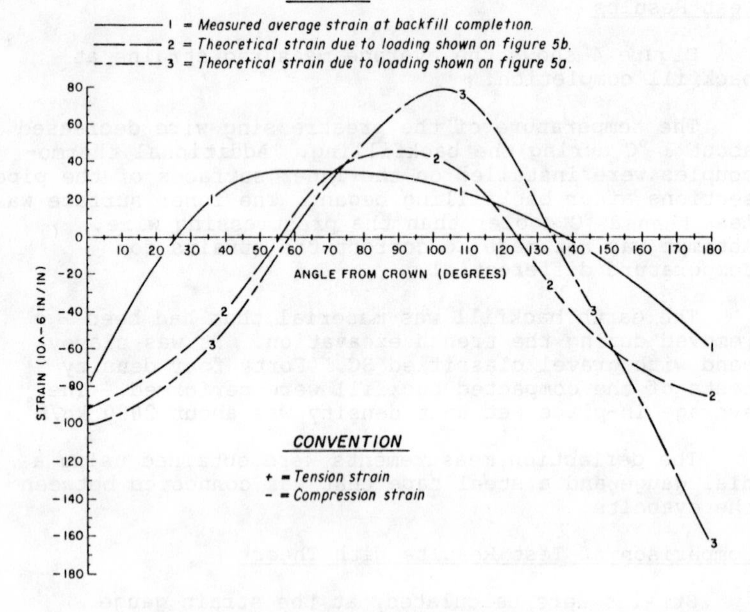

Fig. 4. Measured and theoretical calculated stains

on figure 5a. This is the loading which the Bureau of Reclamation has used for many years to design concrete pipe.

The finite element model shown on figure 6 was used to calculate deflections and to calculate moments and thrusts. Strains were then calculated from the moments and thrusts.

The above calculated strains are shown on figure 4. The pipe wall thickness and amount of prestressing wire was designed for the calculated moments and thrusts at the pipe invert. Figure 4 shows that the measured strain is a maximum at the pipe crown, but is only about one-half of the calculated strain at the invert.

The pipe bedding is non-uniform (see Fig. 2). The pipe invert is located on compacted earth backfill and the pipe haunches are on lean concrete. It was assumed that the pipe support might be other than the single-bulb shape shown of figure 5a. Accordingly, supports consisting of two- and three-bulb shapes acting over difference angles were assumed and strains were calculated.

Strains from the three-bulb shape support shown on figure 5b gave the best comparison with the measured strains. These strains are shown on figure 4. The three-bulb shape support has been discussed by (Heger 1988). He has determined, through a soil-pipe interaction finite element program, that a three-bulb support provides reduced thrusts and moments in a pipe wall.

Reclamation accounts for increased soil loading in a wide trench caused by soil settlement along the sides of the pipe. The soil weight is multiplied by the factor shown in equation 1:

$$(1) \quad F = 1 + 0.2H/D$$

where F equals factor, H equals height of soil above the top of the pipe, and D equals pipe outside diameter.

The preceding equation is supposed to account for long term settlement. Since the measured strains were recorded on the last day of the backfilling operations, the above factor was not used in calculating the theoretical strains.

The measured and calculated deflections at the vertical and horizontal pipe diameters are shown in Table 1. The calculated vertical deflection (using the

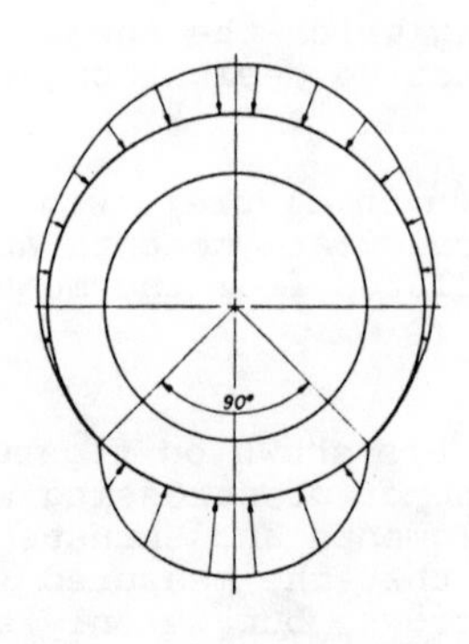

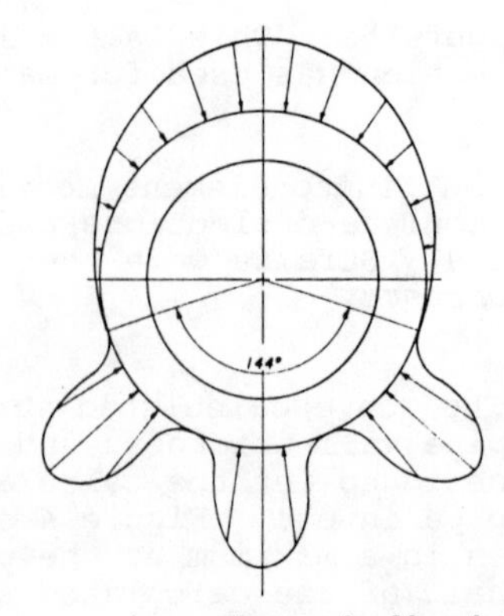

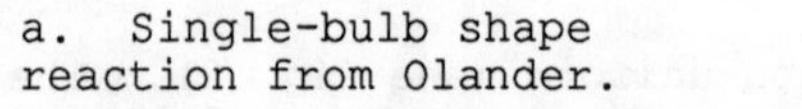

a. Single-bulb shape reaction from Olander.

b. Three-bulb shape reaction from Heger.

Fig. 5. Assumed soil loadings

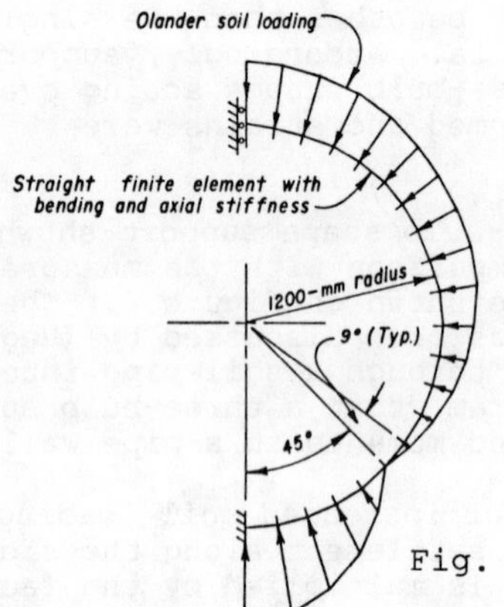

Fig. 6. Finite element model.

Table 1. - Pipe deflections

Location	Average measured deflection (mm)	Calculated Deflection (mm) Fig. 5a Loading	Fig. 5b Loading
Vertical diameter	- 1.02	- 0.94	- 0.64
Horizontal diameter	+ 0.58	+ 0.82	+ 0.56

Notes: + = lengthening
- = shortening

Fig. 5a loading) best matches the measured deflection, whereas the calculated horizontal deflection (using the Fig. 5b) loading best matches the measured deflection.

The deflections are small and the differences between measured and calculated may be due to (1) accuracy of the measuring device, and (2) the soil models do not perfectly represent the actual soil conditions.

Conclusions

The measured prestressing wire strains do not precisely match the calculated strains of either figure 5a or 5b. But upon considering the variable properties of soil and concrete, and that the test was done in a field environment--where conditions can not be as closely controlled as in a laboratory--the results clearly show that the assumed Olander loading gives a conservative design.

Consideration should be given to providing a pipe bedding that will ensure a soft area under the pipe invert with resulting lower strains; therefore, lower pipe costs are achieved.

Acknowledgements

The success of the test program was due to the efforts of the Bureau of Reclamation laboratory and field staff, the cooperation of the general contractor Centric/Jones Constructors, and their subcontractor Ameron who manufactured the pipe.

Appendix I. References

Heger, F. J. (1988) "New Installation Designs for Buried Concrete Pipe", Proceedings of the Conference Sponsored by the Pipeline Division of the American Society of Civil Engineers, Boston, Massachusetts, June 6-7, 1988.

Olander, H. C. (1950) "Stress Analysis of Concrete Pipe" Engineering Monograph No. 6, U.S. Department of the Interior, Bureau of Reclamation, Denver, CO.

AMBIGUITIES AND INANITIONS OF CURRENT PLASTIC PIPE SPECIFICATIONS

LAWRENCE I. ERDOS*
M. ASCE

ABSTRACT:

This paper discusses current specifications, including materials for plastic pipe as compared to other pipe specifications, including rigid pipe materials. It discusses the ambiguities in these plastic pipe specifications as well as the inanitions in these specifications,which can lead the specification user astray, and allow conditions that may result in failure.

More particularly, the issues discussed herein are not only the lack of information, but in fact the existence of disinformation in those specifica tions as they relate to deflection limitations, and any corrective action that may be or is taken to overcome overdeflections. Also discussed is the absence of limitations as to the materials used in these pipes, including the absence of any specif ication as to fillers, impurities and geometric requirements.

MATERIALS SPECIFICATIONS

Current ASTM standards specify, to some extent, the materials that are required to be utilized in the manufacture of various pipes. They also establish some test requirements to determine the composition of these materials. They do not, however, limit the presence of impurities, nor that of fillers.

Typically, the material composition of these plastics, as specified in a pipe specification, refer to other specifications. These specifications typically allow a wide range of chemical composition of the basic material, but sometimes restrict the composition to a select few combinations of the materials. The material specification referenced in the pipe specification is typically selective as to the materials, but is silent as to certain parameters. Some of these inanitions are the lack of limitations as to fillers or to impurities. In formulating the chemical composition of resins, both the resin

***Engineer of Design, City of Los Angeles, Department of Public Works, Bureau of Engineering, Executive Division, Room 900, City Hall, 200 North Spring Street, Los Angeles, California 90012**

manufacturer as well as the pipe manufacturer have a wide selection as to the particular composition to be formulated, as well as to the types and amounts of fillers to be included therein. Also, there are few limitations as to impurities, and other extraneous matter that may be introduced, either accidentally, or intentionally into the resin formulation.

Calcium Carbonate (Limestone) has been utilized as a filler for many years. This is partly due to the characteristic of this material in that it increases the modulus of elasticity of the material when mixed with the other chemicals in the resin formulation. Historically, there was no problem, and the resin formulators utilized only a small percentage of limestone, usually one to three parts per hundred parts of resin by weight. However, it was learned that by adding even more limestone, the modulus of elasticity could be further increased. Even with a reduction of tensile and flexible strength, a pipe manufactured with greater amounts of limestone could, due to its increased modulus, resist deformation and deflection more readily. Due to this physical property, formulators and extruders increased the amount of limestone, and with this increase, decreased the wall thickness of the pipe so as to save on materials, and thereby reduce costs. Reducing costs is, of course, an admirable goal. However, reducing the cost of an engineering material may create unsafe conditions, or shorten the life of the structure. This practice of increasing the limestone content has in recent times seen resin mixes with over ten or twenty parts of limestone to one hundred parts of resin, by weight, or maybe by now, even thirty parts, with investigations continuing to increase the limestone content to sixty parts per one hundred parts of resin in the near future.

The problem of adding such large amounts of limestone lies in the manner in which the limestone relates to the other materials in the resin mix. The limestone does not chemically bond to any of the other elements present; nor is there a physical bond to the other elements present. Thus, the limestone does not have any molecular relationship with the other elements present. Neither has there been any proof of any encapsulation of the limestone by the other chemicals, or of any of the other chemicals by the limestone. Rather, the limestone exists as granules adjacent to other granules of resin. While there may be some cohesion of the granules when the pipe is extruded, there is no evidence of any long term assurance of the continuance of such cohesion, either from a pipe having been installed for many years (25 years or more), nor of any type of acceptable accelerated test. Moreover, various technical organizations, as well as resin manufacturers, extruders and manufacturers have for many years investigated this topic, but have yet to arrive at any conclusion, and more importantly, have been unable to establish any acceptable standard, either a limitation, or a test procedure. Until such time as the exact amount of limestone, or other filler that can be safely added with some degree of scientifically determinable reliability of future performance, it is recommended that all fillers, regardless of the type or purpose, be limited.

The Joint Cooperative Committee (Joint Committee)of the Southern California Chapter of the American Public Works Association (APWA) and the Southern California District of the Associated General Contractors of California (AGCC) has since 1962 established a consensus specification, including both material and construction specifications entitled

"Standard Specifications for Public Works Construction," but more commonly referred to as the "Green Book." These specifications, while primarily used in Southern California, have also seen widespread use in Northern California and in many other western states as well. Presently over 200 public works agencies (Cities, Counties, Special districts, etc.) have adopted the Green Book which is supplemented or revised in its totality every year by the Joint Committee.

The current edition (1988) of the Green Book established the limit of these fillers for all resins for plastic pipe as follows:

> "Additives and fillers, including but not limited to stabilizers, antioxidants, lubricants, colorants, etc., shall not exceed ten parts by weight per one hundred of the resin in the compound."

This limitation of ten parts per one hundred of the resin is more than sufficient to allow the manufacturer and extruder to formulate a number of compounds, all of which have proven themselves for many years as being able to provide for a long term (50 years) design life of a pipeline. It further provides more than sufficient latitude to allow the manufacturerand extruder to include those additives that are necessary to adequately and economically manufacture both the resin and the pipe. And it also provides the installing contractor with a material that can be readily utilized for installation.

Some industry representatives have argued that the resin formulator, the extruder and the pipe manufacturer are better able to determine the exact mix necessary, and should not be constrained in any manner by the establishment of restrictive limitations as to the resin mix. Such arguments are not persuasive. The American Society for Testing and Materials (ASTM), in its material standards, has for decades established limitations for fillers and impurities in all other manufactured, as well as naturally occurring materials. All one needs to do to verify this is to look at the standards for Cement (ASTM C 150, C 595, et. al); Aggregates (ASTM C 289, Reactivity Test); Sand (ASTMC 33; Reinforcing Steel (ASTM A 615); Wire Reinforcement (ASTM A 82, A 185); Concrete Block (ASTM C 90, C 331); Steel (ASTM A 36, A 242, A 440, A 606, A 607, A 446, A607; ad-infinitum, etc. Suffice it to say that these many and varied standards for engineering materials have for decades established standards for materials, fillers, and the limitations for impurities. Plastics should be considered and treated no differently, nor deferentially. They should be required to subscribe to the same high standard as other engineering materials, and be subjected to the same constraints and limitations to ensure quality and long term successful application. Unfortunately, ASTM is silent to the limitation of fillers for plastic pipe.

PIPE DEFLECTIONS, LIMITATIONS AND CORRECTIVE ACTIONS

The subject of pipe deflections, their limitations and corrective actions to be taken upon determination of overdeflections have been given considerable thought and treatment in recent years by many, including this writer. In spite of the considerable amount of discussion and papers written on this subject, it remains a highly controversial item. For all practical purposes, there is today unanimity as to the need for deflection control and limitations. However, there is a total lack of unanimity as to the amount of allowable deflection, how to measure it, when to measure, and what to do when a pipeline is determined to be overdeflected. Unfortunately, much of the disagreement is, on one hand, between pipe manufacturers who seek to market their product and have it installed at the lowest possible cost, and with standards that are, in the opinion of this writer, lacking as to manufacturing standards, installation standards, and inspection, and on the other hand with end users, as this writer, who insists on policies, practices and standards to insure the viability, free from any possible failure of the product and its installation for a period of time of not less than fifty years.

As demonstrated in Table 1, the application of the ASTM standard may, without proper analysis, result in deflections greater than expected. ASTM standards recommend a 7.50 percent initial deflection, and additional factors (including one called a "Tolerance Package"), to increase the allowable deflection. Although almost all ASTM plastic pipe standards contain this same deflection standard, this discussion will refer to only that of "Type PSM Poly (Vinyl Chloride) (PVC) Sewer Pipe and Fittings" per ASTM D3034 as an example. The comments and conclusions, however, are applicable to all other plastic pipe standards.

Table 1 indicates the possible deflections, up to 17 percent, that can occur utilizing the allowable linitations of ASTM D 3034. Long term (50 year) deflections are assumed to be 150 percent of short term (30 day) deflections, a commonly accepted standard, although suspected of being too low, especially for low stiffness pipes.

First, it should be noted that the need to have a deflection limitation and standard remains the most important issue. Without a deflection limitation and standard, the user has no means of determining what the life cycle of the pipeline will be, and therefore the end user will be unable to evaluate various different pipe alternatives having different lives.

Second, as discussed in greater detail below, the deflection limitation itself must be established, and adhered to. Failure to do so would amount to having no deflection standard at all, and can result only in a failure.

Third, it must be established as to what tests to conduct, and to what extent. It is the opinion of this writer that testing must be done for all mainline plastic pipes, for their entire length, without exception. The test is not so much a test of the pipe material, nor the pipe, but rather, a test of the installation. It assures that the pipeline will last the design life. As to the test procedure, there are only two tests possible for deflection testing: the Deflectometer Test and the Mandrell Test. The Deflectometer Test is a test utilizing a custom designed and constructed testing machine which contains computerized measurement devices and records the measurements taken. The deflectometer will pass through the pipeline in all cases, and make actual measurements of the pipe in its deflected state. While a very accurate test method, it is impractical for testing of all pipelines for reason of the considerable amount of time involved as well as the high cost. It is a very valuable tool for research however. The other test, the mandrel test is, on the other hand, a simple, quick and inexpensive test. It consists of obtaining a standard manufactured mandrel, inserting it into the pipeline and pulling it through. The test is in effect a simple "Go-No Go" test. If the mandrel can be pulled through the pipeline (without extreme force), it indicates that the pipeline has not deflected more than a predetermined amount. If the mandrel cannot be pulled through, it conclusively indicates that the pipeline is overdeflected.

This is a rather simplistic approach to the mandrelling concept. However, as with anything else, gamesmanship has been introduced into the test procedure so as to allow the mandrel to pass through an overdeflected pipeline. To overcome this gamesmanship, a number of restrictive limitations must be established.

The first of the restrictions is to establish the maximum long term and short term deflection. For lack of better information, the long term deflection may be assumed to be one hundred and fifty percent of the short term deflection. The Green Book has established the following deflection limitations which are recommended to the user:

PIPE I.D.	SHORT TERM DEFLECTION LIMITATION
All plastic pipes except ASTM D2680.	
a. Up to 12 inch I.D.	5%
b. Over 12 inch, but not over 30 inch I.D.	4%
c. Over 30 inch I.D.	3%
ASTM D-2680	4%

The reason for decreasing the deflection with an increase in pipe size is due to the effect on the street surface that is almost invariably above the pipe. If a sixty inch I.D. pipe had an initial deflection of five percent or three inches at thirty days, the additional deflection due to long term deflection, almost all of which would occur with a few weeks or months, would amount to an additional 2.50 percent or 1.50 inches. The damage to an asphalt or concrete street surface above this pipe, as well as other underground pipelines or other structures above or immediately adjacent to the plastic pipeline that would result from these deflections would be considerable, resulting in a significant failure.

The second of these restrictions is to establish the actual diameter of the mandrel. To do so, it must first be determined what the I.D. of the pipe is. A reading of the various ASTM standards quickly reveals that there are several I.D.'s: The "average I.D.", the "minimum I.D.", the "maximum I.D.", the "base I.D.", etc. Because a plastic pipe extruded on a machine is a fairly accurately fabricated engineering item, it can be required to have certain tolerances, but it must have tolerances due to the manner in which it is manufactured and treated. The only rational basis is to utilize the manufacturing tolerances contained in the specification, and to determine the minimum and maximum I.D.'s.The justification for this method is that the manufacturer may while in complete compliance with the appropriate specification, furnish pipes within those parameters, and the user, if he has not established a narrower range of acceptable dimensions, must accept pipe meeting those tolerances.

The maximum allowable deflection must be based upon the maximum possible I.D., of the furnished pipe, and the value of the deflection determined therefrom. It may come to pass that the pipe furnished is at or near the minimum I.D., and so the manufacturer, as well as the installing contractor must live with it. The manufacturer always has the option to guarantee a narrower range of maximum/minimum I.D.'s, and bear the cost of any extra inspection and testing to overcome the disadvantage of impressing on him the deflection based on the maximum I.D. passed through a minimum I.D. pipe.

The mandrel itself must be designed to assure the requisite diameter. To this end, the mandrel must be an odd numbered leg mandrel. An even numbered leg mandrel can pass through a smaller diameter pipe, even when measured accurately, for reason that an odd legged mandrel has one leg at the maximum extremity of a pipe, while an even legged mandrel does not. Further, to assure that the diameter is controlled, a certain minimum number of legs is required. Three, five or seven legs can allow errors. Obviously twenty-five or more preclude virtually any possibility of error, but that would be extremely impractical and expensive. A good compromise, and one accepted by the Green Book is a mandrel with nine legs minimum. If the contractor or manufacturer wishes to furnish more legs, the user certainly cannot be heard to complain, but more need not be required. Next, the legs must be permanently attached. A mandrel with legs that can be reset, or a "variable mandrel" cannot be accepted for reason that it would have to be recertified each time it was changed or adjusted, or if it slipped , and if the hardware holding the legs onto the mandrel frame were not completely tight, it could loosen, thereby changing the diameter. Suffice it to say that the purpose of a mandrel is to make an objective test; it is not the objective to deliberately pass the pipe, nor to deliberately fail the pipe. Next the material must be of steel or some other hard metallic, non-corrodible, non-pliable material. A mandrel is subject to wear, even through plastic pipe, or even deformation if too flimsy. The material should be able to preclude this from happening. Finally, the mandrel must be measured by a certified laboratory to ensure that it meets the dimensional requirements, and if necessary, before each use. After it has been measured and certified, access to it should be restricted to the user's personnel until the test is completed.

POST MEASUREMENT CORRECTIVE ACTION FOR OVERDEFLECTED PIPE

After a pipe has been determined to be overdeflected, there is invariably much concern, if not outright turmoil and accusations. Typically, the installer seeks a waiver of the deflection requirement, and failing that, to be allowed to cure the overdeflection without reexcavation and reinstallation. As to the waiver of the overdeflection by the user, suffice it to say that that should never be done.

After refusing to waive the deflection requirement, the only possible cure for the overdeflected pipe is to reexcavate the pipeline or the segment of overdeflected pipe, and if the pipe has not been damaged in any manner, reinstall it; otherwise, the pipe must be replaced.

A recent development has been the use of machinery which attempts to reduce the overdeflection to an acceptable amount by "rerounding" or otherwise reforming the pipe with apparatus placed inside the overdeflected pipe. Typically, this machinery operates by a mechanical device exerting forces on the pipe internally over the entire internal circumference in a rotary fashion as it is pulled longitudinally through the pipeline. In so doing, it is usually able to change the ovality of the pipe, and thus possibly reduce, or at least temporarily, the overdeflection.The problems with this process are threefold. First, it presumes that the overdeflection will not recur; the second is that it presumes that there is no damage to the pipe itself; and third, it presumes that the bedding, backfill and compaction is satisfactory.

Addressing the first, it is unknown how long a "rerounded" pipe will retain its new shape or if it will revert to the prior overdeflected position. The primary reason for any overdeflection, other than inadequate pipe stiffness of a pipe, is that the bedding and backfill was improperly placed. By forcing a reshaping of the pipe, it does not necessarily follow that the bedding and backfill were similarly, and permanently reshaped. And to date, there has been no testing program simulating the numerous variables in flexible pipeline design and installation to verify that any such reshaping is in fact permanent. Next, as to the pipe material being subjected to the process, it cannot be conceived that the pipe would not be damaged from the beating of the machinery. If a process has such force to reform not only a pipe, but earthwork above it, that force may be presumed to damage the pipe in some fashion. Finally, as stated above, except for pipes of low pipe stiffness, overdeflection is usually caused by improper bedding and/or backfill. If this is the cause of the overdeflection, how can "rerounding" a pipe correct the bedding and backfill that was improperly placed. The following specification adopted for the Green Book is recommended for use:

FIELD INSPECTION FOR PLASTIC PIPE AND FITTINGS

Installed pipe shall be tested to ensure that vertical deflections for plastic pipe do not exceed the maximum allowable deflection.

Maximum allowable deflections shall be governed by the mandrel requirements stated herein and shall nominally be:

1. 3 percent of the maximum average ID for ABS or PVC Composite Pipe.

2. For all plastic pipe other than ABS or PVC Composite Pipe, the percent age listed of the maximum average ID.

Nominal Pipe Size	Percentage
Up to and including 12 inch	5.0
Over 12 to and including 30 inch	4.0
Over 30 inch	3.0

The maximum average ID shall be equal to the average OD per applicable ASTM Standard minus two minimum wall thicknesses per applicable ASTMStandards. Manufacturing and other tolerances shall not be considered for determining maximum allowable deflections.

Deflection tests shall be performed not sooner than thirty days after completion of placement and densification of backfill. The pipe shall be cleaned and inspected for offsets and obstructions prior to testing.

For all pipes less than twenty-four inch ID, a mandrel shall be pulled through the pipe by hand to ensure that maximum allowable deflections have not been exceeded. Prior to use, the mandrel shall be certified by the Engineer or by another entity approved by the Engineer. Use of an uncertified mandrel or a mandrel altered or modified after certification will invalidate the test. If the mandrel fails to pass, the pipe will be deemed to be overdeflected.

Unless otherwise permitted by the Enqineer in conformance with Subsection 3-1, any overdeflected pipe shall be uncovered, and if not damaged, reinstalled. Damaged pipe shall not be reinstalled, but shall be removed from the work site. Any pipe subjected to any method or process other than removal, which attempts, even successfully, to reduce or cure any overdeflection, shall be uncovered, removed from the Work site and replaced with new pipe.

The mandrel shall:

1. Be a rigid, nonadjustable, odd-numbered-leg (9 legs minimum) mandrel having an effective length not less than its nominal diameter.

2. Have a minimum diameter at any point along the full length as follows:

Pipe Material	Nominal Size (Inches)	Minimum Mandrel Diameter (Inches)
PVC-ASTM D 3034 (SDR 35)	6	5.619
	8	7.524
	10	9.405
	12	11.191
	15	13.849
PVC-ASTM F 679 (T-1 Wall)	18	16.924
	21	19.952
	24*	22.446
	27*	25.297
	30*	28.502
ABS or PVC Composite Pipe ASTM D 2680	6	5.636
	8	7.663
	10	9.584
	12	11.475
	15	14.356

*Optional.

3. Be fabricated of steel, be fitted with pulling rings at each end, be stamped or engraved on some segment other than a runner indicating the pipe material specification, nominal size, and mandrel OD (e.g., PVC, D 3034-8"-7.524"; ABS Composite D 2680-10"-9.584); and be furnished in a suitable carrying case labeled with the same data as stamped or engraved on the mandrel.

For pipe IDs nominally twenty-four inch and larger, deflections shall be determined by a method submitted to and approved by the Engineer. If a mandrel is selected, the minimum diameter, length and other requirements shall conform to the dimensions and requirements as stated above.All costs incurred by the Contractor attributable to mandrel and deflection testing, including any delays, shall be borne by the Contractor at no cost to the Agency.

INANITIONS OF SPECIFICATIONS AS TO GEOMETRIC PROPERTIES

A number of ASTM specifications describe plastic pipe without sufficient dimensions to allow the specifier or user to determine what the dimensions of the pipe, or its tolerances are.

In those specifications, the I.D. is indicated as well as the "Base Wall Thickness". But the primary strength resistance to deformation as well as pipe stiffness is attributable to "ribs" or corrugations located radially beyond the "Base Wall." Those specifications are totally silent as to either the geometric shape, properties or dimensions of these "ribs" or corrugations.

The contention of the proponents of these pipes is that the specification is a "Performance" specification and the manufacturer should not be constrained to provide only that allowed by a specification, but to be free to provide any configuration that meets the "Performance Requirement," and to be allowed to change the geometry at any time without restraint. The proponents further claim that the "Pipe Stiffness" requirement in the specification is all that is needed.

ASTM D 3034, as an example, not only has a "Pipe Stiffness" requirement; it also has very specific geometric properties, dimensions and tolerances. Further, when specifying a wooden or steel beam or column, neither the cross sectional area nor the section modulus of the member is specified; the actual shape and/or dimension of the member is specified, usually referenced to another specification such as those of the American Institute for Steel Construction (AISC). Those other specifications include detailed dimensional properties of the member, including tolerances. Similarly concrete, clay, steel, copper, aluminum, cast iron and ductile iron pipe specifications all include highly detailed geometric requirements, as do many other plastic pipes. And finally, "Profile "steel and aluminum pipe are required to meet very specific geometric requirements as to the size, shape and dimensions of their "Profile" corrugations of which there are several. The inanitions of these few plastic pipes seeking to be accepted on the basis of a"Performance" specification should be revised to clearly indicate the geometric properties so that the designer /specifier /user knows exactly what will be furnished before the pipe is delivered to the job site.

TABLE 1

GEOMETRIC PROPERTIES

PVC PLASTIC PIPE - ASTM D 3034

①	②	③	④	⑤	⑥	⑦
SDR	Nonimal I.D. (inch)	Min. I.D. (inch)	Max I.D. (inch)	ASTM 7.50% Deflection Mandrell Diameter (inch)	Max. Short Term Deflection With ASTM Mandrell (%)	Max Long Term Deflection With ASTM Mandrell (%)
41	6	5.9396	5.9984	5.37	10.48	15.72
	8	7.9514	8.0286	7.16	10.82	16.23
	10	9.9423	10.0337	8.93	11.00	16.50
	12	11.8354	11.9446	10.62	11.09	16.64
	15	14.482	14.618	12.98	11.21	16.81
35	6	5.8824	5.9476	5.31	10.72	16.08
	8	7.8772	7.9628	7.09	10.96	16.44
	10	9.8490	9.9510	8.04	11.17	16.75
	12	11.7188	11.8412	10.51	11.24	16.71
	15	14.3506	14.5014	12.86	11.32	16.98

① SDR = I.D./ Wall Thickness.

③ Minimum I.D.= Avg. O.D.- Tolerance on O.D.-(2x Min. Wall Thickness)-(2x Wall Thickness Tolerance).

④ Maximum I.D.= Avg. O.D.+Tolerance on O.D.-(2x Min. Wall Thickness)+(2x Wall Thickness Tolerance).

⑤ ASTM 7.50% Deflection Mandrell Dia. Per ASTM D 3034.

⑥ Max. Short Term Deflection With ASTM Mandrell= 1.0- ASTM 7.5% Deflection Mandrell Dia./ Max. I.D.

⑦ Max. Long Term Deflection With ASTM Mandrell= 1.5x Max. Short Term Deflection With ASTM Mandrell.

BURIED STRUCTURAL CAPACITY OF INSITUFORM®

Reynold K. Watkins [1]
L. Phillip Sorrell [2]

INTRODUCTION

In July 1988, a test was conducted at Utah State University to determine the structural support that can be provided to a failing rigid pipe system through the use of the Insituform process. The test was designed to determine the composite ring strength of the concrete/Insitupipe system.

Supervising the test were Dr. Reynold K. Watkins and Dr. Owen K. Shupe, both Professors of Engineering at Utah State. The university's Buried Structures Laboratory in Logan was the test site.

PROCEDURE

The experiment comprised two parallel test sections in the USU large soil cell shown in Figure 1. The two parallel test sections were 30 inch pipes placed in the soil cell separated by a spacing of 7.5 feet center to center. See Figure 2. The test sections were twenty feet long. The height of soil cover over the tops of the test sections was 3 feet. The bedding was a level plane of well-compacted soil. The pipe zone backfill soil (PZB) was silty sand placed in layers and compacted to a uniform density. A vertical soil load was applied by 50 hydraulic cylinders attached to ten beams. Vertical diameters (deflection) of the test sections were measured after each increment of load.

---

[1] Professor of Engineering, Utah State University, Logan, UT 84322.
[2] Associate Engineer, Insituform of North America, Inc., 3315 Democrat Road, Memphis, TN 38118.

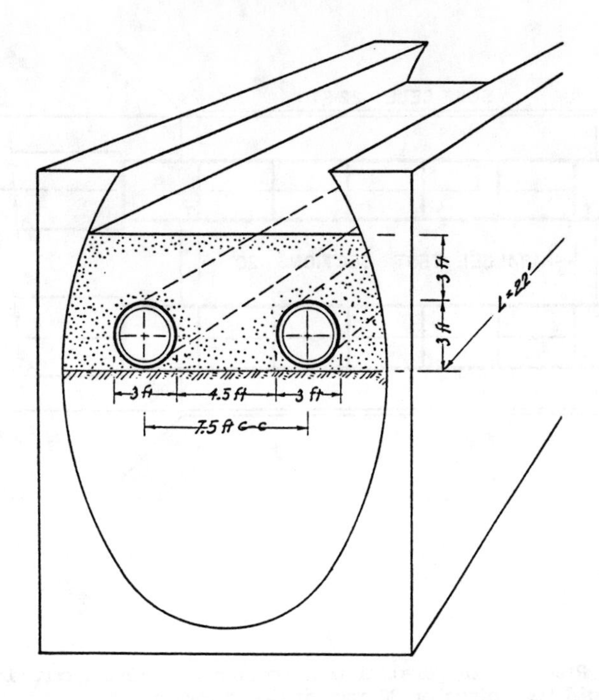

Figure 1. Cross section of the USU large soil cell showing the location of the test pipe sections in the Insituform tests.

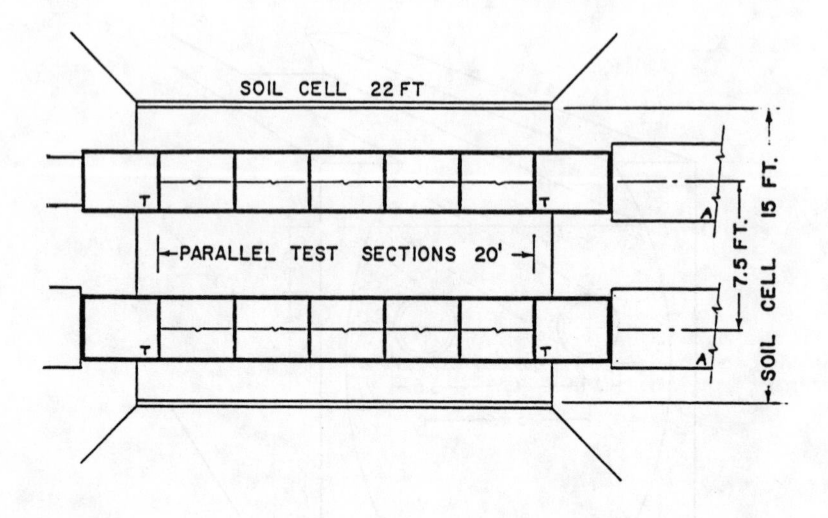

Figure 2. Plan view of parallel test sections for Test 1 comprising broken rigid test pipes of 30 inch inside diameter.

A = access pipes

T = transition sections of rigid pipe each four ft. long

MATERIALS

PIPES

Each of the two parallel pipe sections was made up of 4 foot lengths of unreinforced concrete pipes, 30 inch inside diameter. These were Class 1 pipes with a minimum specified three-edge bearing strength of 3000 lbs. per linear foot. The joints were tongue-and-groove. No gaskets or sealants were used at the joints. Each test section comprised five of these 4 feet long pipe sections for a laid length of 20 feet. Both of the test sections in the first test were broken. An unbroken pipe 4 feet long was placed on each end of each of the parallel test sections to serve as a transition. Access pipes were placed in tandem with each of the transition pipes to provide for entrance of personnel. After backfill was placed, one of the test sections was Insituformed.

The ten broken sections of pipe were cracked in a three-edge-bearing device. The average ultimate load was 3806.4 lbs. per linear foot of pipe. The standard deviation was 398.04 lbs. per linear foot of pipe. Before loading in the three-edge-bearing device, each pipe section was banded with steel bands and stuffed with three 14 inch diameter paper sonotubes to serve as mandrels for holding the circular pipe cross section during transportation and installation in the soil cell.

INSITUPIPE

A 30 inch O.D. by 21mm Insitutube® with standard formulation polyester resin was used to repair one of the broken pipe systems.

SOIL

The pipe zone backfill (PZB) was a cohesionless silty sand of relatively poor quality, locally referred to as blow sand. It was fine sand with about 20% silt, classified as SM according to the Unified Soil Classification System. Plasticity index was zero. This soil was selected because its poor quality represents worst conditions and because it can be compacted to relatively uniform density by mechanical means. Compaction of the PZB is an important factor in the structural strength of rigid pipes -- especially broken rigid pipes. Soil density was measured by a nuclear test device.

INSTRUMENTATION

The basic instrumentation was a micrometer for vertical and horizontal measurement of diameters inside the pipes before and during loading. The locations for diameter measurements were the center of the soil cell and four feet each side of center. The center of the middle rigid pipe was located at the center of the soil cell. These vertical diameters were measured at each increment of soil load. They provided an adequate check on the accuracy of the measurements, and contributed to a meaningful evaluation of the mean and standard deviation of each data point. The resulting plots of data were good enough to be compared with theory by the chi-squared analyses of the data plots.

Thermocouples were located throughout the test sections to check on the thermosetting process of the Insituform.

TEST

The test comprised two parallel test sections, 20 feet long, of 30-inch ID rigid pipes that had been previously cracked by a vertical line load in a three-edge-bearing test device. The cracks occurred approximately at 3:00, 6:00, 9:00 and 12:00 o'clock. The pipes were so oriented in the soil cell that the top cracks in all of the five pipes in each test section were at the top and were in-line. Backfill was placed in 1 foot lifts up to a height of 3 feet above the pipes. After the first lift was placed on the bedding, soil was hand-shoveled under the haunches of the pipes. These 1 foot lifts were compacted only by personnel walking throughout the soil cell. Soil density was measured at each 1 foot lift with a measured average density of 75.7% AASHTO T-99, with a standard deviation of 2.25%. After backfilling, steel plates were placed on the soil surface and the hydraulic cylinders were positioned for loading. One of the two test sections was Insituformed with a 21mm thick Insitupipe. The objective of the test was to provide a direct comparison between the structural performance of two buried broken rigid pipes under increasing vertical soil pressures; one test section Insituformed, and the other non-Insituformed. This test showed the extent to which the Insitupipe provided structural support to a broken rigid pipe under soil load.

Prior to Insituforming, a preliminary vertical soil pressure of 1450 psf was applied with a corresponding pipe deflection of approximately 2%. This configuration was established as the configuration of the broken rigid pipes for Insituforming. Pipe deflections during loading were based on this initial pipe deflection as zero and on this vertical diameter of the broken rigid pipes. Inversion was accomplished using a 21mm thick standard Insitutube. After inversion, the Insitutube was allowed to cure overnight. The loading of the test was performed early the following morning.

Vertical loads were applied to the soil cell in increments of 50 psi in the hydraulic cylinders. This applies a vertical soil pressure at the top of the test pipes of 727 psf which is equivalent to 6.06 feet of soil cover at a unit weight of 120 pcf. After each increment of load the vertical and horizontal ring deflections were measured, in the middle of each test section and 4 feet on either side. The drop in elevation of the soil surface above the pipes was measured. All other pertinent observations were recorded. The maximum pressure applied through the hydraulic cylinders was 8724 psf which is equivalent to 72.7 feet of soil cover at unit weight 120 pcf. Measurements and observations were recorded and the test was terminated.

RESULTS OF TEST

Results of the test are summarized on page 6 as Test Summary which shows the relationship between vertical soil pressures and pipe deflections. Vertical soil pressure is the average pressure acting downward at the level of the top of the test sections. A formula for this vertical soil pressure was available from previous calibration of the test cell. Pipe deflection is defined here as the decrease in vertical diameter of the pipe or Insitupipe divided by the inside diameter of the broken rigid pipes. The Test Summary is averaged and corrected from the basic data recorded. The two plots are a graphic comparison of the Insitformed test pipe and the identical, but unInsituformed, test pipe. Pertinent observations from the Test Summary are as follows:

1. The Insitupipe contributes significant strength to the pipe-soil system. The strength contribution is the result of two phenomena, the reinforcement phenomenon and the stiffener phenomenon as explained in the next paragraphs.

2. From the Test Summary, with less than a vertical soil load of about 2200 psf, pipe deflection was too small to be measured. Pipe deflection is small because of compaction of the sidefill and preloading of the soil cell. Side support is developed by the compaction of the sidefill soil as it falls into place from the conveyor belt and by hand placement of soil under the haunches. The initial preloading of the soil cell was to P = 1450 psf.

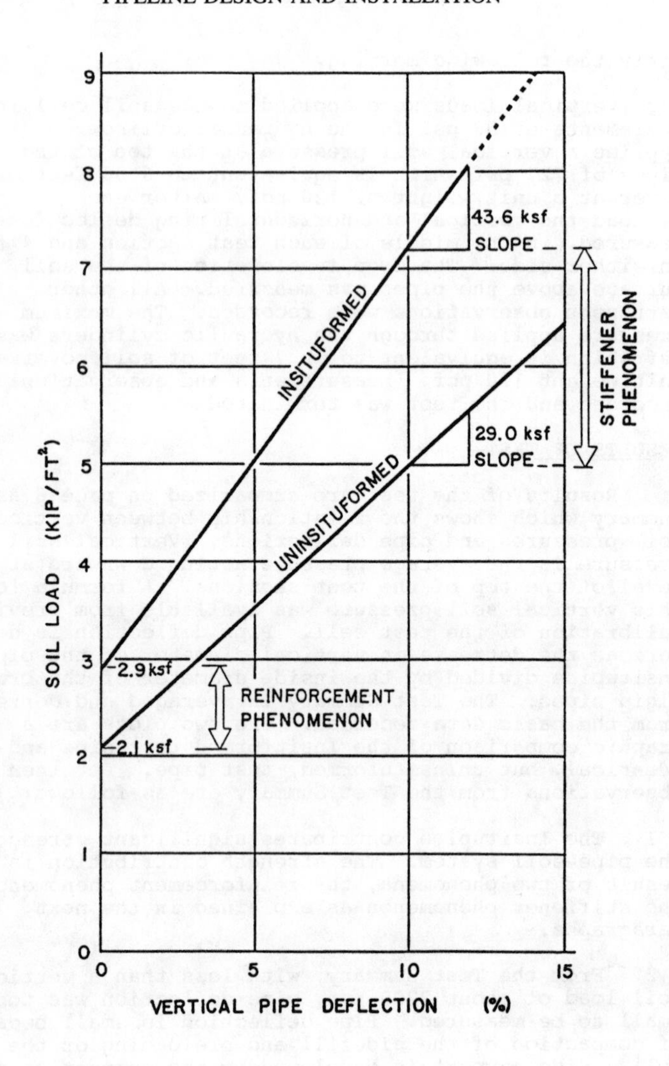

TEST SUMMARY

Load-deflection diagrams for two, buried, broken, rigid, 30 inch ID pipes one of which was Insituformed before loads were applied (Preloading was 1.45 kips per square foot (ksf) before the Insitutube was inverted into the pipes.)

3. As the soil load is increased above P = 2200 psf, the unInsituformed test section begins to deflect. The Insituformed test section does not begin to deflect until the soil load is 2900 psf. This increase in strength is the reinforcement phenomenon. The Insitupipe serves as reinforcement. As cracks inside the rigid pipe widen at 6:00 and 12:00 o'clock, the Insitupipe holds the cracks together. Clearly, the Insitupipe is not perfectly bonded. Nevertheless, bond is not insignificant for it increased strength by a factor of 1.4 in this test. Bond includes 1) adhesive bond to the surface and 2) mechanical bond in the cracks, indentations, joints, etc. In practice, bond may or may not be significant depending on the inside surface of the broken pipe.

4. Above a soil pressure of 2900 psf, the plots are approximately linear up to pipe deflections of more than 10%, but the slope of the Insituformed plot is 1.5 times as steep as the unInsituformed plot, in other words, the Insituformed pipe deflected at a much slower rate. This is the stiffener phenomenon. This increase in strength is the contribution of pipe stiffness by the Insitupipe.

5. The safety factor due to the reinforcement phenomenon is 1.4. The safety factor due to the stiffener phenomenon is 1.5. The two are not cumulative, but clearly the safety factor is no less than 1.5, even if bond is not achieved. These numbers can be interpreted as an increase in strength of at least 50% for the Insituformed section.

6. It is noteworthy that even under the maximum pressure of 8724 psf (equivalent 72.7 feet of soil at 120 pcf), neither of the test pipes collapsed completely. The fact that neither pipe collapsed was due to the support of the side fill soil. However, in real conditions, the severe cracking observed in the unrepaired pipe would result in an in-migration of soil. In time, voids would develop around the pipe, allowing it to collapse. Because the Insitupipe did not break and is jointless, it prevents such infiltration and no voids would occur. This keeps the important side fill soil support intact, adding many years of useful life to the completely fragmented concrete pipe.

Innovative New Drainage Pipe

J. Paul Tullis,[1] Member, ASCE
Reynold K. Watkins,[2] Member, ASCE
Steven L. Barfuss,[3] Associate Member, ASCE

ABSTRACT

Hydraulic tests on corrugated HDPE pipe with a smooth liner show that its Mannings n-value is in the range of 0.009 to 0.015 depending on the smoothness of the liners. With the smooth liners, the friction loss is less than or equal to that of competitive products.

The new product can withstand high external loads without distress to the pipe. The amount of ring deflection is primarily a function of the degree of soil compaction. The polyethylene material relaxes with time, relieving the stress in the pipe and allowing the soil to take its full share of the vertical load.

INTRODUCTION

A new thermoplastic pipe has recently been introduced into the marketplace. It is fabricated of high density polyethylene (HDPE). Externally, it is corrugated; internally it has a smooth liner. The new product combines the corrosion and impact resistance of polyethylene, the structural strength of corrugated pipe, and the favorable hydraulic characteristics of smooth-wall pipe. Tests were conducted to identify the hydraulic characteristics and the structural performance of the pipe as a buried conduit.

----

[1]Professor, Civil and Environmental Engineering, Utah State University, Logan, UT 84322-8200.
[2]Professor, Civil and Environmental Engineering, Utah State University, Logan, UT 84322-8200.
[3]Research Engineer, Utah State University Foundation, Logan, UT 84322-9300.

HYDRAULIC TESTS

Hydraulic tests were conducted at the Utah Water Research Laboratory (UWRL) on nine different pipes which were supplied by three manufacturers. The pipes were 12, 15, 16, and 18 inches in diameter. Additionally, external soil pressure tests were conducted on a 24-inch diameter pipe.

Experimental Methods

The various HDPE pipes were installed in two general configurations. The test sections consisted of a minimum of 100 pipe diameters of the HDPE pipe. The first piezometer was placed at least 20 pipe diameters from the inlet of the test section to allow an adequate length for establishing the hydraulic grade line.

In general, three pressure differentials were measured between P1 - P2, P2 - P3, and P1 - P3. The third differential provided a check on the other two measurements. The pressure drop measurements were taken over a minimum length of 60 pipe diameters with P1, P2, and P3 equally spaced at about 30 pipe diameters. At each of the three piezometer locations, two piezometers were installed on opposite sides of the pipe and were connected in a manifold arrangement.

The HDPE pipe sections were joined using a heat-shrink wrap manufactured by Canusa. This provided a watertight connection at pressures up to approximately 10 psi.

The tests were conducted with full pipe flow at velocities between about 3 and 12 feet per second. At least ten sets of measurements were taken within this flow range to accurately determine the variation of the friction factor with velocity and Reynolds Number. The flow was measured with a Mapco Ultrasonic Flow Meter that was previously calibrated with volumetric tanks traceable to the U.S. Bureau of Standards.

The objective of the testing was to determine the magnitude of Mannings n (in English units) and the Darcy-Weisbach f defined in the following equations:

$$V = \frac{1.486}{n} R^{0.667} S^{0.5} \qquad (1)$$

$$h_f = \frac{fL}{D} \frac{V^2}{2g} \qquad (2)$$

in which V is the mean pipe velocity, R is the hydraulic radius (area over wetted perimeter), S is the slope of the energy grade line, h_f is the friction loss measured over a pipe length (L), g is the acceleration of gravity, and D is the inside diameter of the pipe.

Although the pipe was internally lined, the bonding of the liner to the corrugations made the pipe interior somewhat wavy; thus, the diameter (D) was measured as the average minimum inside diameter of the pipe.

The f and n values are related by the following equation:

$$n = 0.0735\ D^{1/6}\ f^{1/2}\ \text{(D in feet)} \qquad (3)$$

which shows a slight dependence of n on diameter D.

Results

Table 1 summarizes the magnitudes of Mannings n and Darcy-Weisbach friction factor f based on a pipe velocity of five feet per second. The experimental values of n and f are listed in columns 7 and 8. The table also contains calculated values of f using the Wood equation (Wood 1966) defined as:

$$f = a + b\ R^{-c} \qquad (4)$$

in which R is the Reynolds Number and a, b, and c are coefficients based on the relative equivalent roughness of the pipe. The equivalent roughness values in column 3 of Table 1 were selected so that the f calculated from the Wood equation (column 6) agreed with the experimental values (column 8).

Both Mannings n and Darcy-Weisbach f varied with pipe diameter, smoothness of the interior surface, and velocity or Reynolds Number. The experimental data showed that the magnitude of these two coefficients decreased slightly with an increased Reynolds Number until the flow became fully turbulent. When the Reynolds Number was changed by using a larger pipe diameter, the magnitude of n increased with the Reynolds Number.

TABLE 1 Experimental and Computed Friction Data for HDPE Pipe.

Pipe No.	D Pipe Diam. Feet	e Equivalent Roughness Feet	e/D Relative Equivalent Roughness	Re	f Calc Wood Eqn	Measured Values n at 5 fps	Measured Values f at 5 fps	r Actual Roughness Feet	r/D Relative Actual Roughness
1	2	3	4	5	6	7	8	9	10
Experimental Data									
1	1.000	0.000050	0.000050	357,143	0.0148	0.0089	0.0148	0.0013	0.00130
2	1.500	0.000210	0.000140	535,714	0.0155	0.0098	0.0155	0.0052	0.00347
3	1.000	0.002600	0.002600	357,143	0.0266	0.0120	0.0266	0.0104	0.01040
4	1.267	0.014400	0.011365	452,500	0.0405	0.0154	0.0405	0.0130	0.01026
5	1.500	0.004000	0.002667	535,714	0.0266	0.0128	0.0266	0.0104	0.00693
6	1.010	0.000045	0.000045	360,714	0.0147	0.0089	0.0147	0.0049	0.00485
7	1.240	0.000550	0.000444	442,857	0.0184	0.0103	0.0184	0.0049	0.00395
8	1.500	0.021000	0.014000	535,714	0.0435	0.0164	0.0435	0.0158	0.01053
9	1.500	0.041000	0.027333	535,714	0.0563	0.0187	0.0563	0.0192	0.01280

Pipe No.	D	e	e/D	Re	f	n
Calculated Data for Pipe #1, Assuming that e is Constant						
1	1.000	0.000050	0.000050	357,143	0.0148	0.0089
1	3.000	0.000050	0.000017	1,071,429	0.0119	0.0096
1	4.000	0.000050	0.000013	1,428,571	0.0113	0.0099
1	6.000	0.000050	0.000008	2,142,857	0.0105	0.0101
Calculated Data for Pipe #4, Assuming That e is Constant						
4	1.267	0.014400	0.011365	452,500	0.0405	0.0154
4	3.000	0.014400	0.004800	1,071,429	0.0310	0.0155
4	4.000	0.014400	0.003600	1,428,571	0.0286	0.0157
4	6.000	0.014400	0.002400	2,142,857	0.0256	0.0159

Notes: 1. Calculation of f based on Wood Eqn: f=a+bRe^-c
a=0.094(e/D)^0.225+0.53(e/d)
b=88(e/D)^0.44
c=1.62(e/D)^0.134
Re = Reynolds Number
e = Equivalent Roughness Height(Selected so f Calculated=f Measured)
D=Inside Pipe Diameter
Calculations Based on: Fluid Viscosity= 1.4*10^-5 sq-ft/sec
and Pipe Velocity= 5.00 fps

2. Scaling Equation from f to n:
n=0.0735(D)^1/6(f)^0.5
D=Pipe Diameter in feet

3. Pipes 8 and 9 are not Commercially Produced.

It is important to realize that pipe numbers 8 and 9 are not commercially manufactured. These were experimental pipes that were too rough, so the manufacturing process was changed. The data from these experiments have been included to extend the range of experimental data, but should not be used to conclude that this type of pipe has a high roughness factor.

Table 1 contains data in which the experimental values are scaled up to larger pipe sizes to show the magnitude of the dependence of n on pipe diameter. Scaling of the experimental data to a larger pipe size was based on the Darcy-Weisbach friction factor f. First the value of f was scaled, then Equation 3 was used to calculate the corresponding value of the Mannings n.

Figure 1 shows a variation of f with equivalent roughness (e/D), and Figure 2 shows the variation of f with the actual relative roughness. Note that Figure 2 has considerably more scatter indicating that the roughness of the pipe is not merely a function of the height of the pipe waviness.

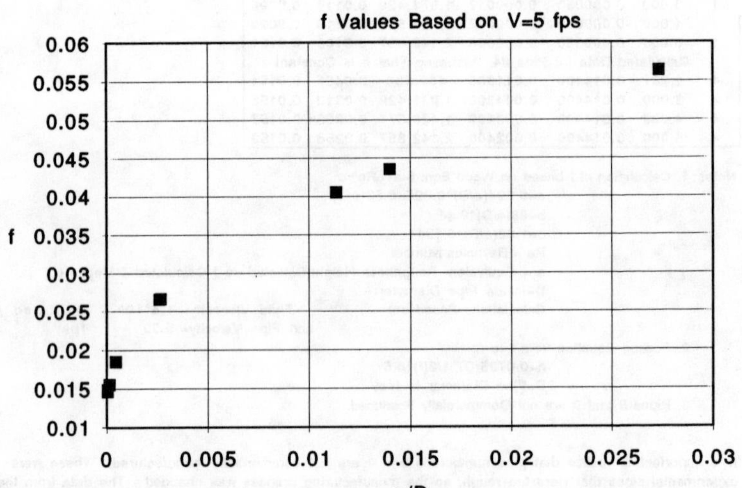

Fig. 1 Variation of f with Equivalent Relative Roughness.

The following procedure is recommended for estimating the Mannings n for a new HDPE pipe: Measure the actual roughness of the pipe, which is the height of the pipe waviness, and calculate its equivalent roughness height using the information contained in Figure 3; use the Wood's equation for the Reynolds Number that the pipe will operate at to determine the friction factor f; and use Equation 3 to convert the friction factor f to Mannings n, if desired.

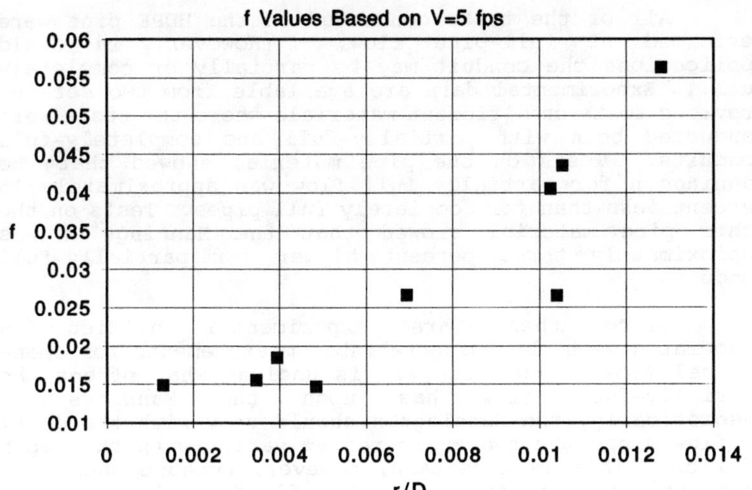

Fig. 2 Variation of f with Equivalent Actual Roughness.

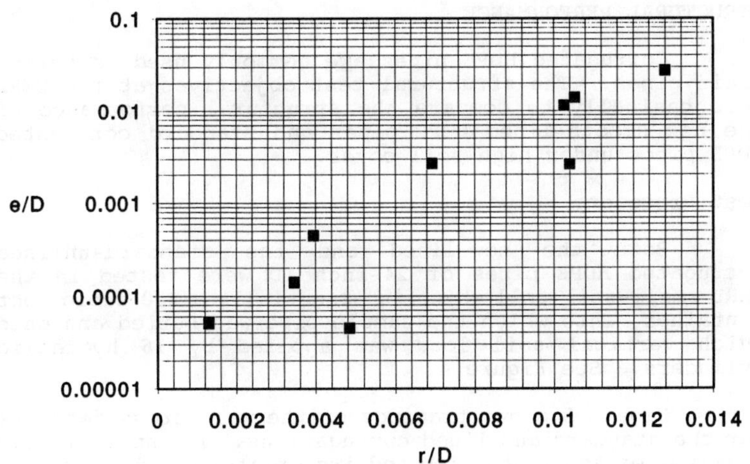

Fig. 3 Equivalent Roughness Versus Actual Roughness.

All of the tests conducted on the HDPE pipe were performed at full-pipe flow. (However, in field applications the conduit may be partially or completely full.) Experimental data are available from two sets of previous tests on different materials where the tests were conducted both with partially-full and completely-full conduits. Tests on one pipe material showed that the Mannings n for partially full flow was approximately 15 percent less than for completely full pipes. Tests on the other pipe material showed that the Mannings n was approximately three percent higher for partially-full conduits.

Since there are experimental difficulties associated with obtaining reliable measurements for open-channel flow in conduits, it is unclear what effect the partially-full flow has upon the Mannings n. Theoretically, the Mannings n should vary with the depth of flow since the hydraulic radius varies with the depth of flow. This is uncertain, however, because there are many unknowns related to the field application of laboratory data. The uncertainty associated with identifying the partially-full Mannings n value is probably greater than the other unknowns associated with the application of these data to field conditions, including leakage into or out of the joints, alignment of the joints, and effect of aging on the surface roughness.

STRUCTURAL PERFORMANCE

Corrugated HDPE pipes are commonly used as buried drain pipes. The structural test objectives at the UWRL were basically to compare the structural performance of the new smooth-lined HDPE pipes and standard corrugated HDPE pipes under high soil cover.

Test Procedure

Both the new-lined and the standard-unlined corrugated HDPE pipes of 24-inch ID were tested in the UWRL small-soil-cell which is essentially a 200 cubic foot container, into which the sample pipe was buried and onto which vertical soil load was applied by 16 hydraulic cylinders. See Figure 4.

Figure 5 shows tracings of the wall cross sections for the standard and lined corrugations. The smaller mean diameter of the lined pipe and its greater weight per foot of length result in greater pipe stiffness for the lined pipe.

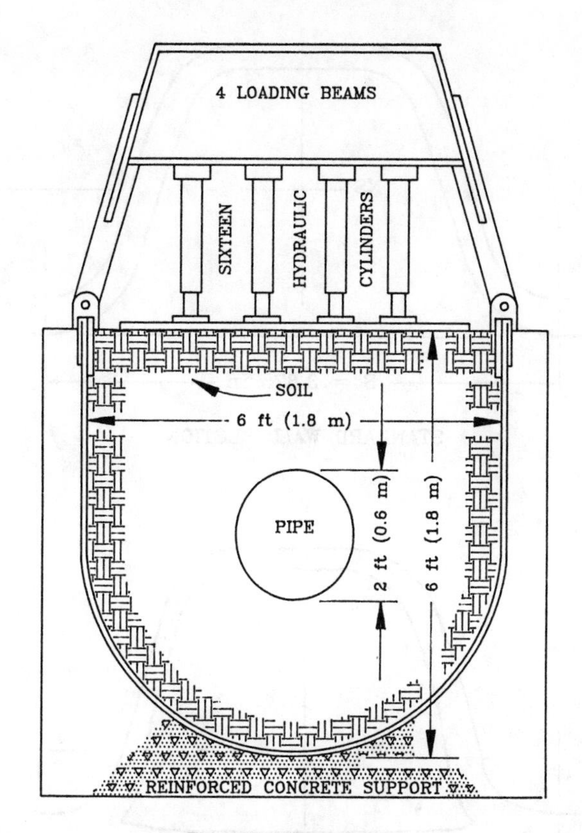

Fig. 4 Soil-Test-Cell.

The observations of performance were ring deflection and visual evidence of distress which turned out to be dimpling of the corrugations inside the pipe at the spring lines. The variables were soil type, soil density (compaction), and the vertical soil load which simulates high soil cover.

The soil type was fine sand, with about 20 percent silt classified as SM according to the Unified Soil Classification System.

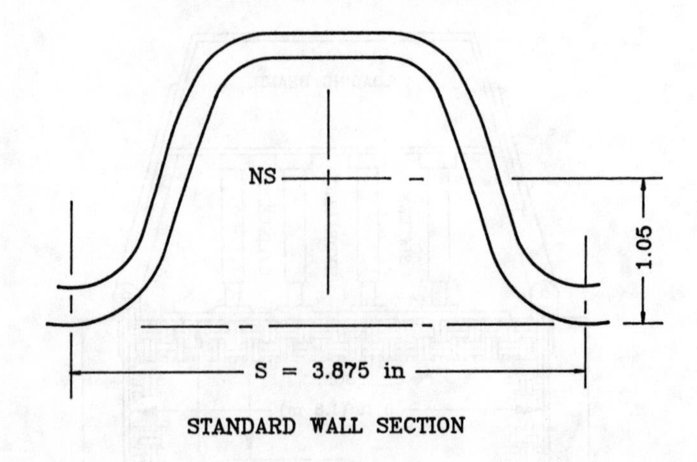

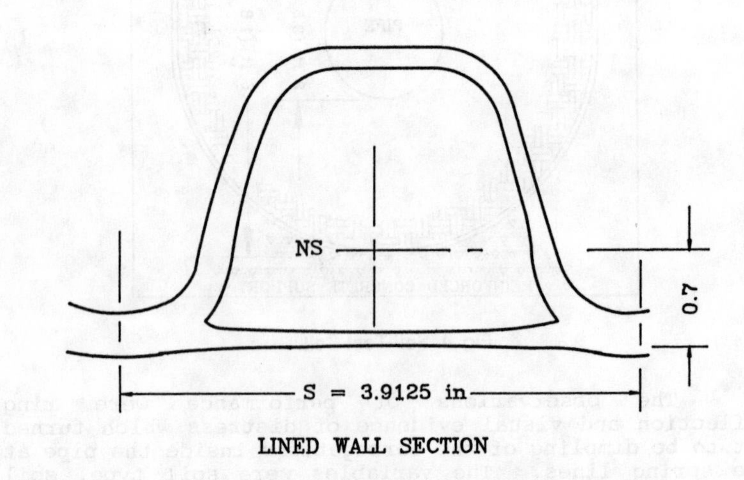

Fig. 5 Corrugations of 24" Pipe.

Test Results

Figure 6 compares the load-deflection performance of 24-inch diameter lined and standard pipes. High loads can be applied without distress to either of the pipes. As would be expected, the stiffer ring of the lined pipe deflects less than the standard corrugated pipe. But with

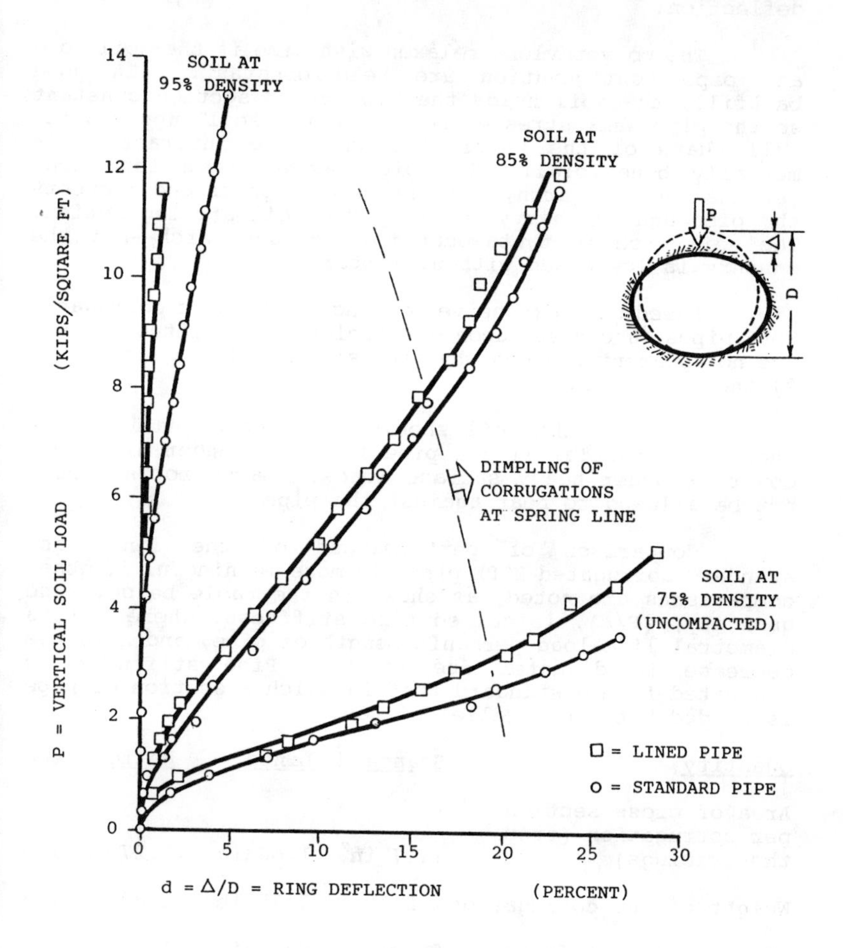

Fig. 6 One typical set of load-deflection curves comparing Lined and Standard HDPE pipes, 24 ID, buried in silty sand at various degrees of compaction AASHTO-T99.

both pipes, ring deflection is primarily a function of the compressibility of sidefill soil. This is most evident when the pipes are buried in uncompacted soil. Clearly, the pipe deflects more in loose soil than it does in dense soil because loose soil compresses more; however, even in loose soil these pipes do not collapse. Dimpling of the corrugations becomes just visible at about 15 percent deflection.

The polyethylene relaxes with time if the soil load and pipe configuration are held constant. In good backfill, the soil holds the pipe cross section constant so the pipe wall stresses relax and the soil supports its full share of the vertical load. The interaction is mutually beneficial. The pipe serves as a form that retains the soil arch, and the soil supports and protects the pipe against heavy loads by the soil arching action. The soil arch is tantamount to a masonry arch--but the masonry is low grade without mortar.

Based upon extensive testing of standard corrugated HDPE pipes, and based upon principles of similitude, there are no apparent reasons why pipe size should be limited to 24-inch diameter.

Clearly, the soil should be of good quality. It should be granular if the pipe is buried under high soil cover or under heavy surface loads. Large rocks should not be allowed to bear against the pipe.

Comparison of performances of the lined and standard corrugated HDPE pipes is more meaningful if other comparisons are noted, as shown in the table below. The quantity, (F/Δ), is called pipe stiffness, where F is a diametral line load per unit length of pipe, and Δ is the decrease in diameter due to F. Pipe stiffness is evaluated from a standard test in which a section of pipe is loaded between parallel plates.

Quantity	Standard	Lined	Ratio
Area of cross section per corrugation (from the tracings)	1.17 in^2	1.48 in^2	1.27 say 1.3
Weight of one corrugation	2.5 lb	3.25 lb	1.30 say 1.3
(F/Δ) = pipe stiffness	53 psi	93 psi	1.75

CONCLUSIONS

Corrugated HDPE pipe with a smooth interior liner has favorable hydraulic and structural characteristics. The Mannings n value varied from .0089 to approximately .015 for the seven pipes commercially produced. The magnitude of n increased with the waviness of the lines pointing out the importance of the manufacturing process used to bond the lines to the corrugated shell. For the smooth lines, the friction loss was equal to or less than that produced by most of the pipes on the market at the present time. Normally, this pipe is used for partial or full drainage. However, the tests showed that when adequate sealing was provided at the pipe joints, the corrugated, lined HDPE pipe worked well for low-head pressurized applications.

The results of the external load tests indicated that structural performance of the lined pipe was as good as, and generally better than, the standard corrugated HDPE pipes now commonly in use.

APPENDIX. REFERENCE

Wood, D. J., "An explicit friction factor relationship," Civ. Eng., 36, December 1966, pp. 60-61.

Short-term Versus Long-term Pipe Ring Stiffness in the Design of Buried Plastic Sewer Pipes

Lars-Eric Janson[1], M.ASCE

Abstract

It is often discussed which pipe ring stiffness should be used when designing the long-term deflection of buried plastic sewer pipes; the short-term value or the long-term value? As the question concerns visco-elastic materials such as PVC, PE and PP, it is easy to understand why it is believed that the stress-dependent creep of the plastic material must be considered in the assessment. This paper evaluates the question, and it is shown that in general only the short-term value is of significance for the long-term deflection of the pipe.

Introduction

It is a well-known fact that a buried plastic pipe undergoes a deflection in the course of time. Through experience it is also known that from a practical point of view the deflection will generally reach a constant value within a period of 2-3 years. In traffic areas, this period is found to be shorter than in areas without traffic. See Fig. 1. Long-term field investigations up to 10-20 years verify these statements, /1/ and /2/.

What happens during this first period of pipe deflection increase is that the pipe is subject to a large amount of short-term load impulses due to traffic, ground water changes, frost action, etc. During each such additional short-term load impulse, the pipe struggles to counteract the deflection increase by virtue of its short-term ring stiffness, which is then determined by the short-term E modulus of the pipe material.

[1]VBB Consulting Ltd
P.O. Box 5038
S-102 41 STOCKHOLM
Sweden

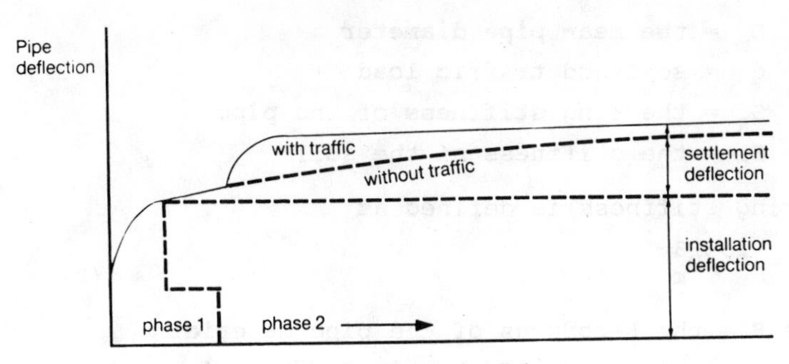

Fig. 1
Principle course of pipe deflection as a function of time. /1/.

It must be emphasized here that the creep phenomenon taking place when loading plastic materials does not mean a weakening of the material in the course of time. For each new loading, the material will always act according to its short-term strength property, independent of how much time has passed since the first loading occurred /3/ and /4/.

The total pipe deflection during the initial soil dislocation period can thus be understood to be the sum of a large number of short-term load impulses. Consequently, the final pipe deflection should primarily be a function of the change of soil stiffness in the course of time (which in turn is dependent on the type of soil and the degree of soil compaction) rather than the creep property of the pipe material.

Analysis

The problem can most easily be discussed based on the Spangler/Molin formula /2/ which in principle determines the relative pipe deflection according to the following equation:

$$\delta/D_m = Cq/(S_R+S_S) \qquad (1)$$

where δ = the reduction of the vertical pipe diameter

D_m = the mean pipe diameter
q = soil and traffic load
S_R = the ring stiffness of the pipe
S_S = the stiffness of the soil

The ring stiffness is defined as

$$S_R = EI/D_m^3 \tag{2}$$

where E = the E-modulus of the pipe material
I = the moment of inertia for the pipe wall

From eq. (1) it can easily be understood that if the left hand side of the equation stands for a momentary deflection increase, caused by a short-term load impulse, S_R must stand for the short-term value of the ring stiffness. The momentary deflection increase must then be explained as a consequence of a momentary decrease in the stiffness of the surrounding soil S_S caused by the actual load impulse.

At first glance, it would seem as though it should be possible to explain the deflection increase also in the following way. The bending stress in the pipe wall caused by the initial deflection is in momentary balance with the soil pressure acting against the outside of the pipe wall. As the bending stress will decrease in the course of time due to relaxation, the force system will successively be brought out of balance and the possibility for the pipe to resist the external load will decrease. However, it can be stated that the pipe will simultaneously recover its resistance force against the further deflection increase each time another deflection impulse occurs (momentarily or continuously), as the actual deflection increase immediately creates an increased bending stress in the pipe wall as a consequence. During the total initial deflection process, the bending stress in the pipe wall will be practically constant, implying that a continuous balance will actually remain between pipe wall stress and soil pressure.

Moreover, for each new deflection impulse the vertical soil column above the pipe will be more or less suspended on the side fill column by friction, giving rise to an increased stability of the soil stored at the side of the pipe. Consequently, the ability of the soil to prevent a further deflection caused by a stress relaxation in the pipe wall (if any) will increase for each new deflection attempt.

It can also be shown by calculation according to eq. (1) that the influence of stress relaxation on the decrease in the ring stiffness S_R cannot explain the deflection increase taking place during the first 2-3 years after installation, as observed by field measurements. See Fig. 2. As an example, it can be shown that the part of the deflection increase from initially 3 % to a more or less constant value of 5 % after 3 years, which could theoretically be referred to a pipe wall stress relaxation, is only 0.3 %. The calculation is made for a PVC pipe with a short-term ring stiffness of 4 kPa, which according to Fig. 3, will have fictively decreased to 2 kPa after 2-3 years. /5/. (Interpolation is made in the diagram between the parameters 2 kPa and 8 kPa.) The soil stiffness in this example is supposed to be constant and determined by a secant modulus of 2,000 kPa.

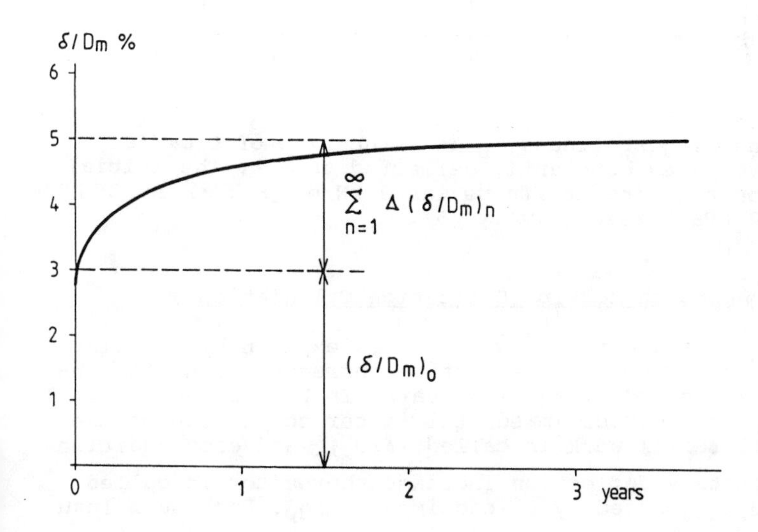

Fig. 2
Pipe deflection increase during the first years after installation.

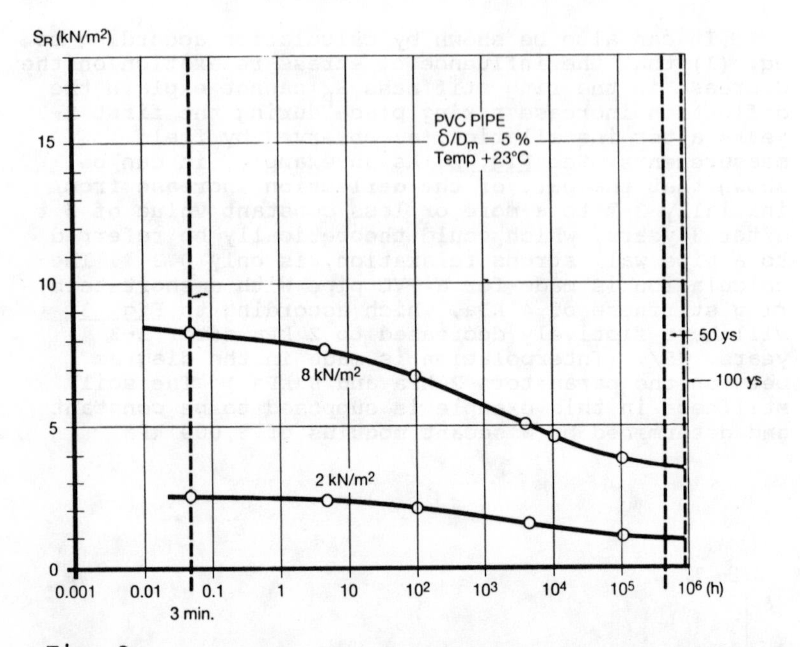

Fig. 3
The ring stiffness S_R as a function of time for PVC pipes constantly deflected to 5 %. The initial nominal ring stiffness was 8 kN/m^2 (8 kPa) and 2 kN/m^2 (2 kPa), respectively /5/.

An Approach to Pipe Deflection Calculation

The question now is how to explain by calculation the observed pipe deflection increase during the initial period of say 2-3 years. In Fig. 2 the initital pipe deflection immediately after completion of the soil refill work is called $(\delta/D_m)_o$ and each additional momentary deflection increase thereafter is called $\Delta(\delta/D_m)_\nu$ caused by a load impulse Δq_ν. Each such load impulse will give the pipe another small deflection contribution; a process that will in turn give rise to an increased soil stiffness $S_{S\nu}$. Each contribution to the increased deflection will then follow the equation:

$$\Delta(\delta/D_m)_\nu = \frac{C\,\Delta q_\nu}{S_R+S_{S_\nu}} \qquad (3)$$

The increased soil stiffness S_{s_ν} can be regarded as the final result of each load impulse, although the immediate reason for the occurrence of the additional pipe deflection increase may be a combination of the load increase and a momentary decrease in the soil stiffness due to soil grain dislocation. However, immediately thereafter, the soil stiffness will have reached a higher value than previously due to the additional compaction work, but also to a large extent as a result of the higher horizontal passive soil pressure reached by the increased pipe deflection. The total process will thus result in a decrease of each additional $\Delta(\delta/D_m)_\nu$ occurring as a result of each additional load impulse. During this 2-3 year process, recognized by a large amount of short-term loading impulses, the pipe has the ability to resist each load impulse with a strength corresponding to the short-term value of the ring stiffness of the pipe.

An equation describing the final pipe deflection reached after some years would consequently appear as follows:

$$(\delta/D_m)_{final} = (\delta/D_m)_o + \sum_{n=1}^{\infty} \Delta(\delta/D_m)_n \qquad (4)$$

The right hand second term stands for an infinite geometric series. As the ratio of the series is less than 1 according to the above, the series will converge which means that the sum of the series has a finite value.

Let us assume that the final pipe deflection has been found to be 5 % after some years, while the initial pipe deflection was 3 %. Then eq. (4) will be satisfied if we as an example assume that the first additional pipe deflection caused by the first load impulse is 0.1 % and the ratio of the series is 0.95.

Discussion

The investigation presented above applies generally to flexible pipes buried in friction soils or firm silty/clayey soils. In the case of loose cohesive soils which show viscose properties when loaded, the pipe material will also creep continuously. When determining the long-term pipe deflection in such loose soils it is recommended to use the long-term ring stiffness together with the long-term soil stiffness. In exceptional cases when it is expected that the soil stiffness will approach zero after a long time, it should, according to eq (1), be the long-term ring stiffness alone, which will determine the final pipe deflection.

In order to prevent buckling of the pipe within the operation time, the pipe stiffness must in such exceptional cases be great enough for the pipe alone to carry the vertical soil load. It should be observed, however, that at precisely the moment when the buckling takes place, it is still the short-term ring stiffness that will be effective according to the expression for the critical soil pressure

$$p_b = 24\ S_R \tag{5}$$

Eq. (5) is valid for a free pipe without soil support while the equation

$$p_{bs} = 5.63\ \sqrt{S_R . E_t'} \tag{6}$$

is valid for a pipe supported by the soil and where the stiffness of the soil is expressed in terms of the tangent modulus of the soil E_t'.

Conclusion

In general, when determining the long term pipe deflection and the buckling stability criterion of buried plastic pipes, it is the short-term ring stiffness which should be used. In order to reach the final pipe deflection after 2 to 3 years, it is more appropriate to use the successive change in the soil stiffness in the course of time than to discuss the creep or the stress relaxation of the pipe material.

References

/1/ **de Putter, W.J. and Elzink, W.J.**1981. Deflection studies and design aspects on PVC sewer pipes. - Int. Conf. Underground Plastic Pipe. ASCE. New Orleans 1981.

/2/ **Molin, J.**1985. Long-term deflection of buried plastic sewer pipes. - Int.Conf.Advances in Underground Pipeline Eng. ASCE, Madison 1985.

/3/ **Janson, L-E.**1987. Hur gammalt kan ett plaströr bli? (How old can a plastic pipe become?). - Report No. 1 from the Swedish KP-Council, Oct.1987. In Swedish. (An English translation is available).

/4/ **Janson, L-E.** Plastic Pipes for Water Supply and Sewage Disposal, Stockholm 1989.

/5/ **Janson, L-E.** Physical aging of buried PVC sewer pipes as affecting their long term behaviour. Int.Conf. Plastics Pipes VII. PRI. Bath 1988.

CASE HISTORIES OF SULFIDE CORROSION

by
Perry L. Schafer,[1] Irene S. Horner,[2] and
Robert A. Witzgall[3]

Abstract

Section 522 of the Water Quality Act of 1987 requires the U.S. Environmental Protection Agency (EPA) to conduct a study of sulfide corrosion and control options in collection and treatment systems, and to report the results to Congress. As input to EPA's report to Congress, Brown and Caldwell Consulting Engineers prepared case histories of five wastewater systems that have experienced sulfide corrosion problems (Brown and Caldwell, 1989). The case histories, which summarize previous studies of systems, are summarized in this paper.

The projects selected for case histories span the United States, with one project located in Lakeland, Florida; two in Omaha, Nebraska; and two projects located in Sacramento, California. This paper includes a brief description of each project area, followed by a description of data collection, field investigations, and predictive models common to all five projects. Recommended solutions to the sulfide corrosion problems for each project are described, and conclusions that can be reached by looking at the five case histories collectively are summarized.

---

[1]Member, ASCE; Managing Engineer, Brown and Caldwell, 723 "S" Street, Sacramento, CA 95814.
[2]U.S. Environmental Protection Agency, 401 M Street, S.W. (WH-595), Washington, D.C. 20460.
[3]Chief Engineer, Brown and Caldwell, 723 "S" Street, Sacramento, CA 95814.

Descriptions of the Projects

Following are brief descriptions of the wastewater collection and treatment systems for which the sulfide/ corrosion control studies were conducted. Each of the studies included field data collection, sulfide and corrosion predictive modeling, and evaluation of control alternatives.

City of Lakeland, Florida, Western Trunk Sewer. The City of Lakeland, Florida, Western Trunk Sewer (Western Trunk), constructed in the early 1960s, receives wastewater discharges from food processing, other industrial, commercial, and residential areas. The collection system (shown schematically on Figure 1) consists of both force mains and gravity sewers. The gravity portion of the Western Trunk consists of 27,300 feet of primarily reinforced concrete pipe (RCP), and some vitrified clay pipe, ranging in size from 24 to 48 inches in diameter with variable slopes. Most dry weather velocities are greater than 2 feet per second (fps) and some reaches have velocities of 7 fps.

The City has rehabilitated or replaced portions of this sewer in recent years because of pipe collapses. The main cause of pipe and manhole deterioration is sulfuric acid corrosion caused by the existence of high hydrogen sulfide gas concentrations in the sewer atmosphere. A study was undertaken in 1988 to assess the existing conditions and to develop a plan to renovate portions of the Western Trunk (Brown and Caldwell, 1988). Odors are also prevalent.

City of Omaha, Nebraska, Papillion Creek Wastewater System. Corrosion and odor problems have occurred in Omaha's Papillion Creek Wastewater System over the past decade. The start-up of the expanded system in the mid-1970s brought new dischargers into the system, and substantially increased transit time, with the service area extending over 25 miles to the Papillion Creek Wastewater Treatment Plant (WWTP). Figure 2 shows the service area and major interceptors. Safety has been a major concern due to high concentrations of hydrogen sulfide in the confined spaces of the interceptors and treatment facilities. A corrosion and sulfide study, which encompassed the interceptor system, as well as in-plant corrosion problems, was undertaken in 1984/85 (Brown and Caldwell, 1985). Adequate velocities are maintained in the interceptors for solids scouring.

City of Omaha, Nebraska, South Interceptor Sewer. In the late 1950s, planning was initiated for collection, diversion, and treatment of raw waste discharges to the

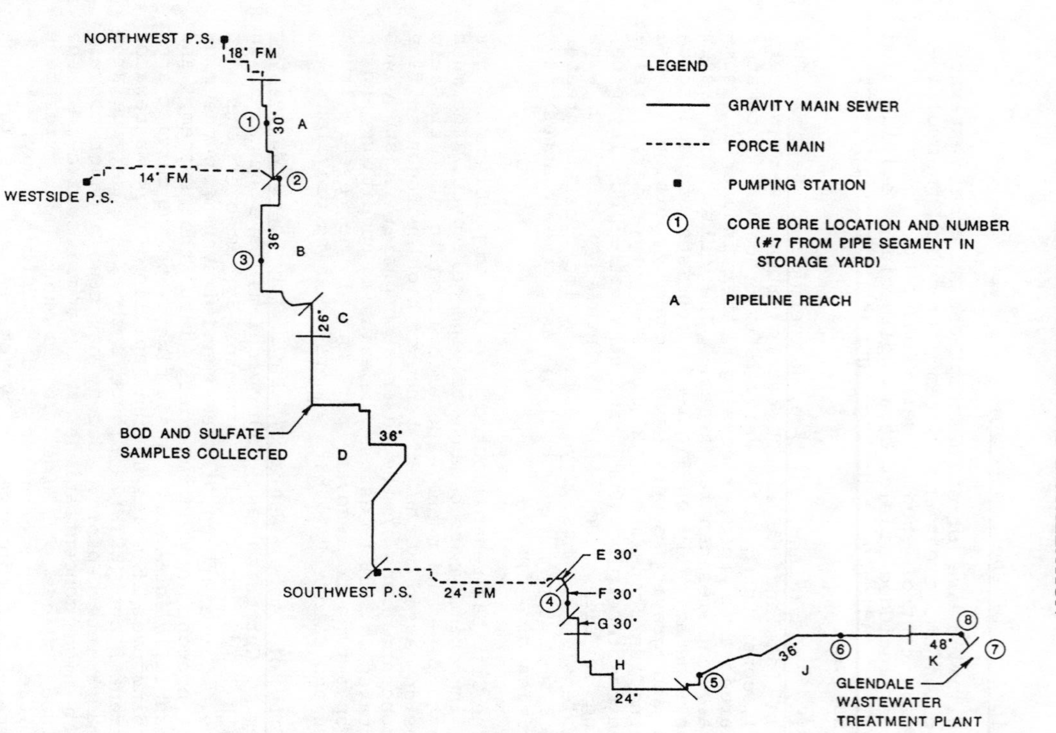

Figure 1. Western Trunk Sewer – System Layout Schematic

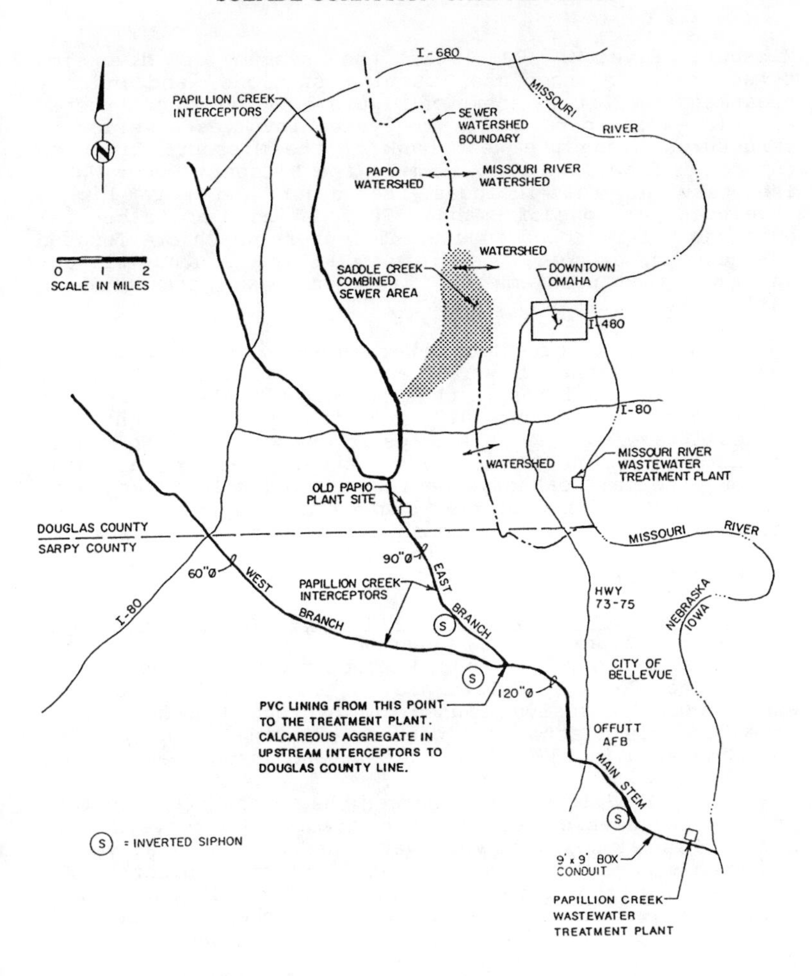

Figure 2. Papillion Creek Sanitary Sewerage System

Missouri River. By 1965, the system of diversion structures, interceptors, pumping stations, and primary treatment facilities shown on Figure 3 had been constructed and placed in operation. The system involves a series of structures along the west bank of the Missouri River to intercept flows and pump them to the Missouri River WWTP. The service area includes the older and more highly developed portions of Omaha. It contains Omaha's central business district and industrial centers which are located adjacent to, or near, the Missouri River. A study of odor and corrosion problems in this system was conducted in early 1984 (Brown and Caldwell, 1984).

The South Interceptor Sewer is 4-1/2 miles long, and conveys dry weather flows of about 20 million gallons per day (mgd). Velocities in this 66-inch-diameter force main interceptor are about 0.7 fps or less at night and typically average 1.2 to 1.6 fps in dry weather. No doubt, solids deposition is occurring under these conditions. Although normal peak daytime flows bring velocities up to about 2 fps, which will keep organic solids in suspension, this does not likely resuspend all of the settled solids. A slow buildup of solids in this pipe is likely under extended dry weather conditions.

Sacramento County, California, Central Trunk Sewer. The Central Trunk Sewer, shown on Figure 4, is a 27-year old gravity-flow, reinforced concrete sewer located in Sacramento County, California. This trunk sewer has experienced corrosion caused by hydrogen sulfide. An intensive field study to identify and assess the problem was completed in 1976 (J.B. Gilbert and Associates, 1979).

The Central Trunk Sewer serves the southeastern portion of the Sacramento metropolitan area. It conveys both domestic and industrial wastewater and, for several years, conveyed sludge from two wastewater treatment plants. The 14-mile-long trunk sewer is comprised of granitic aggregate reinforced concrete pipe, 27 to 60 inches in diameter. Slopes vary from 0.18 percent in the upper reaches to 0.05 percent in the lower end of the trunk.

Sacramento Regional County Sanitation District Regional Interceptor System. The Sacramento Regional County Sanitation District Regional Interceptor System (Regional System), which brought most of the greater metropolitan

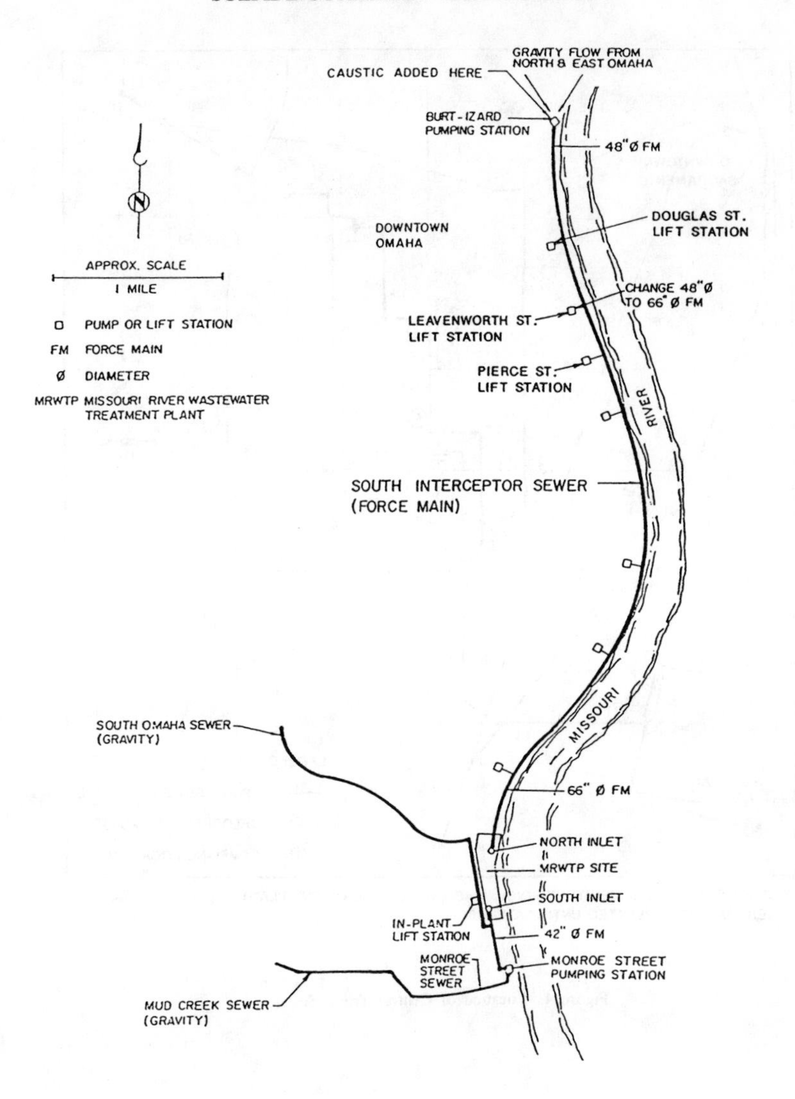

Figure 3. MRWTP Interceptor System, Simplified Layout

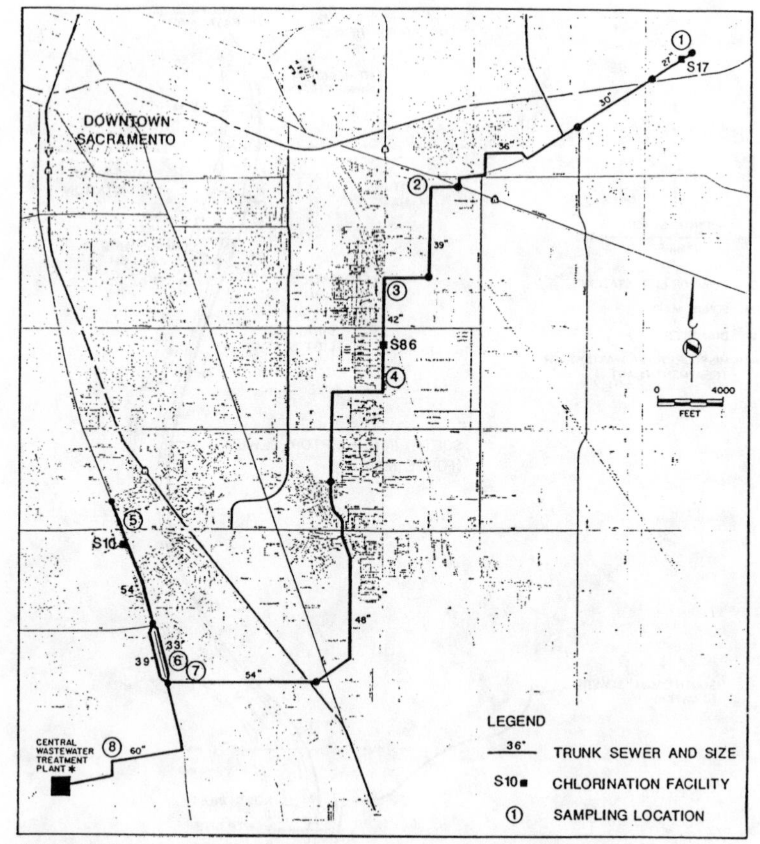

* SITE OF NEW SACRAMENTO REGIONAL WASTEWATER TREATMENT PLANT. CENTRAL WTP OPERATED UNTIL LATE 1982.

Figure 4. Location of Central Trunk Sewer

area of Sacramento flows to a new 150-mgd WWTP, is extensive in scope, with a total capital cost of about $143 million. It was constructed during the 1975 to 1982 period, and encompasses about 62 miles of gravity sewer, with over 25 miles being pipe in the 60- to 120-inch-diameter range. The planning and design phases for much of the Regional System encompassed several years, with considerable attention to sulfide corrosion and odor issues (Sacramento Area Consultants, 1976).

The Sacramento terrain is quite flat, and the climate features hot summers and mild winters. The system extends 20 miles east of the Sacramento River, and for more than 20 miles north to south. The longest interceptor, nearly 30 miles, drops less than 200 feet from end to end. A tributary river, the American, separates the north half of the system from the south. Several pumping stations are included in the system, which is depicted on Figure 5.

Types of Data Collection and Field Investigations

Four of the five case studies were based on projects which were undertaken to investigate existing sulfide corrosion and odor problems. Each of these projects required an extensive data collection and field investigation effort in order to characterize the wastewater, hydraulic, and environmental conditions, and to determine the extent of the corrosion problems. Some of the data collection and field investigation techniques that were employed in some, or all, of these projects are described in this section.

Flow and Wastewater Characteristics. In conducting a sulfide corrosion or odor study, the following wastewater parameters are of interest:

- Biochemical oxygen demand (BOD)
- Portion of BOD that is soluble (definition of "soluble" varies)
- Temperature
- Sulfate
- pH
- Dissolved oxygen
- Oxygen uptake rate
- Total suspended solids
- Oxidation-reduction potential
- Total and dissolved sulfide

Although not a wastewater parameter per se, hydrogen sulfide gas in the sewer atmosphere is also of primary interest.

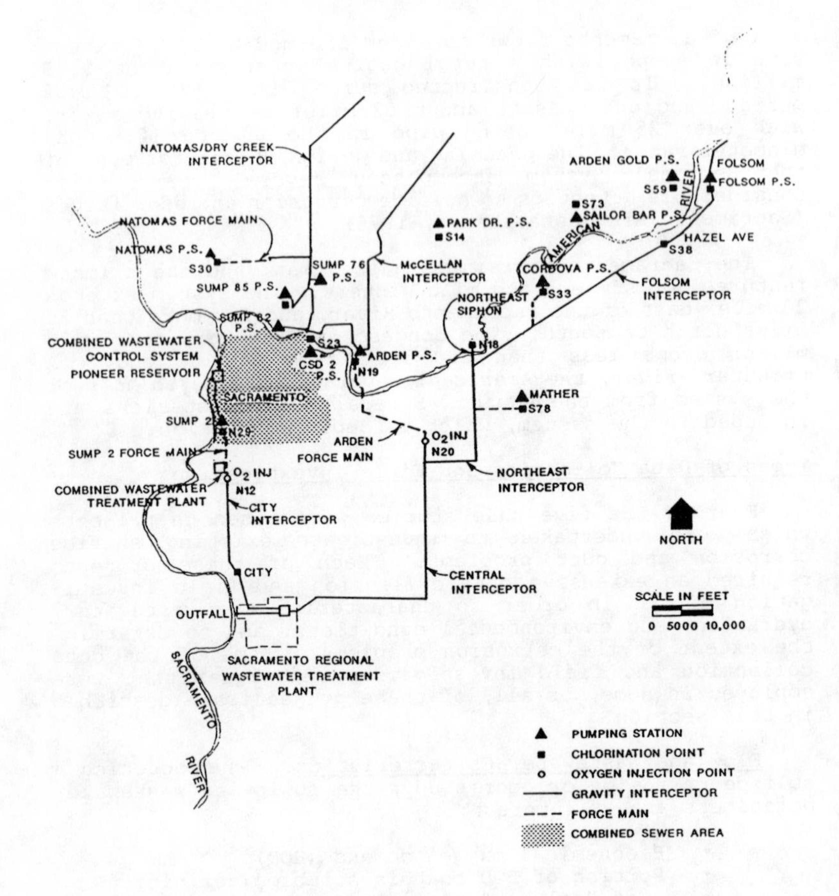

Figure 5. Sacramento Regional County Sanitation District Regional Interceptor System

Hydraulic factors are also important, including pipe slope, velocity, depth of flow, and in-pipe residence times. Full-pipe versus partial pipe flow and sewer gas space characteristics are also critical data.

In the case of the study conducted in 1976 on the Central Trunk Sewer in Sacramento, most of the above-listed data had been collected during the previous 11 years for the system covering approximately 13 miles of the trunk and 28 separate tributary sewers. Maximum BOD concentrations in the Sacramento Central Trunk Sewer ranged from 1,400 milligrams per liter (mg/l) at the upper reaches of the trunk sewer, to 360 mg/l at the downstream end. BOD values were high due to upstream sludge discharges and industrial dis-charges. The percentage of soluble BOD is unknown, but could have been greater than for a normal mixture of municipal wastewater, due to the high soluble portion of BOD from industries such as food processors.

In Lakeland, the major dischargers are mostly food processors, and the discharges tend to contain high-strength BOD concentrations (greater than 500 mg/l) and high percentages of soluble BOD (greater than 40 percent of total BOD). This produces high-strength wastewater in practically all portions of the Western Trunk, with BOD concentrations in the trunk sewer in the 400 to 500 milligrams per liter (mg/l) range on many occasions. Average BOD concentrations in the Western Trunk are closer to 300 to 350 mg/l.

Wastewater temperatures and sulfate concentrations are summarized in Table 1. Summertime wastewater temperatures in all of the systems investigated were relatively high, generally being in the range of 26 to 30 degrees Celsius (79 to 86 degrees Fahrenheit). In all cases, sulfate concentrations in the wastewater are high enough to provide plenty of sulfur for conversion to sulfide, given anaerobic conditions.

Figure 6 shows the variation of total sulfide concentrations in the Sacramento Central Trunk Sewer. The trend is based on weekly sampling for 4 years, and monthly sampling for the subsequent 4 years at one location. The trend is not typical for many agencies since sulfide concentrations remain high in Sacramento through the fall months. This is primarily due to canning season industrial discharges, but is also due to climatic factors.

Table 1. Summary of Wastewater Temperatures and Sulfate Concentrations

Location	Summertime wastewater temperature	Sulfate concentration, mg/l
Lakeland, Florida	30°C (86°F)	115 - 140
Omaha, Nebraska	26°C (79°F)	150 - 200
Sacramento, California	28°C (82°F)	30 - 80

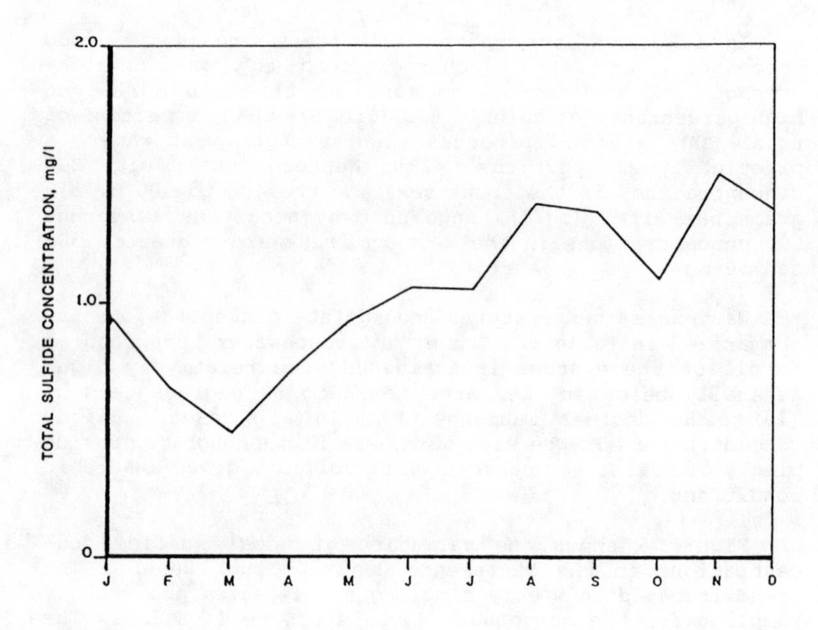

Figure 6. Sulfide Trend – Sacramento Central Trunk Sewer

Field Investigations

Physical inspection of sewer pipelines and manholes was conducted in conjunction with all of the projects. The types of fieldwork conducted included:

- Physical inspection of manholes and portions of sewers.
- Measurements of pH of wall moisture above flow line.
- Measurements of penetration into corroded concrete (i.e., how far a screw driver can be jammed into corroded material).
- Measurement of inside diameter of pipelines.
- Pipe coring and subsequent core analysis.
- Internal television inspection.
- Measurement of clay monitoring plugs installed in pipe.

The pH of the wall moisture was measured by pressing pH paper against the pipe walls. The pH of the moisture on the pipe wall was used as a guide in categorizing the likely rate of corrosion and time of occurrence. If the pH was less than 2, the corrosion was judged to be active. If the pH was 4 or 5 and there were signs of corrosion, the corrosion probably occurred earlier. The average pH on the crown of the Sacramento Central Trunk Sewer, measured at more than 100 manhole locations, was about 2, and always less than 5 in the 1970s.

In the Sacramento Central Trunk sewer, corrosion of about 3/4 inch, with a maximum of 1.5 inches, was observed at both the crown and springline in the 60-inch-diameter reach at the downstream end of the trunk. The penetrations were measured from the face of then-existing corroded material to hard concrete. Also measured in the field were the diameters of the pipe (diameters across the then-existing faces of pipe walls). These data showed that for each pipe diameter, the measured diameters were very close (usually within about 0.2 inch) to the nominal diameter.

In the Lakeland Western Trunk sewer, eight representative locations were chosen for core boring of concrete pipe. Core boring sites were chosen to obtain core samples on each size of RCP, obtain cores primarily at the worst observed points of corrosion (usually about the springline of pipe reaches that were heavily corroded), and one core from an uncorroded piece of pipe. The objectives of the core boring work were to obtain samples whose thickness, remaining cover over reinforcing steel, compressive strength, and concrete alkalinity could be measured and analyzed. Core borings were also checked for possible signs of soil corrosion on the outside surface of the boring. No soil side corrosion was found.

Sulfide and Corrosion Modeling and Correlation

The data collected in conjunction with these projects, in particular the 11 years of data that had been accumulated for the Sacramento Central Trunk system as of 1976, provided a good basis for evaluating sulfide and corrosion prediction models. The data were particularly useful for evaluating the corrosion prediction equations where the history of sulfide conditions must be known to complete an accurate correlation.

In each of these five studies, sulfide and corrosion modeling was conducted. The primary reason for doing the modeling was to be able to evaluate future conditions. Other reasons were: to help determine why corrosion was occurring; and, to determine the likely accuracy of the models. The case histories provide some helpful insight into use and accuracy of the models.

The Pomeroy-Parkhurst gravity sulfide prediction model (Pomeroy and Parkhurst, 1977) was developed in the 1970s through empirical work at the Los Angeles County Sanitation Districts. The authors indicated that the model should be capable of predicting sufficiently conservative values, if appropriate conservative coefficients are used. This has been reiterated in another paper to describe how the coefficients should be chosen to define the degree of conservatism and purpose of the investigation (Kienow, Pomeroy, and Kienow, 1982). Issues that have been raised in these and other case histories include the following:

1. The percent of BOD that is soluble is important to determine when predicting sulfide generation.

2. The models were developed in Los Angeles County and inherently include the local wastewater characteristics, including toxicant conditions that may have existed during the 1960s and 1970s period of development of the model. Those conditions may not compare with other locations.

3. The model assumes normal levels of oxygen in sewer atmospheres. There can be oxygen depression in the atmospheres of sewers, depending on specific local circumstances.

4. The model assumes no sulfide production from deposits, nor considers the impact from a combination of high sulfate concentrations (greater than 100 to 150 mg/l) and solids deposits.

Major issues for the corrosion prediction model (EPA, 1974) are:

1. Input data are extremely critical. pH depression can occur in very long interceptors, which can have dramatic effects on corrosion rates. Long-term information is mandatory.

2. Acid reaction rates with concrete vary widely and depend on factors that are not well understood.

3. The model does not predict corrosion in areas of turbulence, only in uniform pipe reaches away from turbulence. Turbulence corrosion factors have been estimated by others (ASCE, 1989).

4. The model predicts average corrosion around the cross section, not the peak corrosion rate on the cross section.

The comparison of the models with actual data are shown in Table 2. In general, the sulfide model is predicting close to the actual values, within plus or minus 50 percent. Making correlations on the corrosion model is more difficult, since very few wastewater agencies have the history of sulfide input data required to make the comparison. The case studies indicate only a very general level of agreement with the corrosion model, certainly better than order of magnitude.

Solutions to the Sulfide Corrosion Problems

Each of the five projects described in the case histories presented a unique situation relating to sulfide corrosion. Therefore, the solutions, or recommended actions, differed in each case. Taken together, the five case histories provide a comprehensive array of solutions to sulfide corrosion problems. The actions recommended to reduce or control corrosion for each case are described below.

Table 2. Comparison of Model Results to Actual Field Data (Modelling Coefficients are in Parentheses)[a]

Case study	Sulfide model	Corrosion model
Central Trunk, Sewer, Sacramento	Model about 10 to 30 percent higher than actual. (M′=0.32 x 10^{-3} m/hr; m=0.96)	Model about 0 to 30 percent lower than actual. (k=1.0; j=0.5)
Sacramento Regional Interceptor System	Model 0 to 50 percent higher than actual. (Range of M′=0.25 to 0.35 x 10^{-3} m/hr; range of m=0.6 to 0.7)	Field data not accurate enough to compare. (k=1.0; j=0.5)
Lakeland Western Trunk	Model 0 to 30 percent higher than actual. (M′=0.32 x 10^{-3} m/hr; m=0.96)	Model shows generally more corrosion, but very scattered data. (k=0.9; j=0.8)
Omaha Papillion Creek	Model about 50 percent below actual. (M′=0.32 x 10^{-3} m/hr; m=0.64)	Undetermined correlation due to scattered input data. (k=1.0; j=range of 0.5 to 0.8)
Omaha--South Interceptor Sewer (Force Main)	Model ±25 percent of actual. (M=1.0 x 10^{-3} m/hr)	Undetermined due to highly turbulent conditions inappropriate conditions for model.

[a]For definitions of coefficients, see: U.S. EPA, 1974; Pomeroy and Parkhurst, 1977; Brown and Caldwell, 1989; Kienow, Pomeroy, Kienow, 1982. Coefficient "m" here is the same as coefficient "N" in Pomeroy and Parkhurst, 1977.

Lakeland Western Trunk Sewer. The Lakeland Western Trunk Sewer had been in service for less than 20 years at the time of the investigation of corrosion problems in 1988. However, all reaches of the RCP had suffered 1 to more than 2 inches of lost concrete due to corrosive action, and reinforcing steel was found exposed or missing in numerous locations throughout the system. Forty-six percent of the RCP reaches were determined to have no remaining useful life (reinforcing bars were exposed, or worse), and the remaining 54 percent of the total length was determined to have a remaining useful life of 1 to 8 years (i.e., to the point that reinforcing bars would be exposed). There was evidence that historical corrosion rates were higher than corrosion rates in 1988, due to lower flows and higher BOD concentrations, conditions which normally produce more sulfide.

In analyzing acceptable alternative replacement/ rehabilitation options, it was recognized that this system, with its existing slope, variable future flows, high soluble BOD, and adequate sulfate concentrations, would likely cause extensive corrosion of cementitious or metallic materials exposed to the sewer atmosphere. Therefore, the alternatives were narrowed to those which would allow no exposure, or minimal exposure, of corrodible material to this environment. The recommended rehabilitation program consisted mainly of slip lining portions of the existing RCP with high density polyethylene pipe or with fiberglass pipe, and replacing the more deteriorated reaches of RCP with PVC-lined RCP or fiberglass pipe designed for direct burial. Including the rehabilitation of 75 of the 79 manholes, the total rehabilitation program was estimated to cost more than $4.2 million, or approximately $155 per linear foot of trunk line.

Omaha, Papillion Creek Wastewater System. In Omaha, concrete, metal, and instrumentation corrosion in the Papillion Creek WWTP occurred at many locations over the period of 1977 to the mid-1980s. The corrosion was caused by high wastewater sulfide levels, and high H_2S off-gassing rates. High dissolved sulfide levels in the wastewater were partially caused by high influent sulfide concentrations, and partially by sulfide produced within the plant. Solutions to plant hydrogen sulfide corrosion problems are now mostly in-place. These solutions encompassed rehabilitation, process changes, chemical addition, and ventilation improvements. The solutions implemented thus far appear to be working satisfactorily.

The probable causes of corrosion, and best chances to minimize or eliminate corrosion in the Papillion Creek WWTP are summarized in Table 3.

Recommended corrosion control for the Papillion Creek Interceptor included chemical addition. A field testing program adding iron chloride just upstream from the headworks in 1985 resulted in some reduction in dissolved sulfide at the headworks, but the results were not definitive and the chemical feed rates appeared to be high. The test may not have provided good results due to the short travel time (less than 1 minute) between the point of application and the plant headworks. Two field tests of caustic slugging were also completed in late 1985. A truckload of 50 percent strength NaOH was added to the interceptor above the old Papio Plant during each event. The wastewater sludge reached high pH levels (at least 12.5) for about 15 to 20 minutes and was diverted at the Papio Plant into grit/grease/aeration basins which had been emptied to accept and store the high pH slug. These events reduced plant influent dissolved sulfide concentrations from an average of about 3 mg/l to about 1 mg/l, and the technique was considered cost-effective. However, the storage of the caustic slug at the plant was somewhat of a problem, and the technique was not used after these two full-scale tests.

Omaha, South Interceptor Sewer. In the South Interceptor Sewer in Omaha, caustic slugging has provided a cost-effective reduction in high sulfide concentrations since late 1984 when it was initiated. Its performance is variable, primarily because of the solids deposition problem in the interceptor, which is caused by low wastewater velocities. Other sulfide control methods would likely be more reliable and have better overall performance than caustic slugging; however, the costs for other control methods appear to be substantially higher than caustic slugging. The City is investigating the use of iron chloride and chlorine gas.

The second improvement recommended at the inlet to the WWTP was to implement improved ventilation and odor scrubbing to minimize corrosive atmospheres and treat residual H_2S and other odorants more reliably. This system will provide continuous fresh air movement through the bar screen building to eliminate or minimize the corrosion problems within the building. Once improved ventilation

Table 3. Causes of Corrosion - Papillion Creek WWTP

Sulfide Corrosion problem	Immediate source(s) of sulfide	Causes of hydrogen sulfide release	Ventilation of area	Best chances to minimize or eliminate corrosion
Walls of headworks and columns	Plant influent flow and recycle streams.	Large energy drop through headworks, high velocity, and turbulence. Off-gassing of hydrogen sulfide from recycle discharge.	Ventilation limits corrosion to splash and mist zone above water line.	Reduce energy drop and velocity of influent flow. Eliminate Kennison nozzle drop discharge. Reduce influent sulfide. Protect columns.
Upper walls of grit/grease basins	Plant influent flow and recycle streams.	Air stripping and turbulence.	Space above water line too small to keep dry from ventilation.	Reduce influent sulfide and recycle stream sulfide.
Primary clarifiers--launders and splitter box	Plant influent flow, recycle streams, and production in clarifiers	Turbulence and energy loss at launders and in primary clarifier splitter box.	Ventilation keeps most concrete dry. Areas of splash and mist are greatest problems.	Reduce quantity of anaerobic recycle solids. Reduce influent sulfide.
Walls of trickling filters	Plant influent flow, recycle streams, and production in clarifiers.	Wastewater is dropped from circulating arm; hydrogen sulfide is off-gassed and wall is moist from splashing.	Ventilation cannot keep walls dry.	Hydrogen sulfide in filter influent must be reduced.
Supernatant holding tank	Anaerobic digestion process.	High sulfide concentration and turbulence created at, and adjacent to, the tank.	Little ventilation within the tank.	Modify/control turbulence. Reduce solids content of supernatant.
Decant tank	Atmospheric hydrogen sulfide from plant drain system.	See item immediately below.	Limited ventilation pulls hydrogen sulfide from plant drain system into tank.	Eliminate direct connection of decant tank to plant drain system.

Table 3. Causes of Corrosion - Papillion Creek WWTP

Sulfide Corrosion problem	Immediate source(s) of sulfide	Causes of hydrogen sulfide release	Ventilation of area	Best chances to minimize or eliminate corrosion
Plant drain manholes	Digester supernatant, thickener overflow, filtrate, and other streams.	High sulfide concentration, turbulence within system, solids deposition within system.	Little ventilation within drain system Natural ventilation pulls high hydrogen sulfide levels into connected tankage and piping.	Replace with corrosion-resistant material. Control system ventilation.
Drain pipe in primary clarifier building	Atmospheric hydrogen sulfide from plant drain system.	See item immediately above.	Roof vent pulls hydrogen sulfide into pipe from plant drain system.	Replace with corrosion-resistant pipe. Control air movement in drain system.
Sludge storage tanks	Digested sludge, mostly. Raw sludge partially.	Mixing/turbulence, air stripping.	Little ventilation within tanks.	Change to anaerobic storage tank; or protect existing aerobic storage tanks.
Outfall box conduit	Raw wastewater 1974 to 1977; treated wastewater 1977 to 1984.	High sulfide concentration suspected in mid-1970s prior to plant start-up.	Little ventilation in outfall.	Improve effluent quality and eliminate sulfide from effluent. Chlorinate effluent.

and H_2S control are in place, the City will proceed to conduct structural rehabilitation and replace equipment as necessary.

Sacramento, Central Trunk Sewer. The County of Sacramento staff has developed considerable long-term data on the Central Trunk Sewer. These data have been useful for over a decade in evaluating sulfide production in gravity sewers and assessing various corrosion issues. With removal of the sludge discharges and installation of chlorination stations along the Central Trunk Sewer, the sulfide concentration has been reduced in the 1980s, and the rate of corrosion appears to have dropped significantly. High-BOD industrial wastes are still discharged. The pH of wall moisture has been observed to have increased approximately 2 pH units at one location.

Several types of "turbulence" occur at Central Trunk manholes. This turbulence ranges from that induced by sidestream discharges and direction change, to the small amount of turbulence that occurs at manholes without sidestream flows and without direction change. One of the interesting conclusions of the 1976 inspection work was that the almost imperceptible ripples in wastewater moving straight through the bottom of a manhole caused increased H_2S off-gassing and noticeable increased corrosion in the downstream pipe crown for perhaps 30 feet.

At manholes where sidestreams entered, there was considerably more turbulence. In most of these Central Trunk junction structures, noticeable velocity changes occur in the flow, but, overall, the flows are transitioned relatively smoothly. In these junction structures, the increased corrosion in the immediate downstream sections of pipe was more noticeable that in smooth-running manholes.

Sacramento Regional Interceptor System. The planning and design of the Sacramento Regional Interceptor System encompassed several years, with considerable attention to sulfide corrosion and odor issues. The result of careful planning was an optimum combination of passive and active sulfide corrosion control measures. In designing the regional interceptors, a number of passive control measures were implemented, such as:

> A County Source Control Ordinance was adopted, controlling the quality of industrial waste discharges.

- Calcareous (rather than granitic) aggregate was specified for all concrete pipe construction. Calcium carbonate equivalent of 80 percent was required.
- Minimum concrete cover over reinforcing steel was specified as 1.5 inches.
- Junction structures were designed for smooth transitions to minimize wastewater turbulence.
- Junction structures, and other hydraulic elements especially vulnerable to corrosion were lined with plastic.
- Pipe slope and velocity were carefully evaluated to limit solids accumulations within the system.

The selection of active control measures for the Regional System reflected comparisons of cost, safety, effectiveness, and ease of operation. Liquid chlorine was the most common choice for control of sulfide in upstream areas, primarily to control sulfide build-up in the force mains. The availability of high-purity oxygen generated at the Sacramento Regional WWTP made it economical to use a system of gaseous oxygen injection into fall structures for treating the two main interceptor streams. Ventilation of the air space in interceptors also helps control sulfide and odor. Major pumping stations and the two oxygen stations exhaust their wet wells through beds of activated carbon. There are 40 operating carbon adsorbers in the Regional Wastewater System. Both of the main interceptors are ventilated by exhaust systems at the Regional Plant, with odorous air treated by the plant's odor control system.

Findings and Conclusions

Some of the major findings ("lessons learned") from these case histories are:

1. The approach, and level of attention, to sulfide and corrosion control varies significantly between different municipal agencies and engineers. Approaches are often influenced by the degree to

which odor and safety issues are integrated with corrosion issues, and the level of coordination between collection systems planning and design with treatment plant planning and design.

2. Sulfide corrosion problems in operating systems are sometimes not recognized early enough to take corrective action before considerable damage has occurred.

3. Municipal agencies, often due to budget and staff limitations, do not provide sufficient monitoring of wastewater facilities to alert them to the existence of corrosion problems. This is partially due to the difficulty and expense in monitoring out-of-sight sewers and other facilities that can be dangerous for human entry.

4. Few operations and maintenance manuals contain information on what to monitor, or how to monitor sulfide levels and corrosion.

5. There is a lack of accurate concrete corrosion monitoring techniques. Accurate measurements and assessments of concrete corrosion during the early years of system operation are important in defining corrosion rates before long-term problems are created.

6. Historically, collection and evaluation of data needed to make decisions with regard to sulfide and corrosion control has been limited.

7. The principles of sulfide and corrosion assessment and control are not universally applied.

8. Assessment and control of ventilation for sewer atmospheres and the atmospheres of tanks, channels, and other structures in treatment plants is often not adequately handled with respect to hydrogen sulfide-induced corrosion.

9. The causes of corrosion in these five case histories are partly, and perhaps largely, explainable based on current scientific knowledge concerning sulfide corrosion processes. However,

in some cases, basic scientific understanding and/or technology is lacking to define and explain the corrosion problem and to solve it quickly.

Following are several conclusions and concerns about the status of sulfide and corrosion assessments.

1. The Pomeroy-Parkhurst sulfide model seems to be providing reasonable estimates as long as the predictions are based on adequate long-term input data, conservative assumptions are used, and the conditions of the wastewater and the sewer are similar to the conditions under which the equations were developed.

2. There are data to show that the U.S. EPA corrosion prediction model can provide approximate predictive corrosion rates (better than order of magnitude), subject to the same limitations previously stated for the sulfide model.

3. Municipal agencies that need valid sulfide and corrosion predictions should verify these models and the specific coefficients for local conditions. There are sufficient unknowns in the models to warrant caution in using them without local verification. As local verification of the models is completed and the range of appropriate input data are determined, the degree of accuracy and confidence in the models should improve for any given situation. Peaking factors or safety factors must be employed.

4. The significant pH drop in long (several miles long) anaerobic interceptor systems (identified in three of these case histories) does not seem to be widely recognized as a problem. The effect of lower pH on corrosion rates is highly significant.

5. Acid production on pipe and structure walls, acid movement down the walls, and acid reaction rates seem to vary according to factors that are not well identified.

6. The different rates of corrosion through the surface layers of various types of pipe are relatively undocumented. This has often led to errors in corrosion rate evaluations.

7. Corrosion predictions in structures with turbulent wastewater conditions are extremely difficult since the hydrogen sulfide off-gassing rates are so much higher than in nonturbulent conditions. Guidance on this issue is offered by ASCE (ASCE, 1989).

8. Engineers and managers do not appear to be adequately evaluating different operating conditions which may occur during the life of a facility, when assessing sulfide corrosion and the controls needed for a project. Significant changes in flows and wastewater characteristics often occur over the 30 to 100-year lives of wastewater facilities.

9. Many interceptors have solids deposits which generate sulfide. Very little is known about sulfide production in such deposits. In addition, determining where and why solids deposits form is still not well understood. Shear stress requirements to scour deposits are recommended by various authors (ASCE, 1989).

10. The effects of various constituents in raw wastewater (such as metals), and their effects on sulfide production, are not well documented.

11. In general, municipalities and engineers do not typically prepare sufficiently detailed information on their successes and failures in sulfide and corrosion control to allow others to benefit. More well documented case histories on this subject should be encouraged.

Acknowledgements

The case histories summarized in this paper were prepared for the EPA by Brown and Caldwell Consulting Engineers, under a subcontract with E. C. Jordan Co., Portland, Maine. Major contributors to the information contained in the case histories were: Francis Hodgkins, Ed Euyen, Stan Walton, and Jack Wong, from the County of Sacramento, California; John Allison, and Virgil Caballero, from the City of Lakeland, Florida; and William Moore, Jim Swan, David Petrocchi, Ed Tomsu, and Norman Jackman, from the City of Omaha, Nebraska.

References

American Society of Civil Engineers, 1989. Sulfide in Wastewater Collection and Treatment Systems. Manual of Engineering Practice No. 69, ASCE, New York, NY.

Brown and Caldwell Consulting Engineers, September 1989. "Sulfide Corrosion Case Histories." Prepared for U.S. Environmental Protection Agency through a subcontract with E.C. Hordan Co.

Brown and Caldwell Consulting Engineers, April 1988. "Western Trunk Sewer Rehabilitation Investigation," for the City of Lakeland, Polk County, Florida.

Brown and Caldwell Consulting Engineers, February 1985. "Control of Corrosion and Odor--Phase I--Papillion Creek Wastewater System", for the City of Omaha, Nebraska.

Brown and Caldwell Consulting Engineers, July 1984. "Odor Study, Missouri River Wastewater Treatment Plant," for the City of Omaha, Nebraska.

J. B. Gilbert & Associates, April 1979. "A Case Study, Prediction of Sulfide Generation and Corrosion in Sewers", for American Concrete Pipe Association.

Kienow, Karl, R.D. Pomeroy, Ken Kienow, October 1982. "Prediction of Sulfide Buildup in Sanitary Sewers," Journal of the Environmental Engineering Division, ASCE, Vol. 108 No. EE5.

Pomeroy and Parkhurst, 1977. "The Forecasting of Sulfide Build-up Rates in Sewers", Progress in Water Technology, Vol 9, pp. 621-628, Pergammon Press.

Sacramento Area Consultants, September 1976. "Control of Odors and Corrosion in the Sacramento Regional Wastewater Conveyance System", for the Sacramento Regional County Sanitation District.

U.S. Environmental Protection Agency, October 1974. Process Design Manual for Sulfide Control in Sanitary Sewerage Systems, EPA 625/1-74-005.

APPENDIX

ENGLISH UNITS CONVERSION TO METRIC UNITS

1 foot = 0.305 meters

1 inch = 2.54 centimeters

1 ft/second (fps) = 0.305 meter/second

1 mile = 1,609 meters (1.6 kilometers)

Advances in Trenchless Sewer System Reconstruction

Lynn E. Osborn, P.E.[1]
Member ASCE

INTRODUCTION

The new developments and refinements in trenchless reconstruction technologies now permit total sewer system rebuilding of collectors, interceptors, manholes, service laterals, and force mains without excavation. Through the use of cured-in-place, poured-in-place and fold-insert-and-expand technologies, this new generation of reconstruction can help control infiltration and exfiltration more effectively while restoring full structural integrity to deteriorated, leaking storm and sanitary systems. These new advancements include techniques for reconstructing laterals from inside mains to and beyond the property lines, structural rebuilding of existing manholes, and a means of introducing plastic pipe into sewers without excavation and without leaving an annular space between old and new pipe walls.

The problems associated with pipeline rehabilitation are complex and interrelated, with no two projects being exactly alike. Because we all realize the importance of rehabilitation and extending the life of our infrastructure, we must educate ourselves as to the methods and limitations of sewer collection system reconstruction.

EVALUATION

The first step of any sewer reconstruction program should include an inventory and evaluation of the existing storm and sanitary sewer systems. This involves physical inspection of all pipelines and manholes. Also included is an evaluation of the physical condition of

---

[1]Associate Technical Director, Insituform of North America, Inc., 3315 Democrat Road, Memphis, TN 38119.

all structures and pipelines examined. Once the inventory and evaluation are complete, various reconstruction technologies are recommended for deteriorated portions of the system. There are several software packages available that can be used to organize and inventory data collected in the field. Some software packages even evaluate different trenchless reconstruction techniques.

CURED-IN-PLACE-PIPE (CIPP)

Deteriorated sewers in the United States have been reconstructed by cured-in-place-pipe for over 12 years. This proven technology is able to reconstruct pipelines from 4 inches to 96 inches in diameter. CIPP uses a flexible non-woven felt tube coated on the outside with a tough elastomeric layer. This tube is manufactured to the diameter and length required by the pipeline to be reconstructed. The wall thickness is determined by standard buried flexible pipeline equations and can be increased by simply adding more layers of felt. This felt, which has a high content of voids, is vacuum impregnated with a liquid thermosetting resin. A wide variety of resins can be used to fit the application. Different types of resins include polyester, vinyl ester, and epoxy. For standard sanitary sewers, the resin of choice is generally a polyester resin. Polyester resins tend to be the least expensive thermosetting resins and have excellent corrosion resistance to low pH environments. This is beneficial in the sewer environment where hydrogen sulfide gas can lead to the formation of sulfuric acid. The sulfuric acid generated in sanitary sewers has no corrosion effects on the polyester resins used with CIPP. In more demanding applications and/or in pressure pipes, vinyl ester or epoxy resins may be considered.

Since the installation of CIPP essentially plugs the existing pipeline, the portion of pipeline to be reconstructed must be segregated from the collection system. In low flow conditions, this can be accomplished by plugging the sewer just upstream from the reach to be reconstructed. When the flows are higher and plugging the sewer is not feasible, overpumping must be set up. Once the existing pipeline is segregated, it is inspected internally and cleaned. Since CIPP is soft and pliable during the installation phase, any debris left in the pipeline will cause an irregularity in the finished CIPP. Because of this, it is important that the existing pipeline be thoroughly cleaned. When the existing pipeline is segregated and cleaned, CIPP tube vacuum impregnated with a liquid thermosetting resin is hydraulically inserted into the pipeline by the inversion method. This method inverts or turns the tube inside out through energy provided by a column of water. Resin soaked felt is forced tight up against the existing pipe. This allows some resin to migrate into cracks and joints thus locking the CIPP into place once cure occurs. The water used to invert the tube is then heated initiating a cure which converts the pliable tube into a hard CIPP. In large diameter pipes, workers can enter the CIPP to do trim work and open side connections. In non-man entry pipes, the reinstatement of service laterals sealed by the installation process is accomplished by a remote controlled robotic cutter. Installation of CIPP is covered by ASTM F 1216.

SERVICE LATERALS

Reconstruction of service laterals through the use of CIPP can now be performed through modification of the procedure described above. For some time reconstruction of laterals without excavation has been possible in those cases where access through a cleanout is available. Now techniques are available to reconstruct laterals by working completely from within the 8-inch or larger sewer main. Using a robotic launching device and closed circuit TV, a 4 or 6 inch felt tube vacuum impregnated with thermosetting resin is pulled into an existing sewer. The remotely controlled launching device is used to position the felt tube at the service lateral opening into the main sewer. Once all of the equipment is correctly located, compressed air is used to invert the tube up the service lateral. Hot water is then added to initiate cure. Using this procedure, lateral lengths of 20 to 40 feet from the main can be reconstructed without excavation. This method fuses the lateral at the flared opening into the main.

SEWER FORCE MAINS

Force main reconstruction is accomplished using CIPP methods and the recently developed end seal technology. This unique design enables the force main to operate without tracking of effluent between the CIPP and the existing pipe. Tracking is defined as leakage around the end of the CIPP and flow of liquid between the CIPP and the existing pipe. If this occurs, sewage can leak through cracks in the existing pipe.

This new end seal procedure involves creating an annular space where the CIPP terminates. An O-Ring followed by a spacer is inserted into this annular space, and the end of the pipe is then dressed with a grout material. The result is a highly engineered end seal that prevents tracking and allows the CIPP to be terminated without leaking.

RECONSTRUCTION WITH PVC

Today polyvinyl chloride (PVC) is one of the most commonly installed sewer pipe materials in the world for new construction in small diameters. Newly adapted processes now allow this material to be used in trenchless reconstruction for small diameter sewers. The resulting SDR 35 PVC sewer pipe meets or exceeds the performance requirements of ASTM D 3034. In concept, the process is quite simple. A PVC pipe is extruded in a folded configuration. The folded pipe is flexible when hot but rigid at room temperatures. During manufacture, while the pipe is hot and flexible, it is placed on large spools for ease of handling, for transportation, and to facilitate the installation process.

Much like CIPP, reconstruction with the folded PVC pipe requires that the pipeline segment in question be segregated from the collection system. Depending upon the flow through the existing pipe, this can be accomplished by either plugging at a upstream manhole or by overpumping. Since the PVC material is soft and pliable during the rerounding phase of installation, the new pipe will simply take the shape of the existing conduit. For this reason it is important to thoroughly clean the existing pipe before reconstruction.

Spools of the folded pipe are transported to the job site where the material is reheated to make it flexible. While flexible, the material is pulled into the existing pipeline. Once the folded PVC pipe is inside the existing pipe, it is again heated by introducing hot water, steam or hot air. Once the appropriate temperature has been reached, the folded PVC pipe is pressurized internally to unfold and round out to conform to the existing pipe. The new PVC pipe is held under pressure until it cools and becomes strong and rigid.

As with CIPP, the folded PVC technology is designed so that it can be installed in existing sewer pipes without excavation. Manholes provide access for installation. Service lateral reconnections are made by cutting holes in the PVC pipe from within by using a robotic cutting device. When heated and stretched, the PVC pipe dimples at each service so that the connections can be located for internal cutting.

MANHOLES

Along with new methods to reconstruct service laterals, sewer mains, force mains, and interceptors, there are also new methods for reconstructing manholes. Groundwater leakage into manholes can account for a significant portion of the infiltration and/or inflow into a wastewater collection system. One of the new methods is a cast-in-place system that constructs a new manhole inside the existing manhole. After the manhole is pressure washed to remove loose material, reusable lightweight steel panels are lowered through the manhole ring. These panels are hand-assembled with bolts and nuts to form riser and conical sections that fit the interior shape and size while leaving a 3 inch annular space between the existing manhole wall and the new form. Concrete of various specifications is then poured into the void, sealing leaks and eliminating joints. Special inserts placed in active sewers allow uninterrupted flows, even in drop manholes. Where hydrogen sulfide corrosion is a concern, PVC liner material can be used.

After the concrete sets, the panels are removed one by one through the existing manhole opening. All of this work can be completed through a standard 22 inch manhole ring. Finished walls of the new manhole are approximately 3 inches thick. Typical construction times vary from a few hours to one day. The result is a new monolithic manhole within the existing manhole.

SUMMARY

The latest advances in sewer reconstruction technologies now allow total collection system renewal using only manhole access points. This includes service laterals as well as mains, collectors, force mains, interceptors, and manholes. Familiarization with these techniques will allow engineers and municipal officials to make informed decisions when renewing sewer systems.

REVIEW OF PIM (PIPELINE INSERTION METHOD) TECHNOLOGY

Richard J. Scholze, Jr.
Stephen W. Maloney
Ed D. Smith
Prakash M. Temkar

Abstract

The U.S. Army Construction Engineering Research Laboratory (USACERL) conducted the first demonstration of PIM (Pipeline Insertion Method, formerly Pipe Insertion Machine) technology for sewer rehabilitation in the United States in 1987, complete with a battery of physical testing for vibration and soil deformation. The technology, first developed for gas main replacement, uses an impact mole to burst the existing pipe outward into the surrounding soil and replace it at the same rate with HDPE (High Density Polyethylene). PIM is the foremost trenchless technology which can replace existing pipe with equal or larger diameter pipe. The technology is applicable to sewer, water, and gas mains and can be cost-competitive with open trench techniques in specialized circumstances, such as areas with high surface restoration costs: under paved areas, through environmentally sensitive areas, etc. A body of knowledge has developed during the past two years as the number of users has increased. This paper will summarize the existing state-of-the-art of the technology with wastewater collection systems including information on applicability, economics, advantages and disadvantages, and lessons-learned.

Introduction

An "INFRASTRUCTURE CRISIS," cry newspaper and magazine headlines, has hit this country for roads, bridges, sewers, waterlines, treatment plants and

U.S. Army Construction Engineering Research Laboratory (USACERL), P.O. Box 4005, Champaign, IL 61824-4005.

utilities. Wastewater collection systems, while being the largest capital investment by a wastewater utility, traditionally receive little attention until a problem arises; because the components are out of sight, out of mind. Many communities face serious problems with a large percentage of sewers being collapsed or in need of urgent attention. Others experience hydraulic overload due to new development or inflow/infiltration. Repair, renovation, and replacement are the major options for rehabilitation of sewers. On-line replacement, i.e. "no-dig" or trenchless technologies, have recently been receiving increased attention as innovative, cost-effective rehabilitation techniques are sought. One technology which is receiving attention is the pipeline insertion method (PIM) which can replace existing sewers with a larger size pipe. This technology can increase hydraulic capacity while supplying new lines. The proprietary technique is protected with a method patent and marketed in the U.S. by PIM Corporation of Piscataway, NJ.

Technology

The PIM concept and technology were developed in Great Britain, following an original proposal for experimental research in the U.S.S.R. as a method for replacement of cast iron gas mains. Since then it has seen additional use for sewer and water lines and, although primarily European in usage, the process is becoming more widely used in North America. In addition to the applications discussed in this paper, the technology has been used in suburban Washington, D.C., Long Island (New York), and Regina and Edmonton (Canada).

"Impact mole" or "pipe bursting" technology involves installing larger pipes into the place of existing older lines. The technology consists of fragmenting the pipe in place and forcing it into the surrounding soil with an impact mole. A new pipe of high density polyethylene is then pushed into the existing sewage collection route manhole to manhole. Thus, it can be installed without disturbing the surface. PIM has the capability to not only replace size for size but to upsize existing pipe, up to 50 percent greater cross sectional area. The process requires excavation for insertion pits at every other manhole and a reception pit for removal of the mole. In addition, smaller excavation pits are required for every functioning lateral connection.

The bursting of the pipe material is accomplished by the use of pneumatic (Figure 1) or hydraulic bursters. In the hydraulic system, the mole is towed by a winch,

Figure 1. Impact mole just before connection to HDPE pipe, showing airline inside HDPE pipe.

while the new pipe is towed behind the mole. In the pneumatic system for replacing pressure pipe, the mole is guided by a winch and the pipe is tensioned and towed behind the mole. In the pneumatic system for nonpressure pipe, the mole is guided by the winch, and the new pipe is hydraulically jacked behind the mole enabling it to achieve greater distances (Figure 2).

Distinct variants have evolved, a pressure pipe replacement system and a non-pressure pipe replacement system. These are differentiated by the characteristics of the pipes themselves. Pressure pipes (e.g. gas) tend to be of smaller diameter and laid at a more shallow depth than the larger diameter gravity systems. Initial development of the system concentrated on small-diameter gas mains. The gas mains were originally constructed of cast iron, so a PVC-liner pipe is used as a precaution against damage to the HDPE by metal fragments. The PVC liner is first pulled into place behind the impact mole as it breaks up the existing pipe. Then a HDPE pipe is slid into place inside the protective PVC liner. Non-pressure pipes such as found in sewer systems are made of more brittle materials such as vitrified clay which pose less of a threat to the HDPE pipe, thus no PVC liner is used. These pipes also tend to be much larger, making it difficult for the impact mole to both break up the existing pipe and pull in the new pipe. Therefore, a 25 ton pushing machine, placed in the insertion pit, is used to force the new HDPE pipe into the cavity created by the impact mole. This new pipe is connected to the impact mole by a collar which allows the new pipe a limited degree of independent movement.

The pneumatic equipment allows replacement of pipelines from 2 inch to 20 inch although most applications have been at the lower end of this range (6-8 inch). However, it has been used to expand pipes from ten to sixteen inches in diameter. Using the hydraulic option, 6 inch and 8 inch mains have been replaced.

The hydraulically-powered burster differs from the pneumatic burster in that the bursting action comes from sequentially expanding and retracting the top section of the burster as it is pulled through the existing pipe. The forward motion is provided by a simple winch in the receiving pit and the replacement pipe is connected directly to the burster. The expanding section of the burster is powered by an external hydraulic power unit with supply and return hoses running through the replacement pipe.

Figure 2. Diagram of Pipe Insertion Methodology for On-line Sewer Replacement

The high frequency bursting action on dry clay soil in a California application (Jacobs et al., 1988) causes the existing pipe and the surrounding soil to fracture (as opposed to the pneumatic burster which compacts the soil in all directions). The fractured soil collapses on the replacement pipe causing significant friction. The increased friction and limited power available from the winch combine to limit runs to about 250 feet.

The advantage of the hydraulic burster is its short length. With its approximate 3 foot overall length, the excavations for the launch and retrieval pits are much smaller than required for the pneumatic burster. The shorter length allows it to negotiate shorter radius curves. The hydraulic option also is reported to work better in wet environments. The annular space around the replacement pipe is smaller and it appears to be filled with soil fractured by the action of the burster. The disadvantages are the shorter runs and slower speed (1 fpm).

PIM can be used on any brittle, fracturable pipe. Successful use has been with cast iron, asbestos cement, and clay. The process can also be used with some plastics such as ABS and PVC. Improvements such as blades on the front of the mole assist in penetration of plastics. It is not useful for replacing ductile materials such as steel or ductile iron or for plastics such as HDPE which can stretch.

PIM Application for Sewers

Sewer replacement requires the following basic installation technique. A site survey is required which includes locating other underground facilities and an initial closed-circuit television camera survey (accurate within 2 inches) which pinpoints the locations of lateral connections and of any structural defects. Appraisal of redundant laterals is performed, roots and other intrusions are removed as the line is cleaned. Manholes are chosen for both flow diversion and as access and exit points. Section lengths are maximized minimizing excavation of pits and laterals. Effective flow diversion is planned and executed. Efficient operation of the pipe bursting equipment is conducted. (Normal operation is one day's conduct of pipe bursting with appropriate preparation and follow-on service to minimize customer inconvenience). Reconnection of laterals on completion and reinstatement of manholes and excavations and repaving complete the tasks.

The pipeline can be assembled in continuous lengths of up to 400 feet at the roadside. These connections are made by the butt fusion method. The faces of the sections to be joined are heated, then forced together under pressure. When properly fused, the resulting seal is stronger than the pipe itself. The same method is used to make service connections

Benefits of technology

The cost of sewer construction is often dictated by the cost of surface restoration. The actual pipe cost can be 20 percent or less of the overall construction cost. The use of pipe insertion rather than an open trench will allow savings in areas where surface restoration or trench construction is especially costly such as easements, sensitive or high visual impact environmental areas like golf courses or parks, and areas with surface traffic. A major benefit, difficult to measure in terms of dollars, is the avoidance of closed areas such as roads and parking lots. Careful planning of insertion pit locations can avoid all disruptions in surface activity associated with open trench methods. Additional advantages are listed in Table 1.

TABLE 1

Advantages of PIM Technology

Advantages of PIM Technology
Trenchless
Uses existing pipe for subsurface lane
Upsizing capabilities
Capital improvement, not maintenance
Minimizes traffic inconvenience
Minimizes public inconvenience
Cost savings in paving excavation and replacement costs
Useful in areas with high surface traffic
Useful in easements, limited work surface
Good public relations
Possible time savings
Engineering costs are lower than "open cut"

The primary disadvantage is the cost. It is an expensive option, therefore, it must be applied appropriately. There is also the possibility of surface disturbances in dense soils.

Ft. Belvoir Application

At Fort Belvoir, Virginia, the U.S. Army CERL performed the first U.S. demonstration of the technology for sewer upgrade. Two hundred and forty feet of eight-inch main was placed in the path of an existing six-inch main which traveled between two manholes, under two parking lots, one well-traveled street, a retaining wall and a fence. Passage of a pneumatic mole destroyed the old pipe and made a passage for the new high density polyethylene pipe. Traffic was maintained and the wall and fence were not affected.

During the course of the demonstration at Fort Belvoir, Virginia, a series of tests were performed during construction to determine the effect of the mole on adjacent utlities. Standard procedure in the PIM technique is to expose areas where pipes cross the path of the mole and remove the dirt between the pipe being replaced and the crossing pipe. This is a safety precaution, however, engineers often have incomplete plans of buried utilities. Tests were designed to measure potential effects on pipes which were not known and exposed to relieve stress. An experiment was set up with instrumented pipe in place perpendicular to mole direction and 16 inches above it. Three rods were sunk into holes at various depths above the pipe, a pipe (4-inch ductile iron) with strain gages attached was installled 16 inches above the pipe. In addition, geophones were placed on the surface directly over the pipe being replaced, and at least 9 feet from the centerline, to measure vibrations induced by the impact mole. After construction, measurements of deflection of the newly installed pipe were taken to determine if it maintained its roundness.

The results of these tests are detailed elsewhere (Briassoulis et al., 1989) and will only be summarized here. Stress induced on the adjacent pipe produced only small strains, with the maximum strain above the impact mole's path being 200 . Figure 3 shows the longitudinal strain distribution along the crown line of the instrumented pipe as the mole was passing by. The mole was beneath the instrumented pipe. The results were found to be not significant. Maximum strain was developed at the crown of the instrumented pipe just above the sewer pipe. This is an example of classic strain distribution for a beam on an elastic foundation.

Soil displacements measured at different elevations above the sewer pipe as the mole was passing by are shown in Figure 4. The maximum soil displacement was 0.60 inch

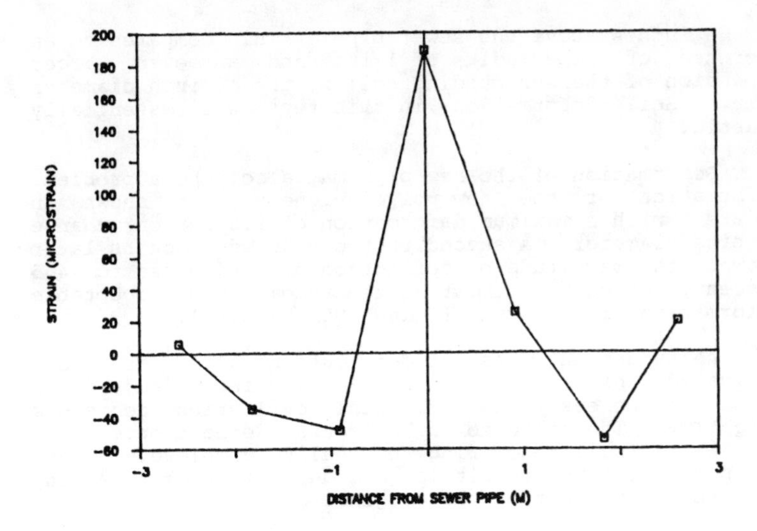

Figure 3. Longitudinal Strain Distribution Along the Crown Line of the Instrumented Pipe When Mole Was Passing By (Mole Beneath Instrumented Pipe).

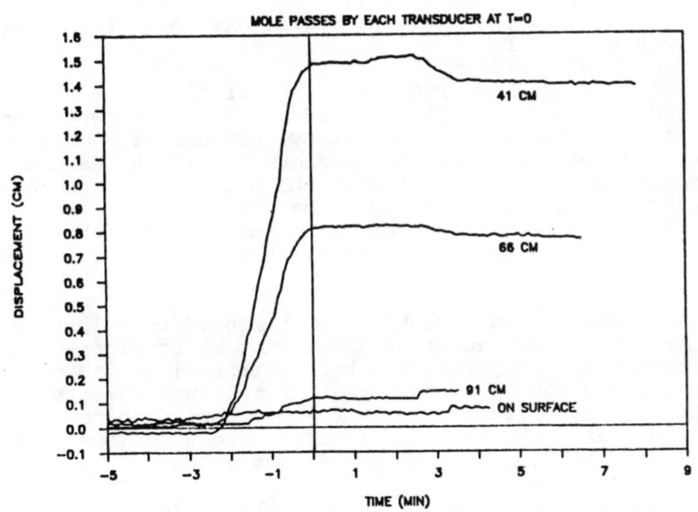

Figure 4. Soil Displacement at Different Depths Above Sewer Pipe as Mole Was Passing By. (Results are normalized so that time zero corresponds to the time when mole was beneath each displacement transducer).

at 16 inches above the sewer pipe. This compares to an increase of pipe radius of 1 inch and an even larger expansion of the surrounding soil by the 10 inch diameter mole. Soil deformation in this test was essentially plastic.

Deformation of the new pipe may also pose a problem. Deformation of the new polyethylene pipe is shown in figure 5 with a maximum deformation of 2.5 percent change of pipe diameter. A second test conducted 6 months later showed the magnitude of deflection had increased to 4.5 percent, which is almost at the recommended acceptable deformation of 5 percent (8-inch SDR 17 pipe).

A subsequent test of deformation after one year indicated that the vertical deflection increased a small amount, whereas the horizontal deflection decreased slightly. The decrease in horizontal deflection may be due to the soil recovering and sealing completely around the pipe. The net result is that deflection is remaining less than 5 percent of the original diameter.

Vibrations were also noted as a potential problem. Vibrations directly over the pipe were found to be significant. The maximum vertical peak particle velocity reached 0.54 in./sec when the mole was beneath the array of geophones (Briassoulis et al. 1989), but damped out quickly (0.06 in/sec) at the 9-foot distance from the mole path centerline.

The results from U.S. Army and British tests were compared (Briassoulis et al, 1989) and were shown to be comparable to vibrations induced by jackhammers. Figure 6 shows a comparison between maximum PPV (Peak Particle Velocity) due to the impact mole and due to jack hammers and trucks. Thus, care should be exercised in locating and protecting (e.g., by excavating and removing the soil between the mole and the critical object) utilities and structures sensitive to vibration.

The conclusion is that, with regard to vibration effects, the PIM technique should not be considered as more damaging compared to other alternatives such as excavation using jack hammers, or to usual sources of vibration like trucks, except for buried pipes and archeological objects located close to the operating mole. Vibrations appear to dampen out quickly.

At Fort Belvoir, the use of the mole technology resulted in minimal surface disruptions. The retaining wall and fence were not affected by the construction and traffic was maintained at all times.

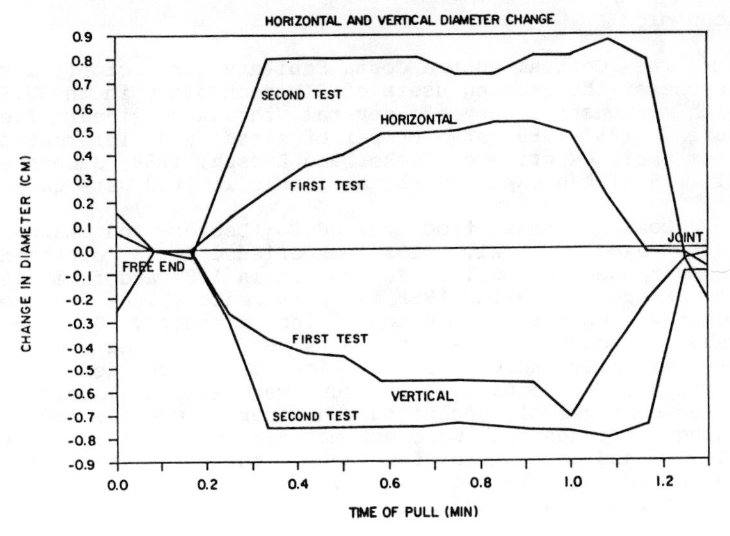

Figure 5. Deformation of Inserted Polyethylene Pipe After Installation (Deformation is elliptical).

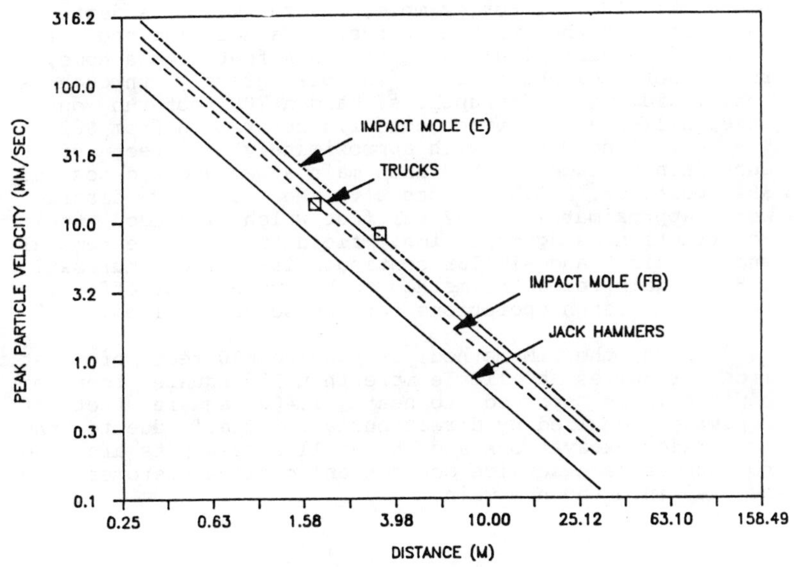

Figure 6. Comparison of Attenuation Curves for Jack Hammer and Trucks and PIM Technique Tests. E=England, FB=Fort Belvoir.

Economics

The Central Contra Costa Sanitary District (CCCSD) is one of the leading users of PIM technology in the U.S. with demonstrations of several thousand linear feet during 1987-1989 at a number of sites and information from their experience (Decker and Larson, 1988; Jacobs et al., 1988) has expanded the available limited data base.

Cost figures from the CCCSD (Decker and Larson, 1988; Jacobs et al., 1988) experience for 8 different runs ranged from $51 to $83 per ft in 1987 and from $69 to $86 per ft during 1988 as approximately 7000 feet of sewer was upsized from 6 to 7.5 inch diameter. Runs were made through a variety of terrains and soils, predominantly adobe clays which are very dense and difficult to compact. The runs were all in easements, often extensively vegetated and steep. Overall savings using the technology were estimated at 6, 11, 15, 16, 28, 28, 44 and -42 percent for each of the 8 runs. It must be noted that the negative savings value was for a run made across a golf course.

CCCSD continued the use of PIM technology during 1989 with their most challenging applications to date (Kurasaki, 1989). For example, one run was at a depth of 20 feet, another within 12 feet of a swimming pool (no damage from vibration), a third ran 4 feet from a home, a run under telephone ducts in pea gravel, through a schoolyard with disruptions, and a 700 foot run on a steep hillside. Contract cost figures ranged from $82 to $99 per linear foot with approximately 3300 feet of 6 inch main replaced with 8 inch main at an average cost of $94 per foot. Overall costs to the sanitary district were approximately $102 per foot which included their construction management time, field time, fence removal and repair, and similar contingencies. They (Kurasaki, 1989) again found the method to be more cost-effective than open-trench operations for the selected sites.

Using the PIM method, replacing 600 feet of cast iron requires a little more than 175 square feet of repaving, as opposed to nearly 1,400 square feet of repaving required by direct burial. That's due to the two major excavations and the small access pits are dug to expose each service connection to allow customers to be tied into the new main.

Lessons-Learned

Lessons-learned from CCSD (Jacobs et al., 1988) experience have established the following minimum specifications for future pipe bursting projects:

* Use of fusion welded saddles for service lateral reconnection

* Removal of all internal butt fusion weld beads

* Repair or correction of all major low spots in existing pipe prior to pipe bursting operation

* Minimum cover of 3 feet in easements and 5 feet in streets

* Final inspection to include low pressure air test, deflection test with 95% mandrel, and CCTV inspection

* One year warranty in installation and materials with deflection test at end of warranty period

They also noted short sags (less than 10 feet) appeared to be corrected by the longer length of the pneumatic burster.

Jacobs et al., (1988) conclude that pipe bursting is cost effective for gravity sewer replacement in easement situations when compared to conventional open-cut methods when the following factors are considered:

* Difficulty of access
* Depth of pipe
* Width of easements and encroachments
* Impact on local residents/commercial activities
* Restoration of surface improvements
* Reduced risk due to unknown conditions
* Improved safety to workers and local residents

Summary

PIM technology is gaining acceptance in the U.S. as a method for rehabilitating and upsizing sewer lines. Experience is being gained for optimal application of the technology and its effect on the surrounding soils.

Acknowledgement: The authors wish to thank Joye Kurasaki and the Central Contra Costa Sanitary District for their information and cooperation.

References

Briassoulis, D., Maloney, S.W., and S.C. Sweeney, 1989. "Static and Dynamic Effects of the Pipe Insertion Machine Technique," Technical Manuscript, U.S. Army Construction Engineering Research Laboratory.

Decker, C. and J. Larson, 1988. "California Sewer Replacement Experience using Pipe Bursting (PIM) Technology," Proc. NO-DIG 88.

Jacobs, E., J. Larson, and C. Decker, 1988. "Addendum to California Sewer Replacement Experience Using PIM Pipe Bursting Technology," Unpublished.

Kurasaki, Joye, 1989. CCCSD, Personal Communication, 8 November.

Maloney, S.W., 1989. "In-Place Replacement of Underground Pipes," in Proc. First International Conference on Underground Infrastructure Research, American Water Works Association Research Foundation, Denver CO.

Maloney, S.W. and D. Briassoulis, 1988. "Wastewater Collection System Rehabilitation Techniques for Army Installations," U.S. Army Construction Engineering Research Laboratory Technical Report N-88/25, Champaign, IL.

INVESTIGATION & REHABILITATION OF SEATTLE'S TOLT PIPELINE

by Walter F. Anton[1] (F.ASCE), Jack E. Herold[2], Robert T. Dailey[3] and William J. Cichanski[4]

ABSTRACT

Seattle Water Department's Tolt Pipeline brings water 24 miles from storage in the Cascade Mountains into city and suburban distribution systems. Twenty-three miles of this pipeline originally consisted of pretensioned concrete cylinder pipe with sizes of 54", 60", and 66". The gravity pressure is as high as 325 psi static.

In November 1987, the pipeline burst during routine shutdown. A pressure reducing station was installed in place of an existing line valve, and a study was begun to determine the pipe's condition. In August 1988, the pipeline again burst when high water hammer pressures resulted from the inadvertent rapid closure of the new pressure reducing valve. A review of design and construction records and field and laboratory data has identified factors that have led to the premature deterioration of this major pipeline. Hydrogen embrittlement of corroded pretensioning rod (with yield strength between 90-100 ksi) has been identified as a key factor contributing to the two pipeline breaks. Approximately half of the pipeline was prone to this type of failure.

( ) Chief Engineer, Seattle Water Department
( ) Manager, Water Engineering Section, Seattle Water Department
( ) Project Manager and Senior Civil Engineer, Seattle Water Department
( ) Branch Manager, Construction Technology Laboratories, Inc.

The rehabilitation program commenced in December 1988 when the decision was made to line about six miles of 66" and 60" pipe with 54" ductile iron and 54" steel pipe, respectively. Construction started in late February 1989 and was completed within three months. Future methods of rehabilitation include replacing the existing pipeline with either the same size or larger pipe or encasing the pipeline with reinforced concrete.

INTRODUCTION

The Seattle Water Department's Tolt Pipeline No. 1 is a vital water supply line that delivers 30% of the system supply from storage in the Cascade Mountains to Seattle and suburban water distribution systems. This pipeline must be in service between the first of June and mid-October yearly. The plan and profile of Tolt Pipeline No. 1 is shown on Figure 1. All but one mile of this 24-mile pipeline is comprised of pretensioned concrete cylinder pipe (PCCP) (originally identified as "modified prestressed concrete cylinder pipe"), varying in diameter between 54 and 66 inches. Pipe joints consisted of steel bell and spigot rings with rubber gaskets. The reinforcing rod is undeformed steel rod which is helically wrapped around the welded sheet steel cylinder. Typical pipe cross-section and range of dimensions are shown on Figure 2. The pipe sections were fabricated and supplied by two prominent pipe manufacturers. All pipe was furnished to a single joint-venture contractor, who was responsible for the entire pipeline installation. The pipeline has been in service nearly 30 years.

In November 1987 and again in August 1988, the Tolt Pipeline ruptured suddenly (Figure 1). Fortunately, on both of these occasions property damage amounted to only limited flooding of a few private properties and debris on roadways and in culverts. There were no injuries, and water transmission was restored in a few days.

The November 1987 pipeline rupture occurred when the pipeline experienced full shutdown pressure while it was being taken out of service in a routine manner. Damaged pipe exposed at that break revealed serious corrosion, prior cracking, and brittle fracture of the spiral reinforcing rod surrounding the pipe's steel cylinder, thereby weakening this segment of pipeline. A pressure reducing station was then installed on an emergency basis at a location roughly midway along the Tolt Pipeline No. 1 route providing some protection for the more densely populated downstream area. The August 1988

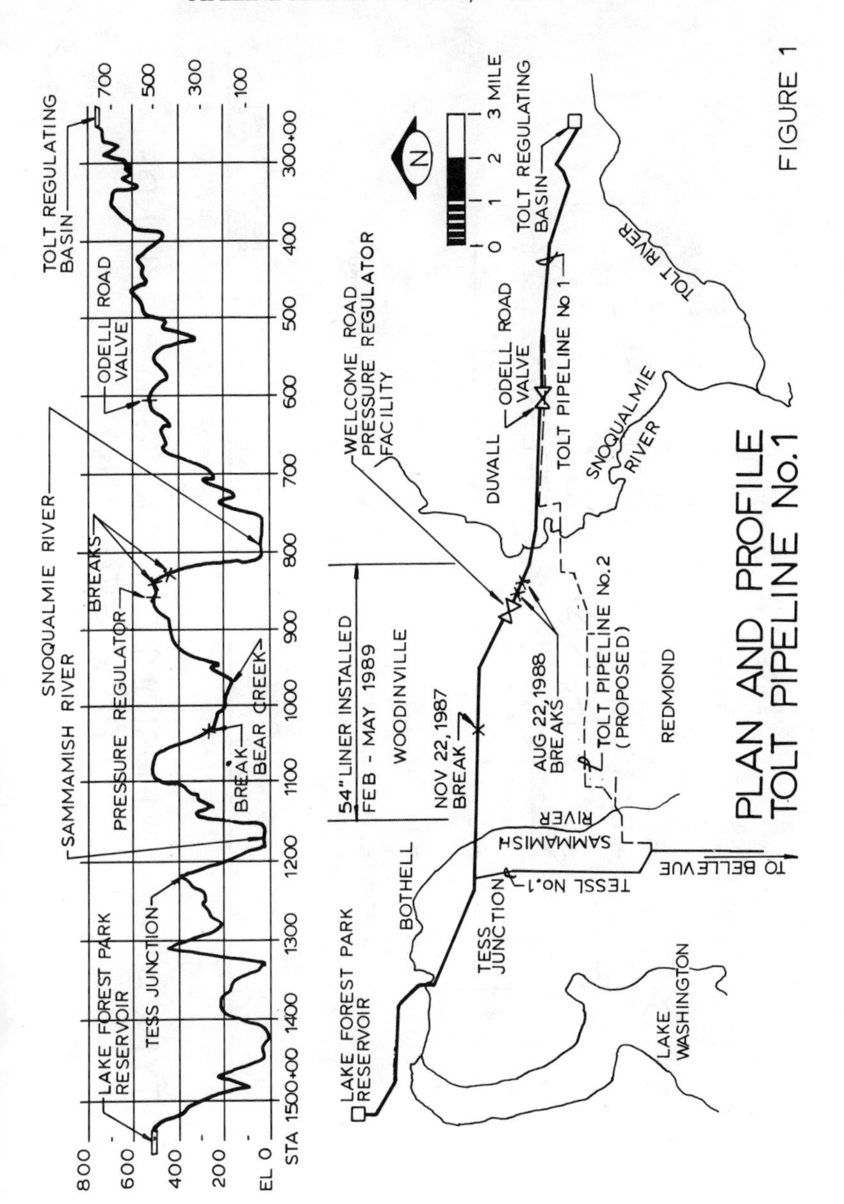

TOLT REGULATING BASIN
ODELL ROAD VALVE
SNOQUALMIE RIVER
SAMMAMISH RIVER
BREAKS
PRESSURE REGULATOR
BREAK
BEAR CREEK
TESS JUNCTION
LAKE FOREST PARK RESERVOIR
800
600
400
200
EL 0
700
500
300
100
STA 1500+00
1400
1300
1200
1100
1000
900
800
700
600
500
400
300+00
WELCOME ROAD PRESSURE REGULATOR FACILITY
N
0 1 2 3 MILE
TOLT REGULATING BASIN
TOLT RIVER
ODELL ROAD VALVE
TOLT PIPELINE No 1
DUVALL
SNOQUALMIE RIVER
54" LINER INSTALLED
FEB - MAY 1989
WOODINVILLE
NOV 22,1987 BREAK
AUG 22,1988 BREAKS
TOLT PIPELINE No.2 (PROPOSED)
REDMOND
SAMMAMISH RIVER
BOTHELL
TESS JUNCTION
TESSL No.1
TO BELLEVUE
LAKE FOREST PARK RESERVOIR
LAKE WASHINGTON
PLAN AND PROFILE
TOLT PIPELINE No.1

FIGURE 1

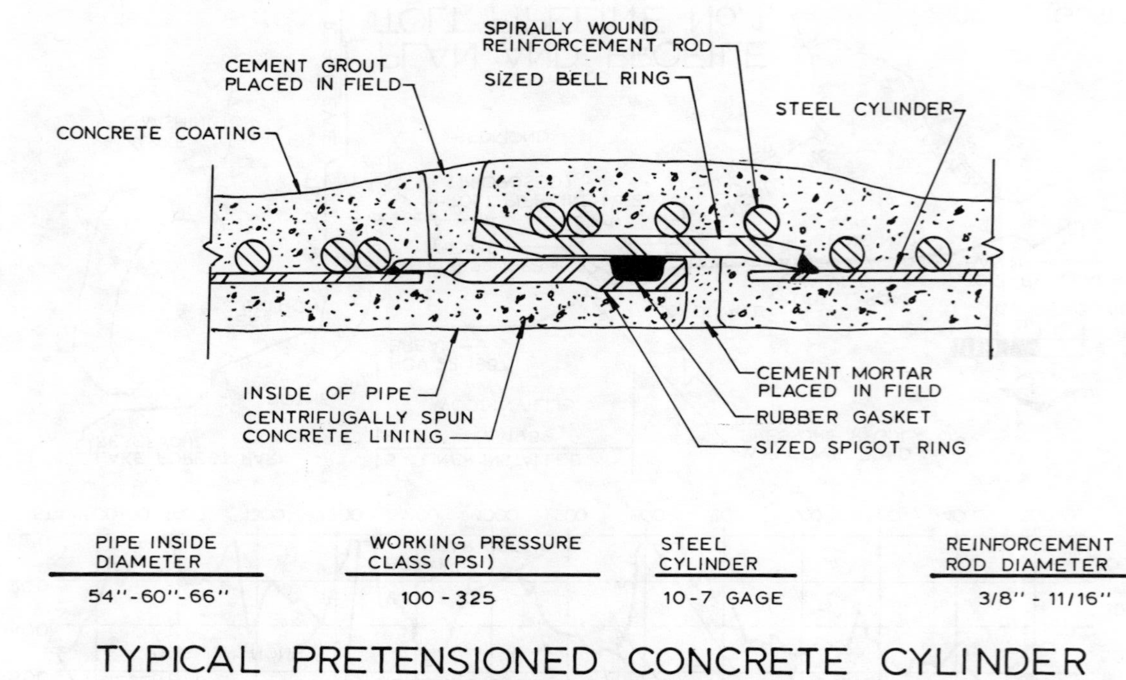

PIPE INSIDE DIAMETER	WORKING PRESSURE CLASS (PSI)	STEEL CYLINDER	REINFORCEMENT ROD DIAMETER
54"-60"-66"	100 - 325	10-7 GAGE	3/8" - 11/16"

TYPICAL PRETENSIONED CONCRETE CYLINDER PIPE SECTION

FIGURE 2

pipeline rupture resulted from a sudden high water pressure surge within the pipeline upstream of the valve caused by a maintenance worker's inadvertent rapid closure of the valve installed at the pressure reducing station. Serious corrosion, prior cracking and brittle fracture of the spiral reinforcing rod was evident again in damaged pipe exposed at this site.

Extensive visual internal and external investigations were conducted by Department staff and specialty consultants. These investigations formed the basis for the extensive rehabilitation and replacement program for the middle half of the pipeline that began during winter 1988-89 and will continue through spring 1990.

PIPELINE EVALUATION AND ANALYSIS

Project Requirements Analysis Phase

The primary goals for the pipeline evaluation were to:

1. Assess the likelihood of additional pipeline failures. It was important to understand whether the two previous failures were isolated incidents, or if the condition of the pipeline was such that further failures could be anticipated.

2. Estimate the remaining service life of the pipeline. This was required to provide input into the fiscal planning for scheduled repairs and rehabilitation.

3. Develop recommendations for extending the service life. Consideration was directed toward (a) potential changes in operating procedures, (b) increased inspection and maintenance, (c) corrosion rate monitoring, and (d) installation of cathodic protection measures.

Scope of Work - Evaluation and Analysis

The scope of work used to direct the evaluation and analysis efforts consisted of the following tasks.

Review Documentation

The Seattle Water Department compiled a comprehensive data package relevant to the Tolt River pipeline project. The data package included project correspondence generated during the design and construction of the line, design data, construction records, pipe fabrication and transportation records, inspection records, soil and groundwater data, operation and maintenance reports, corrosion and corrosion potential reports, and specific reports prepared by other consultants. This documentation was reviewed to gain necessary insight into the history of the pipeline to develop a thorough understanding of the operational requirements for the line, and to form the rationale for specific site excavation and material sampling activities to follow.

Evaluate Failed Pipe Segments

Physical samples were extracted from each of the failed segments, and detailed laboratory examinations were completed on the samples to determine probable causes for the pipeline structural failures. Also, one damaged pipe segment was stulled, protected, and transported to the engineering consultant's laboratory facilities for detailed dissection and examination. Data gathered during these activities was used to identify probable causes for the previous pipeline failures, and to develop a selection criteria for site excavation and further field sampling.

Select Evaluation Sites

All of the data and information compiled during the first two tasks were synthesized to: (1) specify the locations and sequence for excavating and sampling the in-situ condition of the pipeline, and (2) specify the detailed inspection and sampling tasks to be completed at each excavation site.

A total of 30 inspection sites were selected. Many variables influenced the site selection process, not the least of which being that the pipe segments had been supplied by two different pipe suppliers. The primary list of equally weighted variables influencing site selection were:

- Pipe diameter
- Pipe class
- Operating pressure
- Pipe manufacturer (supplier)
- SWD inspection report data
- Proximity to previous failure(s)
- Native soil type
- Bedding and backfilling conditions
- In-situ corrosion potential
- Overburden

Conduct External Inspections and Sampling

The pipeline was excavated and physically inspected and samples were collected to determine the in-situ condition, and to compare it to the conditions predicted during the aforementioned task. In addition to evaluating the pipe structure, both the soil and groundwater were sampled to determine whether their physical and chemical properties constitute contributing factors to the deterioration of the pipeline. Each excavation resulted in a structural condition survey of the pipe section(s) exposed, a geotechnical survey of the surrounding soils and a groundwater analysis.

Pipeline Analysis

The material sampling activities played an important role in evaluating the in-situ condition of the line. Mortar coating samples were subjected to petrographic analysis in the laboratory to check for evidence of carbonation, contamination, and deterioration. Pretensioning rods were nondestructively inspected in the field for evidence of cracking using several techniques (ultrasonic, dry magnetic particle inspection, and wet fluorescent magnetic particle inspection). Pretensioning rod samples were extracted and analyzed in the laboratory to determine chemical and mechanical properties, extent of corrosion area loss and pitting, cracking, and extent of embrittlement.

Analyses of the pipeline were made to evaluate the design conditions, causes of observed damage, and the remaining useful life. Project specifications contain criteria for design of "modified prestressed concrete cylinder pipe". Specifically, the following design considerations were specified for design of the Tolt Pipeline:

1. The required cross-sectional area of steel per linear foot of pipe wall was based on the cylinder and circumferential rod reinforcement reaching the specified minimum yield points simultaneously at the pressure equal to 2-1/4 times the working pressure; thereby defining the elastic limit of the pipe.

2. The required minimum yield and tensile strengths for the steel cylinder were 30,000 psi and 52,000 respectively.

3. The required minimum yield strength for rod reinforcement was 50,000 psi (no upper limit); and hard grade steels were permitted.

Analysis of buried pipelines requires consideration of soil-structure interaction effects. CANDE, a known non-linear finite element analysis program, was utilized for the structural analysis. Using CANDE, a parametric study was conducted to evaluate the effect of back-filling, bedding, and overburden variables on pipe performance. Data gathered during review of Tolt Pipeline construction records quantified the bedding, backfilling and overburden variables selected for this analysis effort. Using the parametric study results, causal relationships for the pipe damage observed in the field were developed. These efforts provided the necessary input for developing a rational damage mechanism hypothesis which closely matched field observations. This analysis was instrumental in quantifying the effects various installation methods may have had on the structural response of the buried pipeline.

Conclusions - Evaluation and Analysis Phase

Allowable pipeline working pressures were generally in conformance with project specifications. No evidence of aggressive soil and groundwater chemistry cause of corrosion was evident. Pipe manufacturing processes appeared, for the most part, to be in conformance with industry standards (circa 1960) and the project specifications. The fundamental cause for failure of the Tolt Pipeline is directly related to corrosion of the pretensioning rods. Both the field data and structural analyses results indicate that a large number of pipe sections have mortar coating delaminated in both the invert and crown regions.

A review of the design and construction records disclosed a number of deficiencies that contributed to concrete coating delamination and the resulting corrosion and premature deterioration of the spiral rod reinforcement. Delamination of the mortar coating destroyed its passivating, corrosion-inhibiting properties to the depth of the steel pretensioning rods. Corrosion of the pretensioning rods commenced and became widespread along the pipeline length. In some pipe sections, delamination spread to the depth of the steel cylinder.

Extensive cracking and delamination of the concrete coating over the rod reinforcement resulted from the excessive radial deformation of pipe segments of much of the Tolt Pipeline. Factors contributing to the excessive deformation included: pipe being not as rigid as assumed; no imported bedding under one-third of length of pipeline -- inadequate bedding elsewhere; inadequate compaction of native soil backfill (particularly in the region bounded by the pipe invert and springline), and premature removal of the interior stulling (bracing). Excessive deformation was observed during an interior inspection of the pipeline following construction, and the Department expressed concern that corrosion would occur. Cathodic protection was considered, but was not installed primarily because the pipe joints were not bonded.

Both pipe suppliers provided pipe which met project specification requirements. However, one manufacturer supplying pipe for approximately 50% of the pipeline provided pretensioning rod having strengths far exceeding minimum specified values. For these pipes, rod yield strengths measured were frequently found to exceed 100,000 psi, with companion ultimate strengths as high as 120,000 psi. The chemical composition of the rods was also found to be highly variable. These rods, subjected to a sustained state of stress and placed in a corrosive environment, were found susceptible to hydrogen-assisted stress corrosion cracking (HSCC). Figure 3 is a photomicrograph taken through one of the pretensioning rods extracted from the Tolt Pipeline. Note the singular, sharply defined crack indicative of hydrogen-introduced stress corrosion cracking.

The American Society for Metals Handbook, Volume 2, describes hydrogen induced cracking as follows: "Most often, fracture occurs at sustained loads below the yield strength of the material. This cracking mechanism depends on the hydrogen fugacity, strength level of the material, heat treatment/microstructure, applied stress, and temperature. For many steels, a threshold stress exists below which hydrogen stress cracking does not occur. This threshold is a function of the strength level of the steel and the specific hydrogen-bearing environment. Therefore, threshold stress or stress intensity for hydrogen stress cracking is not considered a material property. Generally, the threshold stress decreases as the yield strength and tensile strength of an alloy increases. Hydrogen stress cracking is associated with absorption of hydrogen and a delayed time to failure (incubation time) during which hydrogen diffuses into regions of high triaxial stress. Hydrogen stress cracking may promote one mode of fracture in an alloy rather than another form normally observed to benign environments. Thus, all modes of cracking have been observed in most commercial alloy systems; however, hydrogen stress cracking usually produces sharp singular cracks in contrast to the extensive branching observed for stress corrosion cracking."

Photomicrograph Through Pretension Rod Crack - Figure 3

Coating Delamination and Pretension Rod Corrosion on Pipe Invert
Figure 4

Pipe for the remainder of the pipeline was supplied by a second manufacturer. Widespread corrosion of pretensioning rods was also found in this reach of the line (see Figure 4). However, sampled pretensioning rod did not exhibit evidence of HSCC-related failures. Laboratory analysis of rod samples did not indicate that the chemical composition and mechanical properties are such that HSCC-related failures are likely. Compared to the high-strength rods of widely varying composition supplied by the first pipe manufacturer, measured rod yield strengths ranged from 50,000 to 54,000 psi, with ultimate strengths ranging from 81,000 to 84,000 psi. Chemical composition test results, supplemented by mill certificates, also indicated a more consistent manufacturing process.

To date, the measured corrosion in the reach of the line not subject to HSCC has been relatively insignificant, and the observed deterioration has not significantly impaired pipeline function within current operational and environmental parameters. This reach of the Tolt Pipeline has been estimated to have a remaining service life of a minimum of 10 years.

DESIGN/INSTALLATION METHODS VS CURRENT PRACTICES

The pretensioned concrete cylinder pipe installed for the Tolt Pipeline in 1959-61 predated the adoption of the American Water Works Association (AWWA) Standard for Pretensioned Concrete Cylinder Pipe. Compared with the current AWWA Standard (C-303) and recommended design and construction practices:

- Pretensioned concrete cylinder pipe larger than 54 inches probably would not used.
- Only A615 billet steel-grade 40 (intermediate grade, minimum 40,000 psi yield strength) would be used for the steel rod reinforcement; A615 billet steel-grade 60 (hard grade, minimum 60,000 psi yield strength steel) would not be permitted.
- The pipe would not be placed directly on hard trench bottom.
- Pea gravel or other good support material, such as cohesionless fill, would be densified under the pipe haunches, along the sides, and over the pipe.
- Stulls/bracing would be left in place until backfill above the pipe was completely in place.

Joints would be bonded to permit the addition of cathodic protection at a later date.

PIPELINE REHABILITATION - PHASE I

Design and construction of Rehabilitation Phase I was pursued as an emergency project with an absolute need to have the pipeline operational by June 1. Design decisions and construction methods were selected with heavy emphasis on constructibility and common materials.

On December 13, 1988, the decision was made to pursue rehabilitation of the pipe between the Snoqualmie and Sammamish River valleys. This is the reach of pipe which had burst twice, which passed through a once-rural area now occupied by many expensive homes, and which contained embrittled rods coincidental with observed out-of-round conditions. A five-person SWD employee design team was formed.

Many decisions were made early in the project. Lining in lieu of laying new pipe was selected for speed and constructibility. Also, the time required to produce and review an environmental impact statement (EIS) would have made the project impossible to complete on schedule, and it was felt that not completing an EIS was more defensible if lining were performed in lieu of dig and lay. Lining also was judged to be less susceptible to delay due to inclement weather.

Various methods of lining were considered. Polyethelene was ruled out due to pressure limitations and cost. INSTITUFORM was not used due to its pressure limitation and relatively high cost. The split-can steel lining method was given strong consideration, but finally ruled out because of concerns for speed of construction and doubts about the integrity of a field-welded longitudinal seam which could not be x-rayed.

The choice narrowed to ductile iron or steel pipe. The preference was to use ductile iron because it had been previously jacked or pushed in smaller sizes over lengths of 1000 feet to 2000 feet.

The length of steel pipe that could be jacked without damaging the bell-spigot ends with the high axial load was unknown. The disadvantage of using ductile iron pipe is that as a liner it causes a 12-inch diameter (nominal) reduction, and steel could be configured to lose only 6 inches in diameter. For reasons of hydraulic performance the nominal diameter could not be less than 54 inches. Thus, SWD chose to use 54-inch Class 50 ductile iron pipe inside the 66-inch PCCP and 55-inch O.D. steel (1/2" wall weld-bell) inside the 60-inch PCCP. Choosing both steel and ductile iron pipe had the additional advantage that suppliers could meet our schedule demands. Selecting both materials provided a measure of flexibility or choice to react to uncertainties.

Orders were placed for ductile iron pipe on December 23, 1988, and for steel pipe on January 5, 1989. U.S. Pipe and Foundry delivered the first ductile iron pipe on February 3, 1989, and Northwest Pipe & Casing delivered the first steel pipe on March 1, 1989. (Incidentially, each first delivery coincided with a record snowfall.)

Even though this project was pursued on an emergency basis, compliance to City of Seattle rules, policies and procedures was maintained. Purchases and construction contracts were bid competitively. Noise ordinances were not waived. All minority/women owned business requirements were in effect. Paperwork was often hand-carried or expedited, but all rules and requirements were met.

Other design elements of interest included concern for installing the liner through curves which were accomplished in the original construction with deflected joints. Complicating this concern were field measured out-of-roundness in the existing pipe of 2.5 inches or more. (This often occurred as a flattening condition). Figure 5 shows the sketch and computation that were used to determine where curvature would dictate a dig-lay installation vs. liner. By selectively locating insertion pits (and in one case lengthening the pit to 100 feet) the amount of pipe required to be open cut due to curvature was 740 feet. The other open cuts totaled 1310 feet and were made to provide concrete encased steel pipe to act as valve thrust blocking.

Welded steel pipe was used whenever special ductile iron pipe, such as fittings or restrained joints, would be needed. This pipe was rolled to ductile O.D. to simplify fitting to the ductile iron pipe. This was done to assure timely delivery of material and to provide greater flexibility for field fitting and welding where

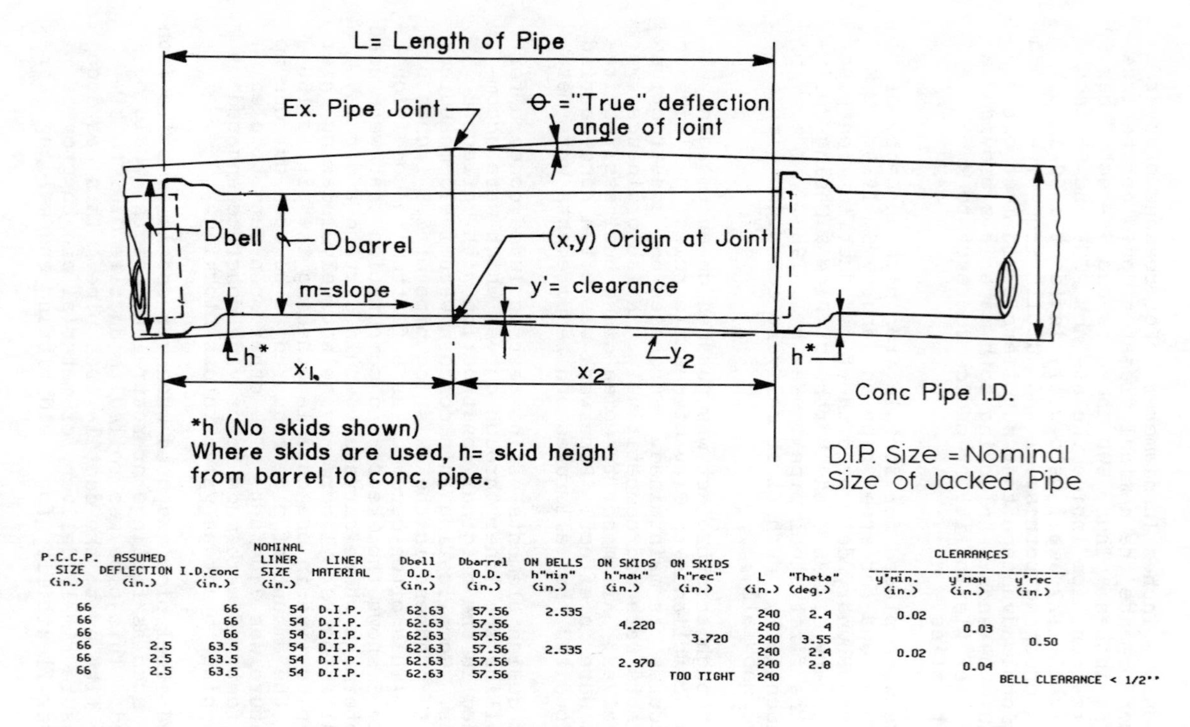

P.C.C.P. SIZE (in.)	ASSUMED DEFLECTION (in.)	I.D.conc (in.)	NOMINAL LINER SIZE (in.)	LINER MATERIAL	Dbell O.D. (in.)	Dbarrel O.D. (in.)	ON BELLS h"min" (in.)	ON SKIDS h"max" (in.)	ON SKIDS h"rec" (in.)	L (in.)	"Theta" (deg.)	CLEARANCES y'min. (in.)	y'max (in.)	y'rec (in.)
66		66	54	D.I.P.	62.63	57.56	2.535			240	2.4	0.02		
66		66	54	D.I.P.	62.63	57.56		4.220		240	4		0.03	
66		66	54	D.I.P.	62.63	57.56			3.720	240	3.55			0.50
66	2.5	63.5	54	D.I.P.	62.63	57.56	2.535			240	2.4	0.02		
66	2.5	63.5	54	D.I.P.	62.63	57.56		2.970		240	2.8		0.04	
66	2.5	63.5	54	D.I.P.	62.63	57.56			TOO TIGHT	240				

BELL CLEARANCE < 1/2"

Relationship Between Liner Clearance and Joint Deflection - Figure 5 connections, outlets, etc., were needed. Also, insertion pits were located where connections, manholes, air valves, and blowoffs were located so that these elements could be done with steel pipe instead of ductile iron pipe.

For simplicity and speed, the corrosion protection and backfill chosen was concrete encasement in the open areas and grouting of the annular space in the PCCP pipe. This choice also enabled ready incorporation of thrust restraint for valves into the design.

Many other design decisions and reviews were accomplished on an expedited basis between mid-December 1988 and mid-January 1989. A construction bid package was put out for bid in mid-January 1989. Bids were opened on February 8, 1989. The low bid for installation of owner-furnished materials over 6.0 miles of pipe was $3.35M. The engineer's estimate was $4.86M. This meant the engineer's estimate was reasonable and the project could be realized.

Activity began in earnest with issuance of the Notice to Proceed to Constructors-Pamco, Inc., on February 23, 1989. Pamco proposed using carriers to put the liner pipe into the existing PCCP pipe instead of jacking a series of pipe lengths into place. The carriers were patterned after the ones used in previous split-can liner projects. In all, four different variations of carrier were used with a total of five carriers constructed. Each represented lessons learned on previous models or was designed to address different conditions of insertion, such as slope, or liner pipe length. (Most liner pipe was in 20-foot lengths, but some steel pipe was in 40-foot lengths.) A carrier is shown in Figure 6.

Each element of construction presented new challenges for Constructors-Pamco, the pipe suppliers, and the Seattle Water Department. Generating enough force to push home the Tyton joints while working inside the PCCP pipe was successfully addressed, but only after a check of pipe on site revealed that the bell mouths fit somewhat too tight. Grinding relieved the fit. Disassembling a partly-seated joint proved an interesting task in such close quarters. Steel pipe assembled relatively easily, but welding the steel joints was time-consuming. More welders were added to the workforce to overcome this problem.

The steel pipe went together well because it was designated for a specific installation location and joints

Pipe Carrier (Battery-Propelled) - Figure 6

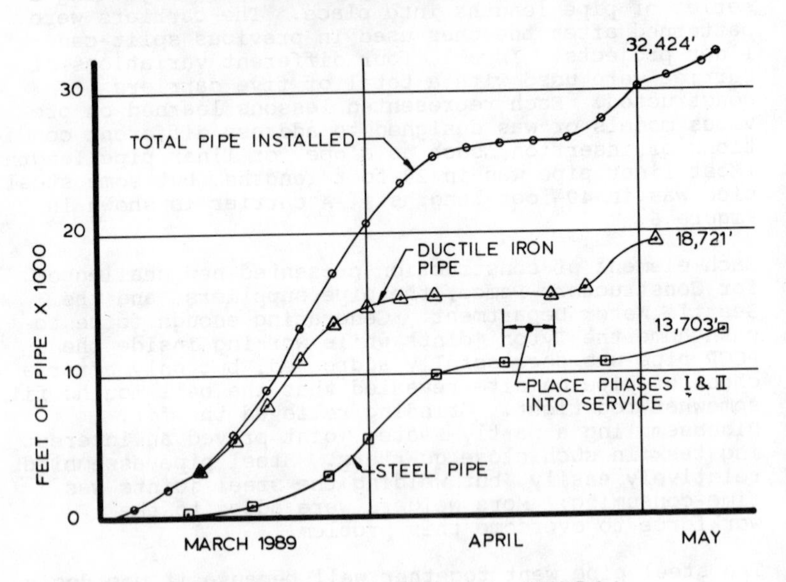

FIGURE 7

were prefit at the plant. Figure 7 shows the rate of pipe installation progress. This designation, plus the fact that all specials and fittings were in steel, not ductile iron, raised the per-foot cost of steel significantly above ductile iron pipe. Costs are summarized as follows:

	Footage	Total Cost	Cost Per Foot
Ductile Iron Pipe	18,721	$ 2,596,664	$138.70
Steel Pipe	13,703	2,692,351	196.48
Construction Contract	32,424	4,333,866	133.66
Other Material & Admin Costs		2,046,177	
TOTALS	32,424	$11,669,058	$359.89

The above totals include a segment of pipe of about 1800 feet added to the original contract by change order. The pipe was fully tested and returned to operation on May 21, 1989, 10 days ahead of the original target date.

Tolt Pipeline No. 1, Rehabilitation Phase I was a success. This success reflects several factors: choices of methods and materials; motivated and ambitious pipe suppliers and installation contractor; prompt delivery of materials; and risk-sharing by the Seattle Water Department. The latter reflects a commitment to pay extra for make-up construction effort as a result of bad weather or material delays. Also, a significant liquidated damages clause was contained in the construction and material supply contracts.

The rehabilitation of Tolt Pipe No. 1 continues in phases as the remaining segments of pipe that contain reinforcing rod subject to hydrogen embrittlement are being replaced. It is not anticipated that remaining pipe will be lined; rather it will be removed and replaced with larger diameter pipe in order to recover some of the capacity lost in the lined segment. Replacement must be done between November and May since summer water demands require the Tolt Pipeline to remain in service from June through October. Approximately 4.8 miles are scheduled to be replaced between November 1989 and May 1990. Another 1.8 miles in the valley reaches may follow at a later date subject to additional investigations.

Column Separation in Pumped Pipelines

Douglas T. Sovern and Greg J. Poole[1]

Abstract

Pressure transients are common occurrences in water supply systems. Pipeline standards and design/testing techniques will normally provide sufficient protection for transmission and distribution systems. However, in pumped systems, "downsurges" or rapid pressure drops can cause column separation in pipelines, can result in severe damage to piping, and may be unrecognized in many water supply systems.

Column separation in water supply systems usually results from power supply or equipment failure. In these instances, control devices such as pump control valves and pressure release valves can be totally ineffective.

Computer modeling can be used to determine the need for prevention of column separation and to size appurtenances such as one-way surge tanks and surge suppression tanks that can prevent column separation from occurring. Two analyses are presented to demonstrate the application of computer modeling. The first case study involves a 600 gal/min (2.1 m^3/min) pump station with a lift of 3,500 feet (1,067 meters) from the bottom of the Grand Canyon to serve the South Rim of the Grand Canyon National Park. The second case study involves use of pressurized tanks in a 25 million gal/day (0.93 million m^3/day) pump station at a pressure of 36 lb/in^2 (2.52 kg/cm^2).

Introduction

Pressure transients are common occurrences in water supply systems. Pipeline standards and design/testing techniques will normally provide sufficient protection for water transmission and distribution systems. However, in

[1]Douglas Sovern, Member, 2420 Alcott Street, Denver, Colorado 80211, and Greg J. Poole, Member, 6771 South 900 East, Midvale, Utah 84047.

pumped systems, "downsurges," or rapid pressure drops, can cause column separation in pipelines and can result in severe damage to piping. They may be unrecognized in many water supply systems.

Column separation is a phenomenon which occurs when the local pressure in a pipeline approaches the vapor pressure of the liquid. Gases in solution begin to come out of solution and dramatically affect the flow behavior. If the drop in pressure is severe enough to cause the local pressure to reach the vapor pressure, then the liquid boils (i.e., cavitates) and vaporizes, forming large pockets of undissolved gases and liquid vapor. This phenomenon is referred to as "column separation" (Watters, 1979).

Much of the population of the arid west is dependent on pumped pipelines for water supply, and, often, a pumped pipeline is the sole method of developing a source of water for a community. A characteristic of pipeline systems is the potential for hydraulic transients ("water hammer") or pressure surges that are created by velocity changes in the pipelines. These pressure surges are propagated throughout the system by the elastic properties of the fluid and the pipe walls. In some systems, severe damage to pipelines and appurtenances have resulted from generation of pressure waves. Pressure surges in pipelines can be generated by a wide variety of conditions. Of those conditions, surges caused by electrical or pump failures present some unique challenges to the designer.

Whether these pressure waves cause problems or not is dependent on the magnitude of the velocity change and the wave speed. It is also dependent on the rate of the velocity change as related to the time it takes a pressure wave to travel from the source of the velocity change to the end of the pipeline and back. The magnitude of the pressure wave is directly proportional to the magnitude of the velocity change. Pressure wave speeds in pipelines often exceed 3,000 ft/sec (914.4 m/sec), and cause both positive and negative pressure changes.

Power interruptions can occur due to external power disruptions and internal disruptions when control systems signal for the immediate shutdown of the pumps. When power is interrupted to the pumps, the flow of fluid immediately in the vicinity of the pump begins to slow. The rate of the velocity change (deceleration) is dependent of the inertia of the pump system in comparison with the operating energy imparted to the pipeline by the pump. An induced pressure wave from a pump or power failure is propagated through a pipeline or a network of pipes at the wave speed of the pipeline. If the pipeline is sufficiently long, the fluid in the vicinity of the pump may come to rest, while at the opposite end of the pipeline, the flow in the pipe is still at steady state.

Column separation is most likely to occur on long pipelines (where the wave reflection does not return to the pumping facility until after the pumps have stopped rotating), at high spots along the pipeline, and/or where there are steep segments in the pipeline route. In each case, column separation will occur if the negative pressure wave induced by the power failure is sufficiently large to cause pressures in the pipeline to occur that are at or below the vapor pressure of the liquid (Figure 1). When the negative pressure wave reaches the end of the pipeline, it will be reflected back as a positive wave that will collapse the vapor cavity formed by column separation. The collapse of the cavity may cause a pressure wave of large enough to damage the pipeline.

It should be noted that column separation is not the only problem that can occur due to power failure in pumped systems. If column separation does not occur, the positively reflected pressure wave can damage hydraulic equipment and pipelines in the pumping facility; however, the case studies that are discussed in this paper will only address the type of problems and potential solutions that occur before the pressure wave is fully reflected back to the pumping facility, namely, column separation.

Two case studies of hydraulic transients induced by pump power failure are presented in this paper. The pumped pipeline that serves Grand Canyon National Park (Indian Gardens/South Rim) rises 3,300 feet (1006 meters) from the Colorado River in the bottom of the Grand Canyon up to the South Rim. The City of Englewood, Colorado, diverts raw water from the South Platte River and pumps it to a water treatment facility located approximately 5,500 feet (1,674 meters) from the point of diversion. The design flow rate is 25 million gallons per day (0.93 million m^3/day) and the total lift is 80 feet (24.4 meters). These two cases were selected to illustrate that the problem can exist on nearly any scale of discharge and pumping head. They also illustrate two different means of preventing the occurrence of column separation. A description of the methodology is briefly presented first, followed by a discussion of the analysis and results of the case studies.

Methodology

The methods used in the analysis of pump power failure induced hydraulic transients in both of these case studies are presented in "Modern Analysis and Control of Unsteady Flow in Pipelines" by Professor Gary Z. Watters (Ann Arbor Science, 1979). The reader is referred to this reference for a more complete narration of the derivation of the methods used.

The magnitude of the pressure wave, H represents the magnitude of the pressure wave caused by a change in velocity. The impulse-momentum equation can be used to develop the following equation:

$$\Delta H = (a/g) \times \Delta V, \qquad \text{Eq. 1}$$

where: ΔH = the magnitude of the pressure wave,
a = the wave propagation speed
ΔV = the change in velocity in the pipeline
g = the acceleration due to gravity.

The speed of propagation of the pressure wave is dependent on the elasticity properties of the fluid and the pipeline materials. Streeter and Wylie (1967) have shown that the equation for the wave speed "a" can be expressed as follows:

$$a = (K/\rho)^{1/2}/[1 + (K/E)(D/e)c_1]^{1/2} \qquad \text{Eq. 2}$$

where: K = the bulk modulus of elasticity of the liquid (for water, approximately 300,000 pounds per square inch),

E = the modulus of elasticity of the pipe wall material (for steel, approximately 30×10^6 pounds per square inch),

ρ = the density of the liquid (for water, approximately 1.94 slugs per cubic foot or 1 kg/liter),

D = the pipe diameter,

e = the wall thickness, and

c_1 = the restraint coefficient = $1 - u^2$ for the case where the pipe is rigidly anchored to prevent axial strain (u is Poisson's ratio for the pipe material, approximately 0.30 for steel).

Figure 1

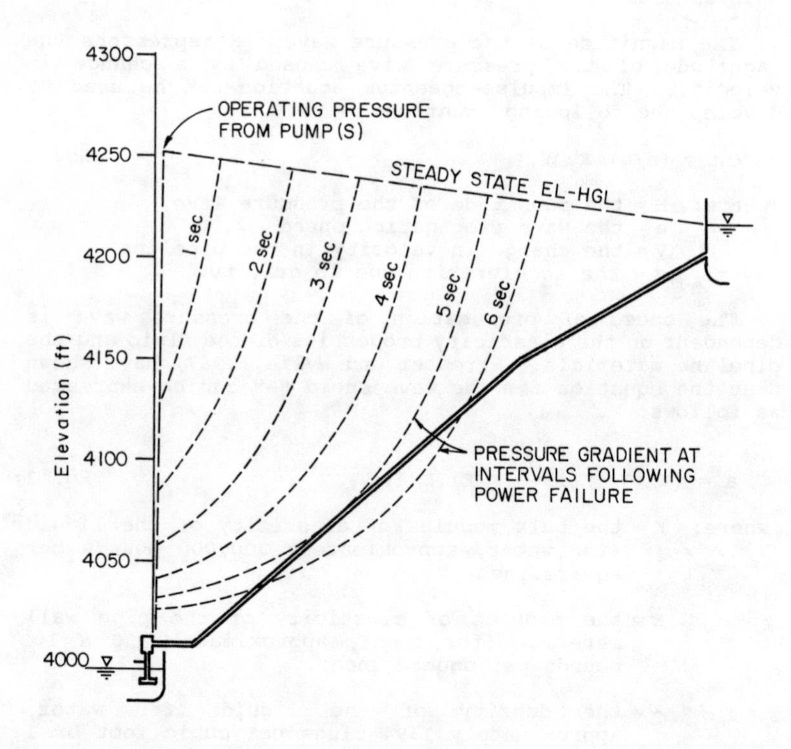

Figure 1. Example of pressure wave propagation from pump power failure.

Equation 1 and Equation 2 can be used to analyze the impact of an abrupt (impulse) change in velocity; however, in the analysis of pump shutdown due to power failure, the rate of change of velocity is important and necessitates an approach taking into account variables that change with time. The Euler unsteady flow equation and continuity equation can be used to derive two simultaneous, independent partial differential equations:

$$dV/dt + (1/\rho)(\partial p/\partial s) + g(\partial z/\partial s) + [f/(2D)][V|V|] = 0 \text{ and} \qquad \text{Eq. 3}$$

$$a^2/(\partial V/\partial s) + (1/\rho)(dp/dt) = 0 \qquad \text{Eq. 4}$$

where: dV = incremental change in velocity,

dt = incremental change in time,

$\partial p/\partial s$ = partial of the change in pressure with respect to distance along the pipe,

$\partial z/\partial s$ = partial of the change in elevation with respect to distance along the pipe,

f = Darcy-Weisbach friction factor, and

g = acceleration due to gravity.

The above simultaneous, independent partial differential equations are amenable to solution by the method of characteristics, which yields the C^+ and C^- characteristics that are represented by the following two ordinary differential equations, with p replaced by (H - Z):

$$C^+: \; dV/dt + (g/a)(dH/dt) - (g/a)V(dz/ds) + (f/2D)(V|V|) = 0 \qquad \text{Eq. 5}$$
only if $ds/dt = V + a$

$$C^-: \; dV/dt - (g/a)(dH/dt) + (g/a)V(dz/ds) + (f/2D)(V|V|) = 0 \qquad \text{Eq. 6}$$
only if $ds/dt = V - a$

For pipeline applications with wave speeds much greater than velocity, the finite difference with rectangular grid method of solution is very amenable to programming and data interpretation (Watters, 1979).

Note that this methodology cannot be utilized to predict the magnitude of pressure surges once column separation has occurred. It is useful to predict the possible occurrence of column separation so that appropriate counter-measures can be taken to prevent it.

Prevention of Column Separation

The first step in preventing damage to pipe systems from column separation is to undertake analyses that will show the potential for occurrence of column separation. Normally, this is undertaken through a computer program utilizing the equations and principles enumerated in the previous section. The two case studies will further explain the procedures for this step. It is important to realize that the potential for column separation cannot always be seen by simple inspection of the pipeline profile.

Once it has been determined that column separation may occur, there are two approaches to protection of a pipeline from being damaged by large pressure surges that are caused by collapse of the vapor pocket. One approach is to limit the magnitude of the pressure wave by the use of such devices as surge anticipator valves. The other approach is to slow the deceleration of the flow by using devices such as air-chambers and surge tanks.

To be effective, a surge anticipator valve must be very near to where the vapor pocket collapse will occur. As a result, the surge anticipator valve has best application where pipelines have no major summits and where the vapor pocket is likely to form in the vicinity of the pumping facility. Neither of the case studies in this paper addresses this situation.

Where vapor pockets are likely to form at some distance from the pumping facility, prevention of column separation is the best approach. Both air chambers and one-way surge tanks retard the deceleration of flow and eliminate the potential for formation of vapor pockets.

Air chambers are similar to hydro-pneumatic tanks used for many water supply systems. Figure 2 illustrates a typical installation. The tank is partly filled with air and partly filled with water, and pressure is maintained by an air compressor. Air chambers are normally located in the pumping facility. When the power to the pumps is de-energized, the air chamber supplies water to the system at the pressure in the tank. The main criterion is to have sufficient water in the air chamber for continuous supply of water to the system until the pressure wave from the end of the system is reflected back to the pumping facility.

Figure 2

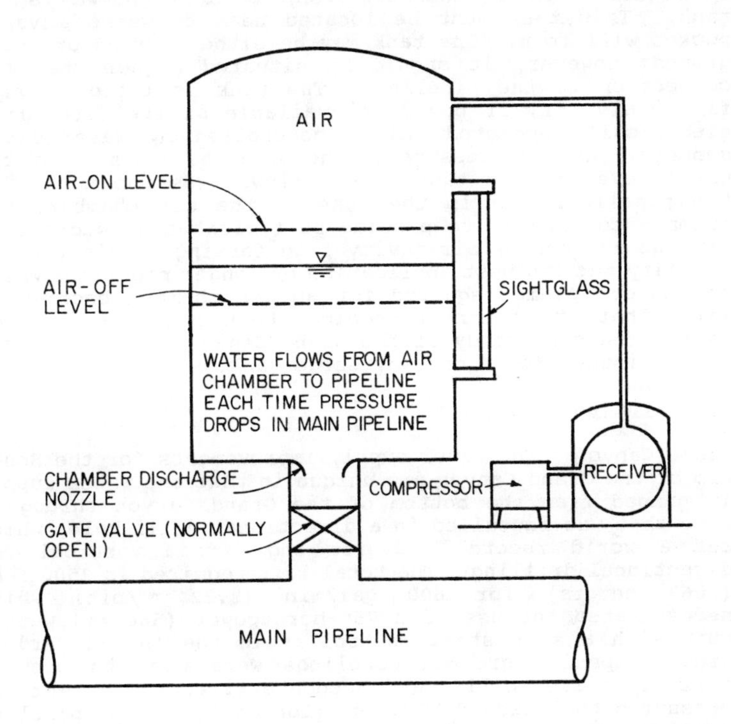

Figure 2. Schematic diagram of an air chamber.

Note that the existence of an air chamber at the pumping facility does not eliminate the need for normal appurtenances that a designer would utilize in the design of the pumping facility, such as pump control valves. An example of this type of system is illustrated in the case study of the City of Englewood pumping facility.

Figure 3 is a schematic diagram of a one-way surge tank. This tank must be located near to where a vapor pocket will form. The tank may be either buried or above ground; however, it should be situated higher than the connection to the pipeline. The tank is filled from a float valve or, if power is available at the site, from electrically operated valves controlled by water level sensors. As the pressure in the pipeline drops below the water level in the tank, water flows from the tank into the pipeline. As in the case of the air chamber, the primary concern is to provide sufficient tank storage to prevent a vapor or air cavity from forming. This type of facility has its best application on longer pipelines where the water volume required for an air chamber becomes so large that it is more economical to use a one-way surge tank. The case study of the Grand Canyon illustrates the design issues for a one-way surge tank.

Case Studies

Grand Canyon. The water supply improvements for the South Rim of the Grand Canyon are unique in that the water supply is pumped from the bottom of the Grand Canyon through a pipe that was installed in a directional drill hole, which set a world record 70-degree angle from vertical for directional drilling. The total lift required is 3500 feet (1067 meters) for 600 gal/min (2.22 m^3/min) which necessitated the use of a 750-horsepower (560 kilowatts) pump with a soft start controller in the Indian Gardens area. Special grooved couplings were used to provide sufficient flexibility and adequate strength for working pressures that exceed 1500 psi (106 kg/cm^2). The pipeline profile is shown in Figure 4.

Analysis of hydraulic transients induced by pump power failure indicates that without prevention devices, a vapor cavity would form at the rim of the canyon about 2.5 seconds after the power failure. The potential exists for damaging high pressures associated with closure of the vapor cavity.

Several alternatives for control of the power failure induced transients were examined, including:

Figure 3

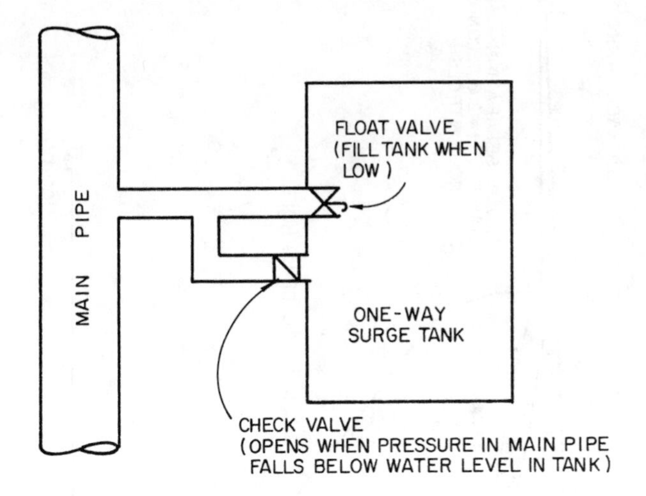

Figure 3A. One-way surge tank schematic

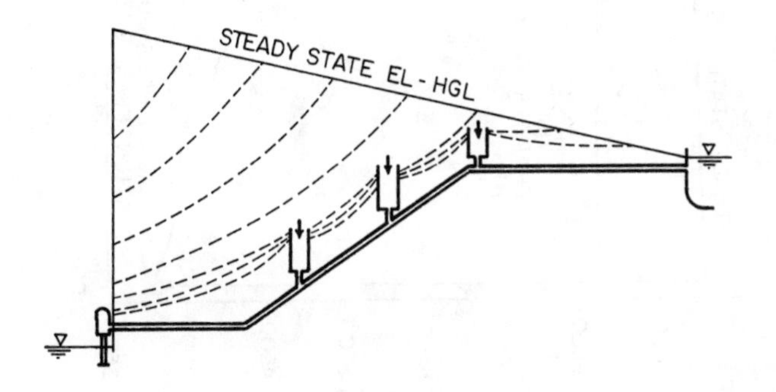

Figure 3B. Typical profile for one-way surge tanks in a pumped pipeline.

Figure 4

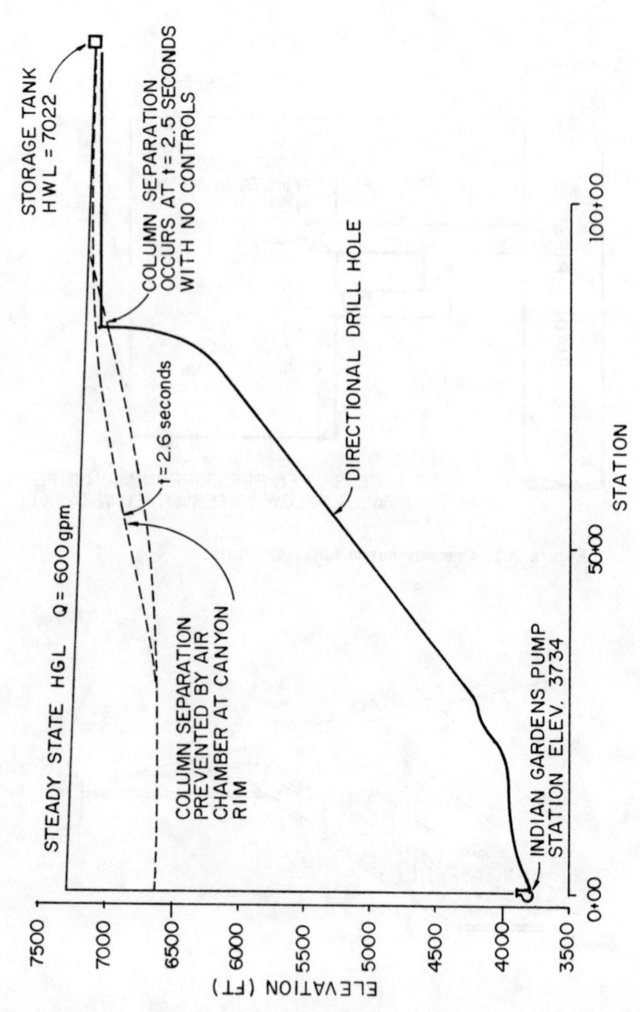

Figure 4. Indian Gardens to South Rim of Grand Canyon

1. An air chamber in the Indian Gardens Pump Station.

2. One-way surge tanks at the canyon rim.

3. An air chamber at the canyon rim.

4. A surge anticipator valve in the Indian Gardens Pump Station.

5. An in-line check valve at the top of the rim with a surge anticipator valve just upstream from the check valve and an air release valve just downstream from the check valve.

Locating an air chamber in the Indian Gardens Pump Station proved to be infeasible due to space constraints in the existing facility. Construction of a one-way surge tank at the rim of the canyon was deemed aesthetically unacceptable that would require a large tank, which would obstruct the view.

The use of a surge anticipator valve in the Indian Gardens Pump Station would result in introduction of air into the pipeline through the air relief valve at the top of the cliff. The addition of air to the line would make restart after power failure more complex and time consuming. Due to the addition of air into the line, both surge anticipator valve alternatives were eliminated.

Of the alternatives analyzed, an air chamber located at the canyon rim was found to be the most practical solution. As shown in Figure 4, an initial air volume of 6 cubic feet adequately mitigated the effects of pump power failure, maintaining positive pressures throughout the pipeline. With the air chamber in place, the maximum pressure at the pump would occur about 6 seconds after pump power failure, with a magnitude of 3,690 feet (1,125 meters) or 1,600 lb/in^2 (113.0 kg/cm^2). The maximum air volume in the air chamber occurred at 14.6 seconds with a maximum volume of 23 cubic feet (0.64 m^3). As a result of the analysis, a 30 cubic foot (0.84 m^3) air chamber was installed at the rim of the Grand Canyon in 1987.

City of Englewood. The City of Englewood, Colorado, diverts water for municipal use from the South Platte River and pumps the water to storage reservoirs located adjacent to the water treatment plant. Initially, the capacity of the pumping facility and 27-inch (68.5 cm) steel raw water pipeline is 31 cfs (0.87 m^3/sec), with an ultimate capacity of 38 cfs (1.06 m^3/sec). The profile shown in Figure 5 illustrates the pertinent physical features of the project.

Figure 5

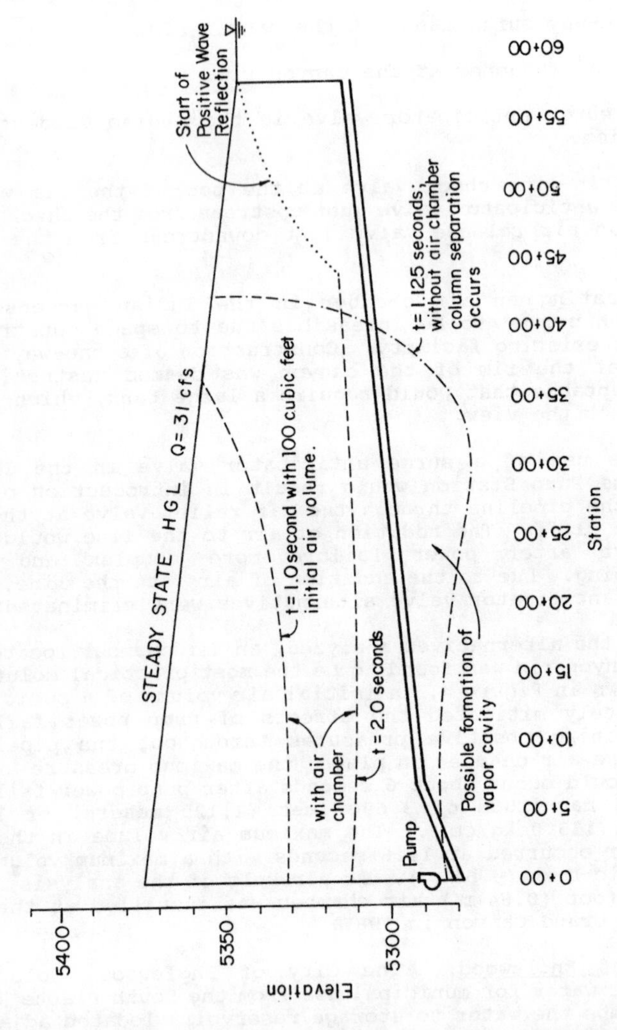

Figure 5. City of Englewood Raw Water Line Capacity = 31 cfs(14,000 gpm)

The profile of the Englewood raw waterline is typical of many pumped pipelines that supply water from a river. The initial segment of the pipeline is steeper than the rest of the profile. The pipeline is about a mile in length. The computed wave speed for the 27-inch (68.5 cm) pipe is about 3,600 ft/sec (1,097 m/sec); therefore, a pressure wave propagating from the pump station will reach the discharge reservoir in about 1.5 seconds and will reflect back to the pump station in about 3 seconds. The analysis shows that a vapor cavity will form at about 1 second after pump power failure, with much of the pipeline experiencing negative gauge pressures. A major concern on a long flat pipeline is the potential for positive pressure waves to propagate from both the upstream and downstream reservoirs, causing the vapor column to close from both sides with potentially large sudden velocity changes and resulting high pressures.

To protect against the potential for vapor cavity closure, an air chamber was designed and placed in the pump station. Two hydraulic grade lines representing 1 second and 2 seconds after power failure, with an initial air chamber volume of 100 cubic feet, are shown on Figure 5. The hydraulic grade line 1 second after power failure demonstrates the propagation of the negative pressure wave from the pump towards the discharge reservoir. The hydraulic grade line at 2 seconds after power failure shows the beginning of the positive wave reflection from the discharge reservoir. The maximum air chamber volume occurs 20 seconds after power failure, with a volume of 420 cubic feet. The air chamber successfully prevents the formation of a vapor cavity and the uncertainties of vapor cavity closure are prevented.

References

Parmakian, John, Waterhammer Analysis, Dover Publications, Inc., 1963.

Parmakian, John, Notes on ... Water-Hammer, N.D.

Streeter, Victor L. and Wylie, E. Benjamin, Hydraulic Transients, McGraw-Hill Book Company, 1967.

Watters, Gary Z., Modern Analysis and Control of Unsteady Flow in Pipelines, Ann Arbor Science, 1979.

COMPREHENSIVE LEAK DETECTION SURVEY AND BENEFIT/COST ANALYSIS

Richard J. Scholze, Jr.
Hany H. Zaghloul
Prakash M. Temkar
Stephen W. Maloney

Abstract

Fort Carson, Colorado was the site of a comprehensive leak detection investigation of the potable water system with the express purpose of quantifying the benefits to be derived by a military installation from use of the technology. Military bases are often the size of a small city and the Directorate of Engineering and Housing has complete responsibility for all real estate (all buildings, all utilities, roads, grounds, etc.) unlike a municipal public works department. The investigation used state of the art noise correlation and computer correlation technology to survey the distribution system mains. This was complemented by a building to building survey covering office and commercial buildings, and family and barracks housing where investigators entered the building and quantified visible leaks in faucets and water closets, etc. A year later, a follow-on survey was performed to again examine all aspects of the system. The result was a complete economic evaluation and benefit/cost analysis of the installation. Representative findings include: the majority of distribution system leaks were at hydrants or similar appurtenances; and family housing was found to be the other major concentration of leaks. However, where the first survey found 80 percent of housing units had leaks, preliminary findings from a second round in 1989 are on the order of 20 percent. Sites where responsibility for correction of leaks was not readily attributable such as communal bathrooms in barracks also

---

U. S. Army Construction Engineering Research Laboratory
P.O. Box 4005, Champaign, IL 61824-4005

showed a high incidence of leaks. Office buildings were found from the first survey to not merit follow-on attention due to limited numbers of leaks. Water-consciousness was raised for both the DEH and individuals in family housing and leak repair was given a higher priority. The paper will present a complete lessons-learned section and also present cost figures and related information showing when and where leak detection surveys are appropriate.

Introduction

While United States Army military installations distribute water much like municipalities, they do not meter the flow of the user, as municipalities do. Military personnel do not pay for their water, they generally have access to an unrestricted supply. Since water is not metered at its terminus, there is no monetary incentive for the user to correct apparently minor leaks and there is no direct way to measure the amount of water consumed, wasted, or lost in transit. Without meters there is no way to gauge water use for each dwelling or building to discover which units may be using excessive quantities. For example, during a short-term study (Matherly et al., 1978), metered water use showed a range of 42 to 248 gallons per capita per day (gpcd) for selected family housing units with increases as the irrigation season began.

Fort Carson is located just south of the city of Colorado Springs, Colorado. The water distribution system contains approximately 115 miles of main comprised of cast iron, ductile iron, asbestos cement, and plastic pipe ranging in size from 6-inch to 20-inch. Water is purchased in bulk from the City of Colorado Springs at a rate of $1.12 per thousand gallons (1987).

Prior to the detailed investigation, a mass balance for minimum overnight water usage was developed and compared to the average daily use. The mass balance calculation was used as a screening mechanism to select potential candidates for implementing a leak detection program. Next, a comprehensive leak detection investigation of the potable water system was conducted to quantify the benefits to be derived by a military installation from the use of modern leak detection survey technology. Two kinds of surveys were conducted: 1) noise and computer correlation equipment was used to survey the distribution system mains, and 2) a two-person team conducted a door-to-door survey through family and other housing units in addition to other types of buildings throughout the installation. This paper will outline the leak detection methodology used, characterize the types and patterns of leaks found, introduce an economic analysis for the entire leak detection process, and finally, provide lessons learned with practical results and implications of the Ft. Carson leak detection project.

Methodology

Mass balance calculations

Since Army installations have little or no internal water meters, a control method had to be devised to quantify the water loss pattern and assess the effectiveness of any water conservation schemes. One method to achieve that objective was to compare minimum night flows to daily average flows. A high overnight flow can indicate continuous leakage or misuse of water during a low consumption period.

Inflow into the Fort Carson water distribution system was determined from three water master meters that serve as intakes from the city of Colorado Springs water utility. The Fort Carson distribution system is configured such that water from three elevated storage tanks would either contribute discharges into the Ft. Carson system - if the water level drops, or withdraw water from the City of Colorado Springs system - if the water level increases while the tanks are filling. Thus, the entire consumption of the Ft. Carson water distribution system is governed by the intake from the three connections to the City of Colorado Springs system, in addition to the fluctuation of the three elevated tanks. The following data items were necessary to perform the calculations:

a. master meters' readings at 2400 hours.
b. all elevated tanks' levels at 2400 hours.
c. master meters' readings at 300 hours.
d. elevated tanks' levels at 300 hours.

One month's data (March) were collected and tabulated in a spread-sheet format. A mass balance calculation for water input into the Ft. Carson system was conducted, using the above data. The first two data sets (a&b) were used to determine average daily flows (in gallons per hour) on a 24 hour basis. The second two data sets (c&d) were used to determine average minimum night flows - between 0 and 300 hrs.

The minimum night flow ratio is that obtained from dividing the outcome of the two above calculations. Certain precautions apply for meaningful application. First, the data have to be collected during a period of little to no overnight irrigation to avoid skewing the overnight flow data. The same precaution should be taken for periods of heavy, irregular daytime activities (e.g. fires) to avoid skewing the daily flow data. Secondly, the correct identification of entry points and their contribution to the water system is imperative for obtaining a meaningful calculation. A minimum night flow ratio of 0.53 was found for Fort Carson using data from 24-27 February, 1987.

A similar mass balance was conducted at two other installations, Fort Lee, VA and Fort McCoy, WI. Fort Lee ranged from 0.13 to 0.27 while Fort McCoy was 0.81. The data on minimum night flow ratio, recoverable costs, cost of water, and other factors were used to select Fort Carson as the prime candidate for conducting a leak detection program. Although Fort Carson did not experience the highest leakage rate calculated, the combination of moderate leakage and high water cost indicated that Fort Carson would benefit most from this type of study.

Water Mains Survey

Two leak detection surveys (Donohue, 1988,1989) of the water distribution system were conducted approximately one year apart. The distribution system leak location survey consisted of a preliminary general listening survey conducted on fire hydrants, service curb stops, and valves throughout the distribution system. Where valves were buried or inaccessible, nearby curb stops were sounded. This general survey was designed to locate areas where leakage was occurring. A detailed location survey using computer correlation technology was conducted to pinpoint leaks in those areas of the distribution system which showed a potential for leakage. A leak sound correlator (Continuous Flow System - CFS), accurate within a few feet, was used to pinpoint leaks.

Approximately 115 miles of water main were surveyed for underground leakage in the Fort Carson water distribution system. This included all water mains in the distribution system. The quantity of leakage at each leak was estimated by an experienced technician using, as indicators, the decibel level and tone of sound created by the leaking water from the pipe and soil conditions at the location.

Building Plumbing Survey

Army installations are often the size and complexity of a small (50,000) city. A variety of residential, commercial, and industrial buildings and activities co-exist within the boundaries of one large installation. Many of these activities are similar to municipalities, however, many are unique to an installation. Additionally, one agency has control over all real estate, buildings, utilities, roads, grounds, etc. permitting a more responsive investigation.

The individual building portion of the investigation began with a survey in 1987-88 which is described below. A follow-on survey (1989) is underway.

The assessment survey of building plumbing systems began with family housing. Eighty-one percent of all units were surveyed. A two-person team surveyed each of approximately 1800 family housing dwellings. An extensive notification and coordination effort supported the project. The procedure required ten to twenty minutes for each family housing unit depending on leakage amount and dwelling size.

The team entered the bathroom(s) first and put several drops of food coloring into the toilet tanks(s). This procedure detects leakage through the seal in the bottom of the tank. Toilet bowls were checked ten minutes later. If color was present, leakage could be identified in the specific tank. After the drops were placed in the tanks, the team checked the bathroom sink and tub fixtures for leakage from spigots, handles, and connecting pipes. The kitchen sink was similarly inspected, as were faucets on the outside of the building. The water heater, washing machine hookups, and other appliances were also checked for leakage. Leaks other than toilet leaks were measured in milliliters per minute. Toilet leakage was classified as slight, moderate, or heavy.

A follow-up sampling of measurements were taken on toilet leakage after completion of the initial study to refine the definitions of what constituted a slight, medium, or heavy leak. The inside width and length of the tank and initial water depth were measured to find the initial volume. Then, the valve to the toilet was turned off. After five minutes the volume of the tank water was remeasured and subtracted from the initial volume. The resulting number was the volume of water lost through the leaking seal in the measured time. Slight leaks were too minute to measure; moderate leaks ranged from 0.1 to 0.5 gallons (0.38 to 1.9 liters) per hour; and heavy leaks were greater than 0.5 gallons (1.9 liters) per hour.

The same method was applied to the remaining buildings on the base including barracks. Leakage proved to be insignificant in these areas. A follow-up resurvey of family housing was performed later in the same year to determine if heavy leaks found in the summer had been repaired. Actual follow up visits were made to one-third of the 227 quarters with the worst leakage. Survey sheets were delivered to all units asking residents to indicate if leaks had been fixed by the base's family housing contractor and if repairs were effective.

Types and Patterns of Leaks

Distribution System Leaks

The initial distribution system survey (Donohue, 1988) was conducted during January and February 1988. During the course of the survey, 35 leaks were located, totaling approximately 435,000 gallons per day. The majority of leakage occurred in vehicle maintenance areas at wash platforms (an estimated 366,000 gpd). Fifteen leaks were on hydrants (approximately 65,000 gpd). Remaining leaks of 4000 gpd were on a service and a valve.

A follow-on distribution system survey (Donohue, 1989) was conducted in February and March of 1989. During that survey 26 leaks totaling 309,000 gpd were identified, again predominantly in vehicle maintenance areas and hydrants. The majority of leakage located occurred on wash platform hydrants (Murdoch valves) and accounted for an estimated 227,000 gpd of leakage. The cause of these leaks could be from improper operation of the wash platform hydrants. If the wash platform hydrants are not tightly closed, water will leak past the seat at the base of the hydrant and wear the seat to the point that the hydrant, when shut, will continue to leak. At that time, only repair or replacement of the hydrant seat would eliminate this leakage. Other possible causes of wash platform hydrant leakage include: in-line surges from opening and closing the hydrants too quickly; or possibly age, corrosion.

Main line hydrant leaks accounted for an estimated 59,000 gpd. Many of the hydrant leaks were stopped or slowed during the survey by tightening the operating nut on the hydrant. This indicates proper hydrant operation should be emphasized to authorized personnel.

Two active services were found to be leaking an estimated 20,000 gpd. The causes of this type of leak can include surges from improper operation of hydrants or valves, surges from pump operations, frost and soil conditions causing movement of the service, poor backfilling or bedding, services weakened from corrosion, and aging or faulty joint material. The remaining leakage occurred on one valve and accounted for approximately 3,000 gpd. This does not appear to be a significant problem at Fort Carson.

Building Plumbing Leaks

During the initial building to building survey, a high proportion of leaks was found in the family housing. For instance, during the initial days in the field, only four of twenty-one quarters studied did not have some sort of leakage. Leaks ranged from a few milliliters per minute to thousands per minute. Approximately 81 percent of all units were surveyed. Upon completion of the survey, 83 percent of the surveyed quarters contained some sort of leakage. Only 246 of the surveyed quarters did not have leakage. The heaviest and most numerous leaks were found in bathroom and kitchen faucets. Over 2300 faucet leaks were discovered. The faucet valves on building exteriors were also faulty in numerous instances. Fewer leaks were found in hot water heaters and washing machine hookups. The leaks in these appliances were small, averaging only a few milliliters per minute. Similarly, only about 130 toilets leaked.

A subsequent building leak survey is underway, again evaluating numbers and types of leaks and having them corrected.

Economic Analysis

The economic analysis for the leak detection project was divided into two aspects, distribution system and individual building. The benefit of reducing the annual water bill cost by leakage repair is compared to the cost of locating the system leakage to produce a benefit-to-cost ratio to guide a utility manager and prioritize any planned system repair list.

Water Distribution System Survey Cost Analysis

The initial distribution system survey (Donohue, 1988) identified 435,000 gallons per day of water loss. Repair of these leaks would save the installation approximately $178,000 per year minus costs of repair and the leak detection survey.

Of the estimated 1.7 mgd average of water metered at Ft. Carson, 309,000 gpd of leakage was accounted for by the subsequent water distribution system survey (Donohue, 1989). This amount of water represents approximately 18% of the total average daily water supplied to Ft. Carson. An annual savings of approximately $126,000 minus survey and repair costs could be realized by the fort from the repair of these leaks. In this case, the savings could pay for the cost of the noise correlation survey in less than three months.

Building Plumbing Survey Cost Analysis

Upon completion of the leak detection project in family housing a cost/benefit analysis was set-up to determine which leaks in family housing would be economically feasible to repair. To develop a range of benefit to cost ratios, conservative leak repair estimates were assumed at $75.00 and $100.00 per repair visit.

Figures 1 and 2 show two graphs which plot the benefit/cost ratio and the total savings from a repair program in family housing, as a function of the number of repaired leaks, respectively. The two curves in each frame represent the variable assumption for the cost of repair visit. The upper curve in each frame is the $75.00 per repair visit, while the lower curve represents the $100.00 cost per repair visit. For determination of leaks economically feasible to repair, the marginal benefit to cost ratio should be greater than 1.0. For the assumptions of this analysis, if the leak repair cost is less than the cost of water saved within one year then the benefit to cost ratio is greater than 1.0. At the $75.00 repair cost it was found that it was economically feasible to repair a total of 227 units. It was estimated that repairing the 227 leaks would reduce total water consumption by 6.3 percent (107,000 gal/day) annually.

Results and Lessons Learned

Residential Leakage Patterns: The leak detection project results revealed a definite correlation between the frequency and magnitude of leaks and the characteristics of the residential building occupants. In senior officer quarters there were virtually no leaks, and there was much less leakage in the officer quarters than in the noncommissioned officer quarters. Few discernable differences were found between family housing neighborhoods. In some neighborhoods leakage inside the unit seemed to be worse, while in other outside faucet

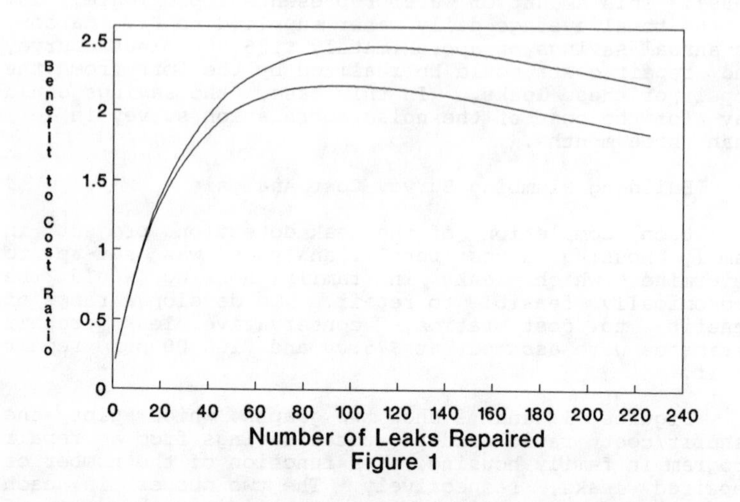

Figure 1

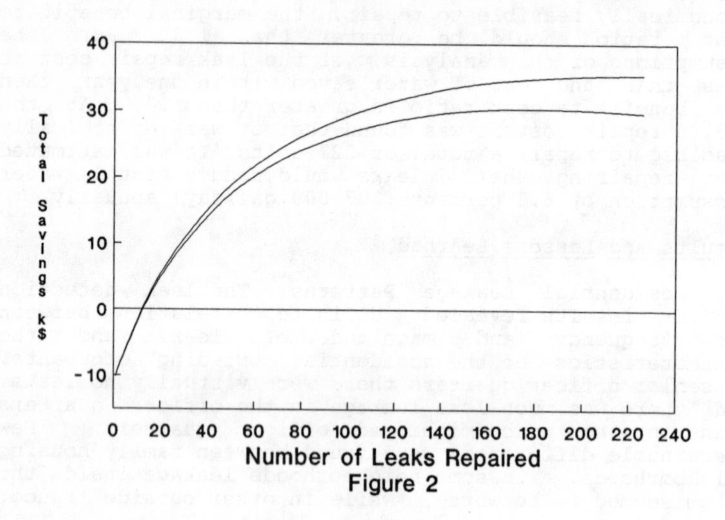

Figure 2

valves were more problematic. Such an observation was attributed to the construction practices and plumbing fixture types prevalent at the time of construction. Aside from family housing, the only serious bathroom leaks were found in barracks with community rather than individual bathrooms. A "commons" problem exists; no one in particular feels obligated to report the leaks.

Metering and Record Keeping: The two major components of unaccounted for water are leakage and meter error. Thus, in addition to periodic leak detection surveys, all meters in the system should be tested periodically for accuracy of registration. Due to the lack of individual service meters throughout the installation, area or district metering programs should be initiated. Metering stations should be located according to area water usage. Therefore, if any jump or unusually high water use in the area is indicated by the meter, the following survey work could be limited to that area. Records for water usage and sewer flow rates should be reviewed regularly for the entire installation. A sudden increase in either of these two records may indicate underground leakage. In addition, a monthly comparison should be made with the previous year's monthly water meter readings. If a significant increase is noted, corrective action may be necessary.

Maintenance Policy Standards: The leak detection project at Ft. Carson has had a significant maintenance policy impact. The priorities for work orders have been redefined. Priority 1 is still for emergencies, priority 2 is now defined as urgent work orders, and priority 3 is considered normal work. Time constrictions for repair remain the same for emergency and urgent work. However, normal work must now be completed within ten days, and the contractor must assess normal work orders within two days of the time it is reported. The current maintenance contract lists different types of work orders and classifies them into priorities. For example, water leaks from faucets are now classified as urgent service orders.

Planning Future Leak Detection Programs: To save time and money, future leak detection research procedures should be altered. Leak detection surveys should select a representative sampling of family housing rather than visiting all buildings. This sample should be large enough to indicate the seriousness of the leakage problem and to determine if further study is warranted. Future and follow-up leak detection survey teams should include an individual with plumbing experience who could fix the leaks immediately upon discovery. Leaks would be fixed, and there would be no future problems with incorrect leakage data due to the lag time between the leak detection survey and the repairs.

Survey Equipment, Type of Leak, and Construction Practices: Survey personnel should have a thorough knowledge of the tools available for detection of leakage and the effectiveness of each tool or method under varying conditions. If the project feasibility is based on the amount of water saved, it should be judged with the understanding that the economics can vary according to the number of small service line leaks, which do not provide high-yield savings (Greeley, 1981). Finally, past construction practices can cause many leaks as a result of insufficient restraint against unbalanced pressures, non-uniform pipe loading, improper bedding or damaged corrosion-proofing. Although it is unreasonable to expect an Army installation to have control over past practice, it should be noted that today's mistakes cause tomorrow's leaks.

Summary

This paper has shown that a comprehensive leak detection survey of the potable water system is a beneficial undertaking for military bases. A demonstration at Fort Carson, CO found substantial economic savings for repair of leaks found in the distribution system mains and for leaks found in family housing. Other categories of buildings did not have significant leakage levels.

References

1. Matherly, J.E., Staub, M.J., Benson, L.J., and R.J. Fileccia, "Water Usage Profile--Fort Carson, CO," USACERL Interim Technical Report N-34, U.S. Army Construction Engineering Research Laboratory, Champaign, IL, 1978.

2. Donohue & Associates, Inc., "Leak Detection Survey - 1988 - Fort Carson, Colorado". Project No. 15555.000, performed under contract to USACERL.

3. Donohue & Associates, Inc., "Leak Detection Survey - 1989 - Fort Carson, Colorado". Project No. 16784.000, performed under contract to USACERL.

4. Greeley, D. S., "Leak Detection Productivity", Water Engineering and Management, Reference Handbook 81, 1981.

PIPELINE LEAK DETECTION AND LOCATION

Chyr Pyng Liou,[1] M. ASCE

Abstract

A real time pipeline leak detection and location system has been developed. This system consists of a numerical model for transient flow, two sets of pressure and flow measurement equipment, a microcomputer, and a data acquisition board. Specific patterns of discrepancy between measured and computed head at pipe ends evolve when a small leak occurs. Usage of these patterns to identify and to locate the leak is explained. Laboratory results are presented to show the concept and the effectiveness of the system.

Introduction

Pipeline integrity is an important and constant concern in pipeline operations. Although pipelines are protected against damage by external impact, internal pressure surge, and corrosion, leaks and line breaks do occur. Such incidents not only interrupt operations and cause economic losses, they can create environmental hazards and endanger public safety. Finding a large leak is easy. As the leak gets smaller, it becomes much more difficult to detect and locate it reliably.

Mass balance over a time interval and transient flow simulation are the two computerized methods used for pipeline leak detection. The mass balance method assumes that the mass storage in the pipe, called line fill, stays constant. The flow rates at the inlet and the outlet of the pipe are compared. A leak is suspected when the difference of the in-flow and out-flow volumes exceeds a tolerance. In the flow simulation method, a numerical model for the transient flow in the pipeline is driven by real time pressure and flow data. The

---

[1]Assistant Professor, Department of Civil Engineering, University of Idaho, Moscow, Idaho 83843

numerical model assumes the pipeline to be intact. When a leak develops, the calculated and the measured pressure and flow at pipe ends soon diverge, thus indicating a leak. This method potentially can detect a leak while the line fill is changing. Several such leak detection systems have been implemented on major oil and petroleum products pipelines (Covington 1979, Goldberg 1979, Huber 1981, Stewart 1983, and Higgins 1983).

One method of locating the leak is to use the fact that the flow rate and the hydraulic gradient are greater at the inlet than at the outlet, and that the projected hydraulic gradient lines (HGL) from the pipe ends intersect at the leak (Seider 1979, Maddox et. al. 1984). The data in Maddox 1984 showed that it is very difficult to locate a small leak by this method. In addition, the leak can only be located after the leak has established and the flow has settled into a new steady state. The leak can not be located at the moment it occurs.

Another method of locating the leak is based on the time of arrival of information at the pipe ends or metering stations neighboring the leak (Stewart 1983). Because the pressure sensors are polled at discrete instants, the arrival times can be over- or under-estimated by one scan period. This method can be fooled by noise and transients in the system.

A leak detection and location method that overcomes the above limitations is presented. The assumptions and the governing equations are discussed first, followed by the numerical solution procedure. The method is explained next. Laboratory test results are then presented to demonstrate its effectiveness.

Governing Equations and Solution Method

Assuming the convective changes in velocity and pressure to be negligible, and the liquid density and the pipe cross-sectional area to be constant except in the definition of the pressure wave speed, the equation of motion and the continuity equation governing one-dimensional transient flow are (Wylie 1985)

$$g \frac{\partial H}{\partial X} + \frac{1}{A}\frac{\partial Q}{\partial T} + \frac{fQ|Q|}{2DA^2} = 0 \quad \text{(1)}$$

$$\frac{\partial H}{\partial T} + \frac{a^2}{gA}\frac{\partial Q}{\partial X} = 0 \quad \text{(2)}$$

in which H = piezometric head, Q = volumetric flow rate, D = pipe diameter, A = cross-sectional area of the pipe, g = gravitational acceleration, f = Darcy-Weisbach friction factor, a = pressure wave speed of the liquid-pipe system, X = distance, and T = time.

Although the simplifying assumptions are acceptable in normal water hammer calculations, their validity is not obvious for detecting a small leak on oil transmission mains where the pressure variation is great, the pipe length is long, and the oil is more compressible than water. The simplifying assumptions yield a constant volumetric flow rate at the steady state. However, the significant pressure difference between the pipe inlet and outlet makes the mass density and the pipe cross-sectional area greater at the inlet than at the outlet. Since the mass flux remains constant, the volumetric flow rate should increase in the direction of flow. Another concern is whether the simplified model will produce the correct pressure for both steady and transient flows.

These aspects were studied by Liou 1988 using more rigorous steady state and transient models in which the pipe cross sectional area and the mass density of the fluid are pressure dependent. Consider a 40 mile 22 inch steel pipe carrying gasoline with a standard specific gravity of 0.7, and a steady state frictional pressure drop of 430 psi. It was found that there is a 0.4% difference in mass density and in volumetric flow rate between pipe inlet and outlet. Therefore, the simplifying assumption of constant mass density may result in a flow error of 0.4%. This error is acceptable since the repeatability of common turbine flow meters seldom exceeds 0.5% full scale. Simulations of transient flows using models with and without the simplifying assumptions generated closely matched pressure and mass flux traces. These results suggest that the simplifying assumptions are acceptable. In addition, the solution to Eqs. 1 and 2 requires far fewer computations than that required by the more rigorous model. The faster speed of the simplified model is an advantage in a real time simulation environment.

The Darcy-Weisbach friction factor is assumed to be applicable to transient flow. This is a generally accepted assumption and verifiable with field data. To facilitate computation, the explicit equation of Swamee and Jain 1976 that relates the friction factor to the Reynolds number and the relative pipe roughness is used.

Denote the steady-state volumetric flow rate by Q_0 and the head rise due to sudden and complete stoppage of the velocity $V_0 = Q_0/A$ by H_0. The potential surge is

$$H_0 = \frac{aV_0}{g} \qquad (3)$$

Let $v = Q/Q_0$, $h = H/H_0$, $x = X/L$, and $t = Ta/L$, L being length of the pipe. Eqs. 1 and 2 take the following dimensionless form

$$\frac{\partial h}{\partial x} + \frac{\partial v}{\partial t} + Rv|v| = 0 \qquad (4)$$

$$\frac{\partial h}{\partial t} + \frac{\partial v}{\partial x} = 0 \qquad (5)$$

where

$$R = \frac{fLV_0}{2aD} \qquad (6)$$

Using the method of characteristics (Wylie and Streeter 1982), Eqs. 4 and 5 are transformed into a pair of total differential equations

$$\frac{dh}{dt} \pm \frac{dv}{dt} \pm Rv|v| = 0 \qquad (7a,b)$$

which are valid along characteristics

$$\frac{dx}{dt} = \pm 1 \qquad (8a,b)$$

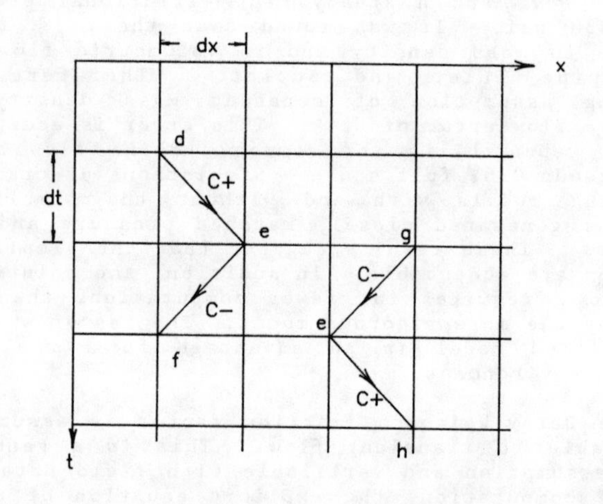

Fig. 1 The characteristics in the x-t plane

Fig. 1 shows the utility of Eqs. 7 and 8. The pipe is divided into a number of computational reaches of equal length dx. The time step dt is computed from dx according to Eq. 8. Eqs. 7a and 7b, called the compatibility equations along the C+ and C- characteristics respectively, are approximated by finite differences as

$$C+: h_e - h_d + (v_e - v_d) + 0.25R(v_e + v_d)|v_e + v_d|dt = 0 \qquad (9)$$

$$C-: h_f - h_e - (v_f - v_e) - 0.25R(v_f + v_e)|v_f + v_e|dt = 0 \qquad (10)$$

$$C+: h_h - h_e + (v_h - v_e) + 0.25R(v_h + v_e)|v_h + v_e|dt = 0 \quad ...(11)$$

$$C-: h_e - h_g - (v_e - v_g) - 0.25R(v_e + v_g)|v_e + v_g|dt = 0 \quad ...(12)$$

Knowing the head and the flow at d and f or g and h, the head and flow at e can be calculated. This scheme is similar to that used in valve stroking (Wylie and Streeter 1982).

Discretization Requirement

Consider the transients created by a sudden and complete flow stoppage at the outlet of a pipe. In non-dimensionalizing the governing equations, all system properties are grouped together as one single constant R (see Eq. 6). This R characterizes a series of similar systems. This fact is used to establish a discretization guideline for general use.

It has been shown (Liou 1988) that Eqs. 1 and 2 yield essentially the same results as those based on a more rigorous formulation where the fluid mass density is pressure dependent. The latter was used to compute the mass imbalance error as a function of discretization. The mass imbalance error is defined as

$$\frac{\text{time rate of pipe mass storage change - net mass influx}}{\text{steady state mass flux}}$$

The error is computed at each time step for two cycles (8 L/a seconds). The root-mean-square values of the errors are presented in Fig. 2 as a function of discretization for R = 0.10, 0.50, 2.50, and 5.0. Suppose that a pipeline has a R value of 2.5 and that a mass imbalance

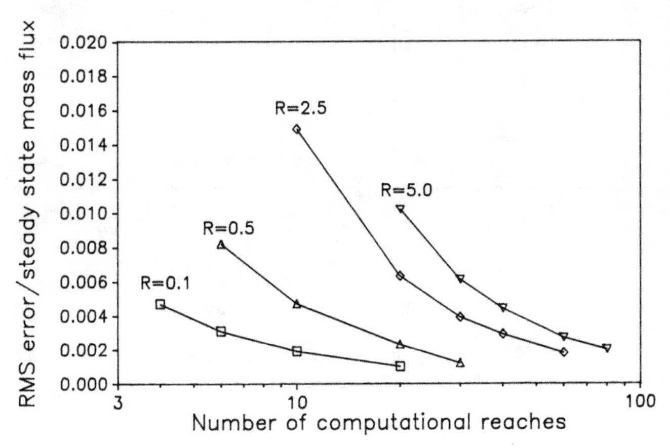

Fig. 2 Mass imbalance RMS error

error of 0.5% of steady state mass flux is required. One can enter Fig. 2 at the abscissa of 0.005 to find 26 as the number of computational reaches. This number is conservative (i.e. high) because Fig. 2 is based on sudden and complete flow cutoff at the outlet which causes the more severe transients.

Methodology

Referring to Fig. 3a, the measured pressures and flow rates at the pipe inlet between time t1 and t4 can be used to calculate pressures and flow rates in the time-distance space confined within t1, t2, t3, and t4. For ease of reference later, call this set of computations U-computations. The calculations are made assuming that no leak of any size exists. In Fig. 3b, the measured pressures and flow rates at the outlet can be used to compute a second set of pressures and flow rates within the space t1, t2, t3, and t4. Call this set of computations D-computations. If there is no leak, the measured pressures and flows should match those computed values at the pipe ends. Should a leak occur, which is not anticipated by the numerical model, the measured pressures and flows will diverge from the calculated ones. This divergence provides the basis for leak detection and location.

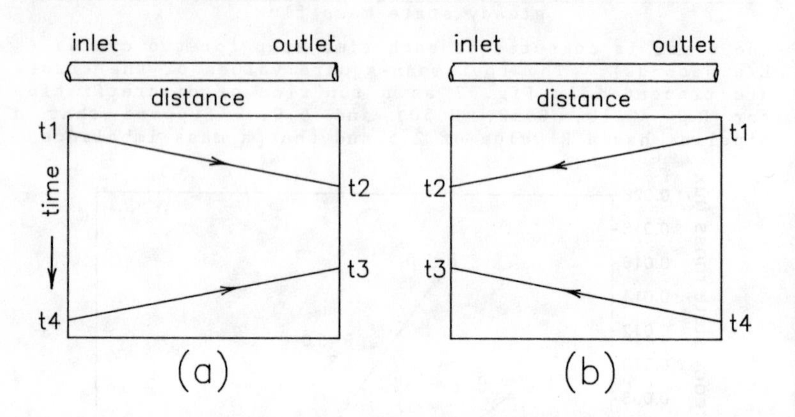

Fig. 3 Zone of dependence

In Fig. 1 and without pipe friction, Eqs. 9 and 10 are used to compute the unknowns at e

$$h_e = \frac{h_d + h_f - v_f + v_d}{2} \qquad (13)$$

$$v_e = \frac{v_f + v_d - h_f + h_d}{2} \quad \ldots\ldots\ldots\ldots\ldots\ldots\ldots\ldots\ldots (14)$$

These equations are used in the U-Computations in the x-t plane shown in Fig. 4. Four computational reaches are used and a leak is imposed at the end of the first reach from the inlet. The leak creates a disturbance which propagates along the C- characteristic to the inlet B and along the C+ characteristic to the outlet J. Let AU and AD represent the sections immediately upstream and downstream from the end of the first reach. The common head and the continuity equation at the node require

$$h_{AU} = h_{AD} \quad \ldots\ldots\ldots\ldots\ldots\ldots\ldots\ldots\ldots\ldots\ldots (15)$$

$$v_{AU} - v_{AD} = q \quad \ldots\ldots\ldots\ldots\ldots\ldots\ldots\ldots\ldots\ldots (16)$$

where q is the dimensionless volumetric leak flow rate. The same relationships apply at EU, ED, GU, GD, and future points at the same location.

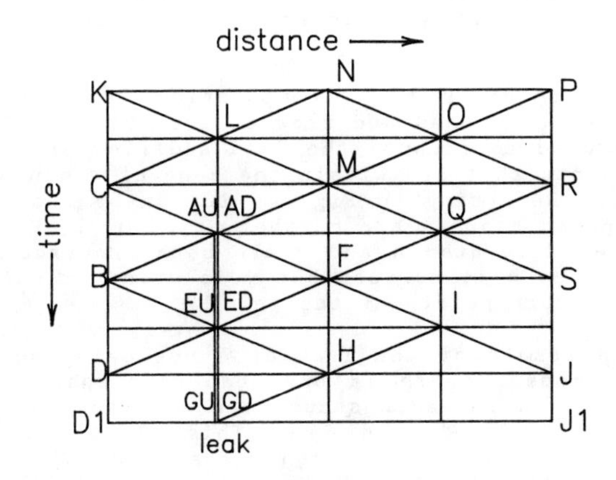

Fig. 4 Computations with leak along AG.

Since the "measured" h and v at B and D contain the effect of the leak, (h_{AU}, v_{AU}) and (h_{EU}, v_{EU}) can be correctly calculated. Note that the model is not aware of the leak and uses the head and flow just obtained at AU to compute the unknowns at F. The results can be expressed by the head and flow at AD via Eqs. 15 and 16

$$h_F = \frac{h_{AD} + h_{ED} - v_{ED} + v_{AD}}{2} \quad \ldots\ldots\ldots\ldots\ldots\ldots\ldots (17)$$

$$v_F = \frac{v_{ED} + v_{AD} - h_{ED} + h_{AD}}{2} + q \quad \ldots\ldots\ldots\ldots\ldots\ldots (18)$$

The last term in Eq. 18 is an error due to the fact that the leak is not included in the model computations. The h_F from Eq. 17 is correct, however, due to the alternating signs of v in that equation. This pattern of error repeats itself at I and J along the C+ characteristic and at all points to the right of the leak location at future times.

The correct h and v can be calculated for L from the "measured" h and v at C and K by using Eqs. 13 and 14. The h and v at M can be computed from the known values at L and AU via Eqs. 15 and 16

$$h_M = \frac{h_L + h_{AD} - v_{AD} + v_L}{2} - \frac{q}{2} \qquad (19)$$

$$v_M = \frac{v_{AD} + v_L - h_{AD} + h_L}{2} + \frac{q}{2} \qquad (20)$$

Because the leak is not modeled, the h_M and v_M just computed are incorrect. By repeating the same algebra, it can be shown that similar errors also exist at O, P, Q, R, and S.

Next, consider the D-computations that start by using the measured head and flow at the outlet. Eqs. 11 and 12 are used to compute the head and flow at the pipe interior points that are within the zone of dependence of the outlet. Referring to Fig. 4, it can be shown that: 1) the computed h and v are correct at O, Q, I, M, F, and AD; 2) the calculated h and v are both incorrect at C; 3) the computed h is correct but v is wrong at points B, D, and all future points at the inlet.

As a result of the leak at A not being accounted for in the model, there is one head discrepancy at the inlet (at C) and three consecutive head discrepancies at the outlet (at P, R, and S). This (1,3) pattern is unique to the leak at the second node (node A), and can be used to locate the leak. There are several other ways to recognize the leak and its location by comparing the head and flow computed from the U- and D-computations at the pipe interior. They all take advantage of the discrepancy patterns outlined above.

In establishing the discrepancy pattern, a threshold is needed to discern the calculated heads from the measured ones. The threshold must lie above the head errors due to model and data inaccuracies and below the differences (between measurement and computation) attributable to the leak. Otherwise, the discrepancy pattern can not be recognized. The model and the measured data have to be sufficiently accurate for a small leak to be detected and located.

After some experimentation, a discrepancy pattern was chosen to recognize and locate the leak with or without pipe friction. Suppose the pipeline is divided into N computational reaches with a reach length dx. The time step corresponding to dx is dt. Attention is focused on a moving time window with a width 2Ndt. Contained in this window are N pairs of measured and computed heads at the inlet and a second N pairs at the outlet. The difference between the measured and the calculated heads at these 2N points inside the window on the x-t plane are compared with a threshold value. When a head difference exceeds the threshold, it is considered a discrepancy. At each of the N time levels (adjacent levels are 2dt apart), the number of head discrepancies at the inlet and, separately, at the outlet are summed up. Shown in Fig. 5 is an example for N = 10. The first discrepancy occurs at the inlet in window 1 and persists for the next two windows. These discrepancies do not match the anticipated pattern and are discarded. Starting at window 6, the number of discrepancies at the outlet increases by one each time. Similarly, starting at window 10, the number of discrepancies at the inlet begins to increase by one each time. The difference of the number of discrepancies between the inlet and the outlet is constant (4 in this example). A leak is suspected once this pattern emerges.

Time Window Number	Number of Head Inlet	Discrepancies Outlet
1	1	0
2	1	0
3	1	0
4	0	0
5	0	0
6	0	0
7	0	1
8	0	2
9	0	3
10	0	4
11	1	5
12	2	6
13	3	7
14	4	8
.	.	.

Fig. 5 Example head discrepancy pattern

The ability to find the leak rate, the location, and the time of occurrence by this method has been evaluated by simulations using examples resembling oil transmission mainlines (Liou 1988). Both steady state and transient flows were considered. The extent to which this method tolerates model inaccuracy due to discretization, data noise, and inaccuracies in system friction factor and wave speed has been investigated. The results suggest that it performs well in the presence of reasonable amounts of inaccuracies and data noise.

The speed of computation on a 16 MHz IBM PS/2 Model 80 microcomputer with disk cache and using compiled QuickBAsic 4.0 is sufficiently fast for real time simulations. For example, consider a 40 mile 22 inch schedule 20 pipe with an R value of 2.22. Fig. 2 suggests that 40 computational reaches are needed to limit the mass imbalance error to within 0.3% of the steady state flow rate. With this discretization, the computation time is only 4% of the clock time for the physical event.

Laboratory Test Facility

A 152.4 m long polyethylene tube is used as the conduit in the laboratory piping system. The average diameter of the tubing is 0.973 cm The tubing is wrapped around a spool 0.366 m in diameter. The core of the spool is a reservoir. Compression fittings are used to insure no leakage at the maximum system operating pressure of 1,000 kPa. The bursting strength of the tubing is rated at 2,068 kPa which is considerably higher than the maximum system operating pressure. The margin ensures the elastic or near elastic deformation of the tube during transients. Mounted on the base of the spool stand is a high head low capacity rotary pump that recirculates water through the pipe coil.

Three in-line quarter-turn ball valves are used in the piping system. Valves No. 1 and 2 are partially open and are located near the pipe inlet and outlet respectively. These two valves are used as orifice meters for flow measurement. A cam and screw mechanism locks each of these valves at pre-set openings where flow versus pressure drop relationships have been calibrated. Valve No. 3 is located 0.914 m downstream from Valve No. 2 and just before the pipe empties into the reservoir. This is the flow control valve used to adjust the flow in the system for different steady state flows and for generating transients. A fourth quarter-turn ball valve is located at a branch line 91.4 m downstream from the inlet. This valve is opened to simulate a leak. Otherwise, it remains closed.

The Reynolds number ranged between 5,000 and 10,000 in this study. The small pipe diameter and the need to create substantial head drop across the valves limit the Reynolds number to the stated range. The head loss across the partially closed valves and the friction for the piping system are determined for several steady state flows. The flow rates are determined by weighing the discharged water over a 2 minute period. The accuracy of these measurements is within 0.3%. The valve head drop coefficients and the system friction factor, all Reynolds number dependent, are given in Liou 1988.

The pressure wave speed for the system is obtained by timing the travel time of a pressure wave between transducers No. 2 and No. 3 spaced 152.4 m apart. The pressure wave is generated by suddenly closing valve No. 3. The wave speed so obtained is 251.5 m/sec with an estimated error of no more than ± 10%.

Instrumentation and Data Acquisition System

Four dynamic pressure transducers are used to obtain the differential pressure across valves No. 1 and No. 2 and local gauge pressures. All transducers were calibrated on the test loop at the same time under hydrostatic conditions. The precision of the pressure readings is within ±3.45 kPa. These four measurements provide the measured pressure (or head) and flow at the inlet and the outlet of the pipe.

The outputs from the four transducers are fed into a data acquisition board. The board uses 12 bit analog to digital signal conversion and can handle 4 channels of analog inputs at a conversion rate of 100 sampling points per second per channel. This rate is considered adequate for the study. The data acquisition board is connected to an IBM PS/2 model 80 microcomputer through its serial port. A transient flow simulation model runs on the microcomputer. At each time step (clock time), it acquires the four channels of analog data, converts them into pressure and flow data at the pipe ends, and performs transient flow simulations in real time.

Real Time Simulations

The timing of acquiring data and of numerical simulation is regulated by a scheme using the real time clock of the computer. At simulation time t, a 100-millisecond window is projected at real time t + dt, where dt represent the time step in the simulation. The four channels of data are acquired for simulation use as soon as the clock time falls inside the window. The simulation of the flow between t and t + dt then takes place. The simulation will be completed well within time period dt. A new 100-millisecond window is then projected at t + 2dt which lies ahead of the clock time. There is an idle period until the clock time falls into the newly projected window. At that moment, the same process starts again. This way, the simulation lags behind the clock time by dt but keeps up the same pace as the physical events in the pipeline indefinitely.

The choice of the 100-millisecond window is affected by the approximately 60-millisecond inaccuracy of the real time clock of the microcomputer. This timing inaccuracy limits the number of computational reaches that can be meaningfully used in the simulation. In the laboratory tests, only 4 computational reaches were used. The corresponding time step was 0.1513 seconds or 2.5 times the maximum timing error. In applications where the pipe lengths are in tens of kilometers instead of 152.4 m, this timing error becomes insignificant.

Test Results

Figs. 6 and 7 show a comparison between the measured head and flow at the pipe ends and those calculated by the numerical model in real time. The solid lines represent the measured values. The upright and inverse triangles represent the calculated values at the inlet and the outlet respectively. The transients are created by throttling valve 3 to bring the flow from 7.45E-5 m^3/sec to 4.67E-5 m^3/sec. The leak is imposed by opening Valve 4 slightly during transients at approximately 40 seconds. The leak flow rate during the unsteady flow is not directly measurable with the present instrumentation. However, by re-opening Valve 3 while leaving Valve 4 untouched, the leak flow rate was measured to be 6.5% of the initial steady state flow.

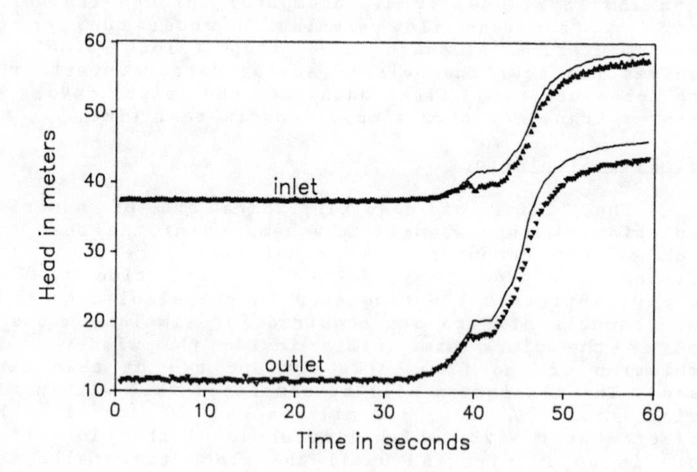

Fig. 6 Comparison of measured and calculated heads

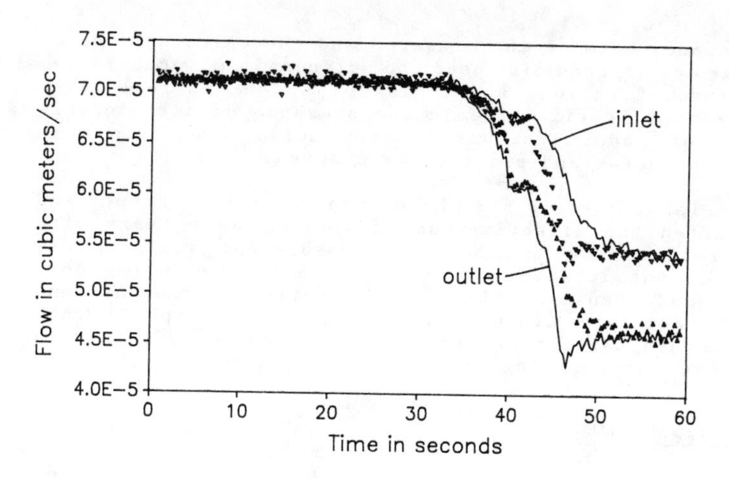

Fig. 7 Comparison of measured and calculated flows

It is seen that the calculated head and flow at the pipe ends match the measured values closely prior to the onset of the leak. The departures between the measured and the calculated head traces are clearly noticeable just before 40 seconds. This leak was detected instantly. Its location was indicated at computational section 3 which is the nearest computational section to the leak. The flow rate and timing of the leak were all correctly computed.

Leaks with smaller flow rates were also tested. The algorithm detects the occurrence and location of a leak correctly if the leak occurs during steady state, but gives either wrong leak flow rate, location, and/or timing when the leak occurs during transients. For even smaller leaks, a head discrepancy threshold and thus the leak can not be found. With the present instrumentation, the smallest leak that can be detected and located correctly is about 4% and 5% of initial steady flow rate for steady and transient flows respectively. In all the tests, the algorithm has never given a false alarm. This desirable feature is attributable to the fact that only leak induced disturbances fit the discrepancy pattern for leak announcement.

Conclusions

A microcomputer based system to monitor pipeline integrity is developed. It uses a transient flow numerical model, driven by measured head and flow rate at the pipe inlet and outlet. A specific pattern of discrepancy between the measured and the calculated heads at the pipe ends is used to discern the occurrence, magnitude, location, and timing of a leak in real time. This method is effective in the presence of a reasonable degree of model inaccuracy, data noise, and errors in friction factor and pressure wave speed.

Laboratory tests have proven the concept and quantified the effectiveness of the method. Leaks with flow rates close to the flow measurement accuracy are detected and located correctly in real time during both steady and transient flows. With better instruments and a more accurate timing scheme, it is anticipated that this method can detect and locate a very small leak with flow rate approaching the repeatability of turbine flow meters.

Acknowledgement

This study was funded by the Engineering Foundation Grant RI-A-87-11 - Chyr Pyng Liou. The idea of using a algorithm similar to that in valve stroking was discussed with Dr. M. A. Stoner.

References

Covington, T. M., 1979, "Transient Models Permit quick Leak Identification", Pipe Line Industry, August, pp. 71 - 73.

Goldberg, D.E., 1979, "Transient Leak Detection for Liquid Pipelines",Pipeline and Gas Journal, Vol 207, No, 8, July, pp. 65 - 67.

Higgins, P., 1983, "Advances in Leak Detection Systems, Pipe Line Industry, Part 1: March, pp. 29 - 32. Part 2: April, pp. 55 -58.

Huber, D. W., 1981, "Real-Time Transient Model for Batch Tracking, Line Balance and Leak Detection", American Petroleum Institute Annual Pipeline Conference, April.

Maddox, R. N., Shariat, A., and Moshfeghian, M., 1984, "Program for TI-59 calculator Helps Locate Position of Leaks on a Liquids Pipeline", Oil and Gas Journal, Dec. 10, pp. 104 - 110.

Seiders, E. J., 1979, "Hydraulic Gradient Eyed in Leak Location", Oil and Gas Journal, Nov. 19, pp. 112 - 125.

Stewart, T. L., 1983, "Operating Experience Using a Computer Model for Pipeline Leak Detection", Journal of Pipelines, Vol. 3, pp. 233 - 237.

Swamee,P. K., and Jain, A. K., 1976, "Explicit Equations for Pipe-Flow Problems", Journal of Hydraulics Division, American Society of Civil Engineers, Vol. 102, no. HY5, pp. 657 - 664.

Wylie, E. B., and Streeter, V. L., 1982 , Fluid Transients, McGraw-Hill Book Co., New York, Republished by FEB Press, Ann Arbor, Michigan, 1983.

Wylie, E. B., Closure to discussions on proceeding paper 18707 - "Fundamental Equations of Waterhammer", Journal of Hydraulic Engineering, Vol. 111, No. 8, pp. 1197-1200.

VALVE CONTROL TO AVOID COLUMN SEPARATION IN A PIPE

Faiq Pasha[1] and Dinshaw Contractor[2]

Abstract.

Maximum pressures in a pipeline can be reduced significantly by closing the valve in an optimal manner. Such an optimal valve closure policy may result in some situations in minimum pressures below the vapor pressure of the fluid. Thus, column separation will occur with consequent high pressures that were not taken into account in the derivation of the optimal valve closure policy. Hence, it is desirable to minimize the maximum pressure (H_{max}) in a pipeline, subject to the constraint that the minimum pressure (H_{min}) in the pipeline be greater than or equal to a pre-selected pressure, e.g., the vapor pressure of the fluid (H). This is accomplished by redefining the problem as the minimization of the objective function $(H_{max} - H_{min})$, with the stipulation that when $H_{min} > H_\upsilon$, the objective function becomes $(H_{max} - H_\upsilon)$.

The method of characteristics is used to simulate fluid transients in a pipe and the simplex method is used to optimize the objective function. A numerical example is provided to illustrate the valve-closure policies obtained when different objective functions are used, e.g, minimizing H_{max}, maximizing H_{min}, and minimizing $(H_{max} - H_{min})$. It is shown that a valve-closure policy can be found that minimizes H_{max}, while avoiding water column separation.

Introduction.

Several investigators have explored methods to control waterhammer pressures in a pipeline by closing the valve in an appropriate manner. Streeter (1963) proposed a technique called "valve stroking" to determine a valve-

[1]Graduate Student and [2]Professor, Civil Engg. and Engg. Mechanics Dept., The University of Arizona, Tucson, Arizona 85721

operating policy that limits the maximum pressure to a preset value, while preventing flow reversal and having no transients in the pipe on valve closure. Driels (1975) studied minimizing waterhammer pressure by operating the valve in two linear stages. A simplex search technique was used to minimize waterhammer pressures for a given valve closure time. Contractor (1983) used dynamic programming to determine a valve closure policy to minimize the waterhammer pressure rise at the valve. Contractor and Conway (1986) applied this technique to a series pipeline in which the junction location, friction factor and closure period were varied. Azoury, et al. (1986) studied, for a simple pipeline discharging into air, the effect of valve closure policy on waterhammer pressures and presented a chart that can be used to determine the valve schedule for minimum pressure. Goldberg (1987) proposed a time-optimal valve closure procedure called Quick Stroking. El-Ansary and Contractor (1988) used the simplex method for minimizing the stresses in a pipe due to waterhammer.

In many of the procedures mentioned above, it is quite probable that the minimum pressure, H_{min}, may fall below the vapor pressure, H_v, of the fluid. The resulting column separation could generate pressures exceeding the maximum pressure, H_{max}, obtained in the optimization procedure. Column separation would thus defeat the purpose of optimization. It is thus necessary to specify that the minimum pressure obtained in the optimization procedure be greater than the vapor pressure of the fluid. This is accomplished by redefining the problem as the minimization of the objective function (H_{max} - H_{min}) with the stipulation that when $H_{min} > H_v$, the objective function will be (H_{max} - H_v). The optimal policy thus obtained accomplishes two objectives; first, it minimizes H_{max} and second, it avoids column separation.

Methodology.

The method of characteristics was used to simulate waterhammer pressures and velocities in a pipeline, Wylie and Streeter (1978). The optimization technique used was the simplex method proposed by Nelder and Mead (1965). This procedure is an extension of the simplex method by Spendley, et al. (1962). The simplex method minimizes an objective function of n variables, with or without constraints. Derivatives of the objective function are not required. The geometric figure formed by a set of n+1 points in n-dimensional space is called a simplex. In the method, the simplex adapts itself to the local landscape, elongating down inclined planes, changing direction on encountering a valley at an angle and contracting in the neighborhood of a minimum. This is accomplished by steps referred to as reflection, expansion, contraction in one direction and contraction in all directions. The minimization is formally defined in the following manner:

$$\text{Min } F(\tau) = W_1 * H_{max} - W_2 * H_{min} \quad (1)$$

subject to

$$H_{min} > H\upsilon \quad (2)$$

$$0 < \tau_1 < 1, \; i = 1,n \quad (3)$$

$$\tau_o = 1 \text{ and } \tau_{n+1} = 0 \quad (4)$$

where τ is the dimensionless valve opening, $(C_dA_v)/C_dA_v)_o$, and the weights, W_1, W_2 can be made equal to 1, 0 or -1.

Previous experience with the simplex method has indicated that when n is greater than 10, the convergence is slow. Thus, in this case, the τ-time curve was optimized at 10 points. These 10 points were connected continuously by a natural spline function for the purpose of interpolating between the points. A FORTRAN program was written to minimize the objective function F. The program was checked for errors and its results verified using the results of the step-closure policy by Contractor (1985). Fig. 1 shows that the results of this program match the results of the step-closure policy very well.

Computer Implementation.

A simple pipeline with a constant head reservoir at the upstream end and a valve at the downstream end is chosen for application of the model. The pipeline data is presented below:

Length of pipe, L	600 m
Diameter of pipe	0.5 m
Reservoir head	150 m
Darcy-Weisbach friction factor	0.018
Wave speed in pipe, a	1340 m/s
$(C_dA_v)_o$	0.038
Velocity in pipe	7.80 m/s

The program was run for three different cases, each case representing a different objective function. The purpose was to study the effects and characteristics of each objective function.

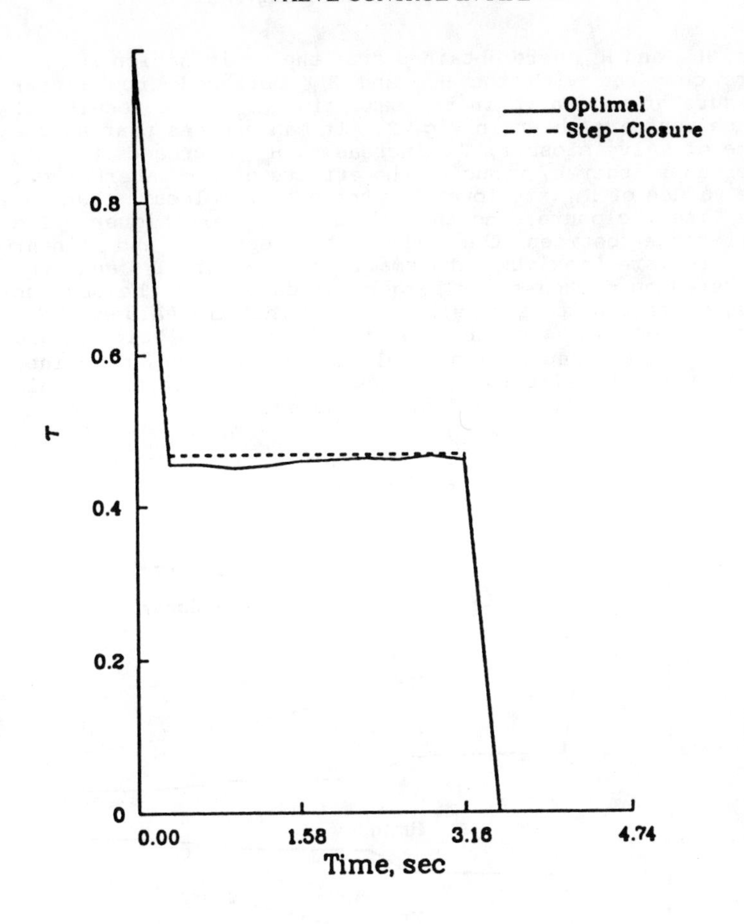

Figure 1. Comparison of Optimal Policy with the Step-Closure Policy of Contractor (1985).

Case I.

The objective function in this case was the minimization of the maximum pressure, H_{max}, i.e., with W_1 = 1 and W_2 = 0 in equation 1. No constraint was imposed on the minimum pressure, H_{min}, i.e., equation (2) was omitted. The program was run several times with the valve closure time, T_c, ranging from 2L/a secs. to 8L/a secs. For each run, H_{max} and H_{min} were obtained from the optimization run and were compared with the H_{max} and H_{min} obtained from linear closure of the valve in the same time, T_c. The results of these runs are shown in Fig. 2. It can be seen that as the time of valve closure, T_c, increases, H_{max} decreases and H_{min} increases, thereby reducing the effects of the waterhammer. The values of H_{max} are lower for the optimal closure than for the linear closure, and the values of H_{min} are higher. The difference between the effects of optimal and linear closure are maximum at small values of T_c and the differences decrease for larger values of T_c. The obvious disadvantage of this case is the fact that the values of H_{min} for both optimal and linear valve closures are below H_v and hence column separation would occur in the pipeline, nullifying the effects of reduction of H_{max}. To solve this problem, Cases II and III were studied.

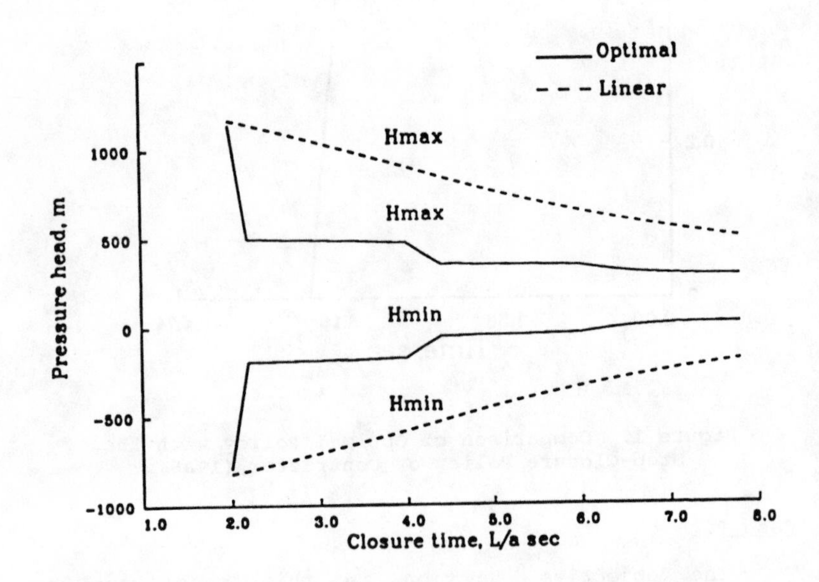

Figure 2. Maximum and Minimum Pressures for Linear and Optimal Valve Closures as a Function of Valve Closure Time, Case I.

Case II.

The objective function in this case was the maximization of H_{min}, i.e., $W_1 = 0$ and $W_2 = 1$ in Eq. 1. The constraint in Eq. 2 was omitted. All other conditions in the pipeline were kept the same as in Case I. The time of valve closure, T_c, was varied within the same range, 2L/a to 8L/a secs. The results of these runs are presented in Fig. 3. It can be seen that the value of H_{min} for the optimal conditions has been increased to the minimum steady state pressure in the pipeline. Since this pressure is greater than H_υ, column separation will not occur. However, the maximum value of H_{max} in this case is much higher than the value of H_{max} in case I, even though it is still less than H_{max} for linear closure.

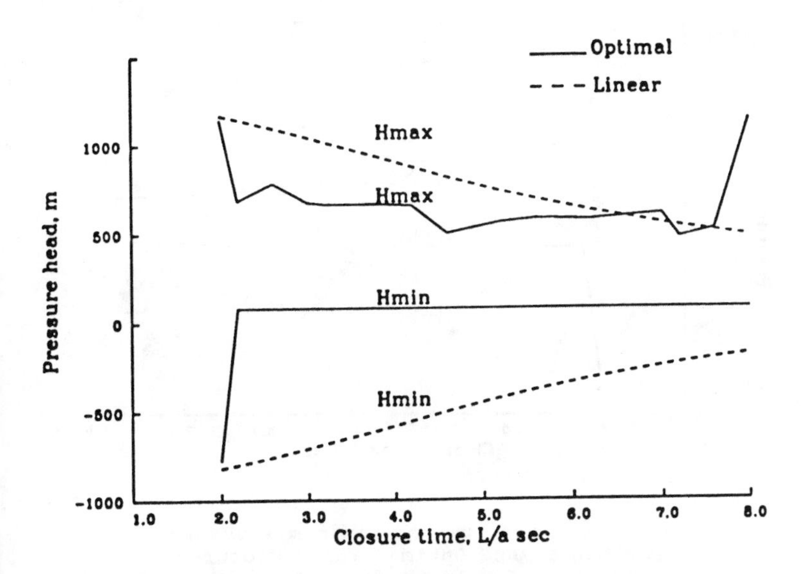

Figure 3. Maximum and Minimum Pressures for Linear and Optimal Valve Closures as a Function of Valve Closure Time, Case II.

Case III.

The objective function in this case was the minimization of $(H_{max} - H_{min})$, i.e, $W_1 = W_2 = 1$, with the stipulation that when $H_{min} > H_\upsilon$, the objective function would become the minimization of $(H_{max} - H_\upsilon)$. The vapor pressure of the liquid was considered to be equal to -10m. The constraint in Eq. 2 was specified to be valid in this case. All other conditions in the pipeline were kept identical with those in the previous cases. The results of these runs are presented in Fig. 4. It can be seen that H_{min} is held at the vapor pressure -10m, while H_{max} is minimized. The value of H_{max} in this case is much lower than the value of H_{max} in Case II and only slightly higher than the H_{max} in Case I. The maximum reduction in H_{max} from the linear to optimal closure occurs at T_c = 2·1 L/a secs and amounts to 52% reduction in the dynamic pressure, i.e., pressure above steady-state pressure.

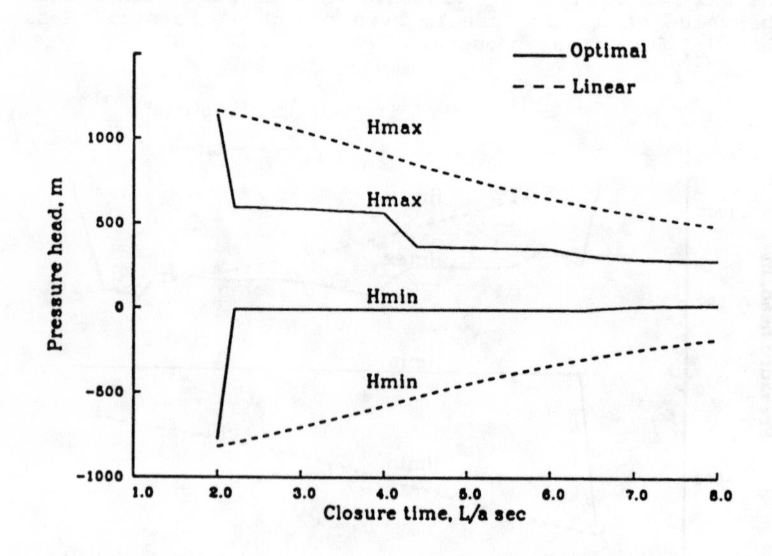

Figure 4. Maximum and Minimum Pressures for Linear and Optimal Valve Closures as a Function of Valve Closure Time, Case III.

Fig. 5 shows the optimal valve-closure policy to minimize H_{max} and avoid column separation (Case III). Thus, the valve must be closed rapidly to τ = 0.21 and kept at that opening for 2L/a secs and then closed rapidly to τ = 0.0. The pressure-time histories at the valve for optimal and linear valve closures are given in Fig. 6. It can be seen that H_{max} for optimal closure is about 50% of that for linear closure. Also, H_{min} for optimal closure is held at -10m and does not go below it, thus avoiding column separation.

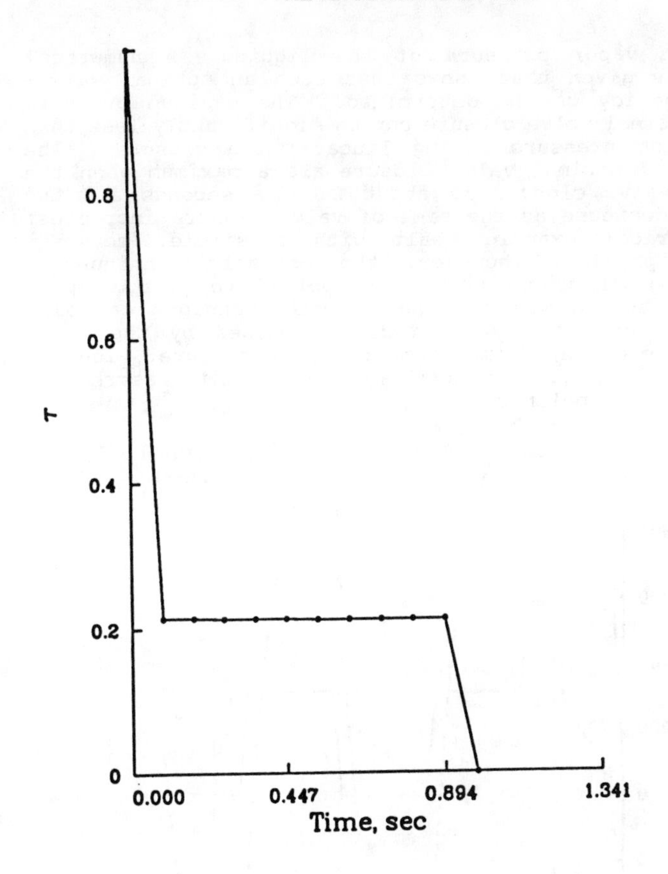

Figure 5. Optimal Valve Closure Policy, τc = .984 sec. (2.2L/a). Case III.

Conclusions.

The control of waterhammer transients can be accomplished by appropriate control of the closing of a valve. In trying to minimize the maximum pressure due to waterhammer, column separation very often occurs, resulting in very high pressures that were not taken into account in the optimal calculations. Thus, it becomes necessary to impose a constraint in the optimal problem that the pressure in the pipe will never go below a pre-set value, e.g., the vapor pressure of the liquid. A numerical example is given that shows that such an optimal valve-closure policy can be determined. The maximum pressure during optimal valve closure can be significantly less than the maximum pressure during linear valve closure. The benefits of optimal valve closure are a maximum when the time of valve closure is about 2·1 L/a seconds and the benefits decrease as the time of valve closure increases. The numerical example dealt with a simple, constant diameter pipeline; however, the optimal technique is general enough that it can be applied to more complex pipelines and networks. The optimal technique can also take into account any constraints imposed by the valve actuator, e.g., a maximum rate of valve closure. Thus, a powerful technique is available to limit waterhammer pressures in pipelines.

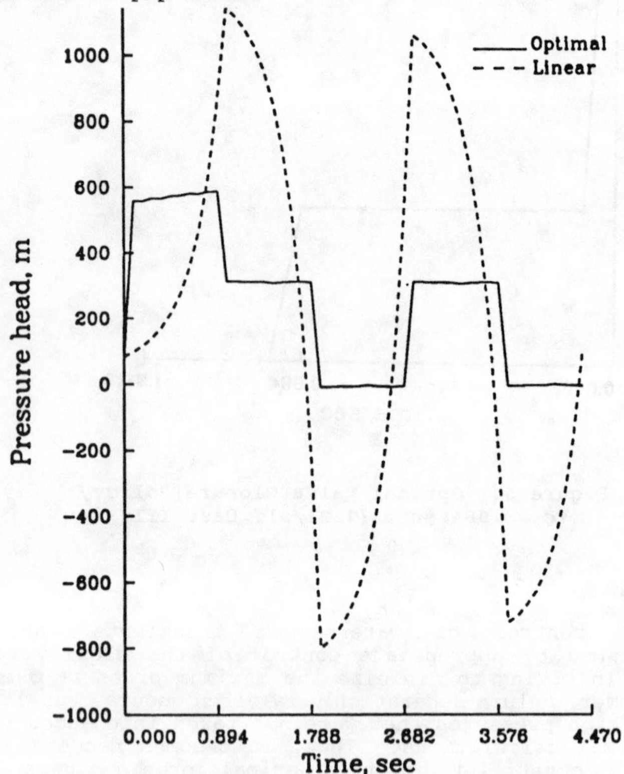

Figure 6. Pressure Variation at the Valve for Optimal and Linear Valve Closures τc = .984 sec. (2.2L/a). Case III.

REFERENCES

Azoury, P.H., A.M. Bassiri, and H. Najm, "Effect of Valve-Closure Schedule on Waterhammer," ASCE, Journal of Hydraulic Engineering, Vol. 112, No. 10, October 1986.

Contractor, D.N., "Two and Three Stage Valve-Closures to Minimize Waterhammer Transients," ASME, Vol. 98-7, 1985.

Contractor, D.N., "Valve Stroking to Control Waterhammer Transients using Dynamic Programming," ASME Symp. Num. Methods in Fluid Transient Anal., Fluids Engineering Division, Houston, Texas, Jun. 1983.

Contractor, D.N., and A.B. Conway, "Minimizing Waterhammer Transients in a Series Pipeline," Forum on Unsteady Flow-1986, ASME Winter Annual Meeting, Anaheim, California, Dec. 7-12, 1986.

Driels, M.R., "Predicting Optimum Two-Stage Valve Closure," ASME, Winter Annual Meeting, Houston, Texas, Nov. 1975. WA-75/FE-2.

El-Ansary, A.S. and D.N. Contractor, "Minimization of Axial Stresses and Pressure Surges in Pipes using Nonlinear Optimization," ASME Vol. 140, Pressure Vessels and Piping Conference, Pittsburgh, Pennsylvania, Jun. 1988.

Goldberg, D.E. and C.L. Kerr, "Quick Stroking: Design of Time-Optimal Valve Motions," ASCE, Journal of Hydraulic Engineering, Vol. 113, No. 6, Jun. 1987.

Nelder, J.A. and R. Mead, "A Simplex Method for Function Minimization," Computer Journal, Vol. 7, No. 4, 1965.

Spendley, W., G.R. Hext, and F.R. Himsworth, "Sequential Application of Simplex Design in Optimization and Evolutionary Design," Technometrics, Vol. 4, No. 4, 1962.

Streeter. V.L., "Valve Stroking to Control Waterhammer," Journal of the Hydraulics Division, ASCE, Vol. 89, No. HY2, Proc. Paper 3452, Mar. 1963.

Wiley, E.B., and V.L. Streeter, Fluid Transients, McGraw-Hill, Inc., 1978.

STRESS RELAXATION CHARACTERISTICS OF THE HDPE PIPE-SOIL SYSTEM

Larry J. Petroff[1], Associate Member, ASCE

ABSTRACT

All piping materials and soils are subject to viscoelastic behavior such as stress relaxation and creep. The combination of soil and pipe viscoelastic properties determine the time-dependent behavior of a pipe-soil system. Generally, the more conforming the pipe is to the soil's response, the more favorable the system's response. For instance, buried, high density polyethylene (HDPE) pipe may undergo a relief in load with time due to creep deflection. The largest portion of creep deflection for HDPE pipe occurs within 30 days after installation.

INTRODUCTION

Time-dependent deformations of soils cause structures to settle and pressures against buried structures to change with time. A common form of time-dependent deformation is primary consolidation, which occurs as excess pore pressure forces porewater from clayey soils. After the excess pore pressure has dissipated, the deformation continues. This continued deformation, called secondary consolidation, usually results in small settlements, occurring over long time periods, that can be attributed to creep within the clay mineral structure. Secondary consolidation due to creep also occurs in granular soils.

Normal geotechnical design practice accounts for viscoelastic behavior. For instance, Schmertmann (1970) gives a method for calculating the settlement of footings on sand that includes a correction factor for creep. The soil pressure against a retaining wall with a wet clay backfill may increase with time due to

---

[1]Engrg. Supervisor, Spirolite Corp., 4094 Blue Ridge Industrial Parkway, Norcross, GA, 30071.

creep along internal failure planes. In this case, the design pressure is considered to be the at-rest pressure rather than the active pressure (Mitchell, 1976).

Likewise, pipe designers should consider how the viscoelastic behavior of soil affects the long term performance of buried pipe. For instance, stress relaxation of soils can lead to the soil pressure against rigid pipe increasing with time. The more rigid and less conforming to the soil's movement, the more pressure is accrued to the pipe. Plastics exhibit rheological behavior similar to soils. Pipes made from plastics may conform to viscous displacements in the soil, which lead to a reduction in bending and thrust loads applied to the pipe with time. Thus, plastic pipes tend to undergo small, continuous deflections which can relieve stress.

VISCOELASTICITY OF SOIL

Viscoelasticity is the tendency of a system to respond to stress as if it were a combination of a viscous liquid and an elastic solid (Mruk, 1989). A simplified model of viscoelastic behavior is the combination of spring and dashpot elements shown in Fig. 1. The spring deflection is independent of time and proportional to the applied load (elastic). The dashpot deflects at a time dependent rate proportional to the applied load (viscous). The dashpot behaves like a cylinder with a leaky piston. When load is applied to the system, deformation is instantaneous in the springs, but is retarded by the dashpots. As fluid slowly escapes from the dashpots, they contract and the overall system deforms. Viscous deformation includes both recoverable and non-recoverable deformations.

Creep is deformation that occurs with time under constant load. See Fig. 2. A soil creeps due to deformation within the microstructure of its fabric, where stress concentrations break down the fabric or cause rearrangements and compression of the particles. Creep occurs in both fine and coarse grained soils. Although dry soil will creep, water acts as a lubricant: wetter soils tend to creep more. A detailed discussion of creep in soils is given by Mitchell (1976).

If a fixed deformation is applied to the system shown in Fig. 1, the springs exert a constant force, but the force exerted by the dashpots decreases with time. Therefore, the load required to maintain a constant deformation decreases or relaxes with time. This is called stress relaxation. See Fig. 3.

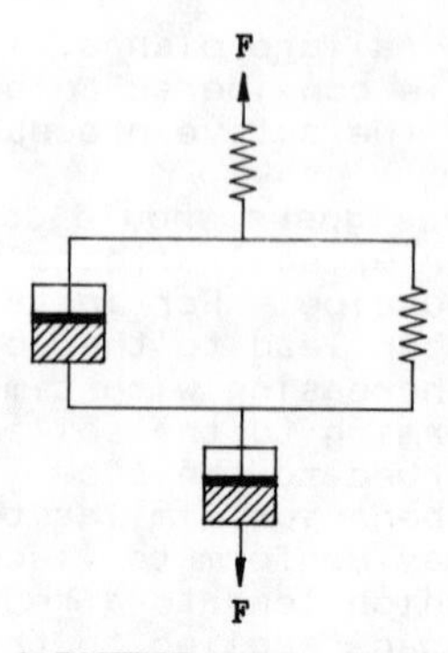

FIG. 1. SPRING & DASHPOT MODEL OF VISCOELASTIC SYSTEM

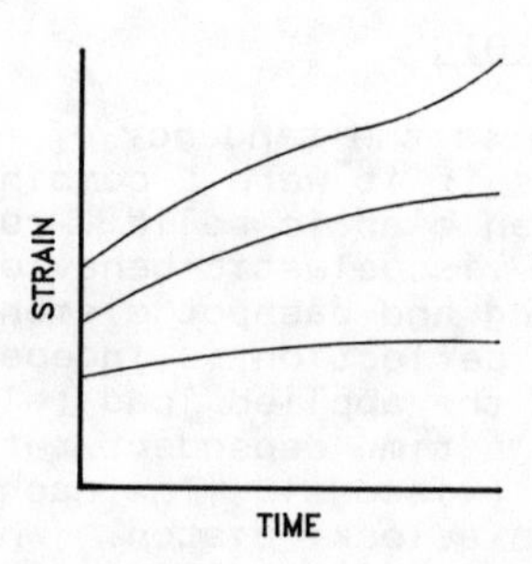

FIG. 2. CREEP (STRAIN UNDER CONSTANT STRESS)

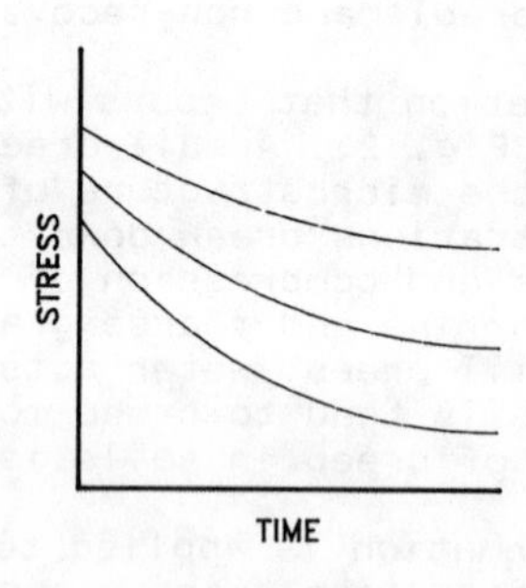

FIG. 3. STRESS RELAXATION (STRESS UNDER CONSTANT STRAIN)

The ratio of the applied stress to the measured strain at any given time is the relaxation modulus. For most materials at strains below yielding the relaxation modulus is given by the power law in time as:

$$E(t) = E_1 t^{-m}$$

where E_1 = the initial modulus; t = time; and m=the power law exponent (Chua and Lytton, 1987). Typical values of the power law exponent, m, for soils and piping materials are given in Table 1 for time in minutes. The elastic and viscous response are inseparable, but where a value for the elastic modulus is desired it is customary to use the value of the relaxation modulus at one minute.

TABLE 1. VISCOELASTIC PROPERTIES OF MATERIALS FROM CHUA & LYTTON (1985,1989)

Description	Relaxation Modulus @ 1 min. (psi)	Power Law Exponent
Clay (CH)	7,048	0.077
Sand	*	0.020
HDPE	131,173	0.083
Concrete	3,410,000	0.028
PVC	328,233	0.031-0.06
RPM	2,590,000	0.048

* Function of confining pressure. Use hyperbolic modulus value.

VISCOELASTICITY OF PIPE MATERIALS

All pipe materials are viscoelastic. Table 1 gives values of the initial modulus and power law exponent for concrete, PVC, RPM, and HDPE. Under an applied soil load plastic pipe creeps. The stress relaxation curve for HDPE is given in Fig. 4.

A common misconception is that plastic pipe loses strength as it creeps. Creep does not damage the plastic. When unloaded and allowed to relax, plastic pipe displays its original strength characteristics, provided it has not been loaded beyond its yield strain. For instance, tests conducted on polyethylene gas distribution pipe after 18 years in buried pressure service showed no evidence of material deterioration (Frantz, 1984). Nor does creep alter the material's modulus. When load is first applied to a pipe, it is met with the pipe material's initial modulus. After the pipe has undergone creep,

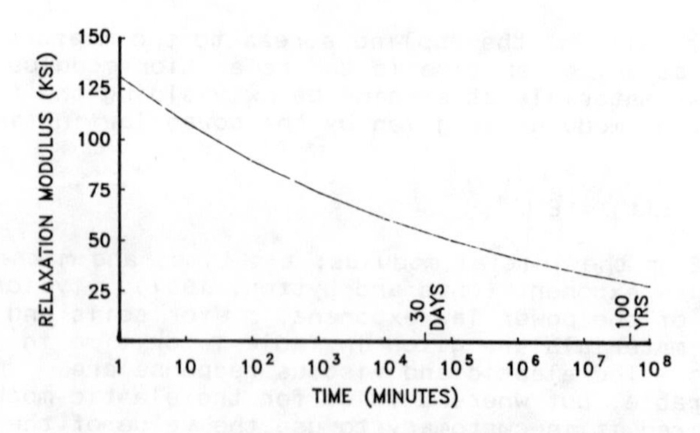

FIG. 4. STRESS RELAXATION MODULUS VS. TIME FOR HDPE

additional applications of load are also met with the material's initial modulus.

Creep deformation can continue forever, but the total creep deformation that occurs after construction is typically small. For example, after construction most buildings settle due to soil creep, but rarely does this cause distress. The same is true for most buried pipe. The stress relaxation curve for HDPE in Fig. 4 shows that a large portion of the total creep deformation for HDPE structures occurs within a few weeks after initial loading. To illustrate this, the creep deformation for HDPE in terms of percent of the total deformation occurring in 100 years under an applied load is given for various time intervals in Table 2.

TABLE 2. HDPE VISCOELASTIC DEFORMATION VERSUS TIME

Time after Load Application	Percent of 100 yr. Deformation
1 day	68%
1 week	76%
1 month	81%
1 year	89%
10 years	95%
100 years	100%

Table 2 shows that 30 days after load is applied, 81% of the creep deformation that will occur in the first 100 years has occurred. Were the pipe's creep to control the time-dependent deflection, an HDPE pipe having 5% deflection at 30 days would deflect only an additional 1.2% in the next 99 years and 11 months. Hence, on a pipeline project taking months to complete a large portion of the creep deformation in the pipe will have occurred by the end of the project.

VISCOELASTICITY OF PIPE-SOIL SYSTEM

The application of load to a pipe-soil system initiates a reaction/counteraction effect between soil and pipe. The specific response of the system is determined by the viscoelastic properties of soil and pipe. The following sections will show that for a flexible, HDPE pipe, installed in granular embedment, the pipe gradually deforms under the soil load until there is an accompanying relief of load on the pipe. This in turn decreases the pipe's reaction until a balance is reached between the soil acting on the pipe and the pipe trying to creep away. Thus, the pipe will shed load due to the viscoelastic response.

The viscoelastic behavior of a pipe-soil system has been mathematically analyzed by modifying the existing methods used for calculating the elastic response (Chua and Lytton, 1987a). Chua and Lytton (1989) have transformed Hoeg's (1968) elastic solution for rigid and flexible pipe to a viscoelastic solution using the correspondence principle. They applied the same transformation to a factorial analysis of CANDE (Katona et al., 1976) to obtain a set of design equations called TAMPIPE (Chua and Lytton, 1987b). TAMPIPE closely matches CANDE's predictions and can be used to determine time-dependent behavior of flexible pipe.

LOADS ON RIGID AND FLEXIBLE PIPE

The terms "rigid" and "flexible" are somewhat arbitrary when applied to pipe. Generally, rigid pipe are pipe that undergo only minute deformations under soil load. Pipe such as clay and concrete are considered rigid. Rigid pipe are usually much stiffer than the soil in which they are installed. Flexible pipe deflect under load and are usually less stiff or only slightly stiffer than the surrounding soil. Plastic pipe are commonly thought of as flexible pipe, but whether a pipe behaves as a rigid or as a flexible pipe depends on the relative stiffness between the soil and the pipe. For instance, the pressure distribution around a plastic pipe installed in very soft ground might resemble that normally associated with a rigid pipe. Another example is thin wall concrete pipe, which can be considered flexible when its stiffness under load is near that of the surrounding soil.

The load that ultimately reaches a buried pipe depends on the relative flexibility between the pipe and the soil. The trench load consists of the dead load weight of the backfill and any surcharge. What portion of the trench load that reaches the pipe embedment zone or a horizontal plane at the pipe crown depends on the shear strength of the soil, its stiffness and the trench width. The trench load reaching the pipe embedment zone is proportioned between the pipe and the embedding soil based on the relative stiffness between pipe and soil. Where pipe and soil are of equal stiffness, the trench load is spread uniformly over pipe and soil. Where the pipe is less stiff than the soil, which is usually the case with flexible pipe, the pipe carries proportionately less load. This is a consequence of the pipe's deflection and the internal shear resistance of the soil. As the pipe deflects the soil above the pipe displaces downward to accommodate the deflection. Internal shearing stresses within the soil develop and resist this downward movement, thus reducing the amount of load reaching the pipe. The load remaining in the soil above the pipe is shunted around the pipe into the soil beside the pipe. Consequently, the soil acts as an arch and carries a larger load than the pipe.

Arching of soil load around flexible pipe is well documented. For instance, Lefebvre et al. (1976) showed the occurrence of arching by measuring the variations in soil pressure around a flexible culvert. Arching can occur in all soils that have an angle of internal friction greater than zero. This includes all granular soils and most fine grain soils in the drained state. Arching comes about by the grain-to-grain contact of the soil particles. It is a form of shear resistance and is as stable and permanent as other forms of shear resistance. For example, footings founded on sand are supported by the sand's shear resistance.

Where the pipe is stiffer than the surrounding soil, reverse arching occurs. This often happens with rigid pipe because it undergoes only small deformations when loaded. These deformations are often smaller than those required to mobilize the shear resistance of the embedding soil. As the sidefill or soil beside the pipe deforms, part of the load from the column of soil directly above the sidefill is transmitted by soil shearing stresses to the column of soil directly above the pipe. This increases the downward force applied to the pipe and the rigid pipe ends up with more load acting on it than the prism of soil above it. This condition is also referred to as negative arching.

Viscoelastic behavior of soils and pipe promote arching. If the pipe creeps under load faster than the surrounding soil, the soil picks up some of that load through arching. There follows a decrease of stress with time in the pipe. Conversely, if the soil creeps faster than the pipe, the stress will increase with time. To illustrate this, the change of bending moment in the pipe wall with time was found for a 48" HDPE profile pipe under two different backfill conditions using the aforementioned TAMPIPE method. See Fig. 5.

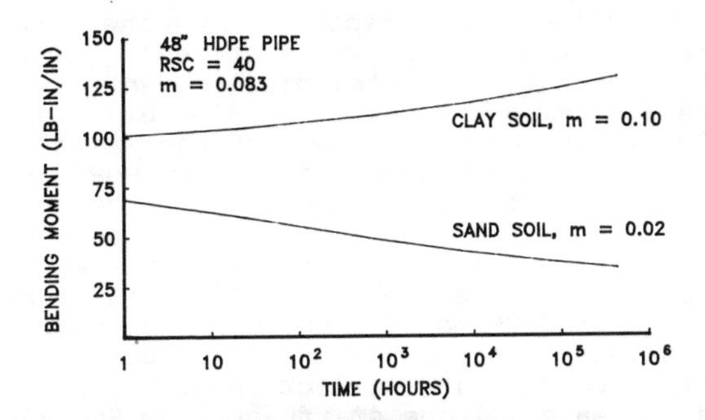

FIG. 5. CROWN BENDING MOMENT VS. TIME FOR HDPE PIPE IN CLAY AND IN SAND

The two backfill conditions were (1) sand backfill and (2) clay backfill. The sand was modelled with a power law exponent equal to 0.02, while the clay model had a power law exponent equal to 0.10. The pipe's exponent of 0.083 falls in between that of the two soils. The stress in the pipe is shown to decrease with time for the sand backfill, which undergoes less creep than the pipe. Conversely, the pipe sees increasing stress when installed in the clay backfill, which creeps at a faster rate. McVay and Papadopoulos (1986) produced similar results with centrifuge tests. They reported that viscous consolidation of a silty clay increased the load on a long-span, metal culvert with time.

Likewise, stress acting on rigid pipe may increase with time due to viscoelasticity. Havell and Keeney (1976) have reported significant increases in trench load with time on buried clay pipe. On the other hand Bacher et al. (1984) reported "negligible" increases with time in the effective soil density around concrete pipe buried in deep embankments. Examination of the figures from Bacher's report shows several cases where slight increases were observed. These increases were on the same order of magnitude as the increases in radial pressure predicted by Chua and Lytton (1989) using a viscoelastic version of Hoeg's equation to calculate long term loading on concrete pipe.

Pipe can be designed to take advantage of the viscoelasticity of the soil, if it can be made more flexible than the soil. This can be accomplished with HDPE because it is a ductile material with a high strain capacity (design strain equals 0.042) and thus can deform to relieve stresses.

VISCOELASTICITY AND HDPE PIPE RING DEFLECTION

Ring deflection is characterized by the vertical displacement of the pipe crown relative to the invert with a simultaneous horizontal displacement of the pipe sides at the springline. For flexible pipe, ring deflection is controlled by the combined stiffness of the pipe and the surrounding soil. When HDPE profile pipe or large diameter corrugated metal pipe are installed in compacted embedments, the embedment soil offers significantly more resistance to deflection than the pipe. For example, load-deflection curves calculated by TAMPIPE for 48" HDPE profile pipe show that an increase in soil stiffness from 85% Proctor to 105% Proctor reduces ring deflection much more than a four-fold increase in pipe stiffness from RSC 40 to RSC 160. See Fig. 6. (Although the pipe contributes little to resisting ring deflection, the pipe stiffness should be considered for buckling resistance. For design considerations for buckling of HDPE profile pipe, see Chau, Chua, and Lytton (1989).)

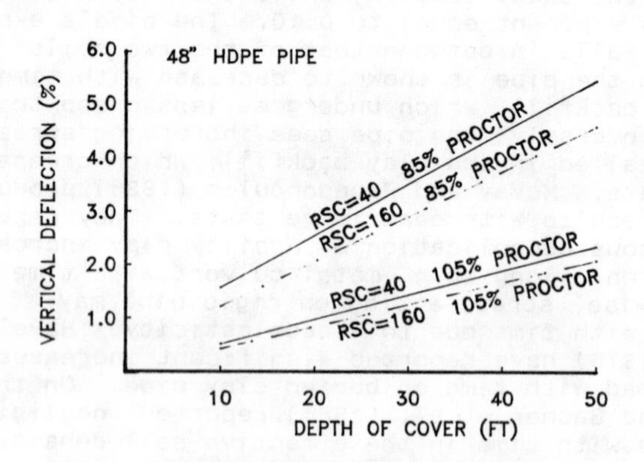

FIG. 6. COMPARISON OF PIPE STIFFNESS AND SOIL STIFFNESS FOR RESISTING DEFLECTION

Where the soil's stiffness, and not the pipe's stiffness, is the major determinant of deflection, stress relaxation of the soil has a significant effect on the pipe's long term ring deflection. This is because changes in deflection due to soil creep are like changes occurring due to a reduction in the soil's stiffness. Therefore, as the soil relaxes the pipe undergoes deflection.

Plastic pipe will undergo long term ring deflection due to the creep of the plastic as well as the soil creep. Where the pipe is installed in a granular embedment this component of deflection is slight, since the pipe contributes little to resisting deflection. As the pipe creeps it can not deflect without displacing the soil, which is stiffer.

Where the soil properties control deflection which is the case for most HDPE pipe-soil systems, the creep rate of the soil virtually determines the rate of deflection of the pipe with time. As shown previously, a large portion of creep deformation occurs within a short time after loading. Table 3 shows the percent of the 100 year deformation at different time intervals occurring in sand with a power law exponent of 0.02.

TABLE 3. VISCOELASTIC DEFORMATION OF SAND VERSUS TIME

Time after Load Application	Percent of 100 yr. Deformation
1 day	45%
1 week	56%
1 month	64%
1 year	77%
10 years	89%
100 years	100%

Chua (1986) has indicated that the power law exponent of granular embedment soil is approximately equal to sand. Therefore, the rate of deformation shown in Table 3 is a first approximation of the time-dependent deflection behavior of most flexible pipe, whether metal or plastic, installed in conditions where the in situ soil, backfill, and embedment are all granular. Where creep occurs according to Table 3 a pipe having 5% deflection at 30 days would deflect an additional 2.8% in a 100 years. For this case, the creep-deflection factor, which is defined as the ratio of the 100 year deflection to the 30 day deflection, equals 1.6.

Where the embedment soil differs from the in situ

soil, the creep rate is determined by the combined effect of the embedment soil, the in situ soil and the trench backfill, which may be remolded in situ soil. For instance, the creep-deflection factor is lower for clayey or silty in situ soils than for granular in situ soils.

The order of magnitude of the creep-deflection factor has been confirmed by field measurements of the vertical deflection of HDPE profile pipe. Deflection data for two projects are plotted versus time in Figure 7 and Figure 8. Each data point represents an average of 30 deflection readings on the 48" pipe and 50 deflection readings on the 60" pipe. Description of the pipe properties and the soil properties for each project are given in the figures. A straight line has been fit to the data on the semilog scale and extrapolated to 100 years. The ratio of the 100 year deflection to the 30 day deflection, or the creep-deflection factor, is 1.5 for the 48" pipe and 1.3 for the 60" pipe. These ratios are of the same order of magnitude as that approximated above for a granular soil and, incidently, close to the lag factor value given by Spangler (1951).

A more precise method to determine the change of deflection with time is TAMPIPE. TAMPIPE considers the embedment zone, the in situ soil, and the backfill soil as separate materials. The TAMPIPE deflection versus time lines are plotted on Figs. 7 and 8. Hyperbolicsoil and power law parameters used in the TAMPIPE analysis are noted on the figures, where me = power law exponent of embedment; mb = power law exponent of backfill and in situ soil; KI = modulus number for in situ soil; KB = modulus number for backfill soil; and KE = modulus number for embedment soil. The deflection values predicted by TAMPIPE's viscoelastic analysis coincide closely with the measured deflections and give 100 year deflection values only slightly higher than those obtained by extrapolation of the data. Two conclusions can be drawn from these figures: (1) the long term deflection observed on the two referenced projects was due to the viscoelastic behavior of soil and pipe, and (2) both the field data and the TAMPIPE calculations show that creep deflections are small and bounded within the service life of HDPE pipe. Thus, as a practical matter, a line that passes a 30 day inspection will not fail by creep within a 100 years.

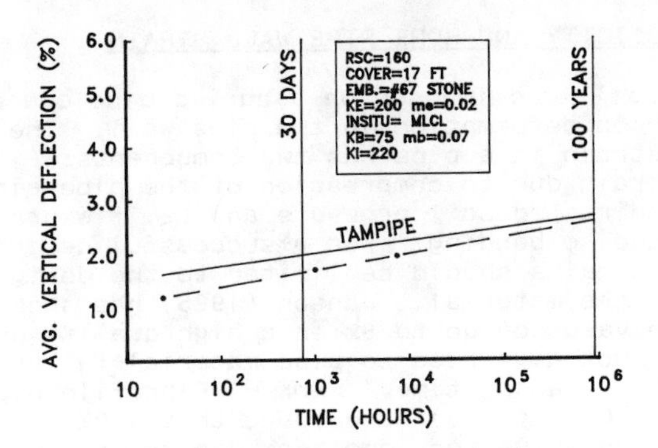

FIG. 7. VERTICAL DEFLECTION VS. TIME 48" HDPE PROFILE PIPE

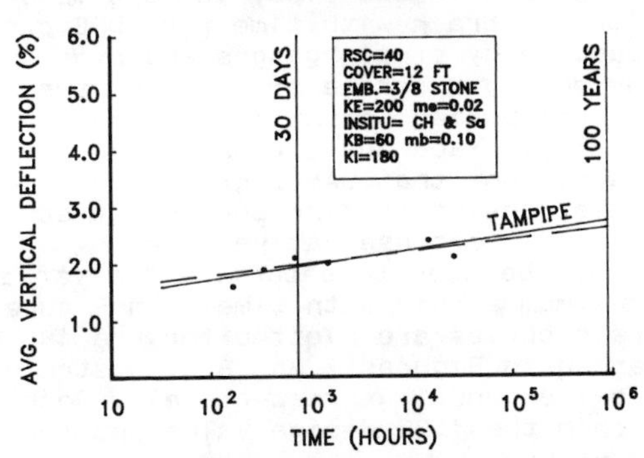

FIG. 8. VERTICAL DEFLECTION VS. TIME 60" HDPE PROFILE PIPE

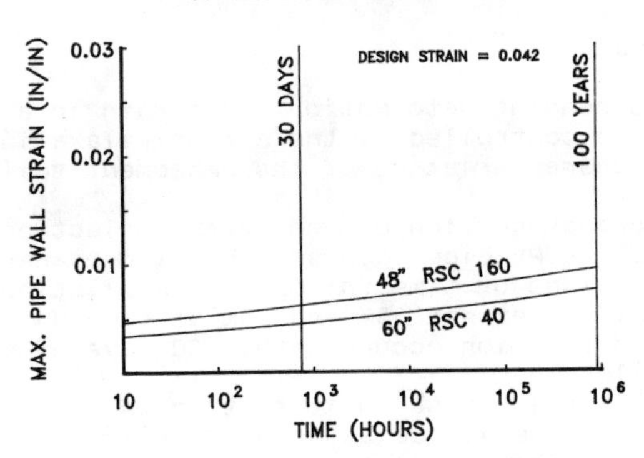

FIG. 9. HDPE MAXIMUM PIPE WALL STRAIN SEE FIG. 7 & 8 FOR PARAMETERS

VISCOELASTICITY AND HDPE PIPE WALL STRAIN

The soil load imposed on a buried pipe causes a ring or hoop deformation in the pipe wall. The hoop or ring strain in a pipe has two components: (1) thrust strain due to compression of the pipe ring by radially directed soil pressure and (2) flexural strains due to bending. For a successful design the combined strains should be limited to the design strain of the material. Janson (1985) has shown that "a strain value of up to 8% in a high quality HDPE pipe will not give rise to pipe material failure prior to 50 years loading time." For HDPE profile pipe Lytton and Chua (1985) have found that 4.2% is a conservative value for long term design strain.

Since the wall strain is dependent on the deflection and load in the pipe, it is time dependent. The variation of strain with time for HDPE profile pipe was measured by strain gauges and by a profilometer during a series of soil box tests conducted at the Buried Structures Laboratory of Utah State University. Yapa and Lytton (1989) used these measurements to show that CANDE predictions and thus TAMPIPE predictions of bending and thrust strains are almost always on the conservative side.

TAMPIPE can be used to determine the variation of pipe wall maximum strain with time. In Figure 9 TAMPIPE strain curves are plotted for the 48" and 60" pipe referenced in Figures 7 and 8. The strain at 100 years equals 1.9% and 1.5% respectively. Both values are well within the 4.2% design value and far below the 8% safe value.

CONCLUSIONS

1. Time-dependent deformation and strain in a buried pipe are controlled by the stress relaxation and creep characteristics of the embedment soil and pipe.
2. The largest portion of long term deflection of a flexible HDPE pipe occurs within a comparatively short amount of time following installation. Typically, between 65% and 80% of the 100 year creep deflection occurs within 30 days after installation.
3. The flexural stress and thrust in the pipe wall depend on the viscoelastic properties of pipe and soil. Whether the load remains constant, increases, or decreases with time depends on the relative values of the relaxation moduli of soil and pipe materials. Typically, the load on rigid

pipe will tend to increase with time. The load on flexible pipe could increase or decrease depending on the relative stiffness between the pipe and the embedment soil. For HDPE pipe, embedded in compacted granular material, the highest flexural stresses occur at completion of installation. There follows a decrease of stress with time.

4. The deflection in flexible pipe continues minutely forever, but the rate at which deflection changes with time is so slight that changes occurring over decades are barely measurable. The rate of deflection change is similar to the rate of long term settlement of buildings founded on sand. For HDPE pipe the consequences of creep deflection are insignificant, when compared to the benefit gained from stress relaxation with its incumbent load reduction.

REFERENCES

Bacher, A. E. (1984). "Design of Thin Wall Reinforced Concrete Pipe Using Dimension Ratio (DR)." Proc., Pipeline Materials and Design, ASCE, San Francisco, Ca.

Chau, M., Chua, K. M., and Lytton, R. L. (1989). "Stability Analysis of Flexible Pipes: A Simplified Biaxial Buckling Equation." 68th Annual Mtg., Transp. Res. Board, Washington, D.C.

Chua, K. M. (1986). "Time-Dependent Interaction of Soil and Flexible Pipe." Dissertation Presented to Texas A&M Univ., ph.D.

Chua, K. M. and Lytton, R. L. (1987a). "A Method of Time-Dependent Analysis Using Elastic Solutions for Nonlinear Materials." Int. J. for Numerical and Analytical Methods in Geomechanics, Vol. 11, 421-431.

Chua, K. M. and Lytton, R. L. (1987b). "A New Method of Time-Dependent Analysis for Interaction of Soil and Large-Diameter Flexible Pipe." 66th Annual Mtg., Transp. Res. Board, Washington, D.C.

Chua, K. M. and Lytton, R. L. (1989) "Viscoelastic Approach to Modeling Performance of Buried Pipes." J. Transp. Engrg. Div., ASCE, 115(3), 253-269.

Frantz, J. H. (1984). "Plastic Pipe Service Life Outlook". Proc., A.G.A. Operating Section, American Gas Association, San Francisco, Ca., 69-72.

Havell, R. F. and Keeney, C. (1976). "Loads on Buried Rigid Conduit-A Ten Year Study." J. Water Pollution Control Federation, Washington, D.C.

Hoeg, K. (1968). "Stresses Against Underground Structural Cylinders." J. Soil Mech. and Found. Engrg. Div., ASCE, 94(1), 833-858.

Janson, L. E. (1985). "Investigation of the Long Term Creep Modulus for Buried Polyethylene Pipes Subjected to Constant Deflection." Proc. International Conf. on Advances in Underground Pipeline Engrg., ASCE, Madison, Wi.

Katona, M. G., et al. (1976). "CANDE-A Modern Approach for the Structural Design and Analysis of Buried Culvert," Report No. FHWA-RD-77-51, Federal Highway Administration, Washington, D.C.

Lefebvre, G. et. al. (1976). "Measurement of Soil Arching above a Large Diameter Flexible Culvert." Canadian Geotech. J., 13.

Lytton, R. L. and Chua, K. M. (1985). "Laboratory Test Report-High Density Polyethylene Pipe Material." Research Report, Texas A&M Univ.

McVay, M. and Papadopoulos, P. (1986). "Long-Term Behavior of Buried Large-Span Culverts." J. Geotech. Engrg. Div., ASCE, 112(4), 424-442.

Mitchell, J. K. (1976). Fundamentals of Soil Behavior. John Wiley & Sons, Inc., New York, N.Y.

Mruk, S. A. (1989). "Plastic Piping Terms and Definitions". Eleventh Plastics Fuel Gas Pipe Symposium, San Francisco, CA.

Schmertmann, J. H. (1970). "Static Cone to Compute Static Settlement over Sand." J. Soil Mech. and Found. Engrg. Div., ASCE, 96(3), 1011-1043.

Spangler, M. G. (1951). Soil Engineering. International Textbook Co., Scranton, Pa.

Yapa, K. A. S. and Lytton, R. L. (1989). "An Analysis of Soil-Box Tests on HDPE Pipes." Res. Report, Texas A&M Univ.

IN-SITU EVALUATION OF 30-IN. (762-mm) DIAMETER PRESTRESSED CONCRETE LINED-CYLINDER PIPELINE

D. M. Schultz[1], R. G. Oesterle[2], J. J. Roller[3]

Abstract

The results of an extensive investigation to evaluate the in-service condition and future serviceability of an existing lake water make-up pipeline are briefly described herein. The pipeline consisted of 30-in. (762-mm) diameter prestressed concrete lined cylinder pipe with a length of approximately 5 miles (8047-m). During the 6-year period since installation, the pipeline had experienced five separate ruptures.

As a result of the ruptures, an in-depth investigation was performed by the authors. Conclusions and recommendations were based on observations and data developed as a result of field investigations, pipe specimen inspections, full-scale performance testing, pipe dissection, material property tests, evaluations of design and manufacturing parameters, and other background information. The net result of the investigation completed in July, 1986 was the complete replacement of the pipeline.

Introduction

An electric power generating plant is located approximately five miles west of Lake Michigan. Makeup water is pumped from Lake Michigan to the power plant

---

[1]Consultant, Construction Technology Laboratories, Inc., 5420 Old Orchard Road, Skokie, Illinois 60077.
[2]Project Manager, Structural Engineering Department, Construction Technology Laboratories, Inc., 5420 Old Orchard Road, Skokie, Illinois 60077.
[3]Engineer, Structural Development Section, Construction Technology Laboratories, Inc., 5420 Old Orchard Road, Skokie, Illinois 60077.

to replace water lost by evaporation and blowdown at the cooling tower. The lake water makeup line consists of 30-in. (762-mm) diameter prestressed concrete lined-cylinder pipe designed in accordance with AWWA C301-72 running in an east-west direction. Figure 1 illustrates typical pipe wall construction of the prestressed concrete lined-cylinder pipe. Figure 1 also shows the proprietary joint protection material used for the pipeline. This material consisted of a continuous polyurethane foam loop impregnated with portland cement. Table 1 gives the pipe component dimensions. Four alternate designs were provided using No. 6 or No. 8 gauge wire and Class III or Class IV wire.

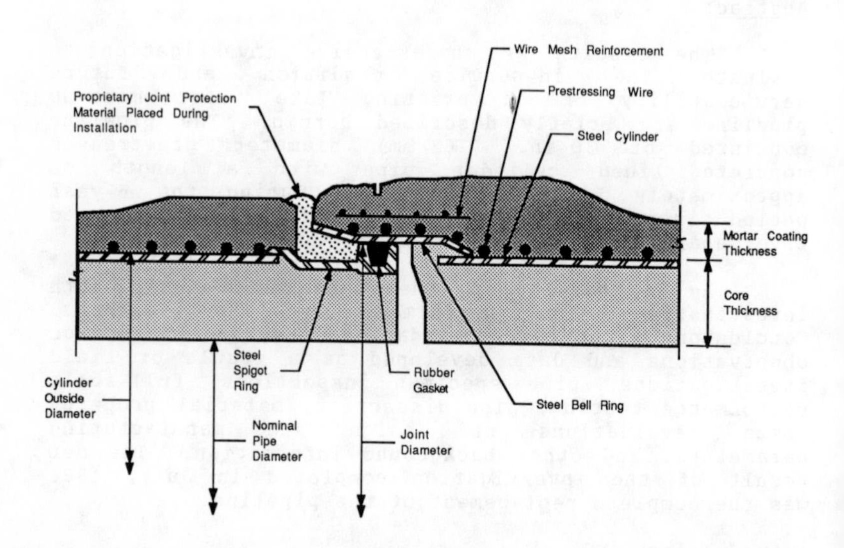

Figure 1. Wall Construction of the Prestressed Concrete Lined Cylinder Pipe

The specified normal working pressure was 160 psi (1103 kPa) and the specified working plus surge pressure was 300 psi (2069 kPa). Also specified was a test pressure of 300 psi (2069 kPa). The required design cover was 15 ft (4.57 m). Because of the high surge pressure specified, the design pressure class rating for the pipe was 215 psi (1482 kPa). The design P_o was 268 psi (1848 kPa).

Since the power plant went into commercial operation in July, 1980, five ruptures of this pipeline

Table 1 Dimensions of Pipe Components as Provided in the Manufacturer's Submittal Sheets

Component	Dimensions
Nominal Diameter	30 in.
Nominal Core Thickness (including cylinder)	1-7/8 in.
Minimum Mortar Coating Thickness over Steel Cylinder	13/16 in.
Minimum Cylinder Thickness	0.0478 in.
Nominal Outside Cylinder Diameter	33.75 in.
Prestressing Wire Size (Class III or Class IV)	No 6 or No. 8 Gauge
Wire Area (A_s) (Class III)	0.395 or 0.381 in.^{2}/ft
Number of Wire Wraps (Class III)	13.62 or 18.51 per foot
Wire Area (A_s) (Class IV)	0.352 or 0.339 in.^{2}/ft
Number of Wire Wraps (Class IV)	12.15 or 16.47 per foot
Zinc Coated Joint Rings: Spigot Ring (width) Bell Ring (thickness x length)	 4-1/2 in. 3/16x4-1/2 in.
Joint Depth	4-1/4 in.
Average Laid Length	20.02 ft.

1 in. = 25.4 mm; 1 in.^{2} = 645.16 mm^2

had occurred. The dates of the five ruptures were July 31, 1983; August 6, 1983; June 17, 1984; October 2, 1984; and October 15, 1984. The ruptures occurred at random locations along the entire length of the pipeline, with each causing shutdown of the pipeline for at least a day.

The ruptures began three years after initiation of service with all five occurring with a 15-month period. This suggested a failure mechanism initiating as soon as the pipeline was put in service. As a result of concern by the owner, an investigation was performed by the authors to evaluate the future serviceability of the existing lake water makeup pipeline.

Details of Investigation

Eight full-length sections and three half-length sections of lined-cylinder prestressed concrete pipe were removed from the makeup pipeline between the power plant and Lake Michigan. Removal of the sections required a near complete shutdown of the power plant. The locations were selected to sample as many field conditions as possible given that only three excavations could take place in the 72 hour shutdown period allowed. Each of the selected pipe sections was subject to an in-place inspection prior to removal. During this field inspection, trench conditions, pipe joint condition, pipe mortar coating cracks, and pipe mortar coating defects were documented. Soil conditions were examined and soils were sampled at each excavation. After inspection and removal, the pipe sections were transported to CTL for testing and evaluation.

Upon arrival at CTL, all eight full-length pipe sections were thoroughly cleaned and inspected once again to establish if any damage had occurred in handling and transport. Four of the eight full-length pipe sections were tested under internal hydrostatic pressure to evaluate mortar coating crack performance and to compare actual pressure-strain relationships to those assumed in design. One of the eight full-length pipe sections was tested under the application of a three-edge bearing load to determine W_o. Pipe performance during testing was assessed based on evaluation of mortar coating cracking, measured pressure versus wire strain relationships, and measured three-edge bearing load data using performance criteria developed within the prestressed concrete pressure pipe industry and accepted standard specifications.

All five of the tested pipe sections, one of the three remaining full-length pipe sections, and one of the three half-length pipe sections were dissected and pipe components were examined. Material properties of core concrete, mortar coating, prestressing wire, and cylinder steel were obtained for the five tested pipe sections and compared to design assumptions and specification requirements. Testing of prestressing wire included bend and split tests. Petrographic examination of core concrete and mortar coating was also conducted for the five tested pipe sections. The single untested full-length pipe section and the single half-length pipe section were only partially dissected to examine prestressing wire beneath existing cracks at the spigot end. A petrographic examination of a crack surface and the proprietary joint filler was performed on samples taken from the half-length pipe section.

Material properties of core concrete, mortar coating, prestressing wire, and cylinder steel were also obtained from pipes stored in a warehouse at the power plant. Among these pipes were the remains of four of the five pipe sections that had previously failed in the field. These four failed pipe sections were thoroughly inspected to determine the probable causes of the failures. The fifth failed pipe section was not available for inspection. Petrographic examination of core concrete and mortar coating was also conducted on selected samples from these pipe sections.

Results of pipe tests, dissections, inspections, material property tests, and material examinations indicated that manufacturing defects due to inadequate quality control had been the major cause of the poor pipe performance observed. Results of soils analyses indicated that the environment played no unusual accelerating role in the observed deterioration, nor did pipe bedding or surface loads.

Results of the pipe tests indicated that, in general, the pipe was adequate for the design conditions used by the manufacturer. These design conditions included a design pressure class rating of 215 psi (1482 kPa) and a working plus surge pressure of 300 psi (2069 kPa). However, less than adequate pipe performance for these design conditions was observed during one of the pipe tests in an area which was later found to incorporate a cylinder dent. Also, measured performance in local areas of the pipe wall of two other tested pipe sections did not correspond to the expected design performance. These local areas of the pipe wall were later found to also incorporate cylinder dents.

Dissections performed on pipe sections revealed defects as well as instances where the different pipe components did not conform to the industry standard specification. These included brittle wire fractures on shiny wire, splitting wire, cylinder dents with depth in excess of 1-in. and with widths in excess of 12-in., kinks in prestressing wire over dents and cylinder weld seams, mortar coating thickness and core concrete thickness which did not meet minimum requirements, foreign debris including wood and mud on cylinder steel beneath wire wraps, cylinder weld seam leaks, and excessive out-of-roundness of the cylinder. In many cases, the pipe sections did not conform to the manufacturer's own quality assurance guidelines. These defects were found on both the exhumed pipes and the pipes that had ruptured while the pipeline was in service. A finding of particular significance but unrelated to manufacturing was the widespread end mortar coating cracking and wire corrosion near the pipe ends. This deterioration was observed on the majority of the dissected pipe.

Results of the petrographic examinations indicated that a long-term durability problem existed in the pipeline. Mortar coating samples taken from pipe sections were found to incorporate a porous layer at the level of the prestressing wire. At the level of this porous zone, the mortar coating did not provide a dense, durable encasement for the prestressing wires. Paste in this porous zone was partly carbonated, indicating that passage of water and/or air could occur. The examinations also indicated that calcite deposits existed along the surface of many of the mortar coating cracks found on pipe specimen ends during the field investigation. These calcite deposits indicated that mortar coating cracks had been allowing percolation of water for some time. Petrographic examination of the proprietary joint filler sample indicated that this material was porous and allowed percolation of water through the material.

As indicated by the end cracking and serious deterioration of the prestressing wire at pipe ends, the proprietary joint filler used for this pipeline did not adequately seal the bell and spigot ends of the pipe sections from intrusion of water and any deleterious material carried by water.

Probable Failure Modes

The following probable failure modes were developed based on data obtained from field and laboratory, and analytical investigations carried out by CTL. Failure herein is defined as a condition of the pipe which is leading rapidly to a loss of adequate or intended function and/or rupture of the pipe. Two broad categories of failure modes were identified. They were 1) failures associated with cylinder dents and 2) failures associated with mortar coating end cracking of the pipes.

Cylinder Dents

Based on inspection of the remains of the ruptured pipe sections, three of the five were directly related to the presence of dents that existed in the steel cylinder at time of pipe manufacture. The remains of the core concrete from the failed area of these three pipe indicated excessively large depressed areas resulting from dents present in the steel cylinder at the time of casting of the core concrete as indicated in Fig. 2. The core thickness variation at the depressions was extremely large. The minimum measured core thickness in the largest depression was one inch less than the designed core thickness of 1-7/8 in. (47.6-mm). Each of these failed pipe sections had a weld seam running along the edge or through the center of the depressed area in the core concrete. Remains of the core concrete from the pipe that failed on July 31, 1983 were not available for inspection by CTL. However, observations of the remaining steel cylinder indicated that the rupture was instigated by a loss of water tightness in the steel cylinder at a

Figure 2. Spigot End of Failed Pipe Section Found on October 2, 1984 with Large Depression and Cracking in the Core Concrete

point near a weld seam. Remains of the pipe that failed on June 17, 1984 were not available for inspection by CTL.

There were two possible causes for the ruptures associated with these large dents in the steel cylinders. Both start with the following progression of events.

1. Dents occur in the steel cylinder from handling during manufacture, probably after the hydrostatic test but before the centrifugal casting of the core concrete. Denting of the thin cylinder was recognized as a quality control problem by the manufacturer, and dents in excess of 1/8-in. deep over an area greater than 2 in. x 2 in. were to have been repaired.
2. Because of the increased relative stiffness of the cylinder at the location of the weld seams, the weld seams in a dented region were located along the edge of a dent. The high local bending in the steel around the edge of a dented region damaged the weld seam.
3. The dents in the steel cylinder resulted in local reduction of concrete core thickness.
4. With the reduced core thickness, the concrete was overstressed at the time of wrapping resulting in increased elastic and creep strains in the core concrete. These increased strains caused a local region in the core with a low value for the decompression pressure, P_o.
5. With pressurization of the pipe during operation, the precompression in the core concrete at these local regions having a low P_o was overcome. Therefore, the core concrete cracked and water was allowed to pass through the cracked core concrete. At this stage, a thinner wall section consisting of the steel cylinder, prestressing wire, and mortar coating became the pressure resisting element. Local bending would occur in this thinner wall section within the region of the depressed area after the prestress in the core concrete was overcome.
6. From this stage on, one of two scenarios lead to failure of the pipe. The first scenario considered that the leak tightness integrity of the cylinder steel was disrupted by flexing of the cylinder steel in the vicinity of weld seams. This could occur at initial pressurization of the pipeline. With the start of a leak through

the steel cylinder, and water flowing from the inside leaking outward under pressure, corrosion of the prestressing wire started and progressed until fracture of the wires occurred. As wires fractured, the cylinder steel yielded and bulged outward until a general fracture of the cylinder steel occurred resulting in burst of the pipe section.

7. The second scenario considered local cracking of the mortar coating in the vicinity of the depressions associated with flexing of this region. With the mortar coating cracked, ground water could have penetrated from the outside-inward and induced corrosion in the prestressing wire and cylinder steel. Local corrosion of the cylinder steel may have disrupted the leak tightness integrity of the cylinder thereby providing an additional source of water for wire corrosion. Wire corrosion and eventual fracture of the wire again lead to a burst of the pipe section.
8. Both scenarios for failure were hastened by the fact that there were high local strains in the "kinked" regions of the wire around dent locations and that the wire used in the manufacture of these pipe exhibited splitting and a potential for brittle fractures.

Observations by CTL on the remains of the ruptured pipe included erosion of the steel cylinder around the apparent starting points of the cylinder failure and the presence of corrosion products in a band completely around the pipe on the inside surface of the steel cylinder and the outer surface of the core concrete. These observations indicated a "loss of water-tightness" of the steel cylinder located near a weld seam prior to burst of the pipe section. It cannot be discerned from the evidence examined which of the two scenarios is the actual failure scenario. The primary conclusion, however, is that the failures resulted from the presence of cylinder dents.

The dents associated with the early-age bursts which had occurred to date were large, randomly located dents in the steel cylinder. Large random dents found in the pipe specimens dissected by CTL, some of which had been pounded out, indicated that these dents occurred with a relatively high frequency. It was the authors' opinion that, there were a number of large dented regions in the existing pipeline still in service. It was probable that the pipe wall in the region of each of these dents was undergoing a process of deterioration that would have eventually lead to failure. It was not possible, with available data, to predict the timing nor the number of failures to expect in the future.

In addition to the occurrence of large random dents, observations of pipe specimens inspected by CTL indicated that there was a large number of "girth band" dents present in the existing pipeline. A girth band was used in manufacturing these pipe and was clamped around the circumference of the cylinder midway between joint rings to add stiffness to the cylinder prior to centrifugal casting of the core concrete. Girth band dents did not produce the extremely low thickness of core concrete in the depressed areas associated with the observed large random dents as illustrated in Fig. 3. However, during one of the pressure tests strains measured in the prestressing wire wrapped over one of the girth band dents demonstrated a low measured P_o. A low P_o is associated with the potential for premature cracking of the mortar coating. With the integrity of the corrosion protection system disrupted by the local cracking of the mortar coating, it was possible that, in time, the presence of these girth band dents would have lead to a large number of pipe failures. Similar to the scenarios for failure from the large random dents, "kinking" of the prestressing wire producing high local strains at the dent location and the tendency towards splitting and a potential for brittle fracture of the prestressing wire, hastened the probability for eventual failure resulting from the presence of the girth band dents. The use of this type of prestressing wire was defined as an extreme accelerator to the deterioration mechanism. It was not possible, with the available data and scope of work carried out by CTL in this investigation, to quantify the potential for failure resulting from girth band dents.

Mortar Coating End Cracking

In evaluating the mortar coating end cracking, two basic scenarios were considered. The first scenario considered that the mortar coating cracks existed prior to wire corrosion and corrosion of the prestressing wire resulted from external penetration of solutions through the cracks. The second scenario considered that the wire corrosion began as a result of some other means and the mortar coating end cracks were a result of the wire corrosion. Each of these scenarios are discussed below.

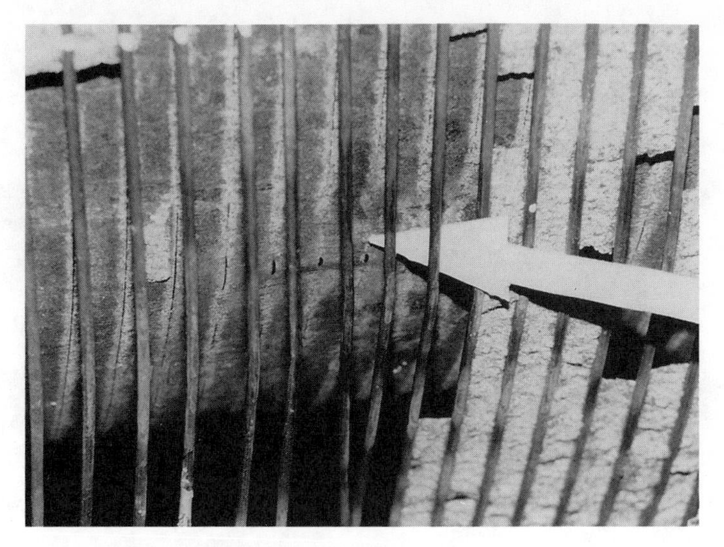

Figure 3. Girth Band Dent in Cylinder Steel. Note "Kinked" Prestressing Wires that Bridged Across the Dent

End Cracking Prior to Wire Corrosion

It was known by the manufacturer that minor mortar coating cracks at the ends of lined-cylinder pipe can occur. These were not considered detrimental by the manufacturer. Several pipe with cracks in the mortar coating at the ends of the pipe were found during the field investigation, as indicated in Figs. 4 and 5. Since the proprietary joint filler did not seal these cracks, it was most probable that solutions penetrated these existing cracks to the level of the prestressing wire to varying degrees along the pipeline. This, in turn, caused corrosion of the prestressing wire, followed by further deterioration of the mortar coating, as indicated in Fig. 6.

Wire Corrosion Prior to End Cracks

The dissections and petrographic analyses determined that a weakened zone of high porosity existed in the mortar coating at mid-level of the prestressing wire. This zone is a direct result of the mortar coating application technique used by industry, whereby mortar is applied by spray-up from spinning brushes while rotating the pipe. At this level, the

Figure 4. Mortar Coating Crack Found at the Spigot End of Pipe Section 3+68

Figure 5. Mortar Coating Cracks Found at Bell End of Pipe Section 95+14

Figure 6. Prestressing Wire Corrosion Beneath Mortar Coating Cracks at End of Pipe

mortar coating did not provide a dense, durable encasement as specified in AWWA C301-72 for the particular wire used by this manufacturer. Paste in this porous zone was partly carbonated, indicating the passage of water and/or air through the mortar coating at this level was possible.

The proprietary joint filler used during the installation of this pipeline did not prevent solutions from penetrating to the joint rings or to the observed delamination between the joint rings and the mortar coating, as shown in Fig. 7. This occurred either as a result of poor contact between the irregularities in the edge of mortar coating or as a result of the porosity of the joint material itself. As a result, solutions gained access to this porous zone in the mortar coating at mid-level of the prestressing at the pipe ends. This led to corrosion of the prestressing steel and subsequent deterioration of the mortar coating in the form of longitudinal cracking due to expanding corrosion products.

Considering the difficulty of checking for looped O-ring gaskets during installation as a result of the use of the proprietary joint filler and recognizing this pipeline also had a history of looped

Figure 7. Delamination Between Spigot Joint Ring and Mortar Coating

gasket failures, internal water may also have gained access to the porous zone as a result of gasket leaks.

Extent of Deterioration

The end cracking failure modes described most probably had occurred in conjunction with one another. The net results were 1) significant deterioration had already occurred at the ends of a large number of pipe sections in the pipeline; 2) deterioration was continuing with time; and 3) deterioration to varying degrees was probable in a large number of pipe sections in the pipeline. The use of wire which was prone to splitting and embrittlement was defined as an extreme accelerator to the deterioration mechanism. Conditions where the pipeline was near or below the groundwater table were also accelerating the deterioration.

Conclusions and Recommendations

Numerous manufacturing flaws and significant end deterioration were found on all of the pipe sections examined in this investigation. Based on these findings, it was deemed probable that there were a number of pipe sections still existing in the pipeline that had significant manufacturing flaws. The pipe sections were undergoing significant deterioration either due to the defects or due to conditions at the ends of the pipes specifically related to the use of the proprietary joint material. These deteriorating pipe sections had a potential for sudden failure at any time. Considering that this pipeline was known to be buried in close proximity to populated areas, and that continued operation of the pipeline at any pressure could produce a sudden failure, the 30-in. (762-mm) diameter makeup pipeline providing Lake Michigan water to the power plant was taken out of service and replaced.

Final Remarks

There are many steps that can be taken to help ensure the long-term performance of a pipeline. Obviously, it is important to start with a high-quality product. Proper design and selection of materials are the primary factors affecting long-term serviceability. Quality control in manufacture and installation is also essential. If a pipe is manufactured with poor quality control or is damaged by improper handling or installation, its service life may be significantly reduced. Like all structural components, concrete pressure pipe must be properly designed, manufactured, and installed in accordance with specific criteria to help ensure long-term durability.

It may be necessary to conduct periodic in-service inspection of concrete pressure pipe. Localized excavations can be performed to inspect the exterior of the pipe for cracks and deterioration. Visual inspection of the pipe interior can be done during shutdowns in areas accessible by manholes.

When significant deterioration is discovered, it is important to conduct a more thorough investigation to determine whether the deterioration is an isolated incident or an event that could jeopardize the entire pipeline.

COMBINED LOAD TESTING
OF PRESTRESSED CONCRETE CYLINDER PIPE

By Armand W. Tremblay [1], Member, ASCE

Abstract

A series of combined load tests were run on prestressed concrete cylinder pipe in order to evaluate a proposed new method of design for pipe made to AWWA Standard C301. These tests involved the simultaneous application of external loads and internal pressures in various combinations. The test apparatus was designed and constructed with several unique features, which permitted visual observation of the interior pipe surface at the invert and measurement of vertical deflection during the test. In addition, strain gauges were used to monitor strain levels in the concrete core, the mortar coating, the steel cylinder, and the prestress wire.

Introduction

Recently, the firm of Simpson, Gumpertz & Heger (SGH), in conjunction with the American Concrete Pressure Pipe Association (ACPPA), developed a Unified Design Procedure (UDP) for prestressed concrete cylinder pipe (PCCP). Manufacture and design of this pipe is currently covered by American Water Works Association (AWWA) Standard C301.

Development of a new design method for this type of pipe involved extensive review of previous tests of the product and its constituent materials, previous research work by noted authorities, and state-of-the-art design procedures for other prestressed concrete structures. Results of this background research and the development of the design criteria for the proposed UDP have been presented earlier in various technical forums. This paper will present the unique methods used to conduct a series of tests on PCCP to verify the proposed design method.

----

[1]Manager of Engineering and Technical Services, Pressure Pipe Division, Price Brothers Company, P.O. Box 825, Dayton, Ohio 45401

Test Criteria

As the development of the design method neared completion, SGH recommended a series of combined load tests. These tests involved the simultaneous application of external load and internal pressure to duplicate the combination of circumferential bending moments and thrusts to which a buried pressure pipe is subjected in service. In this case, the external load was applied by a three-edge test machine. The criteria established by SGH for this test program were:

1. One size of lined cylinder pipe (LCP) and one size of embedded cylinder pipe (ECP) to be tested.
2. Three different strength classes of each pipe size to be tested.
3. Test pipe to be 2-3 months old when tested.
4. Test sections to be at least 9 feet (2.7M) long.
5. Each of the four components of the pipe (prestress wire, concrete core, steel cylinder, and mortar coating) to be strain gauged for monitoring strain changes during testing.
6. Test equipment to be designed to supply negligible support to the pipe.
7. Test equipment to allow for visual observation of the pipe interior at the invert throughout the test.
8. Test equipment to have a means of measuring pipe deflection during the test.
9. Test equipment to safely allow internal pressures up to 500 psi (3445 kPa).

Test Pipe

The test program was conducted by Price Brothers Company under the sponsorship of ACPPA and consisted of three pieces of 48-inch (1220 mm) LCP and three pieces of 60-inch (1524 mm) ECP. The 48-inch (1220 mm) LCP, shown in Figure 1, had a 16-gauge (.06 inches; 1.5 mm)

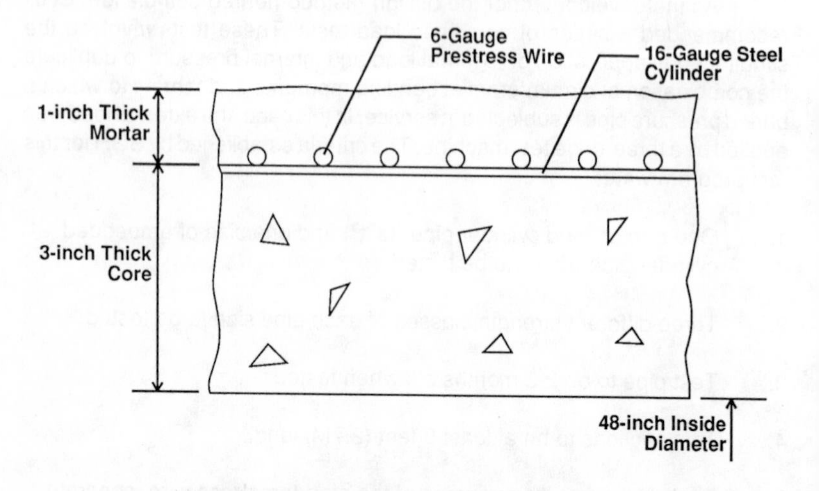

Figure 1. Cross section of 48-inch LCP

steel cylinder, a 3-inch (76 mm) centrifugally placed concrete core lining the cylinder, a uniformly spaced wrapping of 6-gauge (0.192 inch diameter; 4.9 mm) prestress wire and a 1-inch (25 mm) exterior mortar coating. The 60 inch (1524 mm) ECP, shown in Figure 2, also had a 16-gauge (0.06 inches; 1.5

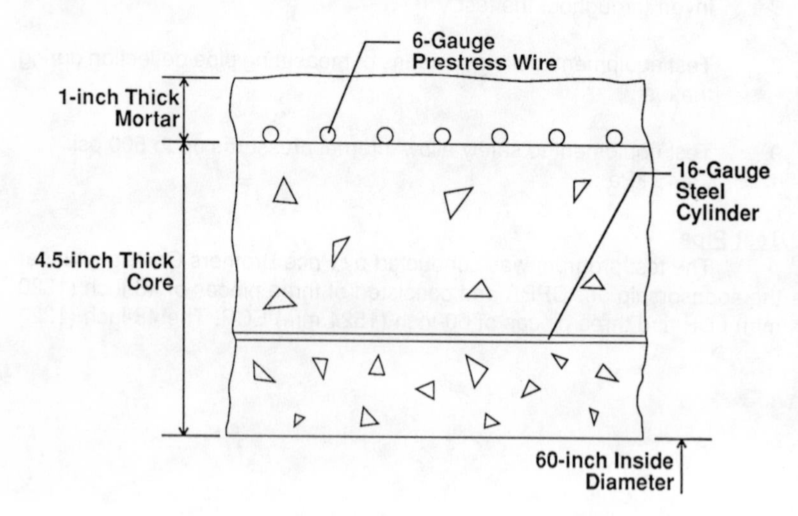

Figure 2. Cross Section of 60-inch ECP

mm) steel cylinder, but the cylinder for this pipe is embedded in a vertically cast 4-1/2-inch (114 mm) thick core. The core is wrapped with prestress wire and mortar coated in the same manner as the LCP.

Test Equipment

In order to allow observation of the pipe invert during the test, a displacement mandrel system was designed with two 6-inch (152 mm) x 12-inch (305 mm) observation windows. The 500-psi (3445 kPa) pressure required these windows to be made with 2-inch (51 mm) thick plexiglass.

End seals between each end of the mandrel and the pipe had to be water tight up to 500 psi (3445 kPa), but transmit negligible three-edge load from the pipe to the mandrel. Previous combined load tests on PCCP required that a correction factor be used on the applied three-edge load to account for the support provided by the bulkheading arrangement. To eliminate the need for this correction factor, the bulkheads for this test were specially designed. The end of the mandrel referred to as the stationary end is shown schematically in Figure 3. The combination of the large-diameter rubber gasket, the

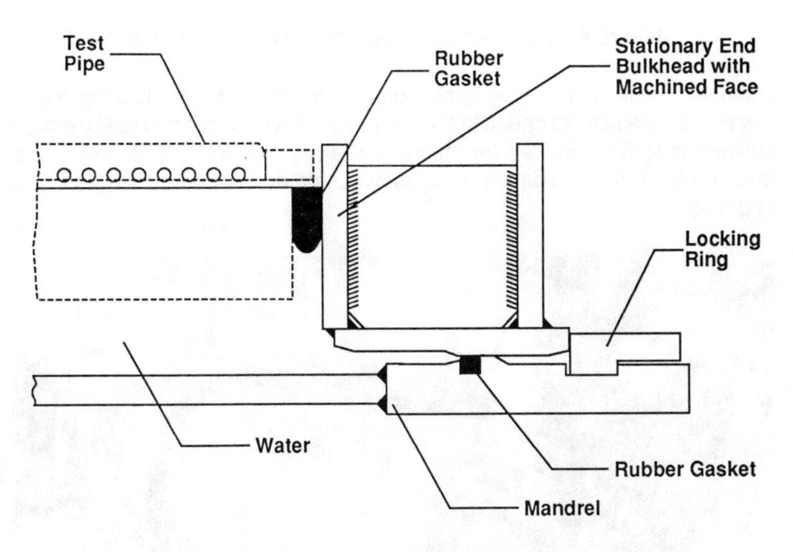

Figure 3. Cross section of stationary end of test mandrel

smoothly machined vertical face of the bulkhead, and a special silicone lubricant transferred less than 125 pounds (556 N) of three-edge load to each end of the test mandrel. An end ring, locked behind the bulkhead, restrained the end thrust which was in excess of 300,000 pounds (1335 kN) at the maximum internal hydrostatic pressure stated in SGH's test criteria. The bulkhead arrangement at the other end of the mandrel, called the piston end, is shown in Figure 4. A hydrostatically pressurized piston built into this

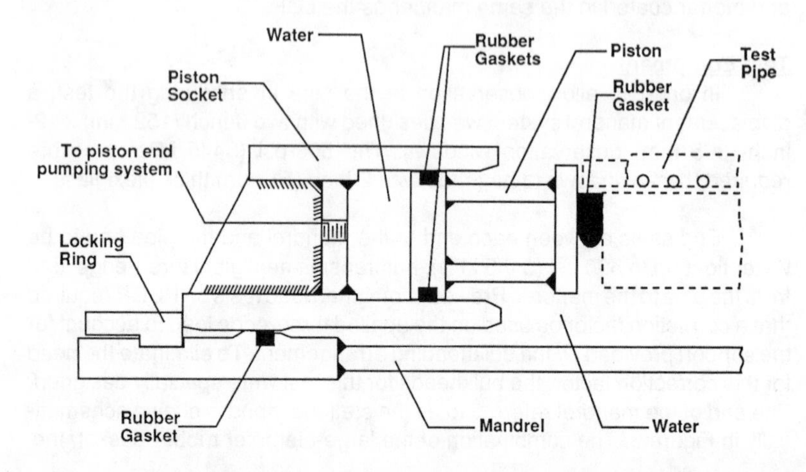

Figure 4. Cross section of piston end of test mandrel

bulkhead applied the sealing force required for both ends. The locking ring at this end, as well as the one at the stationary end, was a split ring to allow both bulkheads to be removed for installing the mandrel into, or removing the mandrel from, a test pipe. Figure 5 shows the mandrel being inserted into a test pipe.

Figure 5. Test mandrel minus end bulkheads being inserted into test pipe

Instrumentation

In order to measure changes in pipe diameter (deflection) during a test, special bolts with spring loaded plungers (see Figure 6) were installed at the

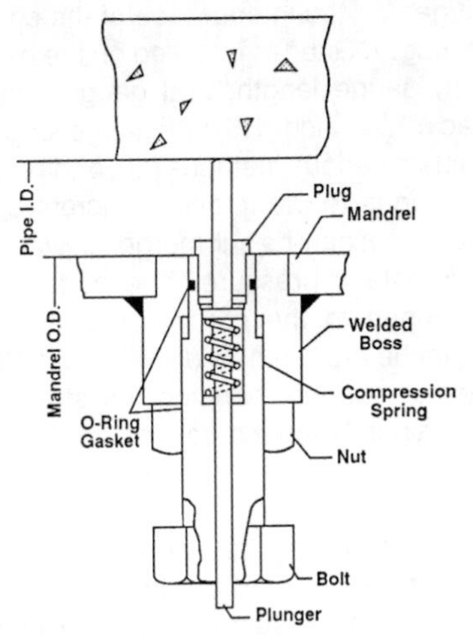

Figure 6. Cross section of deflection indicator

crown and invert of the test mandrel. The outboard ends of the plungers were in contact with the pipe at crown and invert, and the inboard ends were in contact with a rod and attached dial indicator (see Figure 7) inside the test mandrel. Increases or decreases in diameter to the nearest .001 inch (.025 mm) were read on the dial indicator.

Figure 7. Deflection indicator bolt at inside crown of test mandrel connected to dial indicator for measuring vertical deflection

Strain gauging of the pipe components provided one of the biggest challenges, particularly the gauges inside the pipe. More than 30 strain gauges were installed on each of the six test pipe in order to provide various levels of redundancy. Gauges on the interior face of the concrete core had a 4-inch (102 mm) gauge length. Gauges installed on the exterior mortar surface had a 2-inch (51 mm) gauge length, and gauges on the steel cylinder and prestress wire had a 1/16-inch (1.6 mm) gauge length. Laboratory tests were conducted to evaluate various methods for bonding and waterproofing before attaching strain gauges to the interior concrete surfaces of the test pipe. These gauges would not only be submerged in water, but would be subjected to substantial hydrostatic pressure. The method eventually selected for bonding these gauges to the concrete surfaces consisted of abrading aluminum tape 2 inches (51 mm) wide x 4 mils (.10 mm) thick, bonding the tape to the concrete surface, bonding the strain gauge to the tape, and waterproofing with a butyl rubber recommended by the strain gauge manfacturer.

For the LCP, all strain gauges were mounted after the test pipe was completely manufactured. During the mortar coating operation, special blockouts were attached to the prestressing wire to form openings for later installation of strain gauges on the wire and the steel cylinder (see Figure 8).

Figure 8. Blockout in mortar coating of LCP test pipe and strain gauges attached to wire and steel cylinder

For the ECP, it was necessary to install the strain gauges on the steel cylinder in the pipe plant prior to casting the concrete core (see Figure 9). All

Figure 9. Cylinder for ECP test pipe with strain gauges and lead wires attached

other strain gauges for the ECP were installed after the pipe was completely manufactured. Lead wires from the strain gauges on the pipe interior were carried through the pipe wall using a special factory assembled gland unit which provided a high pressure seal around the wires.

Two of these special glands, one near each end of the pipe, were installed in 3/4-inch (19 mm) threaded outlets built into the pipe (see Figure 10).

Figure 10. Special high pressure gland with factory assembled lead wires from strain gauges inside the pipe

Another 3/4-inch (19 mm) outlet in each test pipe was used to install a transducer for measuring the internal pressure inside the pipe. A digital readout unit (see Figure 11) connected to the transducer displayed the pressure to the nearest 1 psi (6.89 kPa). Backup was provided by a standard Bourdon tube-type dial pressure gauge.

Figure 11. Digital readout of internal pressure inside test pipe

The strain gauges were monitored using five switching and balancing units and three strain indicators with digital readouts (see Figure 12). A personal

Figure 12. Equipment and personnel recording strain gauge readings

computer with an electronic spreadsheet was used during each of the tests to input strain and deflection measurements (see Figure 13). Output from this computer was displayed on the screen or plotted by an A/B plotter to monitor trends in deflection and strain gauge readings.

Figure 13. Computer and display screen used to track strain readings during test

Test equipment was designed with two separate pumping systems; one to pressurize the piston end of the test mandrel which activated the end seals and one to pressurize the annular space between the pipe and test mandrel (see Figure 14). The system for pressurizing the pipe also had a subsystem

Figure 14. Separate pumping systems for pressurizing test mandrel bulkhead and test pipe

for circulating the water in the pipe through a filtering unit to insure that the water was kept clear for good visual observation of the invert through the plexiglass windows. A special long focal length microscope was used to measure inside core crack widths once they were observed through the windows. The width of visible cracks in the mortar coating were measured with a standard 40x microscope. Exterior and interior views of a fully instrumented pipe ready for testing are shown in Figures 15 and 16.

Figure 15. Test pipe in 3-edge tester ready for test

Figure 16. Interior of test mandrel showing plexiglass observation windows, long focal length microscope, lighting unit and vertical deflection bar

Test Machine

Three-edge load was applied by a Forney PT-250 test unit capable of testing pipe specimens up to 16 feet (4.9 M) in length and applying external load up to 500,000 pounds (2225 kN). Testing was conducted inside a specially designed and constructed enclosure built around the three-edge tester to provide shelter from the weather (see Figure 17).

Figure 17. Enclosure in which combined load tests were conducted

Test Procedure

During the combined load test, various internal pressure and external load combinations selected by SGH were applied. At each combination, deflections and strains were recorded, and the pipe was carefully inspected for visible cracking in the core and coating (see Figure 18). When cracks did

Figure 18. Dr. Frank Heger & Dr. Mehdi Zarghamee of SGH checking pipe during test

appear, their widths were measured with microscopes and recorded. All tests were conducted at the Price Brothers plant in Perryman, Maryland, and were witnessed by representatives of ACPPA and SGH. The results of these tests will be presented in a separate paper by Dr. Mehdi S. Zarghamee of SGH.

Summary

Combined load tests on the two types of prestressed concrete clyinder pipe commonly used for pressure lines were conducted by the American Concrete Pressure Pipe Association to confirm design theory established by Simpson, Gumpertz & Heger. The end bulkhead arrangement of the test apparatus was designed to eliminate the need to correct the external load. In addition, the pipe components were successfully strain gauged, waterproofed, and monitored even on internal surfaces subjected to hydrostatic pressure. Plexiglass "windows" built into the test frame permitted visual observation of the pipe invert where tensile stress from combined loading is highest. A unique apparatus was also built into the test mandrel to allow vertical deflection to be measured.

Appendix

1. "Combined Internal-External Load Tests on Concrete Pipe", Engineering News Record, February 16, 1956.

2. Heger, F. J., Zarghamee, M. S., and Dana, W. R., "Limit States Design of Prestressed Concrete Cylinder Pipe, Part A: Criteria", submitted for publication to ASCE Journal of Structural Engineering.

3. Kennison, H. F., "Tests of Prestressed Concrete Embedded Cylinder Pipe", ASCE Journal of Hydraulics Division, November, 1960.

4. Zarghamee, M. S., Heger, F. J., and Dana, W. R., "Predicting Prestress Losses In Prestressed Concrete Pressure Pipe", Concrete International, October, 1988.

5. Zarghamee, M. S., Heger, F. J., and Dana, W. R., "Experimental Evaluation of Design Methods for Prestressed Concrete Pipe", ASCE Journal of Transportation Engineering, November, 1988.

6. Zarghamee, M. S. and Dana, W. R., "A Step-By-Step Integration Procedure For Computing State of Stress In Prestressed Concrete Pipe", presented at ACI Convention, Atlanta, Georgia, February, 1989.

7. Zarghamee, M. S., Heger, F. J., and Dana, W. R., "Concrete Creep and Shrinkage and Wire Relaxation In Buried Prestressed Concrete Pipe", accepted for publication by ACI Structural Journal.

INNOVATIVE CONCEPT FOR PRESTRESSED PRESSURE PIPE

A. T. Ciolko[1] and T. E. Northup[2]

Abstract

A new design and manufacturing concept was developed by General Atomics for prestressed pressure pipe. With this new concept, patented in the U.S. and in the major industrialized nations, a cylindrical concrete core is cast against the interior surface of a high-strength sheet steel cylinder. However, instead of achieving prestress by helically wrapping the concrete core/steel cylinder with high-strength steel wire, a cementitious grout is pressure injected into the interface between the concrete core and steel cylinder. The pressurized grout material produces circumferential compressive stress in the concrete core and a circumferential tensile stress in the steel cylinder.

Introduction

This paper will describe a prototype development program (sponsored by General Atomics) undertaken by CTL to develop a commercially viable production process which could be used to manufacture this new type of pressure pipe. In this program, a preliminary prototype pipe was successfully constructed and prestressed. The 4-ft (1.22-m) diameter, 10-ft (3.05-m) long prototype pipe, shown in Fig. 1 consisted of an exterior cylindrical steel membrane fitted with injection inlets, an interior concrete core and special interface end seals. A schematic of the prototype is shown in Fig. 2. Following casting of the concrete

----

[1]Manager, Structural Development Section, Construction Technology Laboratories, Inc., 5420 Old Orchard Road, Skokie, IL
[2]President, General Atomics International Services Corporation, P.O. Box 85608, San Diego, CA

Figure 1. Prototype Pipe After Prestressing

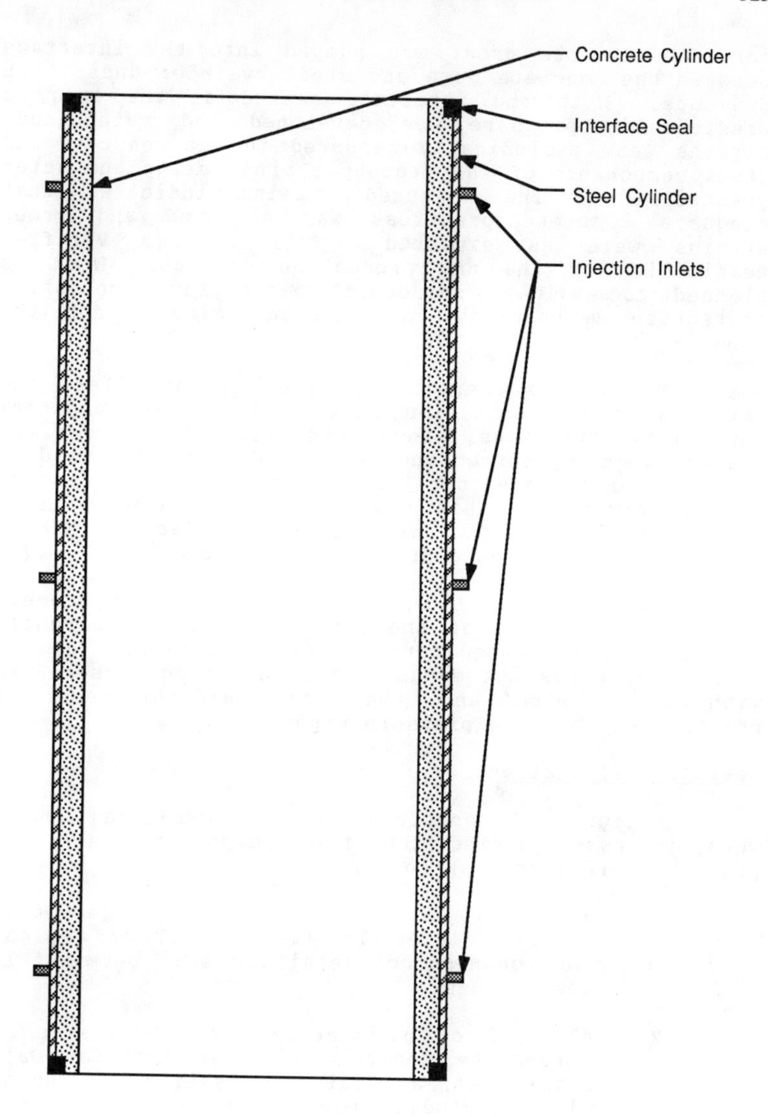

Figure 2. Prototype Pipe Schematic; Joint Rings Omitted for Clarity

core, a flowable grout was pumped into the interface between the concrete core and steel cylinder under high pressure. With the interface sealed at the ends, a prestressing pressure was developed and maintained. Strains were periodically measured in the concrete and steel components of the prototype pipe during and after prestressing. The measured strains indicated that adequate concrete prestress was achieved and creep strains were as expected. This program verified feasibility of the new production concept. Work is planned to evaluate structural performance, corrosion protection methodology, and design criteria for the pipe.

The unique prestressed pipe concept offers the potential for fabrication cost savings and long-term durability advantages, since it eliminates the need for high-strength prestressing wire and the wire-winding operation necessary for current prestressed pressure pipe fabrication. This concept improves the mobility of prestressed pipe manufacturing plants, making job-site prestressing a practical consideration.

It is believed that the manufacturing process will eliminate some of the production effort currently needed for prestressed concrete pressure pipe. The new concept combines the manufacturing advantage associated with steel pipe and the cost effectiveness of prestressed concrete pressure pipe.

Design Considerations

Design and construction considerations which were addressed during the development of the first prototype are listed as follows:

1. Selection of steel cylinder geometry, injection inlet locations, bell and spigot ring connection details, and permissible steel stress.

2. Selection of a concrete core thickness, an appropriate concrete mix, an interior wall form system, and suitable casting procedures which included high slump concrete and external vibration.

3. Details for an end-sealing system which incorporated a soluble form material, an injected rubber sealant, and development of an end-seal construction procedure.

Prototype Construction

Construction details for the prototype are described in detail in this section.

The first component fabricated for the pipe system was the steel cylinder. The cylinder included injection inlets for prestressing, bell and spigot rings, and steel end rings required for sealing the end of the pipe during injection prestressing. The steel cylinder was fabricated from 2-ft (0.61-m) wide, type 4130 steel strips. To start the fabrication, five 2-ft (0.61-m) long cylinders were fabricated by butt welding a 157-in. (3.99-m) long steel strip to form a closed ring. The 10-ft (3.05-m) long cylinder was constructed by butt welding five 2-ft (0.61-m) long rings together. The cylinder wall was 0.125 in. (3.18 mm) thick and the outside diameter was 50 ± 1/16 (1.27 m ± 1.59 mm) inches.

Next, a specially shaped extruded polystyrene ring was glued to each steel end ring. The inlet holes were covered with square mylar sheets attached to the interior surface of the steel cylinder using conventional masking tape. This prevented the core concrete from leaking through the injection inlets during casting. The steel cylinder was positioned vertically prior to erecting the formwork for casting the core. A Sonotube cylindrical form having a 43 in. (1.09 m) outside diameter was used as an inner form. This form was varnished and secured to a plywood base platform at the bottom end of the prototype. An internal bracing system, consisting of several plywood disks, was erected inside the Sonotube to counteract the inward radial pressure produced by the weight of the concrete during casting. The disks were spaced on 6-in. (152.4-mm) centers from the bottom to the 7-ft (2.13-m) level. Since hydrostatic pressure from the concrete cast was low above the 7-ft (2.13-m) height, bracing above this level was not provided. The top of the Sonotube was secured to prevent movement during casting. An anchor rod was placed through a top steel hold-down plate and connected to the laboratory floor to prevent floating of the Sonotube during casting.

A steel hopper was placed around the top opening of the annular concrete core space to facilitate casting. A concrete mix having an 8-in. (203.2-mm) slump was placed with a bucket and chute. External vibrators were mounted on wood studs that contacted the exterior surface of the steel cylinder. The vibrators were moved in 2 ft (0.61 m) vertical increments as the concrete casting progressed upward. This procedure concentrated the vibration energy in the vicinity of each new lift of concrete. After completing the cast, the Sonotube remained in place for seven days to ensure adequate curing conditions.

Five days after casting, the pipe was positioned horizontally and the extruded polystyrene rings were dissolved by pouring xylol solvent through openings in the steel end-seal rings. The pipe was rotated 180° and xylol was again poured through the opening to ensure that the extruded polystyrene was completely dissolved. Next, grease fittings were screwed into the openings in the steel end-rings and a liquid rubber compound was injected into the annular void left by dissolving the extruded polystyrene. An air powered grease gun was used to inject the rubber compound into the void at each end with sufficient pressure to ensure that the annular voids were completely filled. The liquid rubber compound cured for two weeks prior to prestressing to ensure that it had achieved a stiffness sufficient to resist forces produced during injection prestressing.

The properties of the hardened rubber were carefully selected to create a compression seal during prestressing. When the grout was injected into the sealed interface under high pressure, a circumferential tensile stress was produced in the steel cylinder and a circumferential compressive stress was produced in the concrete core. As the diameter of the steel cylinder enlarged from the tensile strains, the gap between the concrete and steel components widened. At the same time, the steel cylinder length decreased due to the Poisson's strain effect and each rubber end-seal was compressed by the steel end ring as it moved longitudinally inward. Compression of the rubber into the interfacial gap between the steel and the concrete sealed the ends of the pipe from grout leakage during prestressing.

Trial Specimens

Prestressing Grout Mixture

A highly flowable grout was made with Type I portland cement and water. The grout was mixed at a water-to-cement ratio of 0.50 using a Shar high-shearing mixer. A solution containing 1.94% Kelco Xanthan gum (by weight) and water was mixed using a Waring high speed blender. The Kelco Xanthan gum increased the grout viscosity and minimized grout penetration into the exterior concrete core surface. It also provided a slippery quality that allowed the grout to flow easily in a small gap. By slowly adding the Xanthan gum into the water, a highly viscous, homogeneous mixture was obtained. This solution was combined with the grout in the Shar mixer.

Construction of Trial Specimens

Using trial specimens, it was possible to test grout mixes and become familiar with the injection prestressing operation. Three trial specimens were constructed and tested. Each trial specimen was constructed using a 12-in. (304.8-mm) diameter, schedule 40 PVC pipe filled with concrete. All specimens were 6 ft-6-in. (1.98-m) long and had a 1-in.(25.4 mm)-diameter injection inlet at the center of the pipe length. A 6-in. (152.4 mm) square void centered within the pipe cross-section ran throughout the length of the interior concrete. The interface between the core concrete and the PVC pipe was not sealed at the ends of the trial specimens. Trial specimens were used to evaluate grout suitability for the prestressing operation prior to prestressing the prototype pipe.

Grout Injection Trials

After constructing the trial specimens and developing a pumpable grout mix, grout injection trials were conducted. A supply-return circulation loop began at the grout pump and was ended at the pump reservoir. A single grout line connected this loop to a closed ball-valve located at the injection inlet on the trial specimen. Once a pressure of 250 psi (1.72 MPa) was achieved in the circulation loop, the ball valve at the injection inlet was opened. A high pressure stream of grout immediately discharged from the ends of the trial specimen.

Leakage at the ends of the trial specimens indicated that the grout mix had flow characteristics most probably suitable for successful prestressing of the prototype. This trial grout injection was used as the basis for selecting the final grout mix used for prestressing the prototype.

Prototype Test Setup

The setup for prestressing the prototype involved installation of strain gages to measure prototype strains during prestressing, selection of prestressing equipment, and specific procedures to achieve the prestress in the prototype.

Instrumentation

After completing construction, the prototype pipe specimen was instrumented with twenty weldable strain gages located on the steel cylinder as shown in Fig. 3. The strain gages were oriented to measure strains in the circumferential direction. Strains at gage locations were measured using a digital data acquisition system during the prestressing operation. Prestress losses were monitored using a strain meter in the days following prestressing. At each strain gage location, two Whittemore mechanical strain gage points were placed on the interior concrete core surface and oriented to measure circumferential strain. The gage length was 10 in. (254 mm). The movement between Whittemore points was measured using a Whittemore strain instrument with a least reading of 1/10,000 in. (1/25.400 mm) or a strain of 10 millionths.

During the prestressing operation, a Digital Data Acquisition System (DDAS) provided immediate values of strain measured by the weldable strain gages attached to the steel cylinder. This information was used to assess the uniformity of the steel stress within the pipe system during prestressing.

Prestressing Setup and Procedures

The procedure used for prestressing was as follows:

1. A supply-return circulation loop began at the pump outlet and ended at the pump reservoir. The loop included a 12-branch manifold, a pressure gage, and a valve, placed respectively between the supply and return lines. Each branch of the manifold was connected to an injection inlet on the prototype. The valve was used to adjust the flow rate in the circulation loop. By adjusting the flow rate in the circulation loop, a uniform pressure was produced in all lines connected to the prototype.

2. A ball valve and a quick-connect fitting were placed at each injection inlet on the pipe. A valve at each inlet was available for quick use of an injection inlet, if necessary.

3. Strain gages were connected to a DDAS. The DDAS scanned the 20 strain gages. Millivolt readings from each strain gage were then

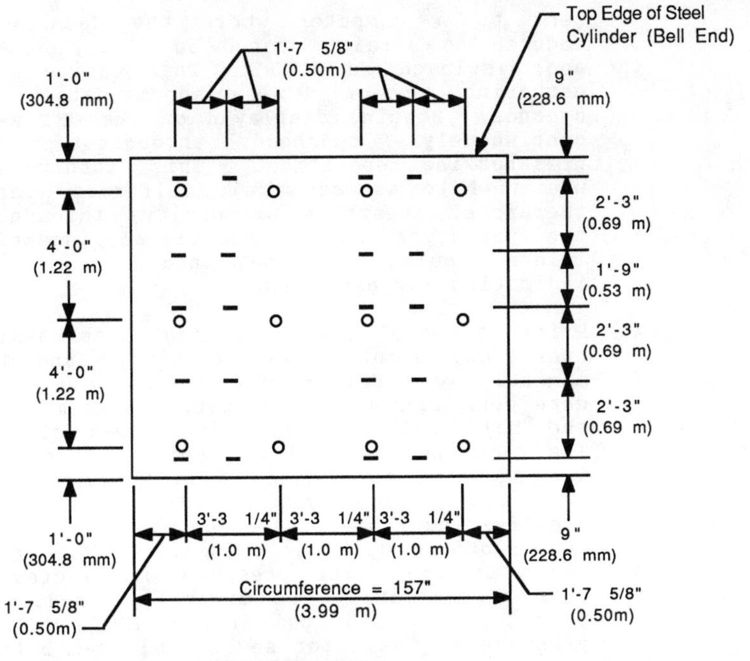

Figure 3. Prototype Pipe Instrumentation

sent to a computer where the data were reduced to strain, stored on floppy disks, and displayed on a CRT. This entire data gathering cycle took approximately [illegible] seconds. Strains displayed on the CRT were continuously updated throughout the prestressing operation. This information was used to assess strain uniformity (and, therefore, prestress uniformity) throughout the prototype wall. Additional injection inlets could then be opened in areas indicating low prestress.

4. Prior to the start of the prestressing operation, grout lines, fittings, and the manifold were filled with grout. All lines were connected to the prototype specimen and the ball valve at each inlet was closed. The regulator valve in the circulation loop was adjusted to produce a line pressure of 100 psi (0.69 MPa). To start the prestressing, one injection inlet at the center of the pipe was opened. As the grout began to flow, the pressure was increased over approximately one minute to produce a pressure of 250 psi (1.72 MPa). This pressure was held for several minutes before closing the valves at each injection inlet. This ensured that all elastic deformations of the concrete core had occurred.

Results

After the first injection inlet had been open for several minutes, steel strains were measured. Lower values of strains were measured diametrically opposite the open injection inlet. To achieve a more uniform prestress, the injection inlet on the opposite side of the pipe from the open inlet was opened. Steel cylinder tensile strain in the barrel of the pipe ranged from 2465 to 3468 millionths. Additional inlets at the mid-height level of the prototype were subsequently opened, but no significant change in steel strains occurred. This indicated that by this time (approximately 45 minutes after the first inlet was used), the grout had stiffened and opening additional inlets was ineffective. At this point, all inlets were closed.

After closing the injection inlets, the seal at the bell end (the top end of the prototype) was visually inspected. A circumferential crack developed between the steel end ring and the concrete core. A small longitudinal displacement was observed to occur between the steel and concrete core at the crack location. This indicated that the Poisson effect occurred and compressed the rubber end-seal at the ends of the pipe. Losses in prestressing pressure due to leakage were not observed during prestressing or thereafter in either the concrete core or at the end seals.

On the day of grout injection, the concrete compressive strength was 8.3 ksi (57.2 MPa) and the modulus of elasticity was estimated to be 5200 ksi (35.4 GPa). Using curves obtained from steel coupon tests, the steel yield strength was 51 ksi (351.6 MPa) and the modulus of elasticity was 30,000 ksi (206.9 GPa). Strains in the concrete core and the steel cylinder measured one day after prestressing ranged from -222 to -382 and 2389 to 3319 millionths, respectively. At twenty-five days after prestressing, concrete and steel strains measured at similar locations ranged from -412 to -572 and 2245 to 3196 millionths, respectively.

To estimate an average 25-day prestress loss in the concrete core, the change in the average steel strain was calculated over a 25-day period that started immediately after grout injection and ended 25 days later. The average steel strain at the beginning and at the end of the 25-day period were each calculated from a total of twelve strains measured at three gaged levels located at the mid-height section and approximately 2-ft (0.61 m) above and below the mid-height section. These three levels were selected for calculating the average steel strain because the restraining effect of the bell and spigot rings generally produced lower measured strains at the gaged levels located at the top and bottom of the pipe. The average steel stress immediately after grout injection was 50 ksi (344.7 MPa) based on curves from steel coupon tests and the average of the corresponding measured steel strains. The change in the average measured steel strain that occurred over the 25-day period was -293 microstrain (one microstrain equals one millionth and a negative sign indicates a decrease in strain). This decrease in the average steel strain represented an average loss in the steel prestress of 8.8 ksi (60.7 MPa) based on the initial modulus of elasticity determined from steel coupon tests.

Conclusions

The success of the prestressing operation clearly demonstrated that the new production concept can be applied to full size pipe production. In addition, significant information has been obtained that will improve future development. At this time, the following conclusions can be drawn from the construction and prestressing of the first prototype:

1. A pressurized injectable end seal system can work in conjunction with the Poisson effect to contain grouting pressures up to 250 psi (1.72 MPa). No leakage of this system occurred during or after prestressing.

2. At least a two-inlet injection system is needed to approach a uniform prestress in a 4-ft (1.22 m) diameter, 10-ft (3.05-m) long pipe. The inlets should be located across from each other at the mid-length of the pipe.

3. A high-slump concrete mix utilizing uniform external vibration throughout the pipe length is necessary to achieve a dense, void-free concrete core. Standard industry casting techniques should be adequate for core fabrication.

4. Standard bell and spigot rings provided sufficient restraint against the radial expansion of the steel cylinder in the vicinity of the end seals. With the end seal system and the bell and spigot ring arrangement used for this prototype, prestress was achieved throughout the pipe length. Although, standard bell and spigot rings could be effectively utilized in the construction of the prototype pipe, specially designed rings will most probably be required during optimization of a production pipe.

5. A flowable grout mix made using Type I portland cement, water, and Xanthan gum was successfully used for prestressing.

6. Injection inlets for prestressing must be prelocated and used simultaneously. If inlets are not used simultaneously, a nonuniform prestress may be produced due to grout stiffening which can occur within a few minutes after initial prestressing. This is especially important if the exterior surface of the concrete core is allowed to dry subsequent to casting.

7. An average steel tensile stress of 50 ksi (344.7 MPa) was achieved immediately after grout injection. An average loss of 8.8 ksi (60.7 MPa) occurred in the steel tensile stress during the 25-day period that followed the grout injection. The measured decrease in steel strain over 25 days and curves from coupon tests were used to estimate the loss in the steel tensile stress.

GEOMETRICAL NONLINEARITY IN OFFSHORE PIPELINES

Bulent A. Ovunc[1]

Abstract

Pipelines are very flexible structural systems which are subjected to various boundary conditions. For instance, during the laying process, the pipes roll on the guidance tracks on the barge, slides freely on the stinger and repose on the floor of the sea. The loads acting on the deformed configuration of the pipelines are its own weight, tension applied at the barge, the buoyancy force varying with the depth of the water and loads due to water currents. Due to the large deformations in the suspended submarine pipelines, the geometrical nonlinearity must be considered in their structural analysis. For the geometrically nonlinear analysis, the differential equations of equilibrium are expressed on the deformed configuration in Eularian coordinates. The two independent displacement functions are reduced into a single one by the proper selection of the independent variable. The displacement functions are obtained through a set of iterative processes satisfying the equilibrium of the members and the entire suspended pipeline on the deformed configuration. The iterations consist of finding a set of fictitious loads on the deformed configuration which become the actual external loads on the deformed configuration, keeping in mind that the buoyancy forces are acting all the time on the deformed configuration. Since the fictitious external loads are applied on the undeformed configuration, thus the iterations to be performed are on the given externally applied loads without

[1] J. and J. Achee Chance Professor of Civil Engineering University of Southwestern Louisiana, P. O. Box 40172 Lafayette, Louisiana 70504-0172

requiring any intermediate deformed shape. Therefore, the same standard structure stiffness matrix is used during the entire iteration process so that the computation time is reduced tremendously. The free sliding of the pipeline on the stinger are realized through a geometrical consideration. Illustrative examples are provided for the practical applications of the method.

Introduction

Depending on the oil reserves under the ocean bed, the installations of the new platforms for the exploration and production of oil are shifting from the shallow to deeper water depth. Therefore, the flexibility of the submarine pipelines are increasing with the increase in the depth of the water. The increase in the flexibility producing large displacements leads to higher nonlinear behavior in the pipelines. There are various types of externally applied loads on the pipe: their own weight including the weight of coatings and the weight of the liquid in the pipe, if any, the tension force applied to the pipeline at the barge, the interaction forces developed between the pipe and the stinger while the pipeline slides freely on the stinger, the buoyancy forces depending on the depth of the water between a point on the pipe and the free surface of the water, the forces due to the winds, waves and currents, and the reaction of the soil to the pipe at the sea floor beyond the touchdown point of the pipelines. The complexity in the boundary conditions, the variety in the external loads and the increase in the flexibilty due to the increase in water depth, leading to higher nonlinear behavior require that the submarine pipelines be given special attention during their design and laying process.

The types of the loads acting on a submarine pipeline are more numerous than than those acting on many other structural systems. The external loads acting on a submarine pipeline are classified and listed according to their types and source (Small 1970). The lift, drag and inertia forces exerted by the waves and the currents are investigated. Relationships are determined between lift and drag coefficients and Reynolds number (Jones

1978). The effect of the angle of incidence of the wave forces on the on the submerged pipelines is examined (Denson and Priest 1974). It has been shown that the Morison's equation predicts the measured wave force with the same degree of accuracy as that of the normal cylinder, provided that the force coefficients appropriate to each yaw angle, Reynolds number and Keulegian-Carpenter number are used (Morison 1950, Sarpkaya and Isaacson 1981). The reliability of various force coefficients used in calculating the hydrodynamic loadings and the validity of Morison's equation are still questioned by some researchers (Akten 1982). The range of the validity of the various wave theories such as: Airy, Cnoidal, Stokes has been given in terms of the length, heigth of the waves and the water depths (Dawson 1983). Random waves have been generated for the model testing. The relationship betwen the spectral peakedness and the correlation coefficients between adjacent waves has been determined through an extensive analysis of field records of swell waves displaying a narrow peaked spectral shape (Goda 1983).The motion of the lay barge due to sea waves produces additional dynamic loads. The motion of the barge has a measurable effect on pipes with large diameters (Zienkiewicz, Lewis and Stagg 1978). The rigidity, buoyancy and the length of the stinger along with proper tensioning are very important to the protection of the pipelines from overstressing or breaking (Blumberg, Osborn and Taher 1971). The sea floor does not provide a solid, stable support for the pipelines. The forces exerted on the pipeline through scour, wave induced soil instabilities create undesirable stresses in the pipes (Herbich 1977). The pipelines are also subjected to loads stemming from the motion of the soil such as: movement of sediments (Wright 1976), earthquake motion (Wang 1979) and liquefaction (Mes 1976).

There are variety of boundary conditions. On the barge, the pipe rolls on the guidance tracks. The horizontal or inclined curves towards the end of the barge to provide the pipe a prescribed lift off angle (Wilkins 1970). The pipe slides freely on the stinger. the stinger may be a continuous element or composed of articulated elements (Ovunc and Mallareddy 1970). For continous stinger, the ratio of the rigidity of the stinger to the pipe is very important so that the pipe can slide continuously on the stinger without lifting off the stinger along one or more intermedite intervals. Lengthy separation of the pipe from the stinger can produce undesirable additional stresses in pipes. The pipe must separate tangentially from the stinger before or almost at the end of the stinger in order to prevent an exceesive shearing force reaction exerted to pipe by the stinger. The pipeline touches tangentially the sea floor and reposes elastically on the sea floor (Ovunc 1982,2). The modulus of elasticity and the characteristics of the soil along the pipeline route such as: mudslide, liquefaction, sand wave migration, faulting, mudlump are the parameters which appear at this end of the pipelines (Audibert 1979).

The submarine pipelines are very flexible structural system with nonlinear behavior (Wilhoit and Merwin 1967). Various methods of analysis have been applied to determine the deformed configuration and the stress distribution. An iterative process has been used to satisfy the equilibrium equations on the deformed configuration, by means of standard linear analyses applied on the intermediate deformed configurations (Tezcan and Ovunc 1966; Ovunc and Mallareddy 1970). The tangent stiffness matrix which was based on the approximations in the conventional beam-column theory has been given by clearly separating the contribution of large rigid body displacements from elastic and locally nonlinear effects (Oran 1973). For the space structures under large displacements and buckling, the likely failure modes have been obtained by using tangent tangent stiffness matrices near the collapse loads (See and McConnell 1986). A procedure which has assumed small strains and used fictitious forces based on the rotation from the initial confifiguration has been applied to the geometrically nonlinnear analysis of structures (Kohnke 1978). An updated Lagrangian approach has been adopted in the formulation of the nonlinear equilibrium equations considering the large deflection behavior of the space frames (Meek and Loganathan 1989). The material nonlinearity has been added to the geometrical nonlinearity for the large deflection analysis of inealstic structures (Bathe and Ozdemir 1975; Cichon 1984). The inelastic frames have been analyzed by assuming that the members may undergo large displacements while their relative displacements are small. It is also assumed that the yielding may occur only at the ends of the members (1983). A large displacement analysis of structures has been based on the Eulerian formulation in which the member tangent stiffness matrix is constructed with reference to the current deformed configuration (Kam 1989). Long free pipeline spans exposed to current and waves have been analyzed and subjected to field test (Bryndum et al 1989). In the design of offshore pipelines, the boundary conditions on the barge, on the stinger and on the sea floor have been considered in the design of offshore pipelines (Ovunc 1982,2). Various computer programs such as NUPIPE, PIPLN (Alyeska), ANSR, INTR (Audibert 1979), OPLSD2, OPLSD3 (Malahy), SCORES, LINK (Zienkiewicz at al., 1978) as to count some of them are available to perform the structural analysis of pipelines.

Herein, the same loading, boundary conditions and same description of the geometrical nonlineariy which were defined earlier are used in the analysis (Ovunc 1982,2). Except, the iteration process is applied on the external loadings, rather than the intermediate deformed configurations (Ovunc 1989). The geometrical nonlinearity is obtained by finding a set of externally applied fictitious loads such that when they are applied on the undeformed configuration, they yield to the actual external loads on the deformed configuration. The iterations to be performed are on the given set of externally applied loads, without requiring any intermediate deformed configuration. Therefore, the standard structure stiffness matrix generated at the start of the calculations is used during the entire iteration process. The differential equations of equilibrium are expressed on the deformed configuration which include the member axial force as integral part of the derivations. The material nonlinearity has not been introduced in the analysis.

PROCEDURE OF ANALYSIS

It is assumed that the materials are elastic, strains are infinitesimal, although the displacements are finite and Bernoulli-Euler hypothesis is valid. The

Pipeline slides freely on the stinger and reposes elastically on the sea floor. The soil at the bottom of the sea is considered as elasic medium. The bottom of the sea may have any topography, including some faults.

The undeformed and deformed configurations of the pipeline during the laying process is shown in Fig. 1.

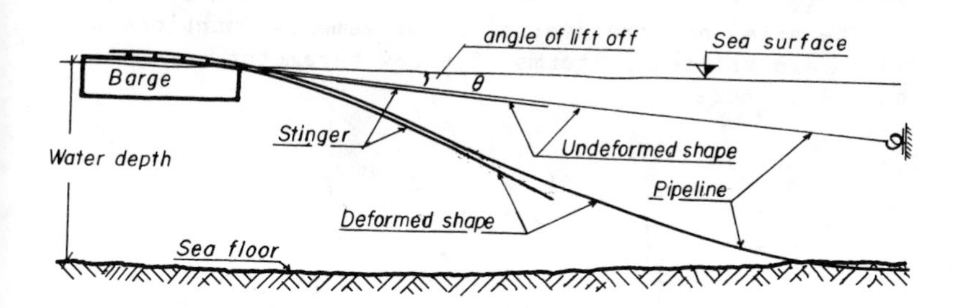

Fig. 1 - Position of the pipeline during laying process

The pipeline and the stinger are divided into members. Smaller member lengths are selected for the pipe on stinger compared to those on suspended part. So that the free sliding of the pipeline on the stinger can be ascertained by a geometrical consideration. To ascertain the free sliding of the pipeline on the stinger, two different nodes are assumed at a point at the interface between the pipe and the stinger. At these points, only the freedoms along the tangential direction on the deformed configuration are different as long as the pipe does not separate from the stinger. If the pipeline and the singer are separated from each other at an interface point, the directions of the freedoms are different for the pipeline and the stinger. For the nonlinear analysis of the system, only a brief summary is outlined below. A more detailed explanation has been given elsewhere (Ovunc 1989).

The system composed of pipeline and stinger is referred to the following axes systems: a global Lagrangian axes system XYZ, a local Lagrangian axes system xyz, an Updated Lagrangian axes system $x_u y_u z_u$ and an

Eularian axes system $x_e y_e z_e$. At a joint, the vector of member end displacements and rotaions {d}, which are obtained with respect to the Lagrangian member axes system xyz, are converted to Updated Lagrangian axis system $x_u y_u z_u$. At an arbitrary point in a member, the strains and the stresses are along the Eulerian coordinates $x_e y_e z_e$, but the displacements u and v, are referred to the Updated Lagrangian coordinates, $x_u y_u z_u$ (Fig. 2). The axial and transversal displacements functions u and v, are written in terms of a new parameter ξ as $u(\xi)$ and $v(\xi)$, where,

$$\xi = y_u + u(\xi) \tag{1}$$

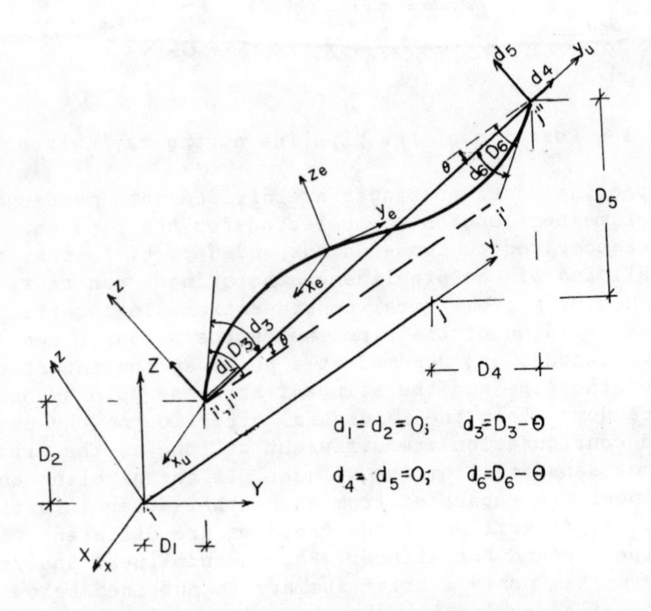

Fig. 2 -Global and local axes systems.

Thus the two independent displacement functions are reduced into one transversal displacement function $v(\xi)$.

For an infinitesimal element on the deformed configuration (Fig. 3), the expression of the strain at a distance η from the neutral axis can be obtained through the differential geometry as,

$$\varepsilon=\varepsilon_0-\eta k_0(1+\varepsilon_0) \tag{2}$$

where,

$$\varepsilon_0=[(1+(v')^2)^{1/2}]/(1-u')-1$$

$$k_0=v''/[(1+(v')^2)^{3/2}] \tag{3}$$

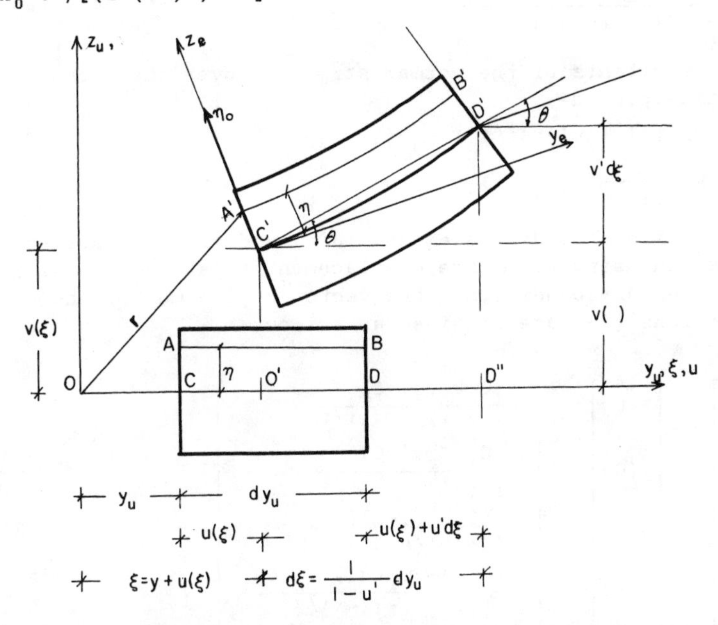

Fig. 3. Deformations of the infinitesimal element

The distributed loads along the direction y_u and z_u being q_y and q_z, the differential equilibrium equations on the deformed shape provide,

$$\frac{dP}{d\xi} + k_o(1 + (v')^2)^{1/2}V + \frac{q_z v' + q_y}{(1 + (v')^2)^{1/2}} \tag{4}$$

$$\frac{dV}{d\xi} - k_o(1 + (v')^2)^{1/2}P + \frac{q_z - q_y v'}{(1 + (v')^2)^{1/2}} \tag{5}$$

$$\frac{dM}{d\xi} - (1 + (v')^2\ ^{1/2}V = 0 \tag{6}$$

The integration of the two simultaneous differential equations 4 and 5 yields to,

$$P = \frac{-v'(C_1 + \xi q_z) + C_2 - \xi q_y}{(1 + (v')^2)^{1/2}} \tag{7}$$

$$V = \frac{v'(C_2 - \xi q_y) + C_1 - \xi q_z}{(1 + (v')^2)^{1/2}} \tag{8}$$

The resultants of the normal stress σ, over the cross section provide,

$$P = EA\ ((1 + (v')^2)^{1/2}(1 - u') - 1) \tag{9}$$

$$M = EI\ \frac{v''}{(1 + (v')^2)}\ (1 - u') \tag{10}$$

By assuming the displacements $u(\xi)$ and $v(\xi)$, in series form and using the force displacement relationships with the boundary conditions, the vector of the member end reactions {f}, are obtained as follows,

$$\begin{Bmatrix} f_1 \\ f_2 \\ f_3 \\ f_4 \\ f_5 \\ f_6 \end{Bmatrix} = \begin{Bmatrix} P(0) \\ V(0) \\ M(0) \\ P(\ell) \\ V(\ell) \\ M(\ell) \end{Bmatrix} = \begin{Bmatrix} \dfrac{-C_1 v'(0) + C_2}{(1+(v'(0))^2)^{1/2}} \\ \dfrac{C_1 + C_2 v'(0)}{(1+(v'(0))^2)^{1/2}} \\ \dfrac{v''(0)EI(1+u'(0))}{(1+(v'(0))^2)} \\ \dfrac{v'(\ell)(-C_1 \ell q_z) + C_2 - \ell q_y}{(1+(v'(\ell))^2)^{1/2}} \\ \dfrac{v'(\ell)(C_2 - \ell q_y) + C_1 + \ell q_z}{(1+(v'(\ell))^2)^{1/2}} \\ \dfrac{v''(\ell)EI(1+u'(\ell))}{(1+(v'(\ell))^2)} \end{Bmatrix} \tag{11}$$

The reasoning behind the present nonlinear analysis is based on the determination of the fictitious external loads. The fictitious loads are such that, when they are applied on the undeformed configuration they produce the actual external loads on the final nonlinear deformed configuration. The fictitious external loads are determined by an iterative process.

The process consists of the iterations on the fictitious external loads $\{F_c\}$. First, the actual external loads are applied on the undeformed configuration and the first set of fictitious loads $\{F_{c1}\}$ are determined through the equilibrium on the first deformed geometry. If the differences between the first set of fictitious loads and the actual loads are negligible the framework does not exhibit any appreciable geometrical nonlinearity and the fist deformed configuration is the final configuration. If the differences between the fictitious and actual loads are not negligible, the first deformed configuration is not the final configuration. Then, the first fictitious loads are applied on the undeformed configuration to obtain a second set of fictitious loads $\{F_{c2}\}$, and so on. Knowing the fictitious loads of last three iterations and the corresponding displacements, the next set of fictitious loads are evaluated from the criteria on the convergence of the fictitious loads. The iterations continue till the difference between two successive sets of fictitious loads is negligible.

The above described procedure to determine the fictitious loads on the undeformed geometry is included in the general purpose software STDYNL for their practical applications (Ovunc 1989). BOSPIP is a simplified version of STDYNL which performs the analysis of the buried and suspended pipelines only, in the main frame machines or personal computers.

APPLICATION

A pipeline laying to a water depth of 277.5 ft is considered as an iluustrative example. The pipeline is in steel with 19 in and 20 in inner and outer diameters, respectively. The pipeline rolls over five rollers on the barge, slides freely on 320 ft length of stinger and reposes on an elastic soil at the sea floor. The soil subgrade coefficient is assumed to be 2100 ksi. The own weight of the pipeline and the stinger are 310 lb/ft and

95 lb/ft. The lift off angle of the pipeline from the barge is 5 degree (Ovunc 1982,2).

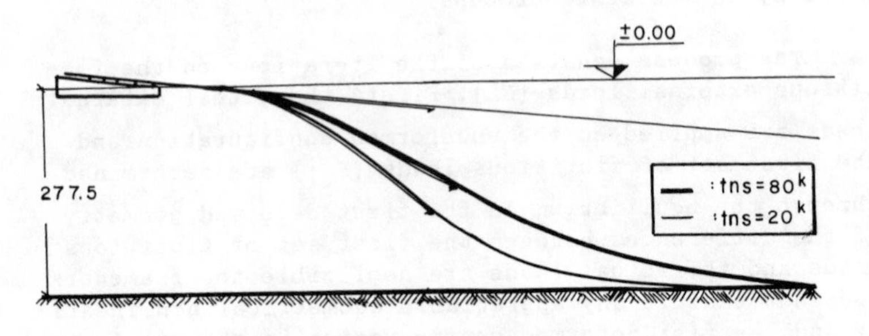

Fig. 4 - The deformed configuration of the pipeline.

The increase in the magnitude of the tension applied to the pipeline at the barge reduces the curvature

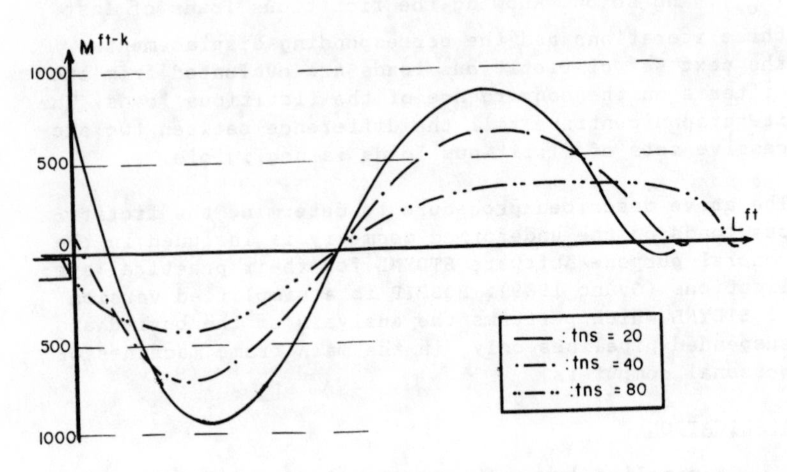

Fig. 5 - Moment diagram of the pipeline along its length

and thus the magnitude of the bending moments along the suspended part of the pipeline. It also increases the the length of the suspended part (Fig. 4).

The variation of the moment diagram with the variation of the tension applied to the pipeline at the barge is shown in the Figure 5. Due to the provision of the free sliding of the pipeline on the stinger, the reactions at the interface are continuously distributed along the interface. Thus, the variations of the bending moment and shearing force along the part of the pipeline supported by the stinger and along the stinger are continuous. A discontinuity on the shearing force may appear at the point where the pipeline leaves the stinger, if the pipeline does not lift off from the stinger before or right before the tip of the stinger. The presence of the discontinuity on the shearing force is undesirable since it may create high shearing stress in the pipe around this point. The discontinuity on the shearing force in the pipeline can be eliminated either by increasing the tension applied at the barge or by increasing the length or decreasing the stiffness of the stinger. So that the pipeline lifts off from the stinger before or right before the tip of the stinger. The magnitudes of the moment and shearing force are very small at the point where the pipeline touches the bottom of the sea. The moments and shearing forces die out very fast beyond the touchdown point. The increase in the ratio of the stiffness of the stinger to that of the pipeline reduces the magnitude of the minimum moment along the part of the pipeline supported by the stinger. But the magnitude of the maximum moment along its suspended length is slightly affected. The increase in the lift off angle from the barge increases the curvature and thus, the moments in the pipes.

The increase in the tension applied to the pipeline at the barge reduces the magnitude of the maximum and minimum normal stresses in the pipes as long as the bending moment is the governing factor. Beyond a limiting magnitude of the applied tension, the tension force becomes the governing factor and the tensile stresses increase with the tension force applied at the barge. An optimum solution for the design of the pipeline can be obtained by the variations of the above described parameters.

CONCLUSION

Most of the static loads or the converted static

loads from their dynamic counterparts are considered in the analysis of suspended pipelines under various boundary conditions. The free sliding of the pipeline on the stinger is taken into account through a geometrical consideration. The continuous free sliding of the pipeline on the stinger provided continuously varying reactions along their interface. Thus, the variation of the shear and moment diagrams along the stinger for the pipeline and for the stinger are continuous.

The new version of the direct integration method is used to detrmine the deformed configuration of the pipeline and the stinger, and the stress distributions within them. The effects of the various parameters on the deformed configuration and on the stress distribution in the pipelines, such that: the tension applied at the barge, the lift off angle, the ratio of the stiffness of the stinger to that of the pipeline, the soil characteristics at the sea floor, etc., are included in the investigations. In the design of the pipelines some of the parameters involved are provided as data, the others need to be determined. The optimum design of a pipeline can be obtained by investigating the variations of the parameters involved in the design.

REFERENCES

Akten, H.T. (1982), -"New Developments in Submarine Pipeline In-Plane Stability Analysis," Proc., First *Offshore Mechanics/Arctic Engineering/Deep Sea System Symp.*, ETCE, New Orleans, LA, ASME, 63-71.

Alyeska Pipeline Service Company, *PIPLIN Computer program Manual*.

Audibert, J.E.M., Lai, N.W. and Bea, R.G. (1979), -"De-Design of Pipelines -Sea Bottom Loads and Restraints," *Pipeline Adverse Environments*, ASCE Vol. 1.

Bathe, K.J. and Ozdemir, H.(1975), -"Elastic-Plastic, Large Deformation, Static and Dynamic Analysis," *J. Compt & Struct.*, 6, 81-92.

Blumberg, R., Osborn, C. and Tahir, S.A.(1971) -"Analysis of Ocean Engineering Problems in Offshore Pipelining," 3rd *OTC*, Houston Texas, Apr. 19-21, OTC 1356.

Bryndum, M.B., Bonde, C., Smith, L.W., Tura, F. and Montesi, M. (1989), -"Long Free Pipline Spans Exposed to Current and Waves - Model Tests," *21rst OTC*, Houston, Texas, May 1-4,

Cichon, C.(1984),-"Large Displacements In-Plane Analysis of Elastic-Plastic Frames," *J. Compt. & Struct.*, 19, 5/6, 737-745.

Dawson, T. H. (1983) , Offshore Structural Engineering, Prentice Hall Inc., New Jersey.

Denson, K.H. and Priest, M.S. (1974), -"Effect of Angle of incidence on Wave Forces on Submerged Pipelines," *ASCE National Meeting on Water Resources Engineering* Los Angeles, Prepint 2120.

Herbich, J.B.(1977),-"Wave Induced Scour around Offshore Pipelines," *9th OTC*, Houston, Texas, May 2-5, OTC 2968

Jones, W.T. (1978), -"On-Bottom Pipeline Stability in Steady Water Currents," *J. Petr. Techn.*, March.

Kam, T.Y., Rossow, E.C. and Corotis, R.B. (1988), - "Large Deflection Analysis of Inelastic Plane Frames," *ASCE ST1*,109, 184-197.

Kassimali, A. (1983), -"Large Deformation Analysis of Elastic-Plastic Frames," *ASCE ST8*, 109, 1869,1886.

Kohnke, P.C. (1978),-"Large Deflection Analysis of Frame Structures by Fictitious Forces, " *Int. J. Num. Meth. in Engrg*, 12, 1279-1284.

Malahy, R., Ocean Pipeline Laying Similation, OPLS2D and OPLS3D, Technical Description, Pipeline Technologists, Houston, Texas.

Meek, S.L. and Loganathan, S. (1989), -"Geometrically Nonlinear Behavior of Space Frame Structures," *Compt. & Struct.*, 31, 1, 35-46.

Mes, M.J. (1976), -"Seabed and Sandbank Fluidization," *Pipeline Gas J.*, Oct.

Morison, J.R., Johson, J.W., O'Brien, M.P. and Schaaf, S.A. (1950), -"The force Exerted by Surface Wave on Piles," *Petr. Trans. AIME*, 189.

Oran, C. (1973), -"Tangent Stiffness in Space Frames," *J. ASCE, ST6*, 99, 987-1001.

Ovunc, B.A. and Mallareddy, H. (1970), -"Stress Analysis of Offshore Pipelines," *2nd OTC*, Houston Texas, Apr. 22-24, OTC 1222.

Ovunc, B.A. (1981), -"Stresses in Offshore Pipelines," *Proc., 3rd Int. Symp. on Freight Pipelines*, Houston, Texas.

Ovunc, B. A. (1982,1), -"Design of Offshore Pipelines,"

J. Pipelines, 2, 285-295.

Ovunc, B.A. (1982,2), -"The Geometrical Nonlinearity of Plane Framework," Proc., *Sino-American Symp.* on Bridge and *Struct. Engrg.*, Beijing, China.

Ovunc, B.A. (1985) , -"STDYNL, A Code for Structural Systems," *STRUCTURAL ANALYSIS SYSTEMS*, The Int. Guidebooks, Ed. by A. Niku-Lari, Pergamon Press, 225-238.

Ovunc, B. A. (1989), -"Geometrical Nonlinearity in the Analysis of Flexible Structures," Proc., *FEMCAD'89, SAS Wold Conf.*, Paris, France.

Sarpkaya, T. and Isaacson M. (1981), - Machanics of Wave Forces on Offshore Structures, Van Nostrand.

See, T. and McConnell, R.E.(1986), -"Large Displacement, Elastic Buckling of Space Structures," J. *ASCE*, *ST5*, 112, 1052-1069.

Small, S.W. (1970), -"The Submarine Pipelines as a Structure," 2nd *OTC*, Houston, Texas, OTC 1223.

Tezcan, S.S. and Ovunc, B.A. (1966), -"An Iteration Method for the Nonlinear Analysis Buckling of Framed Structures," SPACE STRUCTURES, Ed. by R.M. Davies, Blackwell Scientific Publ. 1966

Wang, L.R.L (1979), -"Seismic Design Criteria for Buried Pipelines," *Pipelines in Adverse Environments*, ASCE,1.

Wilhoit, J.C. and Merwin, J.W. (1971), -"The Effect of Axial Tension on Moment Carrying Capacity of Pipelines Stressed Beyond the Elastic Limit," 3rd *OTC*, Houston, Texas.

Wilkins, J.R. (1970), -"Offshore Pipeline Stress Analysis," 2nd *OTC*, Houston, Texas, OTC 1227.

Zienkiewicz, O.C., Lewis, R.W. and Stagg, K. G. (1978), Numerical Methods in Offshore Engineering, John Wiley and Sons.

Technical Work on Flexible Pipe/Soil Interaction Overview - 1990

B. Jay Schrock[1], P.E., M. ASCE

Abstract

Research in the laboratory and field have been ongoing for the past seventy years on flexible pipe/soil interaction. Virtually all of the early work was done on steel pipe. During the past thirty years most of the evaluation work has been done on plastic pipes. More recently additional information has been developed through Finite Element Analysis (FEA) utilizing the computer with various models and assumptions. Literally thousands of experiments have been conducted and observations made in order to isolate and identify various parameters for a better understanding of pipe reactions to loadings. These parameters, their correct use and the pipes reaction and long term performance are necessary for the pipeline designers.

The paper presents an overview and summary of the various results of much of this research and testing.

The International Standards Organization has formed a new committee for developing calculation systems for plastic pipes. Under ISO/Technical Committee 138/Working Group 13/this responsibility will be completed.

General

Flexible metal pipes have been utilized for more than one hundred years. During the first thirty years

- - - - - - - - - - - - - - - - -

[1]B. Jay Schrock, P.E., JSC International Engineering, 1313 Gary Way, Carmichael, CA. 95608

of their use, empirical and/or trial and error procedures were practiced. Not until the early 1920's had much in-depth technical research been accomplished.

Professors Marston (1929) and Spangler advanced their Flexible Pipe Theory on soil loadings through many experiments at Iowa State College. This work had evolved from the Marston experiments published during 1913. Professor Spangler continued his flexible metal pipe experiments throughout the 1920's and 1930's, culminating with the publication of Iowa State College Bulletin No. 153 during 1941. The bulletin illustrates the theoretical development, with supporting experimental data, for the "Iowa Formula".

Professors Spangler (1960) and Watkins continued the work and during 1958, the "Modified Iowa Formula" was developed. At that time the E' (Soil Reaction) value evolved which was substituted for Spangler's earlier "er" value. This relationship provided a more useful formula for the flexible metal pipeline designer. It should be noted that Spangler's original work was predicated upon embankment experiments, utilizing 36, 42, 48 and 60-inch diameter pipes, with pipe stiffness values of 61, 54, 36 and 25-psi respectively. Also, the horizontal deflection was of primary concern and became part of the formula. Each diameter of the culverts were placed in a tamped and untamped test condition. The results of the experiments are shown in Table I. These findings became the basis for the "Iowa Formula".

TABLE I

Nominal Pipe Diameter (in.)	Sidefill Condition	Calculated Horizontal Defl.(%)	Measured Horizontal Defl.(%)	Measured Vertical Defl.(%)
60	T (26)*	1.45	1.68	2.10
	U (12)	2.97	2.98	3.03
48	T (29)	2.02	1.79	2.12
	U (14)	3.33	3.38	3.25
42	T (25)	2.24	1.83	2.00
	U (15)	3.48	3.31	3.12
36	T (28)	2.31	1.86	1.97
	U (13)	3.38	3.53	3.36

* Under the sidefill condition column, the numbers in parenthesis indicate e values for modulus of Passive Pressure ($lbs/in^2/in.$).

Note: T is tamped condition and U is untamped condition.

Spangler assumed that the upper and lower 40° of the flexible pipe developed small movements and that a parabolic load distribution embraced the middle 100° area on each side of the pipe (Figure 1a).

Spangler's test data reveals that for a given soil density (tamped) condition the pipes vertical deflection is greater in magnitude than the horizontal deflection as the pipes stiffness reduces (Figure 2).

The joint Task Committee of the ASCE/WPCF produced a Manual of Practice in 1960 culminating seven years of work. This "Design and Construction of Sanitary and Storm Sewers" Manual discussed structural requirements for pipelines in soils. The various editions included a recommendation that pipe ring stiffness factor, EI/r^3, be maintained at no less than 10 percent of the value of the soil stiffness factor, 0.061 E'. This recommendation still stands but is contested by some manufacturers and practitioners as being unnecessary in good trench installations when using granular material. It is also indicated, in the joint manuals, that "almost the entire performance of a flexible conduit in retaining its shape and integrity is dependent on the selection, placement, and compaction of the envelop of earth surrounding the structure".

Professor Watkins continued his research during the 60's and 70's at Utah State University, investigating flexible steel and plastic pipes. The primary work incorporated the assumption of the interaction of elastic pipe and elastic soil, characterizing maximum possible deflection with soil movements.

Barnard's work (1957) on steel pipe with soil interaction illustrates the pipe loading effects on soil laterally adjacent to the pipe. This relationship was also referred to in the AWWA M-11 Steel Pipe Manual in 1964. The Lauer (1978) and Leonhardt (1979-Fig. 1b & Fig. 2) work was similar to Barnards, however, a detailed interrelationship was developed for the trench wall insitu soil effects on the pipe zone material (Figure 3). This was encorporated into the ATV (1980) calculation method which will be discussed later in this paper.

Luscher developed a buckling and behavior theory (1966) of metal tubes which incorporates a soil support factor B', differentiating it from the Timoshenko "Theory of Elastic Stability" (1961). Gaube, Mueller and others presented pipe buckling relationships during several Kunstoffe Conferences in Germany, from 1963 to 1977.

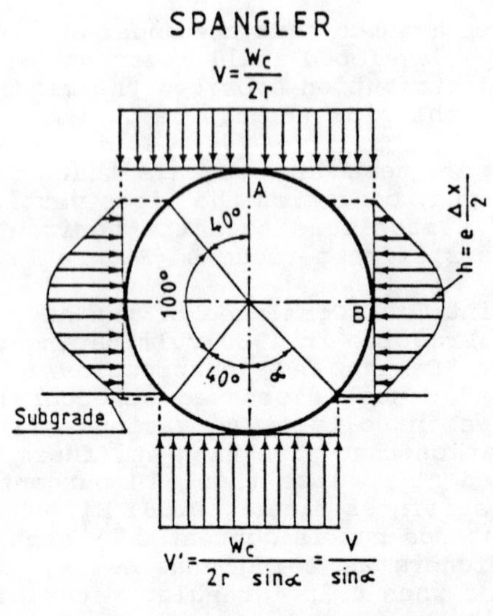

Figure 1a

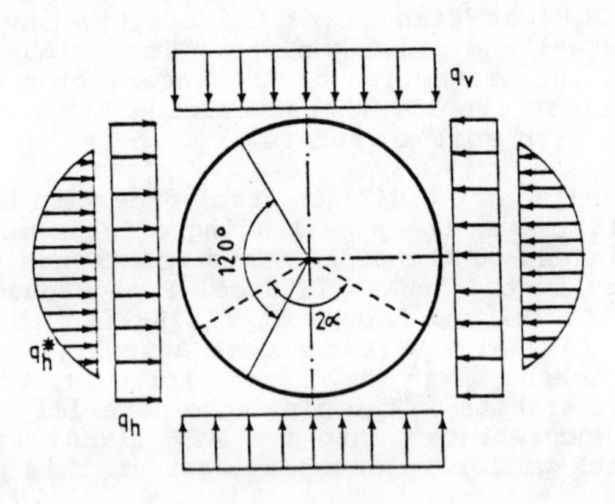

Figure 1b

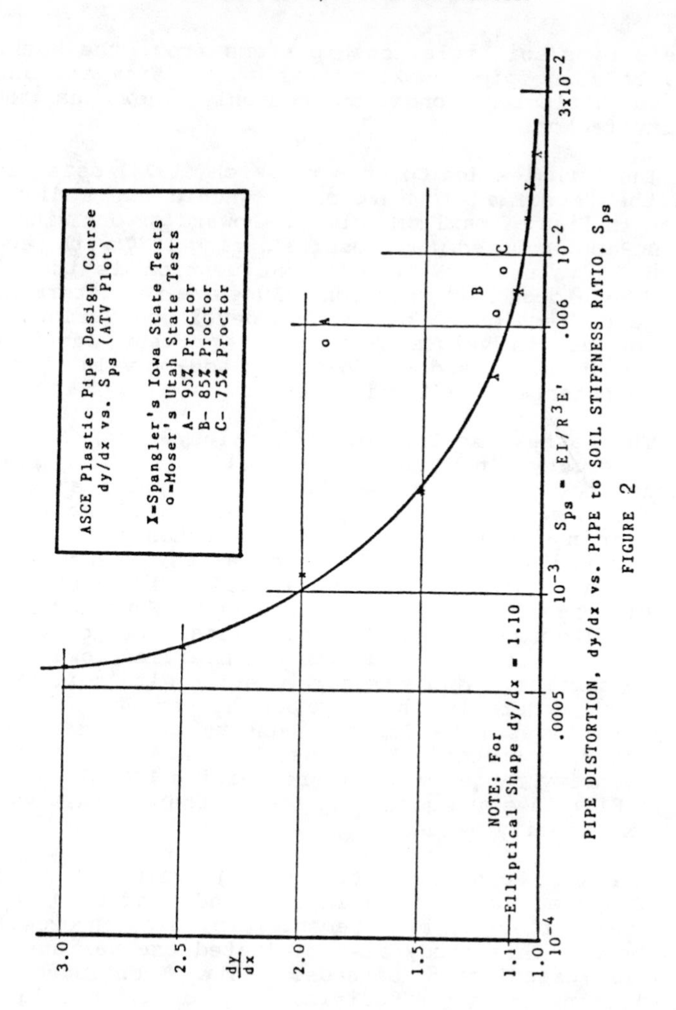

PIPE DISTORTION, dy/dx vs. PIPE to SOIL STIFFNESS RATIO, S_{ps}

FIGURE 2

Gaube's pipe/soil relationship stems from the Watkins work, relating pipe buckling resistance from the anticipated deflected condition commonly known as the ovality factor.

The Parmalee and Corotis research (1974) determined that the "Modified Iowa Formula" should use median E' values in lieu of maximum values. Howard's work with the U.S. BuRec concluded with results during 1977 with recommended E' values. Noteworthy of mention is that the Howard developed soil reaction values are for determining average deflection. Howard also developed various deflection add-on values for the 95 percent confidence levels on average, and an additional add-on value for the maximum potential deflection levels.

The Wastewater Engineering Society, Inc. (ATV) of West Germany published "Guidelines for Static Calculations of Drainage Conduits and Pipelines" during 1980 with a second edition in December 1988. The various soil loadings are similar to the Marston and Marston/Spangler rigid and flexible pipe values, respectively. The ATV calculaton system can be utilized for rigid and flexible pipes, where the "Modified Iowa Formula" is for the flexible pipes only. The soil classification systems of ATV and Howard, as used with the Modified Iowa Formula, are similar. There is some overlap with soil gradation, but generally they are in agreement. The ATV system provides more Proctor Density categories, with each having a specific E' value. The Howard system has wider density ranges with its respective E' values. The Howard E' values are generally higher than the ATV values for like density values.

The Komo Foundation Study (1982) evaluated hundreds of PVC installations with initial and long-term (up to 20 years) deflection measurements providing an excellent data base. This study also evaluated the various deflection calculation methods. They concluded that calculations using the Modified Iowa Formula provide the least correlation to in-ground burial performance. The ATV and the Bossen calculation systems provide the best correlation to actual performance. It should be noted that Howard's modification to the Modified Iowa Formula (1981) provides acceptable correlation for flexible pipes on good installations. Also acceptable correlation when using stiffer pipes on average installations.

Owens-Corning Fiberglass (OCF) culminated a lengthy laboratory and field investigation (1984) on fiberglass pipe performance. These findings have been presented to the appropriate International Standards, ASTM and AWWA

Committees. These various committees have reviewed the information and are incorporating certain salient features into their respective standards. The primary components of these findings have been presented in the form of technical papers by Bishop, et al (1984) and Moser, et al (1985). Carlstroem (1989) compares the Spangler and ATV calculaton systems and contends that ATV produces more correct values when compared to installation results.

Finite Element Analysis (FEA) conducted by Hartley, et al (1987) has provided theoretically developed results for design use on E' (soil reaction) values. This work clearly demonstrates that E' values vary with soil types and depths. This corroborates the work done by Molin (1974) and OCF. Other Finite Element Analysis conducted by Jeyapalan, et al, (1984) reveals that the pipe/soil stiffness ratio is a key element to the successful design of flexible pipes. When this ratio reduces to a certain level, high stresses with accompanying strains can develop adjacent to the crown of flexible pipes. Some pipe materials can tolerate these stresses better than others; therefore, the pipe manufacturer should qualify and quantify material long term strain performance limits.

Howard and Greenwood recent soils work (1989) shows better correlation than previously determined. The OCF research has indicated that pipe stiffness, short-term and long-term, and the use of the Leonhardt Factor with pipe and soil strength reduction, should be used for calculating in-ground performance. Howard and Greenwood have made various adjustments in their previous work which will be the subject of later papers.

Summary

The strain and deflection responses to certain installation irregularities for pipes having various stiffnesses have been observed, monitored and documented. The effects of different types of soils used in the pipe zone embedment immediately around the pipe have been categorized. The effects of trench widths, insitu soil types and densities, the long term pipe stiffness and presence of groundwater all have impact on the pipe's long-term performance. The effect of neglecting soil placement in the haunch areas of the pipe has been investigated and noted as having significant impact on short and long-term performance. Proper haunching has direct impact on the strain levels in buried flexible pipes.

When soil is placed unsymmetrically around the pipe, differing stress levels with accompanying strain have been observed. Also, the lower the pipe stiffness, the more impact on the pipe distortion. As the pipe zone embedment material is placed and compacted, the pipe can experience some initial ovalization. The lower the pipe stiffness, the greater the initial ovalization. This ovalization should be kept to a minimum. When backfill is placed over the top of the pipe, there is a tendency for a certain amount of rerounding to take place, depending upon the level of initial ovalization and soil density adjacent to the pipe. When these levels are high, there is also a tendency for the pipe to flatten at the top, and sometimes at the bottom. This flattening at the pipe's bottom is normally dependent upon the thickness and the density of the bedding material and the pipe's stiffness.

Pipes poorly haunched in compacted silty-sand soils, deflect nearly twice as much as when proper density in the haunch area is attained. Higher strains occur in the pipe when proper haunching is not attained. Also, the higher stiffness pipes show less deflection and strain under the same loads when compared to lower stiffness pipes. Increasing the pipe stiffness tends to dampen out the strain peaks because of better bridging of the bedding irregularities. It should be noted that proper haunching is more readily attained in an embankment or wide trench than narrow or normal trench conditions.

Soil density has a dramatic impact on pipe performance. When pipe stiffness of 20 psi or less is used it is normally recommended that gravel/rock (stone) material be used in the pipe zone. The difference between the horizontal and vertical deflection is greater for the higher density soils. This can be attributed to the effects of the dense materials that are lateral to the pipe, consequentially not permitting horizontal movement, and the greater vertical movement is caused by the vertical loadings. Also, as the density increases, there is a tendency for the pipe to go into a higher mode of deformation.

There is significant evidence that current buckling models can not be used with any degree of reliability to predict buckling resistance for flexible pipes having long-term stiffness values below 12 psi. This mimimum recommended stiffness value pertains to gravity or pressure pipe applicatons experiencing vacuum, traffic loadings, or deep burials.

Recommendations

The most difficult problem confronting the designer of flexible pipelines is the selection of realistic values for the soil modulus and external load parameters required for design. This difficulty arises from the large potential variation in native and pipe embedment soil characteristics. Also, note that E' values vary with soil types and depths.

Until recently, researchers have not emphasized the importance of the insitu trench wall soil character. For decades, however, researchers have emphasized the importance of achieving good compaction of the pipe zone embedment material. With the exception of Barnard's work, emphasis was not given to the importance of materials beyond those in the immediate adjacent backfill envelope until the mid 1970's. The work by Leonhardt emphasized the combined soil modulus developed from the interaction between pipe zone backfill material and the insitu material in the trench wall.

The structural integrity of a buried flexible pipe depends upon the soil load on the pipe, stiffness of the pipe, passive resistance of the soil at the sides of the pipe, time-consolidation characteristics of the soil, and the degree of bottom support of the pipe.

The owner and/or design consultant should provide or arrange for a pipe alignment soils investigation. Also, a competent level of inspection and supervision should be provided during installation.

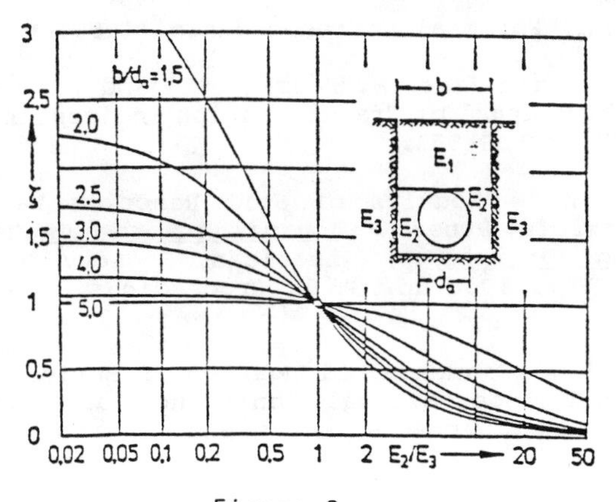

Figure 3

References

American Water Works Association Manual M-11. "Steel Pipe Design and Installation." 1964 and 1988.

Barnard, R. E. "Design and Deflection Control of Buried Steel Pipe Supporting Earth and Live Loads." Proceedings. American Society of Testing Materials, Philadelphia, Pennsylvania. 1957.

Bishop, R. R. and D. C. Lang. "Design and Performance of Buried Fiberglass Pipes--A New Perspective". ASCE Conference, October 5, 1984.

Carlstroem, B. "Technical Report-Part 2-Static Calculation of Underground GRP Pipes". ISO/TC138/WG4/N-127, August 1989.

"Design and Construction of Sanitary and Storm Sewers." American Society of Civil Engineers Manual and Report on Engineering practice No. 37 (Water Pollution Control Federation Manual of Practice No. 9). American Society of Civil Engineers/Water Pollution Control Federation, New York, New York. Editions 1960, 1969, and 1976.

Gaube, E. and others. Kunststoffe 53-67 Conferences. 1963-1977. Several papers.

"Gravity Sanitary Sewer Design and Construction." American Society of Civil Engineers Manual and Report on Engineering Practice No. 60 (Water Pollution Control Federation Manual of Practice No. FD-5). N.Y. 1982.

Greenwood, M. Personal Correspondance (1989).

Hartley, J.D. and James M. Duncan, "E' and Its Variation With Depth", Journal of Transportation Engineering, ASCE, TE Sept. 87, p 538-553.

Howard, A. K. "Modulus of Soil Reaction Values for Buried Flexible Pipe." Journal of the Geotechnical Engineering Division. American Society of Civil Engineers, Vol. 103, No. GT 1, Proceedings Paper 12700. January 1977.

Howard, A. K. "Diametral Elongation of Buried Flexible Pipe." ASCE International Conference of Underground Plastic Pipe, New Orleans, Louisiana. March 30 - April 1, 1981.

Howard, A. K. "The USBR Equation for Predicting Flexible Pipe Deflection." ASCE International Conference of Underground Plastic Pipe, New Orleans, Louisiana. March 30 to April 1, 1981.

Howard, A. K., Personal Correspondance (1989).

Janson, Lars-Eric, "Plastic pipes for water supply and sewage disposal." Neste Chemicals, 1989.

Jeyapalen, Jey K. and Abdelmagid, A. M. "Importance of Pipe-Soil Stiffness Ratio in Plastic Pipe design." ASCE Conference, October 5, 1984.

Kienow, Kenneth K. and Prevost, Robert C. "Pipe/Soil Stiffness Ratio Effect on Flexible Pipe Buckling Threshold." TRB Conference, January 1988.

Komo Foundation, Netherlands. Deflection Measurements on Operational uPVC Sewers. April 1982.

Lauer, H. Statische Berechnung von erdverlegten Entwasserung-skandlen aus PVC hart und PE hart. February 1978. USBR Translation, 1983, "Static Calculation of Buried Pipes of UPVC and HDPE".

Leonhardt, G. "Die Erdlasten bei uberschuteten Durchlassen" Die Bautechnik, 56, No. 11, 361-68, 1979. USBR Translation, 1983, "Soil Loads on Buried Culverts".

Leonhardt, G. "Belastungsannahmen bei erdvertegten." GFK-Rohren. USBR Translation, 1983, "Loading Assumptions for Buried Glass-Fiber-Reinforced Plastic Pipes".

Leonhardt, G. "Soil Loads on Pipes with Different Degrees of Stiffness." Europipe 1982 Conference, Basel, Switzerland. January 18-20, 1982.

Luscher, V. "Journal of the Soil Mechanics and Foundations Division." Proceedings 92. American Society of Civil Engineers. November 1966.

Marston, Anson and A. O. Anderson. "The Theory of Loads on Pipes in Ditches and Tests of Cement and Clay Drain Tile and Sewer Pipe." Bulletin 31. Iowa Engineering Experiment Station, Iowa State College, Ames, Iowa. 1913.

Marston, Anson. "The Theory of External Loads on Closed Conduits in Light of the Latest experiments." Proceedings. 9th Annual Meeting, Highway Research Board. December 1929.

Molin, Jan, "Principals of Calculation for Underground Plastic Pipes - Load, Deflection, Strain", Owens-Corning Fiberglas Technical Report, Brussels 1974.

Moser, A. P., R. R. Bishop, et al, "Deflection and Strains in Buried FRP Pipes Subjected to Various Installation Conditions." TRB, January 1985.

Parmalee, R. A. and R. B. Corotis. "Analytical and Experimental Evaluation of Modulus of Soil Reaction." Transportation Research Record 519. Pages 29-38. 1974.

Richtlinien fur die statische Berechnung von Entwasserungs-Kanalen und-leitungen Abwassertechnische Vereinigung e. V. December 1988. Hardy Translation, "Guidelines for Static Calculation of Drainage Conduits and Pipelines - ATV".

Schrock, B. J., "Plastic Pipe Overview--1983." Europipe 1983 Converence, Basel, Switzerland. June 22-24, 1983.

Schrock, B. J., "Chronology of Technical Work of Flexible Pipe/Soil Interaction." Proceedings of GRP Seminars, April 1988, Abu Dhabi, U.A.E.

Spangler, M. G. "The Structural Design of Flexible Pipe Culverts." Bulletin 153. Iowa Engineering Experiment Station, Iowa State College, Ames, Iowa. 1941.

Spangler, M. G. Soil Engineering. International Text Book Co., Scranton, Pennsylvania. Second Edition. 1960.

Timoshenko, S. and J. M. Gere. "Theory of Elastic Stability." McGraw-Hill, Second Edition. 1961.

Water Research Centre, "Pipe Materials Selection Manual", Swindon, England, U.K., 1988.

Watkins, R. K. and M. G. Spangler. "Some Characteristics of the Modulus of Passive Resistance of Soil: A Study of Similitude." Highway Research Record 519, Washington, D.C. 29-38. 1958.

Flexible Pipe Design Revisited.

Robert C. Prevost [1], M. ASCE.

Abstract. An overwhelming soil factor in the Iowa formula and random deformations at installation affords soil a prominent rôle in flexible pipe; this, together with a maximal deflection rule make obsolete deflection calculation for design purposes.

Buckling sets this maximal deflection at 5% and determines the pipe minimal stiffness. "Dynamic buckling" is the preferred model because it rests upon established rationale and minimal assumptions about soil/pipe interaction, because it is the lower envelope of field and lab experience, and fits well with extant recommendationsi; alternative "composite" models do not fit soil behaviour, and allow too weak pipe in stiff soils.

Classical theory, advanced through "harmonic analysis" and confirmed by experience, evidences the primary rôle of the factors E'/S and p/S, the ratios of the modulus of earth reactions and internal pressure to pipe stiffness. The ratio E'/S (where $S=EI/D^3$) should not exceed a very few thousands, lest "squaring" develops .This with the above maximal deflection rule would keep deformed pipe elliptic and therefore strains under control.

Longitudinal effects bear overwhelmingly on the transverse cross sections, are significant, and alter the conventional loading. Pressure on pipe's shape defects causes significant "secondary" bending stresses as important as, or more, than hoop stresses. Conventional loadings can be admitted however though with some reservations. Superposition of maximal loadings (internal pressure, external loads, pressure's secondary and longitudinal effects) remains the rule as in all other fields of engineering, save, exclusively, for stress design of pipe made out of materials that are not sensitive to strains.

Experiments confirm the above views.

Introduction.

1/ This presention endeavours a synthesis of the available literature, of a series of publications prepared, jointly or separately, with Mr K. Kienow, of documents published in French about 30 years ago, and of a couple of recent papers in final draft form. Principles only will be reviewed with minimal justifications; details and demonstrations are in the references.

The views expressed rest on classical structural design or "strength of materials" principles, available experimental, laboratory or field data, including a series of tests carried out in our home country at the turn of the sixties. The basic references are documents of universal acceptance and availability such as Timoshenko's books and Roark's textbook (Ref. 27, 30, 31), which should facilitate checks and further developments. Yet these views often go beyond, or depart from, the present conventional thinking; some are new; some are controversial.

We review the principles governing flexible pipe design, the theories, the loadings, including longitudinal effects, applied to gravity and pressure pipe, with emphasis in the latter case on the so called "secondary" effects of pressure.

---

[1] Dr. Ir., 2402 Route des Vallettes, 06490 Tourrettes sur Loup, France.

We address design as the process leading to the determination of the pipe characteristics, eg., its stiffness or thickness, given its service conditions and material.

As a matter of principle, we consider all flexible pipe whatever material they are made of as one family, responding to the same basic design rules: their model is indeed the thin, narrow elastic ring, whether corrugated or plain steel, GRP or other plastic. The material characteristics become determinant at the time maximal deflections or strains, or stresses enter the design process.

Gravity Pipe.

The Iowa formula was set forth in 1941 by Spangler (Ref 29) who developped it along the classical methods of structural design or strength of materials:

$$\Delta_x = \frac{D_l K W}{EI/R^3 + 0.061\,E'}$$ We write it:

$$\frac{\Delta_x}{D} = 0.012\ D_l \frac{w}{S + 0.0076\,E'} \quad \text{or} \quad \frac{\Delta_x}{D} = 0.012\,D_l \frac{w}{S} \frac{1}{1 + 0.0076\,E'/S} \qquad (1)$$

to show the pipe stiffness $S=EI/D^3$ and dimensionless expressions.; Δ_x is the horizontal deflection;W is the uniform load on the crown, w the corresponding soil pressures; D_l is the lag factor; the bedding factor K is admittedly 0.1. Spangler indicated how to, but did not, calculate the vertical deflection, which was later done in Ref.24. He admitted that the vertical and horizontal deflections were about equal which experience on the pipe he was using --corrugated steel-- had evidenced.

Equation (1) shows that the main source of strength of flexible pipe is earth passive reactions, since the earth factor 0.0076E' overwhelms the pipe factor S. **The formula is** therefore **of little avail in design** because it yields very low stiffnesses, even absurd negative values, which simply means that soil limits deflection to lower values than admitted.

This fact has widely been aknowledged for sometimes. Yet the Iowa formula is still a basis of plastic pipe design though AASHTO no longer even refers to it in its metal culverts design rules (Ref.1). We should all do the same, but with a deflection limit that AASHTO has not.

Such conclusions applies to all methods inasmuch as they rely on deflection calculation or prediction, whether ATV, standards, applications of the finite elements methods, etc., but this by no means lessens the rôle of Spangler's model in flexible pipe theory, or of the classical methology we shall extensively use below.

The **strain equation** $\varepsilon = D_f \frac{\Delta}{D} \frac{t}{D}$ (2) is generally associated with the Iowa formula. It should be stressed that it **expresses merely the geometry of the deformations;** it is independant of any theory, and of the behaviour of pipe, and its derivation needs one only assumption that plane cross sections remain plane after deformation. Pipe, deformed into the ellipse type shape we advocate to maintain, has a D_f of about 3. Measurements indicate however that D_f may in practice be as high as about 40 which of course may signal problems. Formula (2) may understate the strains as we shall see.

The fact that flexible pipe stiffness has little importance in determining the pipe own deflection leads naturally to considering **pipe of negligible stiffness** (Ref. 18). In such case, there is no bending moment, no bending stresses, no shear, but a constant normal thrust, symbol P:

$$P = W/2 \text{ which becomess } P = w\,\rho \quad (3)$$

in a deformed pipe or a non circular cylinder or arch, ρ being the radius of curvature; the ring compression stress is P/A, A being the ring cross section, which the formula determines given the admitted wall compression stress; plain wall pipe's cross section A is generally redundant .

Pipe in such concept behaves like an arch of no bending strength, a circular one, abutting on its sides; this is typically **pipe arching,** and the very model of AASHTO metal culvert (Ref.1), though not explicitely set forth. This concept may seem contradictory to Spangler's **bending** theory; both however are complementary aspects of flexible pipe behaviour; the former applies particularly well to pipe made of material devoid of strain sensitivity (steel).

Prominence of Installation. As **soil is the main source of flexible pipe strength, installation is paramount**. Further, because flexible pipe lacks strength, installation can cause their deforming more or less than designed (eg. to elongate) and into shapes that may be irregular and far from the ellipse. The more flexible the pipe and the stiffer the soil, the more pipe can be affected. There are thus **"random" deformations, with associated stresses and strains, that are superposed upon, and may be as important as, or even greater than, planned and accepted "design" deformations.**

These random deformations occur as a result of apparently erratic variations of construction processes and soils properties; They are too large to be accomodated through "allowances". **Installation therefore, not "design" in its usual sense, determines pipe's eventual shape; and, as a maximal deflection is to be set, it is pointless in design to further calculate or try to predict deflections** This is not yet aknowledged.

The implications are that the fills around pipe are **"structural" backfills** and that together with the installation procedures (bedding, backfilling, and compaction), they **are integral constituents of design**, which indeed they are in AASHTO. This in turn implies close supervision of work in progress, training of labour and even engineers, guidance by pipe manufacturers, authorization to proceed with construction only upon satisfactory performance (deflection) possibly of test, certainly of completed, sections (see eg. Guidelines 5 in Ref. 21).

Buckling is no more than normal deformation that could proceed in conditions where:

--either the bending moments due to external pressure (formula (5) below, $M_r=-pr\partial$, with negative pressure p) would grow with deformation faster than pipe's reaction (formulas (6) and (7) below) --which is the weak pipe case;

--or vertical deflection would go on increasing while the horizontal "diameter" would stop expanding, even shrink, which suppresses "pipe arching", the very source of flexible pipe strength; this is the "excessive deflection" case. [2]

In such instances, deformation would feed further deformation, which is typical **instability**.

At the onset of buckling, there is thus little difference if any with normal deformation and therefore buckling may not be readily identified as such; it would evolve towards large changes of curvature, flattening, reversal of curvature, very

---

[2] There is a third mode, the local wall buckling of profile wall pipe, which is set by the profile design. We do not address it, because it is entirely the responsibility of the manufacturer to determine wall designs which do not allow local wall buckling of normally deformed pipe.

slowly, over years, decades, and may possibly proceed untill final collapse.[3] Such failure may be attributed to poor installation which may actually be the initial cause, but the failure process is buckling.

In the "excessive deformation" case, the vertical deflection is about the same as, or proportional to the horizontal one until it reaches 15% (Ref. 9); "pipe arching" or soil side support at the spring line abutments breaks down at about 20% (Fig (1)). These facts are aknowledged since the beginning of this century (Appendix to Ref.2,p.546) and noted by Spangler (Ref.29), yet sometimes overlooked nowadays. **With a comfortable and quite current factor of safety of 4, the maximal allowable deflection was set at 5% which should be fully reinstated.**

The theory of the straightforward "weak pipe" case is well over a century old, yet this buckling mode is now the subject of a controversy. There are two main streams of thoughts: the first in which pipe is straightforwardly subjected to pressures from, and constrained by, soil; the second which considers a composite pipe-soil structure.

Ken Kienow's and our's contribution (Ref. 11, 12, 19, 24, 25) has been the **"dynamic buckling"** (DB) model, wherein the classical theory is applied to the 5% normally deformed pipe taken as an 180° arch abutting on the soil on its sides, with fixed ends . The sole assumption about the pipe-soil interaction is that **external pressures are <u>maintained</u> on pipe while buckling possibly proceeds**, as a result of, eg, traffic which shoves pipe and earth to and fro as loads move over.

The **deformation** of the pipe **is taken into account** by assuming that the pipe stiffness is that of a cylinder whose radius is the radius of curvature of the deformed pipe at crown; since the latter varies about three, actually D_f, times as fast as the deformation and the stiffness is function of the corresponding diameter at the third power, **this stiffness is the pipe's stiffness reduced by the factor $1 / (1+ D_f \Delta/D)^3$**, or by 1/3 for 5% deflection. This assumption builts safety into the formula: for this and other reasons, we do not believe it necessary to apply at this stage a safety factor different from 1.[4]

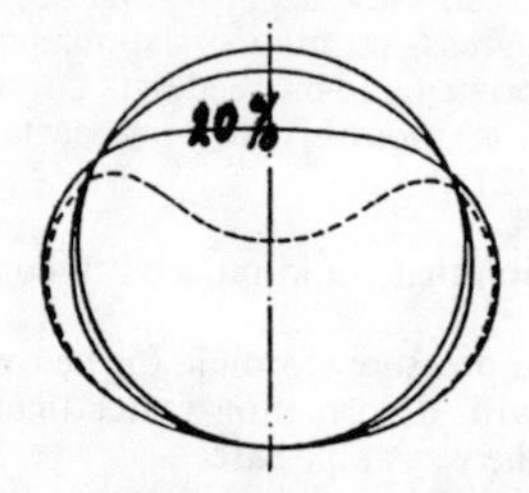

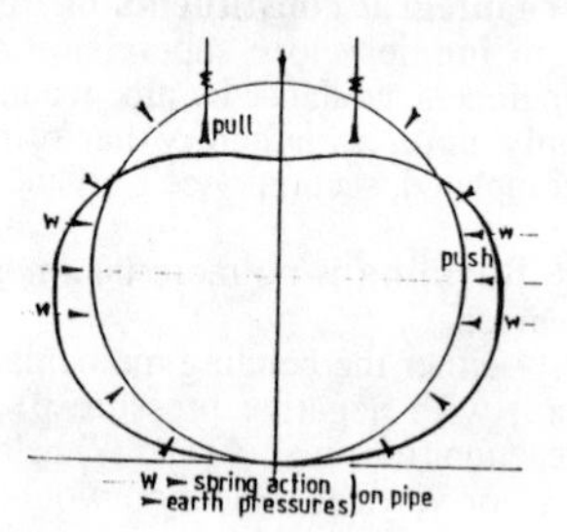

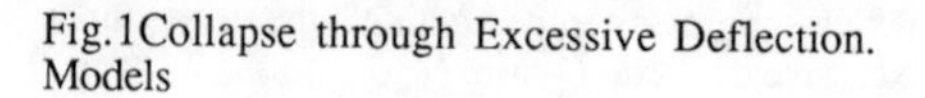

Fig.1Collapse through Excessive Deflection. Fig.2. Sketch of the Composite Models

The **composite models**, which are Luscher's, Winkler's, Moore's, Continuum,

---

[3] Buried pipe buckling may be delayed, even, possibly, prevented: an inadequately installed section may be supported by better installed adjacent ones; the loadings vary from section to section; part of them are transient loads (traffic), and maximal loading may unfrequently occur, pressure stabilizes and reround pipe; but soil may vary with time, work around pipe remove soil support, longitudinal effects redistribute the loadings, overloading sections.

[4] As long as a reduction of the loading through longitrudinal effect is not counted.

consider pipe subjected to external pressure, and bonded to soil (with slippage sometimes admitted though tension bond generally maintained [5]); and soil is implicitely presumed to be a solid to which the theory of elasticity is applied. Springs often represents the pipe-soil interaction as in the image Fig.2. As long as the "springs" are under compression, such image is satisfactory. However, a buckling pipe has a bi-, or multi-, lobe shape. For each outward bulge, which compresses soil or springs, there is an inward bulge which pulls on springs or soil, but **soil cannot withstand** such **tension; and soil is no solid,** but a large number of minute moveable solid particles. This theory is thus flawed.

From a practical point of view, **composite models allow more flexible pipe in stiffer environment and very weak pipe**; their buckling pressure depends indeed from the square root of the product E'S. This is unsafe.

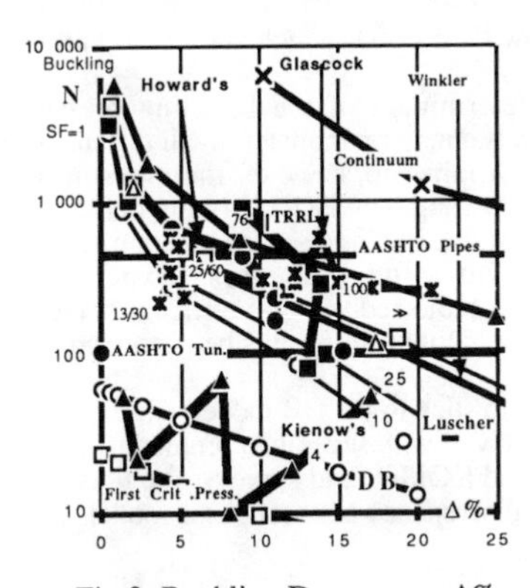

Fig.3. Buckling Data versus Δ%.

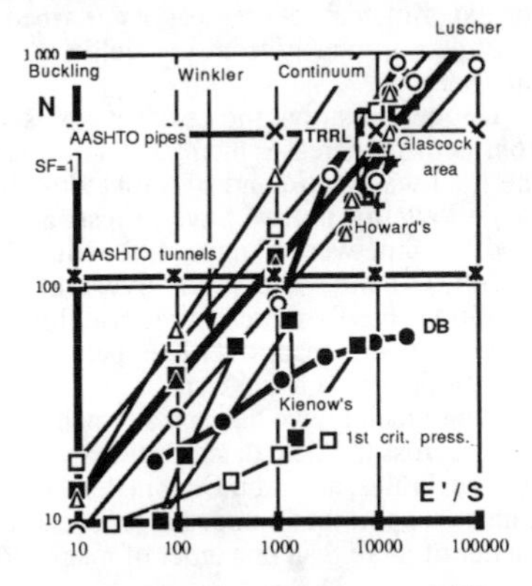

Fig.4. Buckling Data versus E'/S

In any event, the proof of theories lies in actual experience, from the field, and from the laboratory . Such confrontation is attempted in Fig.3 and 4 [6]; it is uneasy and volatile, because data are various, often incomplete, and need some processing, but suffice the purpose.

These scant data can be sorted out into three main groups:

a/ Pr Watkins tests in a collapsable wall lab. cell: about 12 dots virtually

---

[5] There is a third approach to buckling (Ref. 7) in which soil is given a <u>one sided elastic behaviour</u>: where soil is compressed, it behaves as in the "composite" model; where soil is under tension, pulled, pipe is alone in whithstanding inward bulging (eg, around the crown, Fig 1). This removes some of the above shortfall.

[6] Buckling theories quickly lead to indentifying a ratio we called N=w/S, buckling soil pressure to pipe stiffness $S=EI/D^3$ (Ref.10, 24). The first critical pressure is $3R^3/EI$ or N=24. The DB has a N of 64, that 5% deflection reduces to 40. It suggests a minimal long term stiffness of 1500 Pa (PS of 12 psi) in usual conditions, ie 1 to 3 m (3 ft) earth cover and maximal live load plus impact. The composite models have N of at least 500 up to several thousands, for 5% deflection. The theory shows also that $N=8(n^2-1)$, n being the number of lobes of the buckling process. It is unlikely that N, even in the stiffest environment, could exceed about 400, or n exceed 5 or 6, because the difference between the lengths of the chord and of the buckling arc fast decreases under O.3% for higher n, and therefore very small slippage of the pipe within its encasement or settlement of the abuttments would allow pipe flattening.

confounded with the first critical pressure line **sloping down with deflections in accordance with the above factor $1/(1+D_f\Delta/D)^3$**, save for very low deflections, ie, high E', which join the Dynamic Buckling (DB) line: **the collapsable walls would prevent the building up of soil arching**, hence correct evaluation of the vertical loads, and the pipes, through their deformations, would equalize the side pressures to the vertical ones, hence realizing almost the hydrostatic pressures of the theory. This is particularly clear in Fig.4 of Ref.11.

b/ series of dots, scattered among the Luscher's (and Winkler's) formula curves, well below the "Continuum" curve. Only TRRL (Ref. 32) and Howard (Ref. 9), with 4 failure examples each, document their experimental device. In TRRL, soil deformations were suppressed; in Howard's, a confinement effect was noticed. In Glascock et al.(Ref.8, used for an AWWA C950 updating), E' was not measured and the experimental set up was not described.

c/ data primarily from the field ("Kienow's"; 6 dots), which should prevail over lab. ones.

Unquestionably, the test set up is a determinant of the buckling pressures; confinement increases them; and deformation is a major parameter. In all the lab. tests, the load was **static** and of **short duration,** applied to soil over the pipes through **large bearing plates**. Save in tests a/, and, perhaps, in TRRL's (wide cell), actual loads on pipe were Marston's, ie reduced by frictions along walls or within soil (soil arching), hence **the tests overstate the buckling pressures**. Experimental conditions hardly represent the real thing: pipe subjected to static earth and travelling (traffic) "point" loads over long periods of time. Doubts must thus be expressed about the validity of such lab. tests.

Consequently, we have no alternative than to stick to the DB model, whose curve runs accross the field data; further, the DB fits well with several recommendations or practical rules, and Abu Dhabi (Ref.11, 25) and KOMO field surveys. We thus cannot support the "composite" models. The discrepancy between these models is a matter of more than one order of magnitude.

Flexibility factors (FF, MS, or RSC), are still refered to in AASHTO, and, sometimes, in connection with plastic pipe. They are meant to measure "handling and installation rigidity", no more. They are discussed in some details in Ref.18. They are empirical data, have no rational background and appear useless; stangely, such "rigidities" decrease together with the modulus of elasticity of the pipe materials. We suggest to discontinue practicing them.

<u>Longitudinal effects</u> are seldom discussed, yet **sometimes may be more important than those of the conventional transverse loadings**. They disturb the distribution of stains, stresses, deformations. They are caused by **uneven longitudinal pipe loading** as a result of "point loads" ("live loads"); of earth loads variations; and of bedding imperfections, under- or over- dug "bell" holes. Pipe is a beam with a peculiarity: its transverse section can easily be deformed.

It cannot be overemphasized that it is **the transverse sections that are overwhelmingly affected**, because the **shear forces deform them** likewise the transverse loadings, and are superposed upon the latters; shear has the same order of magnitude as, though about half the factors of influence of, the transverse loadings. The effects on the transverse sections are equally suppressed by the side earth passive reactions. This is not in the pipe literature, save a few indications in Ref 23. The study of longitudinal effects replaces the usual "rings" which have no connection with the rest of the pipe, by rings loaded on their faces by shear, and also thrust, but the latter is generally unimportant save in small pipes and very flexible large ones (Ref.20).

The beam effect is significant: pipe, as a "beam on elastic foundation", spreads "point loads" already distributed through Boussinesq on a length of the crown, onto a longer stretch of its bedding (Fig.5). With an earth cover of 1 m (3ft) over a pipe of indefinite length (all welded steel line), the above crown loaded length is about 1.4 times the depth of cover, and the load on the bedding under it, is reduced to between 1/2 and 3/4 of the crown load; but under a pipe 6 m long (20 ft) it is up to about 8/10 under a centered load, with, possibly, some overloading when the load is applied at one of its ends. Bell holes should not cause problems, save in small sizes, and flexible pipe; inadequate foundation support may significantly increase transverse loading in all pipe and very flexible pipe longitudinal strains (Ref.10), and even be devastating for pipes installed on "mounds". **The most loaded area may pass from the invert of the conventional "ring" to the crown[7], which fits many experimental evidences, and, through shear, transverse loadings are redistributed among pipe cross sections, unloading some of them, but overloading others.**

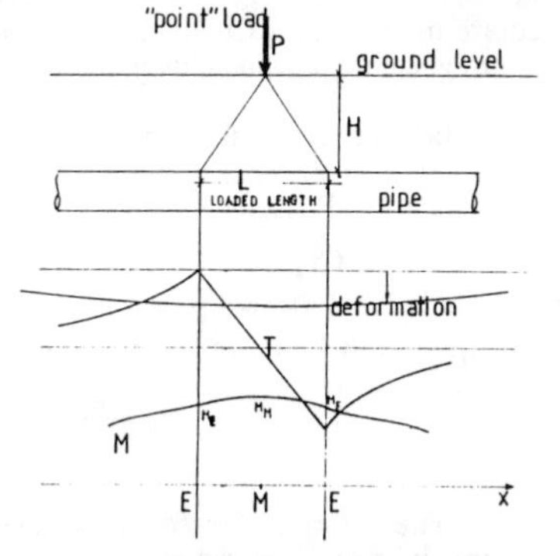

Fig.5. Longitudinal "Point" Loading

For practical purposes, all finite length (6 m) pipe , save the smallest and the most flexible, are longitudinally rigid. Practical rules ensue: no joints between unprotected pipes crossing an heavily trafficked road; pipe barrel designed for axel loads over pipeline crossings, but single wheel load for pipelines along a road; and the usual no hard spot, adequate backfilling everywhere, even bedding, mostly standard rules, sometimes overlooked. The above effects are underestimated, because mainly of two basic assumptions they rest upon: point loads on soil which are in fact always spread on a surface, yet limited; and, more importantly, uniform distribution over the pipe loaded length and pipe diameter of a pressure equal to Boussinesq's maximal intensity. Such assumptions were necessary in a general assessment of the longitudinal effects, but can be refined. As a by-product, we recommend for estimating the live loads on pipe a method more in line with the Boussinesq distribution of pressures in soil and hence better for pipe than the conventional ones.

Because of their complexity and variableness, it would not be feasible to include longitudinal effects into design save in special instances. That they are allways present, and that they are important must be kept in mind. They are usually, but not knowingly, included in the safety factors; however, if they are taken into account to reduce the loadings, safety factors ought to be increased.

[7] Frictions along trench walls may contribute.

Pressure pipe

Most generally, the literature considers solely the main, direct, effect of (internal) pressure on pipe, the "hoop" stress; sometimes, a possible rerounding is taken into account; very seldom are other aspects analysed, such as the pipe's stiffening or shape defects, which effects of however may be as, or more important than, those of the hoop stress. Much is in French papers (Ref.13); some aspects are in Timoshenko (Ref.30, 31); Spangler (Ref.28) touched on the subject and proposed formulas which are used in AWWA C950 (Ref.4); the latters however understate the pressure stiffening by some 50% (save deformations in parallel plate loading), do not include the earth side passive reactions; their rationale is short but crude in adding perpendicular vectors algebraically and omitting the hoop forces. Ref.22 develops the subject in some details.

The hoop stress σ is $\sigma = p\,D / 2\,t$ (4)

It has other expressions, but we keep this one for thin pipe.

The **"secondary effects" of pressure** require more complex mathematical tools than the general pipe literature, which will in turn generate new insights in the gravity pipe and buckling.

The Fourrier transform and the nodal center.readily show that the "rerounding" moment due to the hydrostatic pressure p in a ring of radius R, affected by a **radial deformation** ∂, assumed small compared to R, is:

$$\mathbf{M_r = -\,p\,R\,\partial} \qquad (5)$$

Such rerounding moment is quite important: in a 3% deflected pipe subjected to 100 kPa (1 kg/cm^2 or 15 psi) pressure, it equals the bending moment due to 1 m (3 ft) of fills but no traffic (earth pressure of 20 kPa).

TABLE 1 Samples of Fourrier Series Coefficients

	m2	m3	m4	m5	m6
P-P	-0.212	0	-0.042	0	-0.018
P-Φ1 (invert)					
30°	0.0035	-0.003	0.0034	-0.003	0.0032
90°	0.0244	-0.021	0.0175	-0.014	0.0102
150°	0.0414	-0.031	0.0212	-0.014	0.0091
Shear	-0.106	0.04	-0.021	0.013	-0.009
Spangler's	0.178	0	-0.016	0	0.001
φ1-φ2 (invert--crown)					
30-180°	-0.165	0.028	-0.018	0.010	-0.006
90-180°	-0.144	0.011	-0.004	0.000	0.001
150-180°	-0.127	0.001	0.000	0.000	0.000

The Fourier transform expresses the moments and radial deformations along the "ring" representing pipe, through an indefinite series of cosine and sine of the angle of reference x, counted from the crown; the loading of the originally circular ring and hence its deformations are assumed symmetrical with respect to the vertical axis, so that the sine terms disappear (save in connection with shape defects which may take all sorts of forms):

$$M = W\,R\ \Sigma\, m_n \cos n\,x \qquad (6)$$

and

$$\partial = (W / S)\ \Sigma\, \partial_n \cos n\,x \qquad (7)$$

before applying pressure, where n is an integer equal or higher than 2 (n=1 corresponds to a translation not a deformation), and the m_n are coefficients for each cosine term, to becalculated from the available formulas in, eg, Roark's textbook (Ref.27), through the Fourrier transform .

Samples of the m_n coefficients are in Ref.22 for typical loadings such as parallel plates, vertical uniform loadings, Spangler's parabolic earth side reactions; an excerpt is in Table 1. In general 4 to 8 m_n terms suffice, save for the parallel plates test stresses which require up to or more than 50 terms; the deformations are less demanding. A lag factor may be added within (7) if needed.

The relationship between ∂_n and m_n is: $\partial_n = m_n / 8\,(n^2 - 1)$;

(the factor 8 results from the use of $S=EI/D^3$ instead of EI/R^3)

Including the rerounding moment of formula (5) in the basic relationship between moment and curvature leads, after integration of a linear differential equation, to the following pressure factors m_{np} and ∂_{np} corresponding to the non pressure m_n and ∂_n in formulas (6) and (7):

$$m_{np} = m_n \frac{(n^2 - 1)}{(n^2 - 1) + \frac{p}{8S}} \quad \text{and} \quad \partial_{np} = \frac{\frac{m_n}{8}}{(n^2 - 1) + \frac{p}{8S}} \qquad (8)$$

Pressure moments M_p and radial deformations ∂_p are not linear with respect to p:

$$M_p = WR \sum m_{np} \cos n x \qquad (9)$$

$$\partial_p = W/S \sum \partial_{np} \cos n x \qquad (10)$$

and $\partial_{np} = m_{np} / 8\,(n^2 - 1)$, which copies the above ∂_n.

A "radial" strain or shape factor we call Δ_f to avoid confusion with the diametral D_f is:

$$\Delta_f = \frac{\sum_{n=2}^{n=\infty} \frac{(n^2-1)m_n}{(n^2-1)+p/8S} \cos nx}{\sum_{n=2}^{n=\infty} \frac{m_n}{(n^2-1)+p/8S} \cos n x} \qquad (11)$$

which takes place in a formula applicable to **radial deflections** and similar to (2):

$$\varepsilon = \Delta_f\, \partial/\mathbf{R}\ v/D \quad \text{or} \quad \Delta_f\, \partial/\mathbf{R}\ t/D \quad \text{for plain wall pipe} \qquad (12)$$

Δ_f is a local strains factor, applicable anywhere all along the circonference; it does not make an average over the major diameters, like D_f ; it is what actual measurements show through, eg the chord / offset method.

The practical loading cases (combined PP, crown, invert, and shear loadings) are built up as usual, as in Ref.22 or 24, by adding the elementary loading coefficients of Table 1, modified according to formulas (8) and including them in formulas (9) and (10). Combining vertical loadings with Spangler's requires writing that the earth side reactions with parabolic distribution are proportional to the horizontal deflections; the ratio **E'/S and p/S appear** then **as determinant** in the expression of moments, deformations and Δ_f, per Fig.6 to 9 (Ref.22).

Fig.6 shows the coefficients $-m_2$, m_3,and $-m_4$, for vertical uniform load on the crown (ϕ_2=180°) and an average bedding sector of ϕ_1=90°, in function of the parameters E'/S and p/S. which both encompass, in these figures, a wider range of rigid and flexible pipe stiffnesses, E' values, and pressures than found in normal practice.

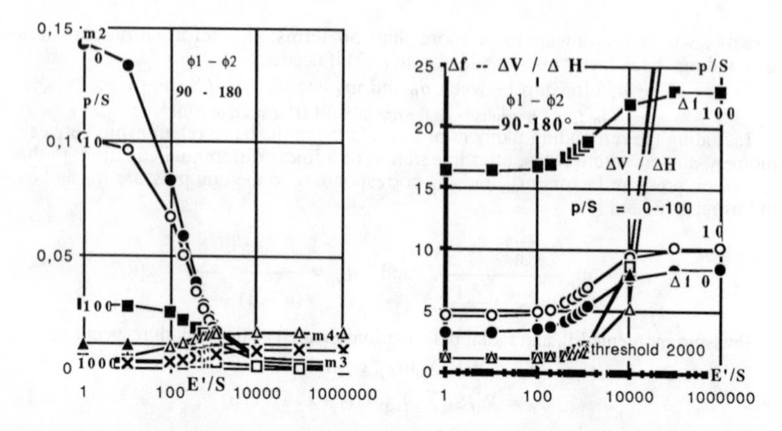

Fig.6. Combined m Coefficients Fig.7. Combined Δ_f and Δ_V/Δ_H against E'/S and p/S.

The effects of both the E'/S ratio and pressure parameter p/S are drastic: the "elliptic" component m_2 (which alone does not define an ellipse) **dwindles down to virtually nothing between E'/S of, about, 10 and 1000, the shape of its curve is not significantly altered, but the whole curve quickly flattened by pressure; the term n=3 is not changed by E'/S,** because there is no third order term in Spangler's loading, **but is suppressed by pressure and disappears altogether with all odd terms when deformation is symmetrical with respect to the horizontal axis; the n=4 term increases in absolute and relative importance with E'/S,** because these terms in the vertical loading and Spangler's have the same sign; it is less suppressed by pressure than terms 2 and 3, and higher order terms lesser also. **There is thus a significant trend towards squaring about E'/S = 1000. This was noted** in earlier papers of ours (Ref.24), and, **experimentally, by Howard** (Ref.9) who indicates that squaring appears about E'/S of 400-500, and is pipe's deformed shape above 2000-2400, with plastic hinges forming diagonally, which fits in with above theory. This concurs also with the recommendation in the early version of Ref.3 (EI no less than 0.1 to 0.15 times 0.061E' R^3, or

E'/S< 8 / (0.1 to 0.15 x 0.061) ie 1000 to 1300)

Fig.7 shows Δ_f and Δ_V/Δ_H. A **threshold** appears: **Δ_V/Δ_H rises steeply beyond E'/S≅ 2000**, whatever the pressure, while the **maximal Δ_f at the invert rises from 3-4 to about 10, and much more under pressure.**

The admitted D_f (3 for the ellipse, 4.27 in PP loading, 6 in AWWA C950) **could thus** apparently **understate the strains. The strains however are determined by formula (12), replacing (2); both include the product of two terms, both affected by E'/S (and the pressure): Δ_f and ∂/R. This product and hence the strains are very significantly suppressed by both E'/S and pressure p/S,** as shown in Fig.8. D_f should not be focussed on alone.

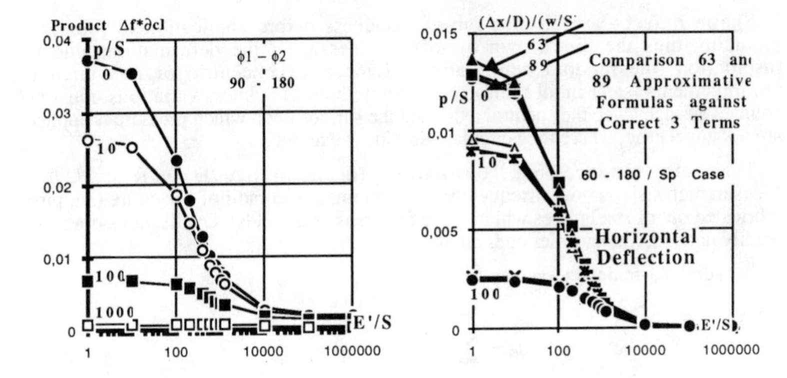

Fig.8. Product $\Delta_f*\partial_{cP}$ versus E'/S and p/S Fig.9.Approximate Formulas:Δ/(w/S)

"First order" or approximate formulas. The use of the above equations is not "handy" because of the number of terms which may be needed. Simplifications may be sought by taking into account only the term of order n=2, the "elliptic" component; this works well but **exclusively for the horizontaal deflection.**and writes (bedding sector of 60°):

$$\Delta x/D = 0.013\ w/S \,/\, [1 + p/24\ S + 0.0077\ E'/S\ ] \qquad (13)$$

which is virtually Spangler's or equation (1), with a pressure term, without lag factor. Fig.8 compares this horizontal deformation, the 1963 one from Ref.23, and formula (10) with 3 terms, for varying E'/S and p/S, with D_l=1. For all practical purposes, these formulas are equivalent. But the ratios Δ_V/Δ_H and the Δ_f of Fig.7 indicate clearly that <u>such simple formulas do not apply to the vertical deflection, nor to the moments, strains and stresses, beyond the threshold</u>. Up to it, vertical deflections are equal, and moments, strains, stresses proportional, to the horizontal deflection.

The stiffening effect of pressure is significant --and implies pipe rerounding--.The pressure and approximate formulas indicate, eg, that, at least with respect to the main "elliptic" component (term n=2), a pressure of 50 psi (3.5 kg/cm^2 or 350 kPa) has the same effect as a stiffness EI/D^3 of 15 000 Pa or a PS of 120 psi.

Rerounding would however **not proceed in all circumstances**: under shallow earth cover, even a moderately pressurized pipe would reround but only in part when transient "traffic" loads are off, and the new pipe shape would to a certain extent consolidate as a result of traffic itself or other causes; but a pipe under high fills would have to exert an upward push of 2 to 3 times its permanent earth load to overcome frictions which soil would increasingly oppose; pressure may not provide that much. Yet, the higher the deflections after installation, the higher the potential for rerounding during pressure tests and operation.

Shape defects [8] are all pipe out-of-roundness before application of pressure, including thus the pipe's own out-of-roundness, and the deformations due to installation, the random deformations. Even non concentric or non circular reinforcements (steel in all reinforced concrete pipe), thickness variations, must be counted as such since the "neutral axis" and the surface upon which pressure is applied are not concentric, where they circular. No pipe is exempt .

There are tolerances, eg, "ovalisation" (order n=2) $\Delta_0/D=\partial_0/R$ of 1/2 % . Measurements show not unfrequently 15 % anomalies in radii of curvature (eg, pipe fabricated out of steel plates which have flat areas near welds). Ovalisations often are readily noticed, higher order ones are not.

A **radial** shape defect ∂_0 may be written

$$\partial_0 = \sum_{n=2}^{n=\infty} \partial_{0n} \cos(nx + \alpha_n)$$

∂_{0n} is the component of order n; no symmetry is to be expected from shape defects, hence α_n, ∂_{0n}, α_n and n may have any values (α_n.substitute the sine terms).

Developped as above the rerounded shape defects ∂ are:

$$\partial = \sum \partial_{0n} \frac{(n^2-1)}{(n^2-1)+\frac{p}{8S}} \cos(nx+\alpha_n) \quad (13)$$ and the corresponding moments are

$$M_r = -pR\partial.$$

To the extent rerounding proceeds, the "elliptic" component of the shape defects is the most corrected; the higher the order, the lesser are defects suppressed by pressure.

The induced bending stresses are: $|\sigma_s| = pR\partial / I/v = 6pR\partial / t^2 = 6p\frac{D}{2t}\frac{\partial}{t}$;

which compare to the hoop stress, $\sigma = pD/2t$ through the ratio σ_s/σ :

$$\sigma_s/\sigma = 6p\frac{D}{2t}\frac{\partial}{t} / (pD/2t) = 6\,\partial/t = 3\,\partial/R\ D/t$$

An **"ovalisation"** (n=2) **$\partial_0/R=0.5\%$, unrerounded and not constrained, causes maximal bending stresses (strains) of 0.015 D/t times the hoop stress (strain)**, ie, a high $\sigma_s = 1.5$ times the hoop stress for a pipe of D/t = 100; this could adversely affect stiff pipe (low D/t or SDR), that reround less. **Free to reround**, the ovalisation drops to that of equation (13), ie, for pipes at the end of the normal ranges, under a 500 kPa pressure (70 psi):

-- for a very thin steel pipe of D/t of 100 (S=17kPa), $\partial/\partial_0 = 0.45$ and

$$\sigma_s/\sigma = 0.45\ 1.5 \approx 65\%$$

-- for a possible plastic pipe of SDR =50 and S= 1500 Pa, $\partial/\partial_0 = 0.07$ and

$\sigma_s/\sigma \approx 5\%$, but

-- " " " " " SDR =25 and S≈12000 Pa, $\partial/\partial_0 = 0.35$ and

$\sigma_s/\sigma \approx 12.5\%$.

These are high stresses, that may often be exceeded in practice, since

---

[8] This subject was developped in the 1960's in Ref. 23 and 26. For further details, see Ref. 22.

shape defects are not all simple 0.5% ovalisations; indeed strain gauges on steel pipes free to reround as they are during pressure tests, currently show stresses and deformations as in Fig.14. With yield strength exceeded, pipes may be permanently but partly rerounded.

The **free rerounding of and the stresses (strains) due to, shape defects can be accurately predicted**, knowing the shape of pipe free of pressure: the Fourrier transform provides the necessary methodology: an example is in Fig.15 to 17

Pipe buckling is a particular case of secondary effects of pressure, when negative pressure equals the denominator of formulas (8) to nought; the series of critical pressures ensues.

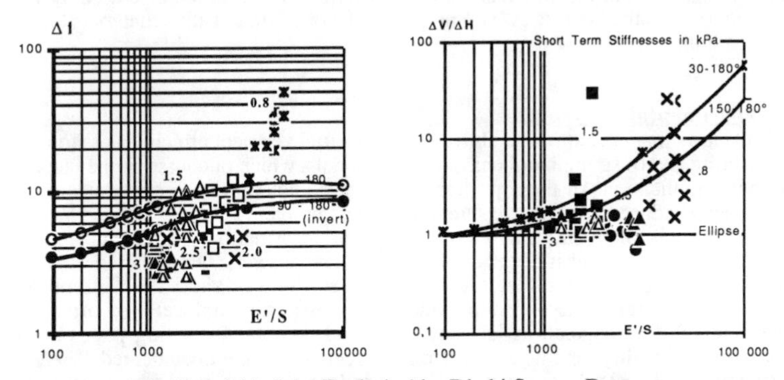

Fig.10 and 11. John Taylor's Abu Dhabi Survey Data.

All the above pressure chapter deals with the parameters of the thin elastic ring; **the random deformations** (and associated strains or stresses) **add their own disturbances** that are affected by pressure; the pipes' shape defects do the same, and the longitudinal effects also, but the latters' action on transverse bending are also suppressed by the earth side reactions. It is thus surprising that, in spite of the above possible disturbances, the Δ_f and $\Delta_V/_H$ theoretical curves run almost right accross the clouds of dots from theJohn Taylor's Abu Dhabi survey (Ref.24), the sole field survey that yields appropriate information (Fig.10 and 11)

Loadings are reviewed in Ref.18. We would stick to the conventional methodology, though with reservations, particularly in connection with **combined loadings.**

The established rule of design is to take into account the worst combination of loadings. It is however infringed in the design of steel pipe, where pressure and external loads are not superposed, exception which was extended to all flexible pipe, but changed by AWWA C950 for GRP.

That "conventional formulae apply only up to a certain strain level and not beyond" (AWWA M11, Steel Pipe Design, Ref.5) is inadequate formulation. The fact that the loadings, pressure and external forces, all are extant cannot be denied; the pipe's material, its flexibility does not change this in any way: pipe is bent, subjected to pressure, shape defects, etc., all superposed (though some of them are not even mentioned in the literature). Yet the above exception is valid.

The fact is that certain materials are not sensitive to **strains**, whether because they yield at a determined stress level, loosing only little of their ductility, like steel; or because they can sustain very high strains, a few % in certain plastics; or through stress relaxation or creep. Though the strains add up algebraically anyway, the stresses do so also but untill yield occurs (steel), where the stresses reach and stay at a ceiling, even under increasing load --to decrease thereafter if loading or deformation subsides. Whenever such kind of stress relaxation can occur, as in pipe bending under external forces or pressure rerounding, these stresses need not be taken into account in "stress design" provided they are relaid by some other mechanism securing pipe steadfastness, such as pipe "arching" over its sides abutments, or pipe rerounding. In any event, pipe undergo another shape which reflects past strains,and can be taken as the "zero" stress state (though it is not), to which are applied the stresses that cannot be erased by yield or creep, such as the hoop stresses. But pipe made of strain sensitive materials bear all their loading.

Experimental Support.

The views that are expressed above rest upon strict classical principles of structural engineering and of strength or behaviour of materials, which of course fit the facts; we extended applications of these principles to aspects not yet developped, as our own experiments induced us when we stumbled accross problems.

Literature is plentiful, and yield a mass of information but, because of the complexity of the matter it does not often yield clear answers. Often there is a lack of basic information on pipe material, --whether eg, long or short term plastic pipe stiffness are referred to-- test parameters, or experimental set up; often also experimental devices such as lab soil boxes and their large loading plates do not represent well reality, or cause confinement effects or are not documented. There are very few field tests or surveys, still less failure reports. In the end very, few experiments can be relied upon to ascertain the fundaments upon which one can build.

We were privileged thirty years ago to proceed with a series of tests which purpose of soon became endowing with a rational and experimental basis the relatively recent --at that time-- American practice in the design of steel pressure pipe which was empirical in its essence. It all started about 1954 when measures within a steel pipe laid under a highway outside Brussels, in anticipation of actual construction of a major distribution main, and therefore accessible, showed deflections more than one order of magnitude smaller than predicted. Spangler had issued his formula in 1941, during the war; AWWA published its M11 Manual in 1964, after a series of Barnard's papers.

There was thus an inducement to a closer look at the design of steel pipe, endeavouring to explain and bring under the aegis of classical theory apparent anomalies which showed up when deflections and strains were measured. Theoretical aspects, experimental devices and instrumentation were improved in successive steps, till they yielded results satisfying our purpose. Building upon a genuin interest in the country about pipes, tests proceeded in Brussels Water Works (CIBE) with publications in French by its senior staff untill the Belgian Water Works Association (Anseau), a water technologies research center emanating from Gent University (Becetel), the Public Works (PW), the Belgian Geotechnical Institute (IGE), and pipe manufacturers (UTM), with assistance from Liège University (LU) joined forces for a final field test near the Eupen Dam operated by PW. LU later continued. I ow to thank all those who collaborated in these enterprises, particularly Messrs De Saedeleer and Meesen (CIBE), Declercq (PWW), Boone (Becetel), and Prs De Beer (IGE and Louvain Univ.), Soete (Gent Univ.) and Dehousse (LU).

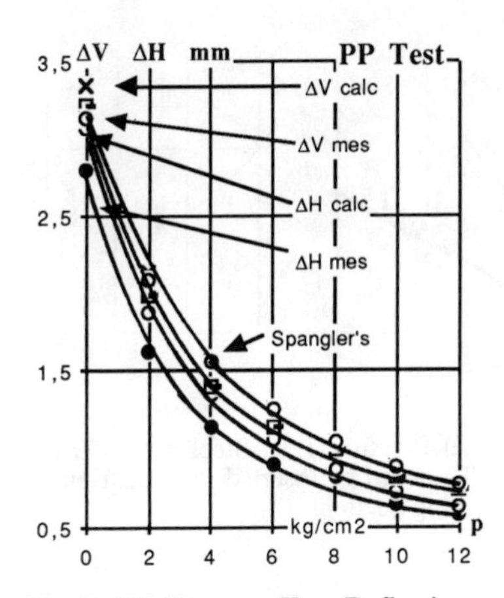

Fig.12. PP / Pressure Test. Deflections
P=250 kN/cm.

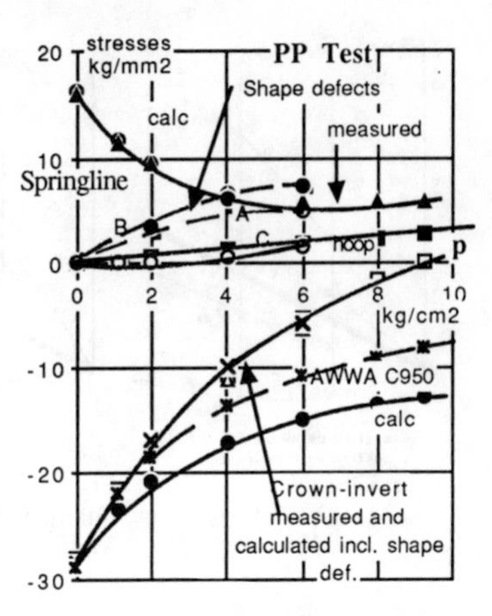

Fig.13. PP / Pressure Test, Stresses.
P=500 kN/cm

There were two types of tests, laboratory tests, mainly "parallel plates" (PP) of pipe specimens under or without pressure, and pipe pressure tests, in which a few well defined parameters (pipe, pressure and, in PP tests, the deflecting force) only interact, thereby allowing accurate and reliable measurements and ascertaining the theory; and, on the other hand, field tests.in trenches mostly under, or worse than, actual service conditions, pressure and traffic. **No** lab. test in earth box was made, because of unavailabiblity of such device, but, above all, because of a lack of trust in its representativeness. Buckling needed not to be investigated. All addressed steel pipe.

A typical PP test was that of a 206 mm (8 in), 60 mm (2.4 in) long, 1.85 mm (and 2.7 mm) thick steel ring, D/t=111.5, stress relieved and then machined to free it best from shape defects. Pressure was applied through a rubber bladder; two dummy rings besides the test ring were deflected through screws by the same amount as the ring itself, to cancel the effects on the ring of the bulkheads and of the frictions with the bladder. Strain gauges about 1 cm long (much for a 20 cm pipe) were glued on the external surface. Fig.12 and 13 compare calculated and measured deformations, and residual shape defects, bending, and combined stresses: the fit is very good, though the tests brought the ring beyond the yield strength (about 25 kg/mm^2); the ring ended 1% permanently deformed. Steel's E appeared to be 2.2 10^6 (31.5 10^6 psi) since the calculated S including the Poison's correction was 13 900 Pa and the measured ones were 14 350 (vertical deflection) and 15 400 (horizontal one), or PS of about 110 psi. Spangler's or AWWA C950 approximation is satisfactory with respect to deflection; it is not to the stresses.

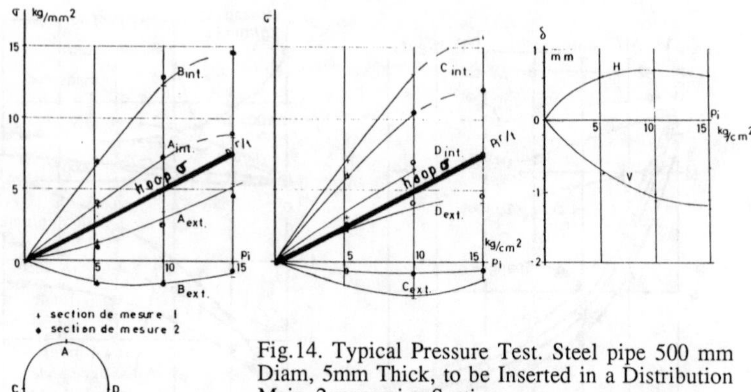

Fig.14. Typical Pressure Test. Steel pipe 500 mm Diam, 5mm Thick, to be Inserted in a Distribution Main. 2 measuring Sections.

All of the several pipe pressure tests show the same picture as Fig.14: shape defects cause maximal stresses often equal or larger than the hoop stresses; the average of the inside and outside stresses at the same spots is in good agreement with the calculated hoop stress. The shape in the measuring sections of the Eupen pipe below was that of Fig 15: the out of roundnesses were for term n=2, 2 mm; n=3, 1mm; n=4, 0.2mm; n=5, 0.7mm; and n=6, 0.1mm, as calculated through the Fourrier transform formulas applied to the measured shape. The predicted (formula (13)) and measured stresses in the two measuring sections are in Figure 16 and 17 ; the fit is also good.

Steel pipes of various sizes, from DN 500 to DN 1000 (40 in) were tested in the field, starting with simple tubes, of, eg 1m and thicknesses of 6, 10, and 15mm, in a normal size trench, loaded through an hydraulic jack abutting on an overhead beam anchored to concrete blocks; at first, deflections alone were measured; then, strain gauges were fitted inside, thereafter outside too; soil tests were added; later, long enough pressure pipes were used. Generally, the test procedure was first PP tests with (where applicable) and without pressure; then installation in a trench; backfilling with dumped or compacted soil, filling in the sometimes deep grooves made by the heavy trucks (usually a 15 T truck, or a 20 T motor crane, both twin rear axles) criss-crossing hundred times the refilled trench; after having completed a test, un-earthing or changing the pipe and resuming the test under different conditions; applying pressure then releasing it, while proceeding with traffic loading; measuring always deflections, most often the strains on the inner and/or outer skin of the pipe and soil densities, this during several days over weeks or months in a few tests.

The last 1963 "Eupen" test (Ref.17), sponsored by the institutions mentioned above, followed thoroughly this procedure: the 15 m long, 900 mm diameter, 8 mm thick fabricated steel pipe with flanges and end plates, was fitted with two 1 m apart central measuring sections, of 12 series of strain gauges, transversal and longitudinal, inside and outside, at crown, invert, spring line and at 45°, with, each, 12 deformations sensors (strain gauges on steel leaf springs) attached to two heavy reference rings resting within the pipe on three screws. Pressure test, then a PP lab test were made; the pipe was burried on sand bedding, while water pressurized to limit "random" deformations, within "dumped" and compacted fills, in a wide trench (3 D) dug into the consolidated fills of a flat area near the Eupen Dam, close to the Belgian-German border. The tests lasted 7 months; besides the usual findings on shorter pipes, **longitudinal bending effects were noticed.**

Two "live" tests were also made: in a transmission main, 1 m diameter, under about 1 MPa pressure, located under a highway, fitted with outside strain gauges only; and in a 500 mm new distribution main, inserting --and removing two months later-- a measuring section fitted with inside, outside, and deformation sensors strain gauges.

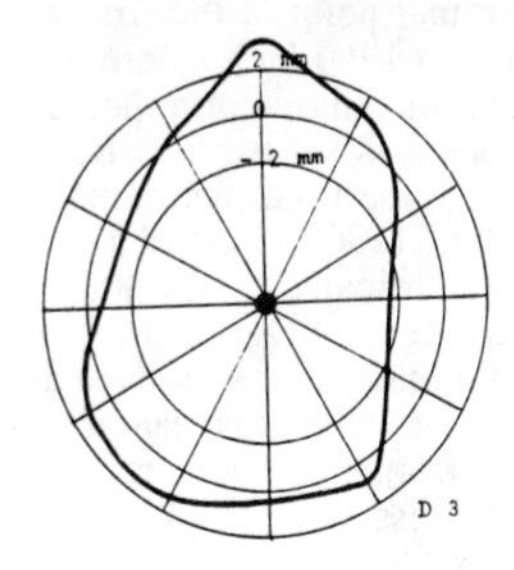

Fig.15. "Eupen" Tests. Average Shape of the Measuring Sections.

Fig.16 and 17. "Eupen" Tests. Predicted (X) and Measured Stresses Due to Shape Defects in both Measuring Sections.

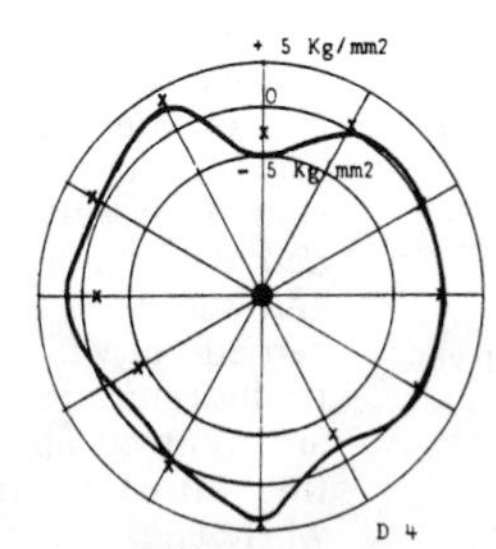

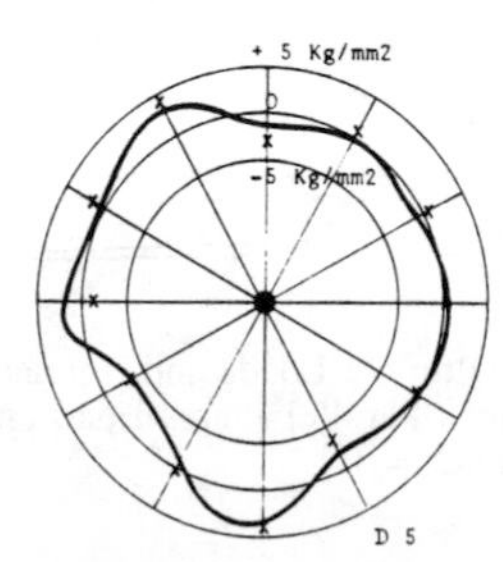

Such field tests are made in **the pipe's very service conditions**, but cannot yield precise values of design parameters; they can and **consistently** did, **confirm the theory and the views presented above**, namely:

-significant but predictable out of roundness effects;

-suppressed bending through earth side support, hence low deformations and stress levels, at the most 1/5 of the predicted ones for pipe without side support; this effect is compounded by pressure;

-presence of stresses and strains due to transverse and longitudinal bending, pressure, and shape defects, measured, superposed, possibly permanently deforming pipe;

-rerounding under pressure, but in part only;

-significant compaction of the sides fills by pipe while deflected by the earth and traffic loads; same stronger effect above the pipe: less than 5 heavy truck passages over a pipe make most of this compaction, which proceeds thereafter very slowly; this effect is permanent; as a result, **flexible pipe becomes quite stiff in a stiff environment;**

-crown instead of invert to be consistently the most stressed area, contrary to ordinary theories.

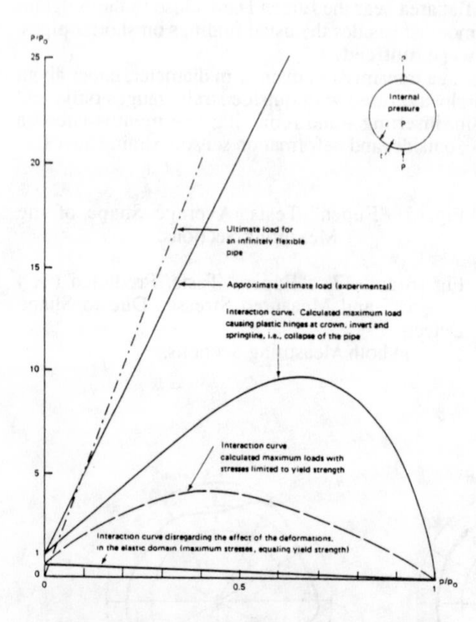

Fig.18. Ultimate Loads and Interaction Curves Four Hinges Parallel Plates Load and Pressure. Steel Pipe, D/t=100.

Liège University proceeded on its own to further PP pressure tests (they were in fact two edges tests) to determine the ultimate load that steel pipe can carry under pressure, when **plastic hinges** appear at the "cardinal points"; the first was made in 1960 on a 2.56m long, 900 mm diameter pipe. The theory was set forth in Ref.14 (1964); the final test, reported in Ref.6 (1966) crushed a series of 4 fabricated tubes 400 mm diameter, 8OO mm long, 4 mm thick, stress relieved and machined with great care to free them from shape defects (out of roundness of 0.25%); the end plates were provided with PE sheets and graphite coated to remove as much as possible frictions during deformations. The limits of the testing equipment were reached, 30 T and deflections of 150 mm, and the tests could not be carried out without damage to the instrumentation. The results are summed up in Fig.18: in this dimensionless graph, where P_0 is the 4 plastic hinges load with no pressure applied, and p_0 the pressure straining steel to its yield strength, the linear interaction straight line is dwarfed by the bulging brought by pressure on pipe at its yield strength (no plastic hinges; D/t=100), which is well below the curves with the plastic hinges, themselves in turn significantly below the experimental and theoretical ultimate curves for an infinitely flexible pipe (ie, $P/P_0 \cong 0.5$ (D/t) (p/p_0), where P_0 is the 4 plastic hinges load)

Conclusions.

The classical theory is reliable; PP and pressure tests evidence that the quantification it provides is satisfactory.

Yet the dominant soil, with its variety of natures and conditions, the possible vagaries of installation with its random deformations, the longitudinal effects, the loadings variations, the shape defects, and the resulting wide uncertainties in connection with buried pipe loadings, blurr the picture.

As a result, in design, the theory's formulas are of little avail for determining bending deflections and strains or stresses, and an essential characteristic of flexible pipe, its thickness or stiffness; flexible pipe's installation is a prominent factor, and "soil design" (AASHTO, Ref.1) fundamental; their directions ought to be enforced through a series of adequate provisions, from close supervision of construction to, possibly, training of all staff involved.

For stability reasons and to help control strains, deflections ought to be limited: 5% is the reasonable, technically and economically affordable maximum.

Maximal deflection being set, and installation being paramount, it is pointless in design to further calculate or predict deformation.

The determination of pipe's stiffness (or thickness) rests thus on two conditions: first, a no buckling condition, which is a subject of controversy. We stick to the "dynamic buckling" model..It derives directly from established classical methods with the very minimal assumptions; it is consistent with field data and several extant recommendations, and safe; it could probably be improved. We discourage using composite models such as Luscher's because they are flawed by considering soil as a solid and endowing it with elasticity in both compression and tension, because they allow very flexible pipe in stiff soils, and have weak experimental support. The immediate problem is by the right choice between models to set the right orders of magnitude. Further experimental buckling research in adequate experimental devices is certainly warranted.

The second condition is a maximal E'/S (with $S=EI/D^3$) that experience and theory sets around 2000; it would keep pipe in an ellipse type shape, with strains under control.

The "harmonic analysis" which is required for the study of the pipe "pressure case", readily shows indeed that the factor E'/S --already in the Iowa formula-- but also the ratio p/S are governing factors. This analysis defines a radial, local, Δ_f, which improves on the diametral D_f; Δ_f however climbs well over 3-5, and, with pressure, to about 20 (within the above E'/S limit but without involving random deflections, shape defects or longitudinal effects); however, it is the product Δ_f or D_f times the relative deflection (respectively ∂/R or Δ/D), that really governs strains, not D_f, so that strains actually decrease with increasing E'/S and pressure.

Pressure considerably stiffens pipe and causes its rerounding, though in part only.

Longitudinal effects are significant and affect primarily the transverse sections.

Conventional loadings are acceptable (earth prism load over flexible pipe, Boussinesq's "point loading"), but should be considered as "steady state" loadings while actual loads may vary widely around, hence the need of adequate safety factors; traffic contributes to fills compaction and incrementally increases deflection in time, which should be taken into consideration. **All loadings**, earth, traffic, together with the effetcs of their longitudinal variations and of the shape defects **actually compound as measurements show, and this reinstates a general rule,** but, and exclusively, for pipe made of non strain sensitive materials (steel, probably certain plastics), only the stresses which persist whatever the deformations, ie, the hoop stresses, need be counted in stress design.

Flexible pipe design would then rest on safe, consistent and comprehensive principles.

References

1 AASHTO Standard Specifications for Highway Bridge AASHTO 1985

2 ARMCO Int. Corp. Handbook of Drainage and Construction Products.Lakeside Press Chicago Ill. 1955 (page 546)

3 ASCE Manual N°37--WCPF Manual n°9 Design and Construction of Sanitary and Storm Sewers WCPF 1970 (4th printing)

4 AWWA Standard C950-81. Glass Fiber Reinf. Thermosetting-Resin Pressure PipeAWWA 1981

with addendum C950a-83 incorporated in sections A2-A4.
5 AWWA Steel Pipe Design and Installation AWWA Manual M11 1964
6 Dehousse N. M. Détermination expérimentale de la charge ultime des conduites en acier doux sollicitées par une pression interne et par deux charges linéaires diamétralement opposées. CRIF Janvier 1966 MT 20
7 Falter B. Grenzlasten von einseitig elastisch gebetteten kreiszylindrischen Konstructionen Bauingenieur 55 (1980) 381-390.
8 Glascock B. C., Cagle L. L. Recommended Design Requirements for Elastic Buckling of Buried Flexible Pipe. Reinforced Plastics/Composites Institute. The Sy of the Plastics Industry 39th Annual Conf., Jan. 1984
9 Howard A. K. Laboratory Tests on Buried Flexible Pipe AWW Journal, October 1972
10 Jeyapalan J.K. Abdel-Magid B.M. Longitudinal Stresses and Strains in Design of RPM Pipes ASCE- Journal of Transportation Engineering, Vol 113, May 87
11 Kienow K. K. Prevost R. C. Pipe/Soil Stiffness Ratio Effect on Flexible Pipe Buckling Threshold A. S. C. E. Journal of Transportation Engineering, Vol. 115 N° 2 Mar. 1989
12 Kienow K. K. Prevost R.C.Stiff Soils - An Adverse Environment for Low Stiffness Pipes ASCE - Proceedings Pipelines in Adverse Environment San Diego -- Nov...1983
13 Luscher U.Buckling of Soil-Surrounded Tubes. ASCE--Journal Soil Mechanics and Foundations Div. 11/00/66
14 Massonet Ch. Dehousse N. M. Roussel J. M. Bonnechère F. Analyse limite d'un tube mince en acier doux soumis à une pression intérieure et à deux charges extérieures linéaires diamétralement opposées Acier-Stahl- Steel N° 10 1964
15 Moore I. D. Elastic Buckling of Buried Flexible Tubes-- A Review of Theory and Experiment. Report N° 024.09.1987 University of Newcastle NSW Australia
16 Moore Ian D. Selig E. T. Haggag A. Elastic Buckling Strength of Buried Flexible Culverts. Presentation at T. R. B. Session 143 Jan 1988
17 Prevost R C Dehousse N. M. Contribution à l'étude du comportement des tuyaux en acier de grand diamètre Revue C n°5 , 1967 (Belgium)
18 Prevost R. C.Alternate Flexible Pipe Structural Design. Pipes and Pipelines International. Vol. 34, N° 5 and 6, Sept--Dec 1989
19 Prevost R. C.Design of non pressure, very flexible pipe. Pipes and Pipelines International.May-June 1983
20 Prevost R. C. Longitudinal Effects in Buried Pipes. in final draft form
21 Prevost R. C.Pipelines for Water and Sewerage World Bank Public Utilities Note TWT / N-5 09/00/81
22 Prevost R. C. Secondary Effects of Pressure on Thin Pipe. In final draft form.
23 Prevost R.C.Calcul des tuyaux d'acier sous pression, enterrés La Technique de l'Eau Decembre 1962-- Janvier 1963
24 Prevost R.C Kienow K. K.Design of Non Pressure Very Flexible Pipe ASCE-Advances in Underground Pipeline Engineering Madison, Aug 85
Also in Pipes and Pipelines Int.,Nov 1985-Jan and Mar 1986
25 Prevost R.C. Kienow K. K.Instability of Buried Flexible Pipe (non pressure case) Pipes and Pipelines International May - June 1988
26 Prevost R.C. Notes relatives au calcul des parois des canalisations enterrées sous pression Revue C - Tijdschrift III N° 5 and 6, 1961
27 Roark R.J.Young W.C. Formulas for Stress and Strain McGraw-Hill 5 th edition
28 Spangler M. G. Secondary Stresses in Buried High Pressure Lines Engineering Report N°23, 1954-1955 I. E. E. S. Iowa State College, Ames, Iowa
29 Spangler M. G. Structural design of Flexible Pipe Culverts Bull. 153 Iowa Engineering Experiment Station 1941
30 Timoshenko S. Gere J. M. Theory of Elastic Stability McGraw-Hill Int. Book Cy 1983. Second ed.
31 Timoshenko S. Strength of Materials. 3d ed. R. E. Krieger Pub. Cy. Malabar Fa USA 1958
32 Trott J. J., G.I. Crabb, A.S. Nagarkatti. Loading tests to failure on buried flexible steel pipes. Pipes and Pipelines International. May-June 1983

NUMERICAL ANALYSIS OF FLEXIBLE CULVERTS IN LAYERED SOILS

P. van den Berg[1] and N.F. Zorn[2]

Abstract

The deformation of flexible culverts is analysed using the finite element method as tool. Results of calculations for installation, traffic and time dependent loads are presented for culverts in homogeneous sand and in layered soft soil embedded in a granular fill and compared with deformation measurements. Finally the ultimate bearing capacity of a culvert under homogeneous conditions and in layered soft soil are compared and the failure mechanisms analysed.

Introduction

Flexible culverts - culverts constructed by bolting together curved, corrugated metal plates - are frequently applied in granular soils. Under these circumstances culverts with spans of more than 10 metres have been succesfully constructed and performed well. In contrast to rigid, concrete structures that provide the load bearing capacity by themselves, the advantage of the use of flexible structures is the increase of the load bearing capacity by activating passive soil resistance. The horizontal culvert deformation generates soil reaction forces which together with the shape of the culvert increase the bearing capacity. The strength therefore must not totally be provided by the culvert cross section itself.

Up to now design models have been primarily based upon rather simplified theories with great reliance on experience. Under standard conditions (culvert buried in homogeneous "elastic" soil) generally these models are sufficient. However, if complex mechanisms are present -

---

[1] Research engineer, [2] Head Fundamental Research, both DELFT GEOTECHNICS, P.O. Box 69, NL-2600 AB Delft The Netherlands

for instance a layered soil profile or soft soil-layers - next to the granular fill, these methods are not always adequate. In such cases and for the case that the installation of such a culvert must be carefully analysed, the finite element method is the only comprehensive approach presently available that is capable of modelling the problem [Duncan, 1979; Selig, 1978]. This method is a very powerful design tool when coupled with good engineering judgement.

This contribution presents the application of a finite element model for homogeneous conditions and compares the calculated results with field measurements. Thereafter it is utilized for a soft soil profile including a sensitivity analysis to quantify the influence of all important parameters. Time-dependent effects such as creep and consolidation of the soft layers and time-dependent settlement of the surface are analysed. Finally ultimate load bearing capacities were calculated to assess the factor of safety. The failure mode of a flexible culvert in a layered profile is compared to the mechanism under homogeneous (granular) conditions.

Finite element model

A cross-section of the coupled soil-structure system is analysed. Due to the axial flexibility effects in this direction can be ignored and a two-dimensional plane strain model can be utilized. The culvert is modelled by two-noded beam elements characterized by bending and normal stiffness. The soil and if included the asphalt layer consist of eight-noded isoparametric quadrilateral (continuum) elements and if geometrically necessary compatible triangular elements. Interface elements may be used to simulate the friction mechanism at the interface between culvert and soil. The general purpose finite element code DIANA [Kusters, 1983] which incorporates the essential requirements needed to analyse a buried culvert as listed by [Selig, 1978] was used.

Loading

The service life of a culvert was considered. The loadcases were subdivided into three groups:
- installation of the culvert
- live loads (surface loading)
- time dependent effects

Installation.

A culvert generally is installed in a pre-shaped bedding. The excavation next to the culvert is then filled in layers, building up the final shape of the flexible culvert. The loads corresponding to this installation procedure can be divided into two groups, dead soil weight and the effect of compaction of the granular fill.

The dead weight is a distributed force, and its effects can easily be calculated with the finite element model by activating the weight and stiffness for only the amount of soil elements that correspond to the specified phase of the installation procedure. Most design models only account for this loading type.

Compaction, however, in granular soils leads to a rearrangement of the particles to a more dense configuration, to an increase of the relative density D_r, which is equivalent to a decrease of the soil volume. The relative density D_r refers to the in situ soil density in % of the range between the minimum and maximum density of the particular soil. The effect of compaction can be modelled as a settlement of the free surface and imposed on the soil elements as a predescribed displacement [Zorn, 1990]. Compaction also influences the mechanical properties of granular soils. Emperical relations exist between the relative density and the stiffness and strength parameters of the soil. In this study general directives proposed by a Dutch Workinggroup on compaction of granular soils are applied [F7, 1987].

First the soil next to the culvert is layerwise compacted, leading to a load of the culvert perpendicular to the design load which acts in a vertical direction. The resulting pre-deformation (the horizontal diameter decreases whereas the vertical one increases) is compensated for when the sand above the culvert is compacted, often resulting in the initial (undeformed) shape. This process is shown in figure 1.

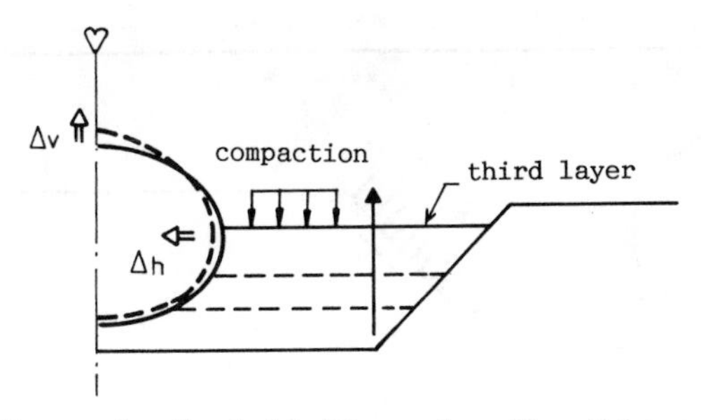

Figure 1. Installation of a flexible culvert

Surface loading.

Surface loads can be applied directly in the model as forces and the resulting (additional) deformations can be calculated. In these cases it is necessary to include the effects of compaction on the in situ stiffness and strength parameters used for sand. These parameters are "relative density dependent" and therefore not the same as the initial values adopted for the installation procedure.

Time dependent effects.

In order to analyse time dependent effects it is necessary to distinguish between homogeneous granular and layered cohesive soft soils.

Generally, if the installation of the culvert has been performed well time dependent effects can be neglected in homogeneous sandy soils.

For soft layered soils the two most important time dependent effects were considered: (a) creep and consolidation of the soft soil layers next to the granular fill and (b) difference between settlement along the boundary of the backfill.

(a) creep and consolidation

Due to the dead weight of the granular fill a stress σ acts on the dividing-line between granular fill and soft soil (figure 2). This stress-state causes a continuing deformation due to creep and consolidation effects in the soft soil. The deformation u increases from u(0) to u(∞) and the vertical deformation of the culvert to $\Delta v(\infty)$. This effect is increased since, due to the deformation of the dividing-line, the volume of the granular fill increases. This leads to a decrease of the relative density and likewise to a reduction of the

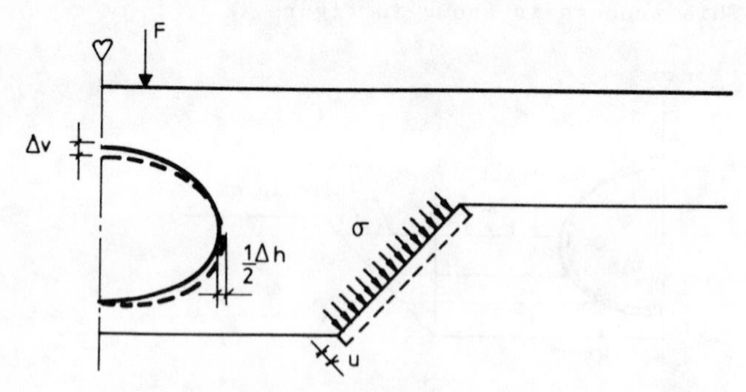

Figure 2. Loading on soft soil layers

stiffness and strength parameters of the fill. The lateral support provided by the soil at a given initial deformation decreases.

(b) settlement

Due to the shape of the granular fill differential settlement may occur. The settlement-gradient results in a time dependent variation of the volume of the fill and the relative density. The latter is related to the mechanical properties of the soil which are responsible for the lateral support of the culvert.

Material models

Soil.

The soilbehaviour was modelled by uzing an elasto-plastic materialmodel. The yield criterium corresponds to the Mohr-Coulomb model including a non associated flow rule [De Borst, 1984]. The material parameters needed are: Youngs' modulus E, Poissons' ratio ν, cohesion c, angle of internal friction ϕ and dilatancy angle ψ. The time dependent analysis is carried out by using a (time dependent) effective value for Youngs' modulus E.

Culvert.

The behaviour of the culvert consisting of curved corrugated metal plates, was described by an elasto-plastic model according to Tresca. The parameters are E, ν and the yield stress σ_y.

Asphalt layer.

An elasto plastic model according to Von Mises was used to model the behaviour of the aphalt layer. E, ν and σ_y depend on the composition and porosity of the mixture and are strongly strain-rate-dependent.

Homogeneous granular conditions

The application of the model for homogeneous conditions had the advantage that measurements were available, which could be used for the calibration of the model. The first step was the simulation of the installation procedure, then a surface loading is added.

Installation

Measurements carried by the Dutch Ministry of Public Works [DMPW, 1987] during the installation of an

Armco MA15 culvert were used to demonstrate the capability of the model when applied to simulate an installation. The procedure can be described as follows: the initial shape of the culvert was measured after the culvert had been placed on the bottom of the excavation (phase 1). After that the backfill was added and layerwise compacted and the shape of the culvert is measured (phases 2 - 6). The height of the fill is given as a distance from the bottom of the excavation to the surface level. The measurements are summarized in table 1. The shape of the structure is characterized by the horizontal (h) and vertical (v) diameter.

Phase	Fill-height(mm)	h(mm)	v(mm)	Δh(mm)	Δv(mm)
1	0, initial	4830	3800	- 5	-15
2	1500	4803	3835	-32	+20
3	2100	4795	3857	-40	+42
4	2700	4787	3858	-48	+43
5	3400	4786	3860	-49	+45
6	3900	4795	3862	-40	+47

Table 1. Deformation of flexible culvert measured during installation

The vertical height is graphically shown in figure 3. It can be clearly seen that the vertical diameter increases up to a fill height equal to about half the vertical diameter of the culvert. After that the shape remains rather constant.

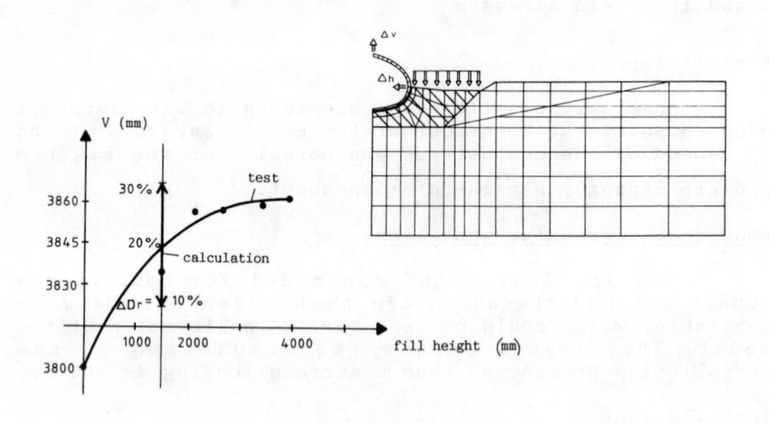

Figure 3. Comparison of measured and calculated deformation of a flexible culvert during installation

Using the finite element model developed the calculated results for phase 2 of the installation (table 1: fill height 1500 mm) are presented in figure 3. Two load types mentioned above, dead soil weight and compaction characterized by an (average) increment of the relative density, ΔD_r, were applied to the model.

Figure 3 shows that the deformation of the flexible culvert strongly depends on the increment of the relative density, ΔD_r, due to the compaction process. As can be seen the calculated size and the measurement are in full agreement in case the increment of the relative density is equal to 20 %. E and ϕ used in the calculation were related to the relative density according to [F7, 1987]. The dilatancy angle was equal to 0°.

Surface loading

Deformation measurements of a full scale buried flexible culvert loaded by surface loading are reported by [Klöppel, 1970]. The results are shown in figure 4.

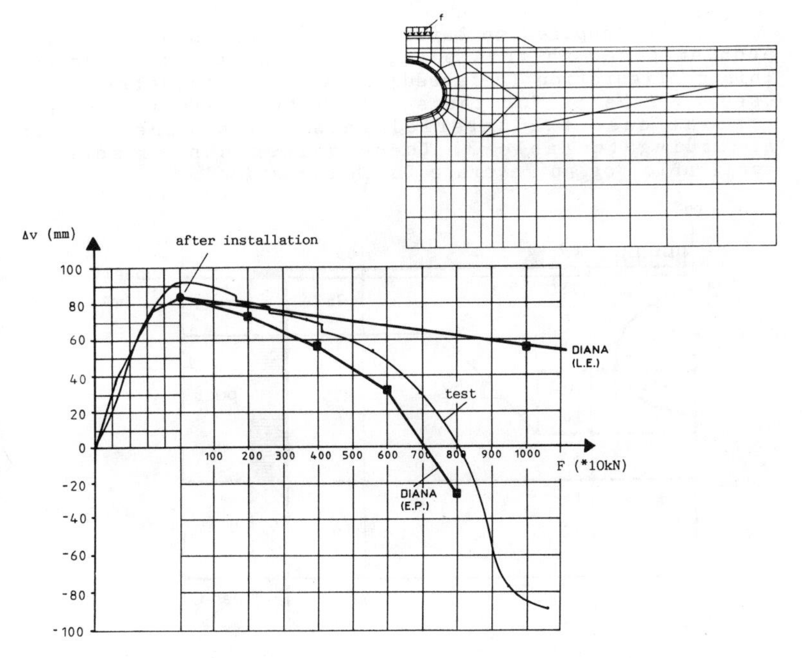

Figure 4. Vertical deformation of flexible culvert loaded by a surface loading (measurement and calculation)

The vertical deflection of the culvert is plotted as a function of the vertical loading. It can be seen that the deformation at the start of the surcharge loading (due to the installation) was about 90 mm in upward direction. Then the load was applied (in opposite direction) and the crown of the culvert moved downward.

The test was simulated utilizing both a linear elastic and an elasto-plastic material model for the soil behaviour. It can be clearly seen that the linear elastic model was accurate only at very low load levels. For higher load levels the load deflection curve is strongly non linear and therefore an elasto plastic must be used. In that case the agreement between the curve calculated and the one which has been measured is fairly good. In this analysis Youngs' modulus E was chosen to 40 MN/m², the angle of internal friction was 40°, whereas the dilatancy angle was equal to 5°.

Layered soft soil conditions

Installation and traffic loading

When applied to layered soft soil conditions the problem posed in this paragraph is shown in figure 5. This configuration (referred to as "basic geometry") was used as a basis for the sensitivity analysis. For this special case the material parameters were chosen according to table 2. These values can be seen as applicable for an "average Dutch situation".

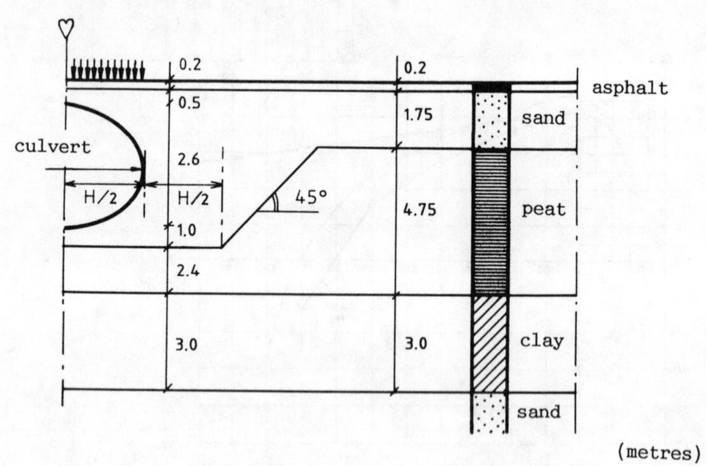

Figure 5. Basic configuration of flexible culvert (Armco, MB6) in soft layered soil

	Sand	Asphalt	Culvert
E (MN/m^2)	20.0	5000.0	210000.0
ν (-)	0.33	0.37	0.10
c (MN/m^2)	0.001	1.25 (σ_y)	13.0 (σ_y)
φ (°)	30.0	0.1	0.1
ψ (°)	10.0	0.1	0.1
γ (kg/m^3)	1700.0	2400.0	7800.0

	Clay	Peat
E (MN/m^2)	0.6	0.4
ν (-)	0.42	0.37
c (MN/m^2)	0.006	0.003
φ (°)	20.0	25.0
ψ (°)	0.1	0.1
γ (kg/m^3)	1500.0	1200.0

Table 2. Material-parameters "basic-geometry"

After simulating the installation and the traffic loading as mentioned a vertical deformation of the culvert equal to 2.1 % of the initial vertical diameter was calculated. The corresponding horizontal deformation was about 0.9 % of the horizontal diameter. Figure 6 shows the deformation of the model due to a (non symmetric!) traffic loading: a heavy truck with three axes. It can be clearly seen that in the case of a proper installation the flexible culvert including the granular fill is pressed into the surrounding soft soil layers. The culvert itself hardly deforms.

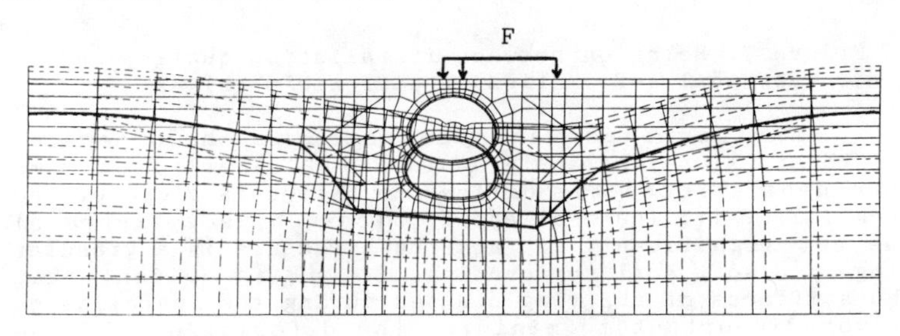

Figure 6. Undeformed and deformed mesh due to a non symmetric traffic loading

The stiffness- and strength parameters of the granular backfill strongly depend on the installation quality: the relative density achieved after installation. To quantify the influence three calculations were performed: (1) bad installation, (2)

normal installation and (3) good installation, referring to the degree of compaction. The result is presented in figure 7. According to this figure it can be concluded that a bad installation quality results in a deformation of the culvert that is about 2.5 times the deformation corresponding to a good installation quality. The compaction-degree of the backfill around the flexible culvert directly influences the lateral support the fill can provide, i.e. the fundamental idea behind the use of this type of structures.

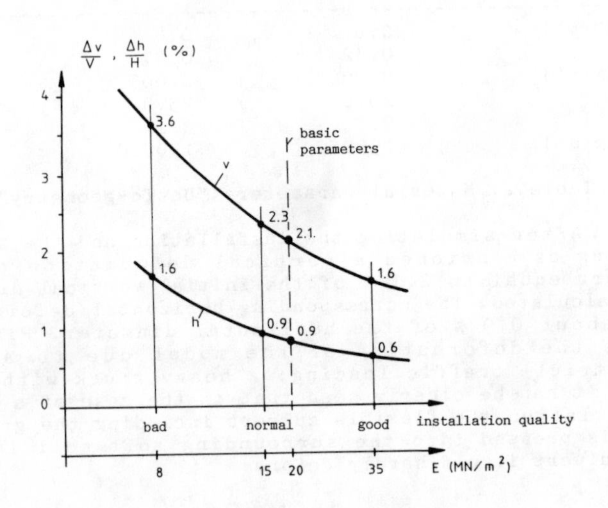

Figure 7. Relation between installation quality and vertical deformation of a flexible culvert

Another important parameter is the dimension (width and slope) of the granular fill. Apart from time dependent effects the difference between a slope of the backfill of 1:1 (case a) and 1:5 (case b, see figure 8) was not significant. In fact the structure is a granular ring in a soft soil surrounding. So, it is evident that the stiffness of the ring created during the installation is very important to minimize the deformation. If the width of the fill is further decreased to about 1 meter the vertical deflection of the culvert increases to nearly 3 % of the initial vertical diameter. This case is referred to as case c (figure 8).

Other parameters having influence on the deformation of the culvert are the height of the soil cover, the stiffness and strength of the soft soil and the stiffness of the asphalt layer. More detailed deformation is given in [Van den Berg, 1989].

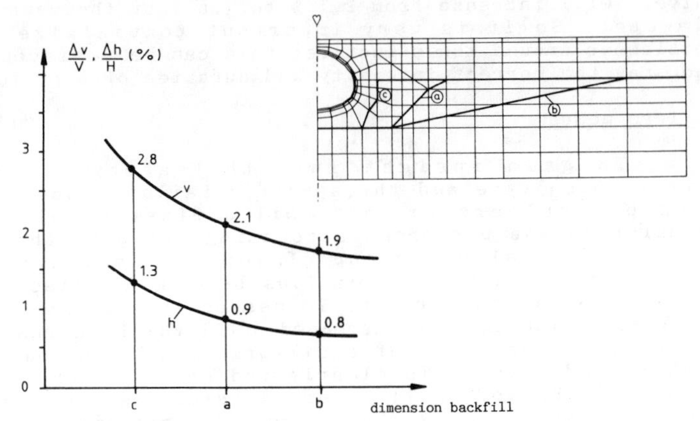

Figure 8. Relation between dimension of granular fill and vertical deformation of a flexible culvert in soft soil

Time dependent effects

The deformations presented in the preceeding paragraph are valid directly after the installation has been completed. Time dependent effects may cause an increase of the deformation and should therefore be added to these initial values.

Creep and consolidation.

A time dependent analysis accounting for creep and consolidation of the soft soil layers was performed for different dimensions of the granular fill referred to as cases a, b and c (see figure 8). The creep-factor, defined as the ratio between the vertical deformation after 50 years of service and the deformation just after installation, resulting from those calculations strongly depends on the width of the granular fill. Going from the largest width (case b) to the smallest one (case c) the creep-factor increases from 1.5 to 3.0.

Settlement.

Due to differences in the loading of the soft soil layers under and next to the culvert differential settlements may occur. The settlement-gradient can be directly related to the relative density of the backfill. Starting from the basic situation (geometry and material parameters) a settlement difference of 25 cm causes a decrease of the quality of the fill from normal to bad (see figure 7). In this case the deformation of the culvert will increase from 2.1 % to 3.6 % of the vertical diameter. So it is very important to minimize the settlement around the structure; this can be achieved by choosing the correct intensity and duration of a preload.

Failure modes

To gain insight into the safety of this construction type and the specific failure mode of a flexible culvert in soft soil a plastic collapse calculation was performed accounting for both the non linear material behaviour of soil and culvert and geometrically non linear relations between displacement and strain in the culvert. To assess the safety factor the surface loading was increased until the load reached its maximum; no state of equilibrium could be found at higher load levels. To clearly indicate the specific effect of the soft soil layers a reference calculation was performed for a flexible culvert in homogeneous sand. For this case Youngs' modulus of the sand was 20 MN/m². For the "soft soil situation" the basic geometry and material parameters were chosen.

The relation between the loadlevel and the vertical deformation of the culvert is shown in figure 9 for both cases. The safety factor in sand is 3.8 referring to the design traffic loading, whereas the factor in soft soil is about 2.4.

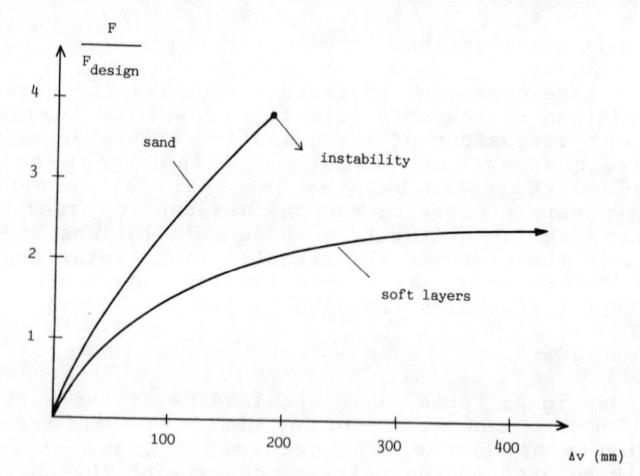

Figure 9: Load-displacement curves following from geometrically non linear plastic collapse calculation

The safety coefficient for a flexible culvert in sand differs from the one calculated for the same culvert in soft soil, but a much more interesting result is the different failure mechanism. In sand the failure is initiated by snap through buckling of the culvert: the incremental deformation concentrates more and more at the crown of the culvert until instability occurs and no equilibrium can be found anymore. This buckling mechanism is initiated suddenly and without prior warnings. Plasticity in the soil is limited to a rather small region just above and next to the culvert.

In soft soil, however, it is not the buckling phenomenon which causes failure of the structure. The deformation of a culvert in soft soil is much larger then the deformation of a culvert in homogeneous granular soil at the same load level. In soft soil when increasing the load it is not the culvert or the granular fill just around the culvert which starts to fail; the maximum shear strength in the soft soil layers under and next to the backfill is reached, so the flexible culvert including the surrounding granular fill is pressed away into the fully plastic soft soil. Both mechanisms are shown in figure 10.

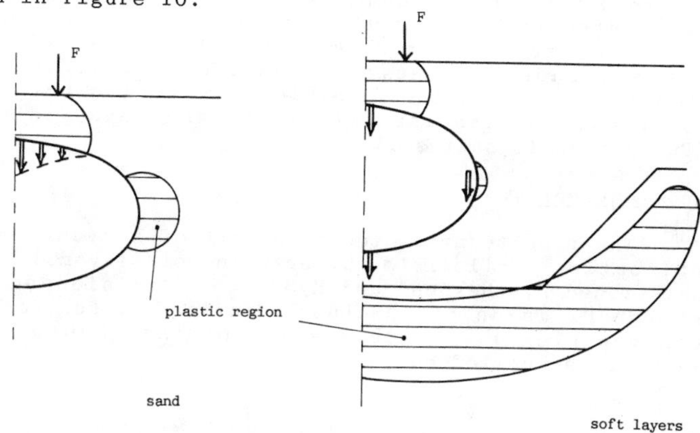

Figure 10: Failure mechanisms in homogeneous granular soil (left) and soft layered soil (right)

Conclusions

- The finite element method is capable of simulating the behaviour of buried flexible culverts. All major features can be represented by the model presented. The installation procedure and effects due to surface loading can be analysed in homogeneous granular soils as well as in soft layered soils. For homogeneous conditions comparison between measurements and calculations show good agreement.

- The most important parameters influencing the deformation of the culvert are: the quality of the installation (compaction level), the height of the soil cover and the stiffness of the asphalt layer. For soft layered soil conditions the dimensions of the granular fill are also very important.

- If the installation of the culvert has been performed well, time dependent effects can be neglected in homogeneous sandy soils. For soft layered soils the time dependent increase of the deformation strongly depends on the dimensions of the granular backfill. Differential settlement along the boundary of the backfill is also of great importance.

- The safety factor for a flexible culvert in sand differs from the one calculated for the same culvert in soft soil, but a more interesting result is the different failure mechanism. In sand the failure is initiated by snap through buckling of the culvert. In soft soil the maximum shear strength in the soft soil layers under and next to the backfill defines the bearing capacity, the flexible culvert including the surrounding granular fill is pressed away into the fully plastic soft soil.

Acknowledgment

The results presented form part of a study on the "Application of flexible culverts in soft layered soils" commissioned by the Road and Hydraulic Division and the Research Division of the Dutch Ministry of Public Works and Armco Flexal. The commission of that contract is greatfully acknowledged.

References

De Borst, R. and Vermeer, P.A., Non associated plasticity for soil, concrete and rock, HERON, Vol. 29, No. 3, 1984.

Duncan, J.M., Behavior and design of long-span metal culverts, J. of the Geotechn., Eng. Div., Proc. ASCE, Vol. 105, No. GT3, pp. 399-418, March 1979.

Dutch Ministry of Public Works, Installation measurements of a flexible culvert, personal communication (not documented), 1987.

F7 Committee, a Dutch working group on compaction of granular soils, part 2: stability and density (in Dutch), SCW, 1987.

Klöppel, K., Glock, D., Theoretische und experimentelle Untersuchungen zu den Traglastprobleme biegeweicher, in die Erde eingebetteter Rohre, Darmstadt, 1970.

Kusters, G.M.A. et al., DIANA - a comprehensive but flexible finite element system, 4th Int. Symp. on Finite Element Systems, Southampton, July 1983.

Selig, E.T. et al., Long-span buried structure design and construction, J. of Geotechn. Eng. Div., pp. 953-966, July 1978.

Van den Berg, P., Application of flexible culverts in soft layered soils - final report, Delft Geotechnics report CO-285280/56 (in Dutch), april 1988.

Zorn, N.F., Van den Berg, P., Deformation of flexible pipes and culverts - mechanics, measurements and calculations - Delft Geotechnics Symposium: Pipelines geotechnically discussed (in Dutch), The Hague, October 1987.

Zorn, N.F., Van den Berg, P., The effect of compaction on buried flexible pipes, International Conference on Pipeline design and installation, Las Vegas, Nev., March 1990.

CONCRETE PIPE TRENCHLESS TECHNOLOGY

John M. Kurdziel*, M. ASCE

ABSTRACT: Precast reinforced concrete pipe is the most commonly used pipe material for jacking operations. For over 90 years, concrete pipe techniques, procedures and applications established the state-of-the-art for trenchless technology. Concrete pipe's long history in pipe jacking provides the opportunity to review and evaluate the effectiveness of various pipe jacking principles.

Concrete pipe is frequently installed by jacking or tunneling methods of construction where deep installations are necessary or where conventional open excavation and backfill methods may not be feasible. Concrete pipelines were first jacked in place in North America by the Northern Pacific Railroad between 1896 and 1900. In more recent years, this technique has been applied to sewer construction where intermediate shafts along the line of the sewer are used as jacking stations. Reinforced tongue and groove pipe as small as 18-inch inside diameter and as large as 132-inch inside diameter have been installed by jacking.

Concrete pipe is uniquely suited for jacking applications due to its inherent compressive and shear strength. Jacking of concrete pipe avoids disruption of traffic, either road or rail; allows installation to tight line and grades, and provides a pipe with high strength characteristics to withstand any depths of fill and not buckle during installation.

---

* Director of Technical Services, American Concrete Pipe Association, Vienna, VA 22182.

LOADS

Two types of loading conditions are imposed upon concrete pipe installed by the jacking method: the earth loading due to the overburden, with some possible influence from live loadings; and the axial load due to the jacking pressures applied during installation.

Vertical Loading and Pipe Strength

The major factors influencing the vertical earth load on pipe installed by jacking are:

- o The weight of the prism of earth directly above the bore
- o Upward shearing or frictional forces between the prism of earth directly above the bore and the adjacent earth
- o Cohesion of the soil

The resultant vertical earth load on the horizontal plane at the top of the bore and within the width of the excavation is equal to the weight of the prism of earth above the bore minus the upward friction forces, minus the cohesion of the soil along the limits of the prism of soil over the bore. This earth load is computed by the equation:

$$W_t = C_t w B_t^2 - 2cC_t B_t$$

where

W_t = earth load, pounds per linear foot
C_t = load coefficient for jacked pipe
w = unit weight of soil, pounds per cubic foot
B_t = maximum width of bore excavation, feet
c = coefficient of cohesion of the soil above the excavation, pounds per square foot

The $C_t w B_t^2$ term in the above equation is similar to the equation for determining the backfill load on pipe installed in a trench condition where the trench width is the same as the width of the bore. The $2cC_t B_t$ term accounts for the cohesion of undisturbed soil. For cohesive soils the earth load on a jacked pipe will always be less than on a pipe in a trench installation.

The design values for the coefficient of cohesion range from zero to 1000, where zero is indicative of a very loose dry sand and 1000 descriptive of a hard clay.

Based on clay's large coefficient of cohesion, pipe jacked through hard clay experiences little, if any, earth loading even under high depths of fill.

The required pipe strength for concrete pipe is commonly determined in terms of a D-load. The D-load strength is further classified as either a 0.01-inch crack ($D_{0.01}$) or ultimate load (D_{ult}) in the three-edge bearing test as computed by the equation:

$$\text{D-load} = (W_t + W_L)(F.S.)/(B_f D)$$

where

D-load = 0.01-inch crack ($D_{0.01}$) or ultimate load (D_{ult}), pounds per linear foot per foot of inside horizontal span
W_t = earth load, pounds per linear foot
W_L = live load, pounds per linear foot
B_f = bedding factor
D = inside horizontal span
F.S. = factor of safety

Live loads are determined by prevailing design procedures for load distribution through earth masses for the particular transportation loading. In general, highway live loads are negligible for fill heights greater than 10 feet. Railroad loadings, however, may be effective to depths of 20 feet or greater.

The bedding factor, B_f, is the ratio of the supporting strength of an installed pipeline to the strength in the three-edge bearing test. Since the jacking method of construction affords intimate contact around the lower exterior surface of the pipe and the surrounding earth, an ideal bedding condition is provided. This intimate contact can be obtained by either close control of the bore excavation to the outside dimensions and shape of the pipe or, if the bore is over-excavated, the space between the pipe and the tunnel can be filled with sand, grout, concrete or other suitable material. For this type of installation a minimum bedding factor of 3.0 is recommended. If the bore is slightly over-excavated and the space between the pipe and the bore is not filled a minimum bedding factor of 1.9 is recommended.

Minimum 0.01-inch crack and ultimate D-loads are based on the strength classes covered by ASTM Standards for circular, horizontal elliptical and vertical elliptical pipe.

Jacking of precast reinforced concrete box sections is not as common as circular pipe but is increasing in use and practice. Design of the boxes is in accordance with the procedures described in Section 17 of the American Association of State Highway and Transportation Officials (AASHTO) Committee on Bridges and Structures.

Axial Loading and Pipe Strength

For axial loadings, it is necessary to provide for relatively uniform distribution of the load around the periphery of the pipe to prevent localized stress concentrations. This force distribution is accomplished by assuring the pipe ends are parallel within the tolerances prescribed by ASTM Standards for precast concrete pipe; by using a cushion material between the pipe sections, such as plywood, hardboard or a similar cushion which will not compact to an incompressible material; and by care on the part of the contractor to ensure that the jacking force is properly distributed through the jacking frame to the pipe and parallel with the axis of the pipe.

The cross-sectional area of the concrete pipe wall is more than adequate to resist pressure encountered in any normal jacking operation. A standard 60-inch B wall (6-inch) pipe with concrete strength of 5000 psi and a 3-inch bearing surface around the outer circumference of the pipe can resist a uniform direct compressive load of 1500 tons. Table 1 contains typical thrust requirements for jacking concrete pipe.

It is imperative to have uniformly distributed loads when jacking. Unsymmetric or eccentric loading may result in excessive point loads and may not only damage the pipe but result in alignment problems. Joint cushioning material permits loads to be distributed uniformly around the periphery of the joint, prevents the build-up of high compressive forces which may spall the concrete or crack the joint, and allows for grade and line control.

Table 1 Tonnage Requirements for Pushing Concrete Pipe (3)

Size of Pipe Outside Diameter -Inches-	Sandy Soil with No Excavation in Front of Pipe	Hard Soil with Excavation in Front of Pipe
18"	1.0 tons	.40 tons
24"	1.4	.52
30"	2.0	.76
36"	2.0	.76
42"	2.3	.88
48"	2.7	1.0
54"	3.0	1.1
60"	3.3	1.2
66"	3.6	1.4
72"	3.9	1.5
78"	4.3	1.6
84"	4.6	1.7
90"	4.9	1.8
96"	5.2	1.9
102"	5.5	2.0
108"	6.2	2.3

Note: Table values are multiplied by the total length of pipe in the excavation (in feet)

There are a number of cushioning materials used for transferring the jacking axial loads. Plywood (0.5 to 0.75-inch) is most commonly used, but it is by no means the only acceptable material. In fact, only half of the pipe's wall diameter is used for transferring the jacking thrust forces between joints. If a preformed cushioning material could be developed which permits transfer of the compressive thrusts through the pipe's entire cross sectional area at the joint, the larger cross sectional transfer area would help prevent damage by point loading. In any event, a suitable compressible material must be used to prevent concrete-to-concrete contact.

For projects where extreme jacking pressures are anticipated, as in cases of long jacking distances or excessive unit frictional forces, higher concrete compressive strength may be specified, along with greater care in avoiding bearing stress concentrations. Concrete strengths of 5000 psi or greater can be supplied to meet higher jacking thrust forces. Using higher strength concrete, however, does not necessitate specifying a higher class of pipe. Little or no gain in axial crushing resistance is provided by specifying the larger circumferential steel requirements of a higher class of pipe.

Although larger circumferential steel in the barrel does not substantially assist in resisting axial forces, stirrups in the tongue and groove sections of the joint may provide greater strength against joint spalling and cracking. Use of stirrups should be reviewed on a project-by-project basis and discussed with project area pipe producers. Most reinforced concrete pipe does not require any additional steel in the tongue or groove to be correctly installed and function adequately.

The manufacturing method used for the production of reinforced concrete pipe has no effect on either the structural strength of the pipe or the pipe's jacking properties. Centrifically spun or wet-cast pipe may have smoother surface finishes than dry cast machine-made pipe, however, the benefit of this property is negligible since a lubricant is required on all pipe jacking operations. External lubricants effectively eliminate any benefit derived from smoother surface finishes.

The use of a lubricant, such as a bentonite slurry, will help reduce the frictional forces on the outside surface of the pipe and the axial loading required for jacking. In very cohesive soils, constant lubrication and non-interrupted jacking advancement is necessary to prevent the build-up of large soil frictional stresses. If jacking operations are halted prior to reaching the receiving pit, the frictional resistance between the soil and jacking pipe may become difficult to overcome and prevent the further advancement of the pipe without greater jacking force.

TRENCHLESS EXCAVATION METHODS

There are a number of trenchless excavation methods available for installation of concrete pipe. Commonly used methods include auger boring, pipe jacking, tunneling and micro-tunneling. Other methods such as slurry rotary drilling and directional drilling are occasionally used, but these are normally limited to very small diameter pipelines.

Auger Boring

Auger boring is a process of simultaneously jacking a casing through the earth while removing the spoil within the encasement by a rotary auger. All operations are performed with mechanical equipment without workers being inside the borehole. Line and grade are maintained with the use of steerable cutting heads and grade sensing devices which permit augering to the tightest grade tolerances. The steerable cutting heads permit grade and alignment changes during boring, but these are limited to only corrective measures. Line and grade of the completed bore depends on the original machine alignment and grade. Steerable heads can not compendate for poor boring set-up and alignment.

The application of bentonite near the leading edge of the casing helps reduce skin friction. Bentonite will reduce thrust requirements and provides better grade control by preventing caking around the steering head linkage. It also acts as a sealant and aids in limiting sloughing and cave-ins.

Auger boring of concrete pipe is normally limited to small diameter pipe, 36-inch and under, but pipe as large as 60-inch have been installed by such procedures. Bore lengths for this method range from very short runs to lengths over 200 linear feet. Augering equipment and procedures are constantly being improved with an ever increasing extension of the maximum bore length. The corresponding average tolerance for line and grade for the auger bore method is approximately one percent of the length.

The auger bore method has many advantages in its ability to maintain a tight line and grade and install the concrete pipe as the bore hole is being excavated, thereby avoiding any cave-in and ground surface subsidence. Its only main disadvantage is its limitation on removal of obstacles such as boulders. In general, the auger can remove an obstacle which is a third or less of the casing diameter.

Jacking

Pipe jacking is the most common trenchless excavation technique used for installing concrete pipe. Unlike the auger boring method, pipe jacking requires men working inside the pipe. The excavation methods vary from very basic processes, involving the manual excavation of material to the use of highly sophisticated boring machine processes. All excavation is accomplished inside an articulated shield. Individually controlled hydraulic steering jacks are utilized in guiding the shield. A track in the pit is added to guide the concrete pipe sections.

Jacking pit design and construction is critical due to the thrust forces required. Alignment and grade control are impossible to maintain without an adequate pit floor foundation.

A concrete slab pit floor with a granular base course is advisable if grade control is critical. A granular floor may also be acceptable if stable ground conditions exist in the pit. It is imperative to ensure the guiding frame be properly supported. Steering jacks can compensate for minor discrepancies in line and grade, but cannot make major corrections due to movement of the guiding frame.

The thrust wall should be constructed perpendicular to the designed jacking axis. Otherwise, problems may develop in attempting to maintain line and grade. These problems may be compounded by the effects of uneven force distribution from the jacks about the pipe centerline.

The excavation process takes place inside the shield. Mechanical excavation with tunnel boring machines or full face cutting heads are the most commonly used trenchless excavation technique for pipe jacking.

Pipe jacking has been performed with pipe as small as 30-inches in diameter, but due to the need to have men within the line, jacking operations are normally performed with pipe 36-inches in diameter or greater. In small diameter pipe, augers may be used to remove the spoil material. The jacking length spans may be in excess of 1500 linear feet, and, as with all jacking operations, the use of a lubricant such as bentonite is strongly recommended.

One of the advantages jacking methods have over auger boring is the ability to remove obstacles without closing down the operation and removing the auger. Due to the nature of the operation, an inspector can easily assess the type and size of obstacle and the best method for removal. Jacking methods are very versatile, pipe can even be jacked on a radius.

Tunneling

Tunneling of concrete pipe is very similar to the jacking techniques. The major difference is in tunneling is that liner plate or rib and lagging form an outside lining, and the concrete pipe is jacked in this tunnel. The annular void between the lining and pipe is then filled with a grout.

Tunneling is typically used for concrete pipe 48-inch diameter and greater, and there is no limit to the jacking length because the liner is not pushed through the ground, but is constructed as the shield advances. Excavation of spoils may be accomplished with either manual or mechanical methods.

Tunneling can obtain the highest degree of accuracy for line and grade. Its major drawback is the time and labor required for the operation.

Micro-Tunneling

Micro-tunneling is a method of horizontal earth boring utilizing highly sophisticated, laser guided and remote controlled equipment. This method permits accurate monitoring and adjusting of the auger and allows obtaining precise line and grade. Micro-tunneling, however, is limited to concrete pipe which is 36-inches or less in diameter. There are a number of proprietary machinery used for installing pipe in this manner. The capital cost for the equipment is high, but if tight line and grade tolerances are of extreme importance, this procedure may be a viable alternative.

LININGS AND COATINGS

Concrete pipe is extremely resistant to both interior and exterior chemical attack, and linings and coatings are seldom necessary or warranted. The production techniques used in making precast reinforced concrete pipe result in a dense cement-rich concrete which is highly resistant to corrosion.

External attack of concrete is limited to either aggressive groundwater or soil sulfates. In either case, the strong alkalie content of the concrete neutralizes the aggressive chemicals next to the concrete surface essentially creating a buffer from further chemical attack. If the pipe is located in an area with a fluctuating groundwater table, the aggressive chemicals could be replenished resulting in a renewed attack on the concrete surface. The severity of this condition, however, may be lessened if calcareous aggregate is used in the concrete pipe.

Interior corrosion of concrete pipe is usually limited to acid attack from the generation of hydrogen sulfides from sewage or acids illigally dumped in the sewer system. There are a number of design procedures which may be used to either minimize or eliminate the possibility of hydrogen sulfide generation. Lining or coating protection that depends on adhesion for jacked concrete pipe performs unsatisfactorily. The jacking operations will scratch, peel or totally remove any external protective lining. Interior coatings may also be damaged by the equipment or men operating within the pipe. Mechanically bonded PVC linings would be an exception.

If a lining is scratched or partially removed, it merely acts as a focus for a concentrated attack at the damaged area. The use of coatings or linings for jacked concrete pipe is unnecessary, impractical and costly and therefore, not recommended.

SUMMARY

Concrete pipe trenchless technology has made great advances in the past decade. Competition among machinery manufacturers and contractors has resulted in more efficient and cost effective operations. New methods for controlling line and grade have allowed the installation of a pipeline within a congested underground urban environment.

The methods and procedures for installing concrete pipe are being constantly improved. The design of concrete pipe, however, has not changed substantially over this same period. Although standard B-wall, reinforced concrete pipe has the inherent strength to resist vertical loading and compressive jacking forces, it is not specifically designed for jacking operations. Standard reinforced concrete pipe generally has excessive circumferential steel, which is not required for jacking, and due to the small dead and live loads on the pipe, it is also unnecessary for load strength. Reduction of the steel area would yield a much more economical design.

REFERENCES

1. Concrete Pipe Handbook, American Concrete Pipe Association, 1988.

2. Horizontal Earth Boring and Pipe Jacking Manual, National Utility Contractors Association, 1981.

3. Horizontal Earth Boring and Pipe Jacking Manual, No. 2, National Utility Contractors Association, 1986.

4. Operation and Safety Manuals, McLaughlin Boring Systems, Plainfield, Illinois, 1987.

5. "Pipe Jacking: State-of-the-Art," C. B. Drennon, American Society of Civil Engineers, September, 1979.

6. Specifications (Construction) for Highway Projects Requiring Horizontal Earth Boring and/or Pipe Jacking Techniques - DRAFT Report, D. E. Hancher, T. D. White, D. T. Iseley, July, 1989.

7. Technical Manual - Horizontal Earth Boring Machines, American Augers, Inc., 1987.

DESIGN OF DUCTILE IRON PIPE ON SUPPORTS

By Richard W. Bonds, P.E.[1]

Abstract

Design procedures for ductile iron pipe in normal underground service have been well established. The standard design considers hoop stresses in the pipe wall due to internal hydrostatic pressure as well as bending stresses and deflection in the pipe due to external loads of earth and traffic above the buried pipe.

In some situations, it is necessary or desirable to use supports at designated intervals along pipelines. Aboveground, supported pipe is needed to transport water and other fluids within treatment plants and buildings. Also, pipe on piers is utilized to cross natural or manmade objects. Additionally, unstable soil conditions or other factors sometimes necessitate the installation of pipe on piers or pilings underground.

This paper reviews the pertinent design considerations for both aboveground and underground Ductile Iron pipe-on-supports installations. Specific procedures, recommended design limits, and allowable stresses are outlined. Design tables based on ductile iron pipe data and suggested loads are also provided.

Beam Span For Ductile Iron Pipe On Supports

Ductile Iron pipe is normally manufactured in 18- or 20-foot (5.486- or 6.096 meter) nominal[2] lengths, depending on the pipe manufacturer and pipe size. The most common joints used with ductile iron pipe are the push-on type joint and the mechanical joint. Both of

---

[1]Research & Technical Director, Ductile Iron Pipe Research Association, Birmingham, Alabama.

[2]Ductile iron pipe may be furnished in shorter lengths per AWWA C151 (ANSI/AWWA C151/A21.51, 1986). If exact lengths are required to fit on pre-built piers, this should be specified.

these rubber-gasketed joints allow a certain amount of deflection and longitudinal displacement while maintaining their hydrostatic seal. This makes these pipe joints ideally suited for normal underground installation. The flexibility of the joints reduces the chance of excessive beam stresses occurring. For pipe supported at intervals, however, flexible joints usually require that at least one support be placed under each length of pipe for stability.

Various schemes have been successfully used to obtain longer spans where particular installation conditions presented the need, but these are special design situations and are not specifically addressed in this paper. The design presented herein is based upon one support per length of pipe.

Support Location

System security is maximized by positioning the supports immediately behind the pipe bells. When the support is placed near the bell, the bell section contributes beneficial ring stiffness where it is most needed. This ring stiffness, in turn, reduces the effect of support loads and localized stress. Supports should normally not be placed under spigots adjacent to bells, due to possible undesirable effects on joints.

Saddle Angle And Support Width

Pipe supports should cradle the pipe in a saddle (see Figure 1). This cradling, which should follow the contour of the pipe, minimizes stress concentrations at the supports. It is recommended that the saddle angle (β) of the support be between 90°and 120°. Little or no benefit is gained by increasing the saddle angle more than 120°. With angles smaller than 90°, the maximum stress tends to increase rapidly with decreasing saddle angle (Evces, C. R. and O'Brien, J. M., November 1984).

There are some differences among published theories and data regarding the importance of axial support width for saddles. The most accepted formulas are found to be completely independent of saddle width. Some test data however, show a decrease in measured stresses with an increase in saddle width. There is little effect on the maximum stress when saddle support width is increased more than $\sqrt{2Dt_e}$ (Wilson, W.M. and Olson, E. D., 1941). Therefore, for saddle supports, the minimum width (b) is determined by equation 1.

$$b = \sqrt{2Dt_e} \quad (1)$$

where:

b = minimum (axial) saddle width (inches)
D = actual outside diameter of pipe (inches)
t_e = nominal pipe wall thickness (inches), see Table 1

Support Design

Additionally, supports, piles, and/or foundations should be adequately designed from a structural and soil-engineering standpoint to safely handle any loads transferred from the pipe.

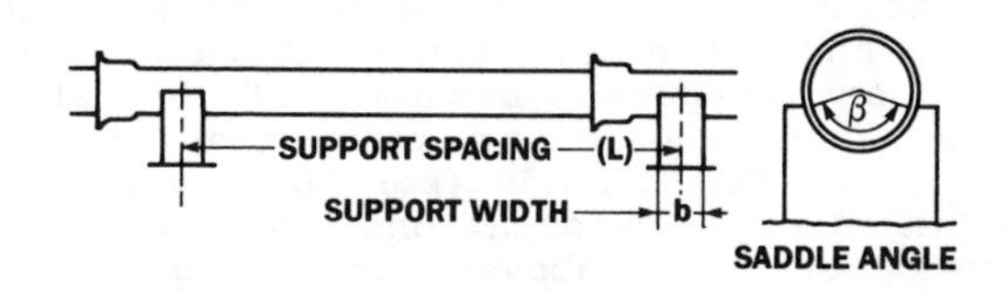

Figure 1 - Saddle Angle and Width

Table 1
Nominal Pipe Wall Thickness of Ductile Iron Pipe

Size in.	Outside Diameter in.	Thickness Class 50	51	52	53	54	55	56
		Thickness--in.						
3	3.96	----	0.25	0.28	0.31	0.34	0.37	0.40
4	4.80	----	0.26	0.29	0.32	0.35	0.38	0.41
6	6.90	0.25	0.28	0.31	0.34	0.37	0.40	0.43
8	9.05	0.27	0.30	0.33	0.36	0.39	0.42	0.45
10	11.10	0.29	0.32	0.35	0.38	0.41	0.44	0.47
12	13.20	0.31	0.34	0.37	0.40	0.43	0.46	0.49
14	15.30	0.33	0.36	0.39	0.42	0.45	0.48	0.51
16	17.40	0.34	0.37	0.40	0.43	0.46	0.49	0.52
18	19.50	0.35	0.38	0.41	0.44	0.47	0.50	0.53
20	21.60	0.36	0.39	0.42	0.45	0.48	0.51	0.54
24	25.80	0.38	0.41	0.44	0.47	0.50	0.53	0.56
30	32.00	0.39	0.43	0.47	0.51	0.55	0.59	0.63
36	38.30	0.43	0.48	0.53	0.58	0.63	0.68	0.73
42	44.50	0.47	0.53	0.59	0.65	0.71	0.77	0.83
48	50.80	0.51	0.58	0.65	0.72	0.79	0.86	0.93
54	57.56	0.57	0.65	0.73	0.81	0.89	0.97	1.05

Table 2

Allowances for Casting Tolerance

Size in.	Casting Tolerance in.
3-8	0.05
10-12	0.06
14-42	0.07
48	0.08
54	0.09

Note: 60- and 64-inch ductile iron pipe is also available. Consult manufacturer for wall thickness and casting tolerance.

Loads On Pipe

For underground design calculations, the total load normally includes the prism earth load plus the weight of the pipe and contents. Since buried pipe is usually installed on supports because of unstable ground conditions, there should, in most cases, be no vehicle loading. Thus, truck loads (per ANSI/AWWA C150) should be used in design calculations only where they are likely to occur. For aboveground design calculations, the total load includes the weight of the pipe and contents.

If the designer expects greater loads to occur on aboveground or underground installations, these loads should be incorporated into the design and are not in the scope of this procedure.

Pipe Wall Thickness Calculations

Design calculations include localized stress at supports, hoop stress due to internal pressure, and flexural stress and beam deflection at the center of the span.

Due to the conservative approach of this design procedure, and in the interest of simplicity, combinations of external load and internal pressure to obtain principal stresses have not been considered. The design engineer may elect to investigate principal stresses due to extraordinary circumstances, e.g., very high internal pressure, etc.

Localized Stress At Supports

The supported pipe is subjected to localized stresses at the support that are a function of the total reaction at the support and the shape (saddle angle) of the support. This maximum stress may be longitudinal or circumferential in nature and is predicted by the following equation proposed by Roark (Roark, R. J., 1975):

$$f_r = K \left(\frac{wL}{t_n^2}\right) \ln \left(\frac{D}{2t_n}\right) \tag{2}$$

where: f_r = localized stress due to support reaction (48,000 psi maximum)
L = Span length (feet)
D = Pipe outside diameter (inches), see Table 1
w = Unit load per linear foot (lb./ft.)
K = Saddle coefficient
t_n = Design wall thickness of pipe (inches), see Table 3

For aboveground applications
t_n = minimum manufacturing thickness of pipe
= nominal pipe wall thickness - casting tolerance

For underground applications
t_n = net pipe wall thickness
= nominal pipe wall thickness - casting tolerance - 0.08" service allowance

Recent research involving ductile iron pipe has established that the function

$$K = 0.03 - 0.00017\ (\beta - 90°) \tag{3}$$

provides excellent correlation between the ring stresses predicted by Equation (2) and the actual stress as measured when β is between 90° and 120° (Eves, C. R. and O'Brien, J. M., November 1984).

The maximum calculated localized stress should be limited to 48,000 psi. This value is equal to the minimum yield strength in bending for ductile iron (72,000 psi) divided by a safety factor of 1.5. It is the same limiting value of bending stress employed in the American National Standard for the Thickness Design of Ductile Iron Pipe, ANSI/AWWA C150/A21.50. (ANSI/AWWA C150/A21.50, 1988).

Hoop Stress Due To Internal Pressure

The net thickness required for internal pressure can be determined by using the equation for hoop stress:

$$t = \frac{P_i D}{2S} \qquad (4)$$

where: t = pipe wall net thickness (inches)
P_i = design internal pressure (psi)
= $2\ (P_w + P_s)$
P_w = working pressure (psi)
P_s = surge allowance (100 psi)
D = outside diameter of pipe (inches)
S = minimum yield strength in tension
= 42,000 psi

If anticipated surge pressures are greater than 100 psi the maximum anticipated pressure must be used.

Flexural Stress At Center Of Span

With one support per length of pipe positioned immediately behind the bells, each span can conservatively be treated as a simply supported beam. The joints being slightly offset from the supports causes some of the simple beam moment and stress to distribute itself from the center of the span to the support. This makes the simple beam approach conservative. The following formula represents the flexural stress at the center of the span of a uniformly loaded, simply supported beam:

$$f_b = \frac{15.28\ DwL^2}{D^4 - d^4} \qquad (5)$$

where: f_b = allowable flexural stress (48,000 psi max.)
D = pipe outside diameter (inches)
w = unit load per linear foot (lb/ft)
L = length of span (feet)
d = $D - 2t_n$ (inches)

Beam Deflection At The Center Of Span

Computations for beam deflection are also based on the simply supported beam concept. This is likewise conservative due to the reality of offset joints. The maximum allowable deflection at mid-span to prevent damage to the cement-mortar lining is limited to:

$$y_r = L/10 \qquad (6)$$

where: y_r = maximum allowable deflection at center of span, inches
L = length of span, feet

Less deflection may be desired. The deflection of the beam may be significant for aesthetic reasons in aboveground installations or possibly for hydraulic reasons in gravity-flow pipelines. Limitations on the deflection, if any, should be determined by the designer as appropriate to a specific installation.

The beam deflection at center span for a uniformly loaded, simply supported beam can be calculated using the following formula:

$$y = \frac{458.4\ w\ L^4}{E\ (D^4 - d^4)} \tag{7}$$

where: y = deflection at center span (inches)
w = unit load per linear foot (lb/ft)
L = length of span (feet)
E = modulus of elasticity (24×10^6 psi)
D = pipe outside diameter (inches)
d = $D-2t_n$ (inches)

Aboveground Installations

For aboveground installations with one support per length of pipe (i.e., a span length of 18 or 20 feet), the minimum thickness class (Class 50)* of ductile iron pipe in all sizes is more than adequate to support the weight of the pipe and water it contains when analyzed in accordance with the suggestions of this procedure.

Other design considerations for pipes supported above ground may include the carrying capacity of the supports themselves, the strength of the structure from which the pipe may be suspended, and/or unusual or additional loads not in the scope of this paper. Such loading may include seismic, frequency or resonance of vibrations, wind, water current, and other special design condsiderations.

It is also necessary to assure a minimum of lateral and vertical stability at the supports for aboveground piping. Deflected pipe joints can result in thrust forces of hydrostatic or hydrodynamic origin, and if not laterally and vertically restrained, unbalanced forces may result in additional joint deflection and possible failure of the pipeline.

Thermal expansion of ductile iron pipelines supported above ground is not usually of concern in

*Class 51 for 3-inch and 4-inch pipe.

correctly designed and installed systems because of the nature of the push-on or mechanical joint. A 100-degree-Fahrenheit change in temperature results in expansion or contraction of a 20-foot length of ductile iron pipe of approximately 0.15 inches. This is easily accommodated by correctly installed pipe and joints. Occasionally, where structures from which ductile iron pipe is to be suspended are expected to have significantly different behavior than the pipeline, special considerations for expansion, contraction, and supports may be necessary. For reference, the following are coefficients of thermal expansion for various materials:

Ductile Iron - 6.2×10^{-6} inch/inch degree-Fahrenheit
Steel - 6.5×10^{-6} inch/inch degree-Fahrenheit
Concrete - 7.0×10^{-6} inch/inch degree-Fahrenheit

<u>Design Procedure</u>

A. Select length of span (18 feet or 20 feet), saddle angle (90°-120°), and pipe diameter.

B. Determine the unit load per linear foot (w) based on the minimum class thickness pipe.

1. For aboveground installations: $w = (W_p + W_w)$

2. For underground installations:

   2.1 No truck loads
   $w = (W_p + W_w) + 12\ D\ P_e$
   2.2 Truck loads included
   $w = (W_p + Ww) + 12\ D\ (P_e + P_t)$

   Note: For D see Table 1
   For P_e and P_t see Table 4
   For $(W_p + W_w)$ see Table 3

C. Determine if the chosen design thickness (t_n) results in an acceptable localized stress less than or equal to 48,000 psi.

1. Calculate the saddle coefficient (K) using equation 3.

2. Calculate f_r using equation 2.

   If f_r exceeds 48,000 psi, increase t_n to the next thickest class and re-calculate starting with Step B. Repeat until the resulting f_r is less than or equal to 48,000 psi.

D. Determine the pipe class thickness required due to internal pressure.

1. Calculate the net thickness (t) required for hoop stress due to internal pressure using equation 4.

2. Determine the total calculated thickness (T) due to internal pressure.

   For aboveground applications
   T = t + casting tolerance
   For underground applications
   T = t + casting tolerance + 0.08

3. Using Table 1, select the standard thickness nearest to the total calculated thickness. When the total calculated thickness is halfway between two standard thicknesses, select the larger of the two.

E. Calculate the flexural stress (f_b) at mid-span using equation 5 and the largest class thickness pipe required in Step C or D along with its corresponding t_n and w values.

   If f_b exceeds 48,000 psi, increase t_n to the next thickest class and re-calculate f_b using the new class thickness and corresponding t_n and w values. Repeat until the resulting f_b is less than or equal to 48,000 psi.

F. Check deflection at mid-span.

1. Calculate the deflection at mid-span (y) using equation 7 and the largest class thickness pipe required in Step C, D or E along with its corresponding t_n and w values.

2. Calculate the maximum allowable deflection at mid-span (y_r) using equation 6. (Note: Less deflection may be desired.)

   If the deflection y is greater than the deflection y_r, increase t_n to the next thickest class and re-calculate y using the new class thickness and corresponding t_n and w values. Repeat until the resulting y is less than or equal to y_r.

G. Choose the largest class thickness corresponding to the largest t_n required in Step C, D, E or F and calculate the minimum saddle width using equation 1.

TABLE 3

PIPE PLUS WATER WEIGHT ($W_p + W_w$) AND DESIGN WALL THICKNESS (t_n)

Size Inch	Thickness Class	Wp + Ww Lb/Linear Ft.	tn(inch) Aboveground Applications	Underground Applications
3	51	14	.20	.12
	52	15	.23	.15
	53	16	.26	.18
	54	16	.29	.21
	55	17	.32	.24
	56	18	.35	.27
4	51	19	.21	.13
	52	20	.24	.16
	53	21	.27	.19
	54	22	.30	.22
	55	23	.33	.25
	56	24	.36	.28
6	50	32	.20	.12
	51	33	.23	.15
	52	35	.26	.18
	53	36	.29	.21
	54	38	.32	.24
	55	39	.35	.27
	56	40	.38	.30
8	50	50	.22	.14
	51	52	.25	.17
	52	54	.28	.20
	53	56	.31	.23
	54	58	.34	.26
	55	60	.37	.29
	56	62	.40	.32
10	50	71	.23	.15
	51	73	.26	.18
	52	76	.29	.21
	53	79	.32	.24
	54	81	.35	.27
	55	84	.38	.30
	56	86	.41	.33
12	50	95	.25	.17
	51	98	.28	.20
	52	101	.31	.23
	53	104	.34	.26
	54	107	.37	.29
	55	110	.40	.32
	56	113	.43	.35
14	50	125	.26	.18
	51	129	.29	.21
	52	133	.32	.24
	53	136	.35	.27
	54	140	.38	.30
	55	143	.41	.33
	56	147	.44	.36

Table 3 - Continued

PIPE PLUS WATER WEIGHT (W_p + W_w) AND DESIGN WALL THICKNESS (t_n)

Size Inch	Thickness Class	Wp + Ww Lb/Linear Ft.	tn(inch) Aboveground Applications	Underground Applications
16	50	157	.27	.19
	51	161	.30	.22
	52	165	.33	.25
	53	169	.36	.28
	54	173	.39	.31
	55	178	.42	.34
	56	182	.45	.37
18	50	192	.28	.20
	51	196	.31	.23
	52	201	.34	.26
	53	206	.37	.29
	54	210	.40	.32
	55	215	.43	.35
	56	219	.46	.38
20	50	230	.29	.21
	51	235	.32	.24
	52	240	.35	.27
	53	245	.38	.30
	54	251	.41	.33
	55	256	.44	.36
	56	261	.47	.39
24	50	316	.31	.23
	51	322	.34	.26
	52	328	.37	.29
	53	334	.40	.32
	54	340	.43	.35
	55	346	.46	.38
	56	352	.49	.41
30	50	465	.32	.24
	51	475	.36	.28
	52	486	.40	.32
	53	496	.44	.36
	54	506	.48	.40
	55	516	.52	.44
	56	527	.56	.48
36	50	653	.36	.28
	51	668	.41	.33
	52	684	.46	.38
	53	699	.51	.43
	54	715	.56	.48
	55	730	.61	.53
	56	745	.66	.58
42	50	868	.40	.32
	51	890	.46	.38
	52	911	.52	.44
	53	933	.58	.50

Table 3 - Continued

PIPE PLUS WATER WEIGHT ($W_p + W_w$) AND DESIGN WALL THICKNESS (t_n)

Size Inch	Thickness Class	Wp + Ww Lb/Linear Ft.	t_n (inch) Aboveground Applications	t_n (inch) Underground Applications
42	54	954	.64	.56
	55	975	.70	.62
	56	996	.76	.68
48	50	1118	.43	.35
	51	1147	.50	.42
	52	1175	.57	.49
	53	1204	.64	.56
	54	1232	.71	.63
	55	1261	.78	.70
	56	1289	.85	.77
54	50	1423	.48	.40
	51	1460	.56	.48
	52	1497	.64	.56
	53	1533	.72	.64
	54	1570	.80	.72
	55	1606	.88	.80
	56	1643	.96	.88

Notes: Approximate pipe weight based on cement-mortar lined, push-on joint pipe and 18-foot-laying-length.

Weight of water based on actual I.D.

60- and 64-inch Ductile Iron pipe is also available. Consult manufacturer for wall thickness and weight.

Table 4

Earth Loads P_e and Truck Loads P_t psi

Depth of Cover ft.	P_e	3-Inch Pipe P_t	4-Inch Pipe P_t	6-Inch Pipe P_t	8-Inch Pipe P_t
2.5	2.1	9.9	9.9	9.9	9.8
3	2.5	7.4	7.4	7.3	7.3
4	3.3	4.4	4.5	4.4	4.4
5	4.2	3.0	3.0	3.0	3.0
6	5.0	2.1	2.1	2.1	2.1
7	5.8	1.6	1.6	1.6	1.6
8	6.7	1.2	1.2	1.2	1.2
9	7.5	1.0	1.0	1.0	1.0
10	8.3	0.8	0.8	0.8	0.8
12	10.0	0.6	0.6	0.6	0.6

Table 4 - Continued

Earth Loads P_e and Truck Loads P_t--psi

Depth of Cover ft.	P_e	3-Inch Pipe P_t	4-Inch Pipe P_t	6-Inch Pipe P_t	8-Inch Pipe P_t
14	11.7	0.4	0.4	0.4	0.4
16	13.3	0.3	0.3	0.3	0.3
20	16.7	0.2	0.2	0.2	0.2
24	20.0	0.2	0.1	0.1	0.1
28	23.3	0.1	0.1	0.1	0.1
32	26.7	0.1	0.1	0.1	0.1

Depth of Cover ft.	P_e	10-Inch Pipe P_t	12-Inch Pipe P_t	14-Inch Pipe P_t	16-Inch Pipe P_t
2.5	2.1	9.7	9.6	8.7	8.2
3	2.5	7.2	7.2	6.6	6.2
4	3.3	4.4	4.4	4.4	4.1
5	4.2	2.9	2.9	2.9	2.8
6	5.0	2.1	2.1	2.1	2.0
7	5.8	1.6	1.6	1.6	1.5
8	6.7	1.2	1.2	1.2	1.2
9	7.5	1.0	1.0	1.0	1.0
10	8.3	0.8	0.8	0.8	0.8
12	10.0	0.5	0.5	0.5	0.5
14	11.7	0.4	0.4	0.4	0.4
16	13.3	0.3	0.3	0.3	0.3
20	16.7	0.2	0.2	0.2	0.2
24	20.0	0.1	0.1	0.1	0.1
28	23.3	0.1	0.1	0.1	0.1
32	26.7	0.1	0.1	0.1	0.1

Depth of Cover ft.	P_e	18-Inch Pipe P_t	20-Inch Pipe P_t	24-Inch Pipe P_t	30-Inch Pipe P_t
2.5	2.1	7.8	7.5	7.1	6.7
3	2.5	5.9	5.7	5.4	5.2
4	3.3	3.9	3.9	3.6	3.5
5	4.2	2.6	2.6	2.4	2.4
6	5.0	1.9	1.9	1.7	1.7
7	5.8	1.4	1.4	1.3	1.3
8	6.7	1.2	1.1	1.1	1.1
9	7.5	1.0	0.9	0.9	0.9

Table 4 - Continued

Depth of Cover ft.	P_e	18-Inch Pipe P_t	20-Inch Pipe P_t	24-Inch Pipe P_t	30-Inch Pipe P_t
10	8.3	0.8	0.7	0.7	0.7
12	10.0	0.5	0.5	0.5	0.5
14	11.7	0.4	0.4	0.4	0.4
16	13.3	0.3	0.3	0.3	0.3
20	16.7	0.2	0.2	0.2	0.2
24	20.0	0.1	0.1	0.1	0.1
28	23.3	0.1	0.1	0.1	0.1
32	26.7	0.1	0.1	0.1	0.1

Depth of Cover ft.	P_e	36-Inch Pipe P_t	42-Inch Pipe P_t	48-Inch Pipe P_t	54-Inch Pipe P_t
2.5	2.1	6.2	5.8	5.4	5.0
3	2.5	4.9	4.6	4.4	4.1
4	3.3	3.4	3.3	3.1	3.0
5	4.2	2.3	2.3	2.2	2.1
6	5.0	1.7	1.7	1.6	1.6
7	5.8	1.3	1.3	1.2	1.2
8	6.7	1.1	1.0	1.0	1.0
9	7.5	0.8	0.8	0.8	0.8
10	8.3	0.7	0.7	0.7	0.7
12	10.0	0.5	0.5	0.5	0.5
14	11.7	0.4	0.4	0.4	0.4
16	13.3	0.3	0.3	0.3	0.3
20	16.7	0.2	0.2	0.2	0.2
24	20.0	0.1	0.1	0.1	0.1
28	23.3	0.1	0.1	0.1	0.1
32	26.7	0.1	0.1	0.1	0.1

Note: 60- and 64-inch Ductile Iron pipe is also available. Consult manufacturer for truck loads.

<u>Nomenclature</u>

b - Minimum saddle width (inches)

D - Pipe outside diameter (inches)

d - Pipe design inside diameter (inches)

E - Modulus of elasticity for Ductile Iron (24×10^6 psi)

f_b - Allowable flexural stress (48,000 psi)

f_r -Localized stress due to support reaction (48,000 psi maximum)

K - Saddle coefficient [0.03 - 0.00017 (β - 90°)]

L - Span length (feet)

P_e - Earth load (psi)

P_i - Design internal pressure (psi)

P_s - Surge allowance (psi)

P_t - Truck load (psi)

P_w - Working pressure (psi)

S - Minimum yield strength in tension for Ductile Iron (42,000 psi)

T - Total calculated pipe wall thickness

t - Net pipe wall thickness (inches)

t_e - Nominal pipe wall thickness (inches)

t_n - Design pipe wall thickness (inches)

w - Unit load per linear foot (lb/foot)

W_p - Unit load of pipe per linear foot (lb/foot)

W_w - Unit load of water in pipe per linear foot (lb/foot)

y - Deflection at center of span (inches)

y_r - Maximum recommended deflection at center of span (inches)

β - Saddle angle (degrees; 90° to 120° is recommended)

References

American National Standard For The Design of Ductile Iron Pipe, ANSI/AWWA C150/A21.50

American National Standard For Ductile Iron Pipe, Centrifugally Cast In Metal Molds Or Sand-Lined Molds, For Water Or Other Liquids, ANSI/AWWA C151/A21.51

Evces, C.R. and O'Brien, J.M., "Stresses in Saddle-Supported Ductile Iron Pipe," Journal AWWA, November 1984

Roark, R. J., Formulas For Stress and Strain, McGraw-Hill, New York, Fifth Edition, 1975

Wilson, W.M., and Olson, E.D., "Tests on Cylindrical Shells", Engineering Experiment Station, University of Illinois Bulletin, 331, 1941

Performance of Pipe Spillways for Dams

David C. Ralston[1], M. ASCE and Jimmy R. Wallace[2]

Abstract

The Soil Conservation Service has used several different kinds of pipe products for spillway conduits of dams. Materials with a shorter estimated life are used for smaller low hazard dams and reinforced concrete steel cylinder pressure pipe with a minimum diameter of 30 inches used for moderate to high hazard dams. Pipe deterioration experienced has resulted from backfill low soil-water resistivity and conveyed water of low mineral content. Maintenance and rehabilitation has been done by replacement and grouting in plastic pipe liners.

Introduction

The Soil Conservation Service has for nearly five decades constructed earth fill embankment dams using pipe materials for the principal (or initial flow stage) spillway. Since the beginning in the early 1940's, it is estimated that there have been over one and one-half million dams built with pipe spillways. The pipes have consisted of both metal and concrete and, in recent years, some have used plastic . The pipes, designed for pressure flow in most cases, have performed very well. The durability and watertightness of the pipe material has been the main factor for determining selection. The more costly dams and those with a potential hazard have required the use of concrete pipe with gasketed expansion joints.

Spillway Material Selection Parameters and Criteria

Structural design for maintenance of the opening and longitudinal continuity is determined by the

---

1 National Design Engineer, Soil Conservation Service, P.O. Box 2890, Washington, D.C. 20013.
2 State Conservation Engineer, Soil Conservation Service, 760 S. Broadway, Salina, KS 67401.

conventional methods for rigid and flexible materials as appropriate. The rigid conduits placed on concrete bedding or cradle are designed and constructed for a positive projection condition to assure that conduits are well embeded in the earth fill. The flexible pipe is designed for a deflection limit of five percent assuming conservatively low backfill compaction consistent with what construction quality control will provide.

The longitudinal continuity of an individual pipe section is normally not a problem unless there are significant differential settlements of the foundation that impose excessive shear forces. Care is taken during foundation preparation to minimize this. However, load concentrations have been found to be a problem when anti-seep collars are used without an adequate reinforced yoke around the pipe to spread the concentrated pressure imposed by narrow collars. The more recent criteria to use a drain diaphragm (properly filtered) for seepage control in lieu of collars should eliminate this condition. The main consideration is the pipe joint which must be able to withstand the shear forces equal to that for the pipe remote from the joint. The joint must also withstand the tension imposed by the soil creep transverse to the embankment centerline by either having enough tensile strength or a joint extensibility length to provide for adequate extension without loss of watertightness.

The need to have a spillway dependably function during flood flows has resulted in establishing a minimum pipe diameter and an inlet configuration that will operate even when trash and debris are excessive. Dams larger than "ponds"[1,2] (hydraulic height to the open auxiliary spillway less than 35 feet and a product of this height, in feet, times the reservoir storage at the auxiliary spillway crest, in acre feet,--known as the height-storage product--less than 3,000) have pipes with a diameter of 18 inches or greater, if the dam is low hazard. The larger the dam and the greater the potential hazard created by the dam, in case of failure, the larger the minimum diameter. A minimum of 30 inches is required for pipes in dams considered to be high hazard. This is considered to be the minimum for purposes of physical inspection and performance of maintenance. With the use of remote control equipment carrying cameras, it is possible to make inspections of smaller pipe, but the satisfaction of first hand examination makes it prudent to retain the current minimum diameter criteria.

Pipe conduits installed in dams have required maintenance and replacement for a variety of reasons. The method of treatment used has also been by a variety of methods, each selected as a result of the size and significance of the dam, the size and kind of pipe, and the availability of the replacement materials and construction methods. Several different situations will be described later in this paper.

To provide greater durability, steel pipe products have been coated, and if soil backfill condition along the outside of the pipe is expected to cause a rapid rate of corrosion, a cathodic protection system is provided. Supplimental treatments beyond coatings have not been provided for corrosive water flow conditions on the in side of pipes.

In July of 1956 SCS issued its first national engineering criteria for principal spillways for dams of any significance (limits were initially less than those cited above i.e. 25 feet of height and 50 acre feet of reservoir water storage). The criteria was simple. Conduits were limited to reinforced concrete, monolithic reinforced concrete, and welded steel, except corrugated metal pipe (CMP) could be used if the life of 30 years was assumed toward its maintenance and/or replacement. The CMP was to be double riveted, asphalt dipped, and with watertight coupling bands. The minimum diameter was set at 18 inches. Trash racks for the inlet were required to protect against clogging. All principal spillway components, *except for the cantilever outlet*, were to be for approximately equal durability. We can see later how the exception for durability of the cantilever outlet has facilitated needed early maintenance measures.

In 1976, issue of new standards, including those for ponds, modified the criteria to require the added use of cathodic protection for steel pipes when the immportance of the structure warrented. By policy, all dams larger than "ponds" warrented such protection. The field conditions to be evaluated were (1) the saturated soil resistivity and (2) acidity. When the resistivity was less than 4,000 ohm-cm or when the pH less than 5.0, a cathodic protection system was required. For a pipe spillway to be installed under these condition, the pipe sections needed to be bonded together to provide continuity and the pipe coated to reduce the electrical current drain. It has since been demonstrated that steel pipe durability is extended when the installation is good.

Also in 1976, use of CMP was expanded for dams when the height-storage product was up to 10,000 [2]. However, added restrictions required (1) a maximum height of fill over the pipe of 25 feet, (2) provision for replacement of the pipe if it was not expected to last the design life of the dam, (3) close riveted and asbestos treated-asphalt coated, (4) structural design for 35 feet of cover, (5) pipe connections bonded, (6) based on soil resistivity and pH measurement, design for a cathodic protection system at less than 4,000 ohm-cm or less than 5.0 pH, (7) if resistivity and pH measurements did not meet the minimum, pipe-to-soil measurements are to be made two years after construction and, if new measurements indicate, a cathodic protection system is to be added.

Performance of Pipe conduits

Approximately 2,300 of the 9,800 dams installed through Soil Conservation Service water resource projects were built using corrugated steel pipe (CMP) for the principal spillway pipe conduit. There are a number of additional dams that were built using concrete pipe in conjunction with a 20 to 40 foot section of CMP at the cantilever outlet end.

In the early 1970's, a field study was conducted on the corrosion of steel pipe by Howard Hall, soil engineer with the Soil Conservation Service in Lincoln, Nebraska. The study was initiated as a result of observed advanced rates of steel pipe deterioration in the eastern portion of the Great Plains area served by the Lincoln, Nebraska office. This report described the need for special pipe treatment when soil and water environmental conditions existed to accelerated corrosion rates.

In 1977 the Soil Conservation Service conducted a brief survey on aging dams with pipe conduit spillways. There was interest being expressed by small watershed project sponsors for the Soil Conservation Service to financially assist them in maintaining their dams in the replacement of spillway pipe conduits. The concern was over deteriorating CMP conduits. The survey determined the number of project dams in place more than 15 years having CMP spillways; their general condition; and how many may need to be replaced within the next few years. The response from 50 states and the Caribbean Area indicated 29 of the states had CMP spillways in 1,206 project dams. Of the 29 states, 16 reported no need for replacement in the near future. The remaining 13 states indicated a wide range of needs for maintenance or replacement. Eight of the 29 states reported 78.5 percent of the total number (four reported 58 percent of the total). Replacement of the CMP spillway was reported as having already been done on eight dams, five of these involved the 40 feet of cantilever outlet section. Many were reported as being in good condition, especially the asbestos treated-asphalt coated installations. Often the asphalt coating was checked and deteriorated, or even off, but the base metal showed only limited rusting. Some comments received regarding the newer spillways installed meeting the more restrictive criteria for coating and cathodic protection substantiated its effect on reduced corrosion rates.

In 1982 during one of the scheduled dam safety inspections for the Ischua Creek watershed in New York, three sections of reinforced concrete pipe of a spillway were found leaking. The concrete pipe wall was so weak that it could easily be cut out with a knife. Tests made on samples indicated manufacture was made without sufficient cement in the concrete mixture. The modification made is described below in the discussion of case histories.

Oklahoma has reported the use of CMP spillways in a large number of dams. The Soil Conservation Service in Oklahoma has observed a variety of pipe deterioration problems due to a variety of causes. The CMP riser (tower) inlet for the spillway has often exhibited severe corrosion at both the weir crest (waterline) and at the base where the riser connects to the conduit (elbow). During the period of 1983 through 1986 a number of studies were made of spillways for dams in several small watershed projects. Some of the findings reported are described. There were 272 dams with the spillway made up entirely of steel CMP and another 150 with only the outlet cantilever section consisting of steel CMP.

In the Oklahoma watershed of Upper Red Rock Creek, five sites constructed during the years 1965 through 1969 using steel CMP conduits had measured water pH ranging from 8.0 to 8.8 and soil resistivity ranging from 1,182 to 1,915 ohm-cm. Three of the sites used reinforced concrete for the riser inlet. Some of the comments made, in abbreviated form, about the condition observed were; CMP risers are "shot", severely corroded and perforated and need to be replaced; concrete risers and trash racks are in good condition; CMP barrels are in good condition; most asbestos bonding is in place and galvanizing and metal is in good condition; only occasional bituminous coating is missing except inside crown of outlet pipe where it is exposed to sunlight; conduit barrels are 18-inch diameter and the risers are 24-inch diameter.

In 1983 an Oklahoma report was made for two dams constructed in the mid-50's in the Double Creek watershed. A summary of the observations made are the spillways consist of a concrete pipe conduit with a 40 feet long cantilever outlet section of CMP; the bottom one-third of the outlet section is rusted out over the projecting visable portion of outlet; a three-foot hole exists in the backfill of one of the dams at the junction of CMP and the concrete pipe conduit.

In 1984 an Oklahoma report was made of the condition of a dam constructed in 1966 in the Upper Black Bear watershed. A summary of the observations reported are the pipe spillway consists of an 18-inch diameter CMP conduit, believed to be asbestos treated-asphalt coated, with a 24-inch diameter CMP riser inlet; the riser section of CMP had a fabricated two-feet long 18-inch diameter stub for conection to the conduit and set in a conconcrete slab for the base; the riser was close-riveted and asphalt coated, but based on memory, not asbestos treated; the outlet conduit appears in good condition; The bituminous coating is gone from the top of the exposed outlet pipe; asbestos is still in place; the as-built drawings indicate the conduit to have a paved invert; the riser for this dam and two others on this watershed were severely corroded and perforated, needing replacement.

In other Oklahoma reports made in 1985 for the Little Wewoka and Big Wewoka watersheds cited some corrosion in CMP spillways for 10 dams with the condition described as fair or in some cases reference made to pin holes and in one case, big holes, in the outlet pipe sections. Lack of information on the age, soil and water condition, and pipe thickness and treatment limits the value of the information in identifying the specific cause for rapid corrosion.

In 1988 a field study was initiated by the Soil Conservation Service in cooperation with Kansas State University to examine the condition of pipe conduit spillways in dams in Kansas. The field work and reporting was performed by Kansas State University. Pipe examined were reinforced concrete (culvert and pressure pipe), steel CMP, and welded steel pipe. A report described the conditions observed on the inside of 38 conduits ranging in age from 10 to 34 years. A video tape was made of each conduit and a suggested rating and schedule for follow-up inspections was developed. The numerical rating system considers pipe condition for cracks, corrosion, lining, and leaking joints based on a description of observed conditions. Based on the rating, the system provides an estimate of the length of time the pipe should function satisfactorily and provides a recommended time until the next inspection. The study concluded that (1) the sample of 38 conduits examined did not provide a data base large enough to refine a rating system or to determine the life span of different construction materials operating in the differing environments, (2) all of the pipe conduit materials inspected appear to have limitations that affect their life span, (3) the life of steel conduits is extended significantly by the use of protective coatings, (4) concrete culvert pipe manufactured to meet ASTM Specification C 76 lacks sufficient strength, without special treatment, to withstand the embankment loadings imposed and, (5) concrete pressure pipe manufactured to meet AWWA Specification C 301 performed the best. The report includes a recommendation that five of the 38 conduits warrent attention for repair and/or replacement.

The 1988 field study by Kansas State University in cooperation with the Soil Conservation Service was extended to add to the data base of conduits examined. The purpose is to develop a more reliable basis for design selection of materials for new construction, repair, rehabilitation, and replacement. The field work of examining 35 additional CMP conduits was completed during the summer of 1989. The report for these additional sites will be completed in early 1990.

Repair, Replacement and Rehabilitation

In the early 1980's several modifications were made of dams by the Soil Conservation Service through one-time construction authorities of the small watershed program. The following cases briefly describe the construction done to correct pipe spillway deterioration problems.

Arkansas - Twenty four dams installed in Six Mile Creek watershed during the period of 1954 through 1957 had CMP with the spillway conduits that had reached the point of unsatisfactory service after about 20 years. The principal spillways consisted of a reinforced concrete pipe conduit with a cantilever section of bituminous coated CMP extending the outlet beyond the toe of the dam. The spillway inlet risers for 12 dams, located within the embankment, used a section of 8-inch CMP to provide an inlet from the reservoir. The other 12 dams have the spillway inlet risers located in the reservoir near its upstream toe. The CMP ranged from 10 to 14 gage steel. In the late 1970's the CMP conduits extending from the spillway inlets to the reservoir were replaced with PVC plastic pipe at a cost ranging from $6,000 to $8,000. The outlet cantilever sections of CMP were replaced with 40 feet of reinforced concrete pressure pipe (AWWA C 301) at a cost ranging from $10,000 to $12,000. The bituminous coating was noted as from the inside the 14 gage CMP conduits and other deterioration ranged from numerous perforations of the invert to complete disintegration of the lower periphery of the pipe. Satisfactory performance of the 14 gage pipe was estimated to be 15 to 20 years. The 12 gage pipe was noted to have moderate to severe corrosion on the conduit interior with perforation ranging from none to numerous. Its length of satisfactory performance was estimated to be 20 to 25 years. A 10 gage 30-inch diameter CMP was used in one dam, there were several perforations in the invert at the time of replacement (25 years). The water flowing through the spillway pipe had a pH ranging from 6.5 to 8.4, a total hardness ranging from 23 to 46 (except one sample was 310), and specific conductance ranged from 78 to 151 (except one sample was 686). Tests were made of the water at five dam sites in 1977 when the pipe corrosion was found to be excessive. The water, inherently soft at four of the five sites tested, had $CaCO_3$ saturation index values indicating protective corrosion scales normally expected would not form. The corrosion evaluation report indicated that a corrosion rate of two to five mils per year (0.051 - 0.13 mm/yr) could be expected.

Arkansas - The concrete interior surface of a reinforced concrete embedded steel cylinder pressure pipe (AWWA C 301) was found, after carrying stream flow for 18 months, to have been seriously deteriorated following installation in 1979. The deterioration was characterized by erosion and etching of the surface, more pronounced in the invert and at the joints. Siliceous particles from the fine aggregate was going into solution, creating voids and producing a honeycombed, sponge-like appearance. Tests on water samples indicated the water to be low in mineral content with a total hardness of 14 mg/l. More tests for pH ranged from 6 to 6.7. A Langelier Index was determined to be about -3.5 . The relatively mineral free, slightly acidic water caused severe chemical attack on the limestone aggregate and to some degree on the cement materials of the concrete. Rehabilitation and protection was accomplished by filling the voids and coating the entire interior pipe surface with an epoxy resin compound.

Kansas - The Riverside Drainage District in Clay County owns and operates four dams that were constructed in 1953 for flood prevention purposes. The inverts of the galvanized CMP conduits rusted through causing a concern for safety of the dams. In 1982, after almost 30 years of service, the first conduit (18-inch diameter and 110 feet long with a canopy inlet) was replaced with a polymer coated CMP at a cost of $7,800. In 1983, the second conduit (18-inch diameter and 85 feet long with a canopy inlet) was replaced with an asphalt coated CMP at a cost of $8,000. In 1985, the third conduit (24-inch diameter and 116 feet long with a canopy inlet) was replaced with an asphalt coated CMP at a cost of $10,800. In 1989, the fourth conduit (36-inch diameter and 144 feet long with a riser drop inlet) was replaced with a polymer coated CMP at a cost of $29,000.

North Dakota - Seven floodwater retarding dams using pipe spillways were built in the Tongue River watershed in 1955. The height of dam emabnkments range from 27 to 85 feet high. The pipe spillways consisted of reinforced concret pipe through the body of the dam with CMP cantilever outlets. The dimensions for each of the spillways is listed below.

The spillway pipe condition consisted of longitudinal & circumferential cracks in reinforced concrete pipe, joints separated and leaking and the CMP outlet sections corroded and in need of repair. The CMP outlet pipes had severe corrosion as a result of soil resistivity less than 2,000 ohms-cm and a pH less than 5.0. Many soils in North Dakota are in this category. Repairs had been previously made consisting of (1) repairing the principal spillway pipe and replacing the CMP foundation drain pipe twice on structure T3-5, (2) replacing the 20 foot outlet CMP section on structure T8-1 in 1968, and (3) replacing the 48 foot outlet section of CMP on structure T3-2 in 1968.

Four alternatives were examined and cost estimates made for purposes of evaluating and selection of the one thought best. The alternatives considered were (1) remove and replace all the pipe (this was judged to be very costly and was not estimated, (2) replace only the outlet CMP sections, (3) line spillway pipe with a polyethelene sleeve, and (4) line spillway pipe with the "Insituform" process. This rehabilitation work, consisting of grouting in a liner of polyethelene pipe and removal and replacement of the outlet CMP pipe section with a vinyl coated CMP, was done in 1984 for about $175,000.

SITE No.	DIAMETER PIPE	LINER I.D.	CONDUIT LENGTH
T2-2	24"	20.2	226 Ft. R/C & 28 Ft. CMP
T2-4	24"	20.2	208 Ft. R/C & 24 Ft. CMP
T3-1	24"	20.2	394 Ft. R/C & 52 Ft. CMP
T3-2	24"	16.9	366 Ft. R/C & 70 Ft. CMP
T3-6	30"	26.2	238 Ft. R/C & 62 Ft. CMP
T7-1	24"	19.9	304 Ft. R/C & 130 Ft. CMP
T8-1	24"	19.9	102 Ft. R/C & 44 Ft. CMP

Wisconsin - In 1979 the reinforced concrete pipe conduit for two small dams were lined with polyethelene pipe as a part of the maintenance. The spillway consisted of reinforced concrete culvert pipe with tongue and groove joints sealed with ribbed compression gaskets. The joints lost their seal due to foundation soil creep opening the joints. The first section of pipe, cast in the wall of the inlet base, cracked due to movements of the inlet riser. The embankment soil piped through the crack and resulted in a surface "sink hole" depression. The cantilever CMP outlet was replaced with polymer coated CMP. The construction was contracted by the small watershed project sponsors based on a design provided by the Soil Conservation Service.

New York - In 1984 a 36-inch reinforced concrete pipe conduit spillway for a dam in the Ischua Creek watershed was lined with a polyethelene pipe having a SDR of 32.5. The 270 feet long lining was installed and the annular space grouted within a 16 day contract period for a total cost of $34,994.

References

1. Soil Conservation Service. (1985). "Engineering Practice Standard Code 378, Pond", National Handbook of Conservation Practice.

2. Soil Conservation Service. (1985). "Earth Dams and Reservoirs", Technical Release No. 60.

3. Koelliker, James K., C. H. Best and A. N. Lin, Inspection and Evaluation of Principal Spillway Conduits in Kansas, Civil Engineering Department, Kansas State University, Manhattan, Kansas, August 1989.

FAILURE OF PRESTRESSED CONCRETE EMBEDDED CYLINDER PIPE

William L. C. Knowles[1], P. Eng., Member ASCE

Introduction

In 1964, the Ontario Provincial Government undertook to supply water to the City of London (current population 300,000). The supply source was from Lake Huron near Grand Bend, Ontario while London is located some 30 miles (48 km) inland.

The supply pipeline is 48 inches (1200 mm) in diameter. Maximum design pressure is 250 psi (1725 kPa). The pipe used was prestressed concrete pressure pipe (AWWA C-301) embedded cylinder-type having a core thickness of 5 inches (125 mm) and a 1-inch (25 mm) coating thickness over the prestressing wire.

The pipeline was constructed over the period from August 1965 to July 1966 and was placed in operation in mid-1967 when the water treatment plant at Lake Huron was commissioned.

On August 16, 1983, a pipe wall blow-out failure occurred in a standard Class 230 pipe (pressure rating 230 psi.). Again, on June 28, 1988 another pipe wall blow-out failure occurred in a Class 250 pipe. The second failure was located some 13 miles (21 km) away from the first failure. At the time of both failures, the pipeline was operating at near design pressures.

Following both failures, MacLaren Engineers Inc. was retained by the Ontario Ministry of the Environment (M.O.E.) to investigate and report on the failures.

---

[1]Vice-President, MacLaren Engineers Inc., London, Ontario, Canada

Pipe and PipeLine Construction

	1983 Failure	1988 Failure
Pipe Specification	AWWA C301-64	AWWA C301-64
Prestressing Wire	ASTM A227-64 Class II	ASTM A227-64 Class II
Design Pressure Class (psi)	230	250
(kPa)	1585	1725
Manufacturer	Canron Ltd.	Canron Ltd.
Date of Manufacture/ Shipping to site	period Nov/Dec, 1965	period Jan. 1966
Date of Laying	March, 1966	June, 1966

The pipe was supplied under separate contract for installation. Consequently, each pipe shipment was inspected by the Contractor as well as by the Engineer. Pipe suspected of being defective was spot-painted 'red' and returned to Canron Ltd. If pipe were not inspected at the time of off-loading, a subsequent inspection was carried out and defective pipe on-loaded on a returning truck from a subsequent shipment.

The construction procedure was as follows:

a) Clear and demark right-of-way which was 100 feet (30.5 m) wide with the pipe centreline 25 feet (7.62 m) from the northerly limit.

b) String pipe along one limit of right-or-way (pipe off-loaded from trucks in the area of the breaks, - pipes were strung after the frost had stiffened the ground to permit truck passage).

c) Topsoil stripping with storage at other side of easement.

d) Trench excavation with 4 cy (3 cu.metre) shovel and D-6 dozer (equivalent); hoe for excavation and dozer to spread and level excavated material along wide side of right-of-way to provide construction roadway and native backfill storage area.

e) Granular bedding material hauled to pipe laying area with some storage. Front end loader and clam assisted operation. A small dozer was used in the trench to spread and level the bedding material.

f) Pipe laying with crane - mortar material for grouting (pointing and external diaper) on stone boat hauled by farm tractor.

g) Backfilling procedure:
- D-6 or larger dozer rip-up of construction road
- D-6 dozer pushed material into trench
- D-6 dozer maintaining ramp of native backfill in trench to ensure uniform distribution and to provide a measure of compaction

h) Topsoil replaced and farm drainage tiles restored.

Investigation Re: 1983 Failure

Observations

The break resulted in a roughly 24-inch (600 mm) diameter hole at the 4 to 5 o'clock position and at about the midspan of a standard length pipe (16 feet, 4.88m). Reported observations of the failure were:

a) The prestressing wires in the zone of the failure broke in a horizontal straight line and showed signs of pit corrosion.

b) The exposed wires on the remaining sections of the pipe away from the hole did not show any visible signs of corrosion.

c) The coating on the undamaged sections of the pipe remained intact. The wires were well embedded in the coating and showed no signs of corrosion when the coating was broken off.

Canron Ltd., obtained one wrap of prestressing wire at the zone of failure, one piece of steel cylinder and some fragments of the coating. The Ontario M.O.E. also made available one piece of coating which was at the zone of failure for Canron to investigate.

Canron's investigations and comments are:

a) The failed prestressing wire was tested for tensile strength and wrap test and meets the

A227-64 standard Class II wire requirements. Samples 1, 2 and 3 were 3'-8", 10'-9" and 12'-8" from one end of the failed wire.

	A227-64	Failed Wire
Diameter	0.192" ± 0.002"	0.191"
Tensile	222,000 psi min.	Sample 1 - 238,000 psi
	251,000 psi max.	Sample 2 - 234,000 psi
Wrap	No breakage	Sample 3 - No breakage

b) The ends of the failed wire showed severe signs of corrosion at the break and showed a necking down of the wire.

c) Corrosion extended for about 24 inches (600 mm) along the wire from the break. The first 12 inches (300 mm) of the wire from the break showed more corrosion and pitting. Corrosion occurred at the top of the wire in contact with the coating. The bottom half of the wire did not show any signs of corrosion even at the break.

d) The steel cylinder was 16-gauge.

e) One fractured fragment of the coating showed heavy rust stains in the grooves which were in contact with the top of the wires. The rust stains extended one-quarter inch into the coating.

f) Another fractured fragment showed heavy rust stains in one groove, no stains in the next two grooves and slight stains in the last two grooves. Again, rust stains extended about one-quarter inch into the coating.

g) The piece of coating from the M.O.E. had a black corrosion product throughout the whole surface between the wire grooves. There were also darker streaks extending across the wires. The coating had a "sulphide" odour.

Chemical analyses show the product to be sulphides which react with hydrochloric acid to release hydrogen sulphide.

h) All outside coating surfaces did not show any signs of corrosion.

Based on the foregoing, we concluded that:

a) The failure of the pipe was due to the corrosion of the prestressing wires.

b) The corrosion would appear to have been localized to the area of the failure. Also, an adjacent pipe was subsequently removed from the pipeline and the cement mortar coating removed by M.O.E. staff. No corrosion of the prestressing wires on this pipe was noted.

Structural Failure of the Pipe

A preliminary structural analysis of the pipe was carried out using 'beam on elastic foundation' theory. This analysis indicated that significant corrosion and pitting of the prestressing wire would need to have occurred over a length of approximately 15 inches (375 mm) or more at near the same point on the circumference to result in catastrophic failure of the pipe under a 220 psi (1500 kPa) internal pressure. It is noted that failure of the pipe would then occur regardless of the condition of the wire beyond this 15-inch $\pm$ (375 mm) section. Further, complete blow-out of the pipe wall would be expected when the longitudinal wire failure reached a total length of 24 inches to 36 inches (600 to 900 mm). Accordingly, this would be the approximate size of the failed area of the pipe.

Mechanism Leading to Failure of Prestressing Wire

a) Prior to, during, or subsequent to installation, the cement mortar coating of the pipe was cracked and delamination occurred between the coating and the core. This would result in an environment suitable for electrolytic corrosion to occur provided water could enter the void formed. It should be noted that the void space need only be of capillary size and therefore would not necessarily be obvious.

b) An electrochemical reaction then set up and pit corrosion of the prestressing wire resulted with the Fe^{++} ions being released from the anode into the electrolyte. The electrolyte cell is completed by the discharge of H^{+} ions at the cathode and nascent hydrogen is formed.

c) The Fe^{++} ions released into the electrolyte react with the surplus OH^- ions resulting from the discharged H^+ ions at the cathode.

d) Sulphates (SO_4) are also present in the electrolyte. The source of these sulphates is either from the cement itself or from the groundwater. A further electro-chemical reaction takes place, i.e. an ion exchange between $Fe(OH)_2$ and $CaSO_4$ in the electrolyte resulting in $Ca(OH)_2$ and $FeSO_4$. Also, an ion exchange between Fe in the electrolyte and Ca in the cement mortar would also occur resulting in the penetration of the rust colour in the coating.

e) The cement mortar coating would prevent the nascent hydrogen from escaping, and being highly reactive reduced the $FeSO_4$ to FeS, e.g.:

$$Fe^{++} + SO_4^{--} + 8H^+ \longrightarrow FeS + 4\ H_2O$$

It is noted that FeS is black in colour and has the familiar sulphide odour.

The foregoing process explains:

i) the form of corrosion;

ii) the presence of rust stains on and in the concrete mortar coating at the interface with the pipe core, and similarly;

iii) the presence of a black stain which exhibited a 'sulphide' odour.

Investigation Re 1988 Failure

Based on initial observations, several similarities between the two failures were noted. Consequently, efforts were directed towards the metallurgical aspects of the prestressing wire failure and the reason for mortar cracking and delamination.

Metallurgical Aspects

Fourteen samples of prestressing wire were taken from the pipe in the location of the failure. A detailed analysis of these was carried out and it was concluded that the wire had corroded in service and was severely weakened by stress corrosion cracks. Thirteen cracks were detected in the fourteen sections examined. The principal conclusions from the study were that:

- corrosion occurred only on the outer half of the wire;
- the failure was due to sulphide stress cracking/hydrogen stress cracking;
- the environment of the pipe would not need to be too aggressive to damage the prestressing wire at its hardness level.

Mortar Coating Cracking

To explain the reason for mortar coating cracking and delamination, files during the construction of the pipeline were reviewed to:

- determine the time when each standard length of pipe was manufactured;
- determine the time when each standard length of pipe was 'strung' along the pipeline easement;
- determine the time when each pipe and special was laid.

The following similarities between these two failures have been drawn:

- Both breaks occurred in the higher pressure classes of the pipeline, i.e. pipe having the highest prestress in the structural concrete core;
- The pipe at both breaks was manufactured in the late fall/early winter of 1965/66, i.e. approximately 18 months prior to being placed in service;
- Concrete under compressive stress 'creeps' with time. In an embedded cylinder pipe, the shortening of the structural core results in a compressive force in the coating, but as well, a 'popping' force tending to delaminate the coating from the core. The larger the duration between initial prestress and the placing in pressurized service, the greater the 'creep';
- Both pipes were subject to winter conditions with extremes of temperature and freeze-thaw action;

Both pipes, it can be assumed, were subject to rougher handling conditions due to winter construction and, perhaps, double and triple handling in the field.

In our opinion, the sum of the foregoing would result in a much higher potential for coating damage, i.e. cracking and delamination from the core.

Conclusions

Firstly, in our opinion, the mechanism of corrosion as determined from the investigation into the 1983 failure is valid. With this mechanism of corrosion, the galvanic cell is contained within the wire-cement mortar coating interface and, therefore, there would be no current through the soil surrounding the pipe from a corroding anode to a cathode somewhere else along the pipeline.

Secondly, when the use of high pressure prestressed concrete embedded cylinder pipe is contemplated, consideration should be given towards minimizing the time lag between pipe manufacture and placing in service. Further, pipe exposure to severe winter conditions should be avoided.

ANCHORAGE SEWER REHABILITATION IN CAMPBELL LAKE

Bruce J. Corwin, P.E., Member, A.S.C.E.[1]

Abstract

This paper discusses the proposed rehabilitation/replacement of the Campbell Lake portion of the Campbell Creek C-5 Trunk Sewer (C-5 Trunk) located in Anchorage, Alaska. The original corrugated metal pipe (CMP) from 36 inch to 54 inch in diameter, installed in the early 1960's, was inspected in 1986 and found to be in a failure mode. Since the portion installed in Campbell Lake was considered to be the most critical to repair, the Anchorage Water and Wastewater Utility (AWWU) authorized the design to proceed in July 1988.

Many unique and interesting challenges have faced the design team for this project including a difficult Environmental Assessment process due to fish concerns, an intensive public involvement program, innovative soils investigation methods, and innovative design techniques which would allow the new pipeline to be installed within the lake.

The C-5 Trunk is one of the largest sewers in the Anchorage area and serves a large portion of the southern part of the community. Design is scheduled to be completed in August 1989. Winter construction of the project is scheduled to begin during January 1990 and be completed by April/May 1990.

---

[1] President, Corwin & Associates, Inc., 1000 E. Dimond Blvd., Suite 205, Anchorage, Alaska 99515

Discussion includes the history and background of the existing sewer, inspection program, analysis of repair/replacement alternatives, value engineering, field investigations, design parameters, and proposed construction techniques.

Introduction

Anchorage, Alaska, like many other United States communities, now finds itself at an age where infrastructure repair and replacement becomes an equal priority with expansion for future development. One such example is the Campbell Creek C-5 Trunk Sewer (C-5 Trunk).

Design service life of our nations sewers depends on a number of factors including materials used in construction, accurate population/development projections, occurrence of unforeseen acts of nature, and the maintenance program utilized to extend service life. If any one of these or other factors has a shortfall, the result will be a need to repair or replace an inadequate sewer line.

The purpose of this paper is to discuss the existing pipeline, the events that have taken place since installation, environmental features, inspection techniques, recommendations for repair or replacement and the proposed design of the replacement sewer in Campbell Lake.

History/Background

In 1958/59, an Anchorage developer, Mr. David Alm, built a dam and spillway at the west end of Campbell Creek to provide a lake environment for his development. Campbell Lake is a private lake devoted exclusively to the owners around the lake. Its primary use is as a float plane base. During 1964, Anchorage was rebuilding from the earthquake, but also constructing infrastructure to service portions of the southern community with sewer service. A private organization named Central Alaska Utilities, Inc. contracted for design and construction of the C-5 Trunk Sewer in 1964/1965.

The material selected for construction was corrugated metal pipe (CMP) from 36 inch to 54 inch in diameter with a smooth flow invert. Pipe was manufactured in approximately 18 foot lengths and bands were used to join sections of pipe. Manholes were also made of CMP and did not contain manhole steps for access. Existing soils along the pipeline route consisted of clays with some traces of silt and sand. Since major development had not yet occurred, installation was fairly easy to accomplish and easements readily attainable from the developer.

Environmental issues were not of much concern at the time and permitting for the project was also a fairly easy task. The pipeline was placed in the logical position along the lake shore except where it actually crossed the lake approximately 800 feet east of the dam itself. Construction was simplified by breeching the dam and lowering the lake after it had frozen over during winter months. The ice left provided an excellent work pad over the otherwise unstable clay soils.

The C-5 Trunk begins just east of the Alaska Railroad tracks, north of Dimond Boulevard and east of C Street in the King Street Industrial area. From there the sewer follows an alignment along the Campbell Creek Greenbelt (low spot in drainage area) to the north shore of Campbell Lake. The existing sewer line then crosses the lake and follows the southern shore of Campbell Creek, below the dam spillway, to the Campbell Creek Sewage Lift Station. Originally, the sewage merely dumped directly into the Cook Inlet, but the lift station was constructed to pump the sewage to the Point Woronzof Wastewater Treatment Plant. (See Figure 1 for location).

Increased development of the Anchorage area in the late 1970's and early 1980's created higher flows to the C-5 Trunk from the southern part of the Anchorage community. In 1982, AWWU moved forward with completion of a long range Wastewater Facilities Plan [1]. At that time, the C-5 Trunk was identified as a sewer requiring construction of a diversion sewer to intercept flows and relieve the excessive flow conditions. Once flow diversion was complete, the C-5 Trunk was to be fully inspected through investigation and rehabilitation accomplished. Upon completion of the West

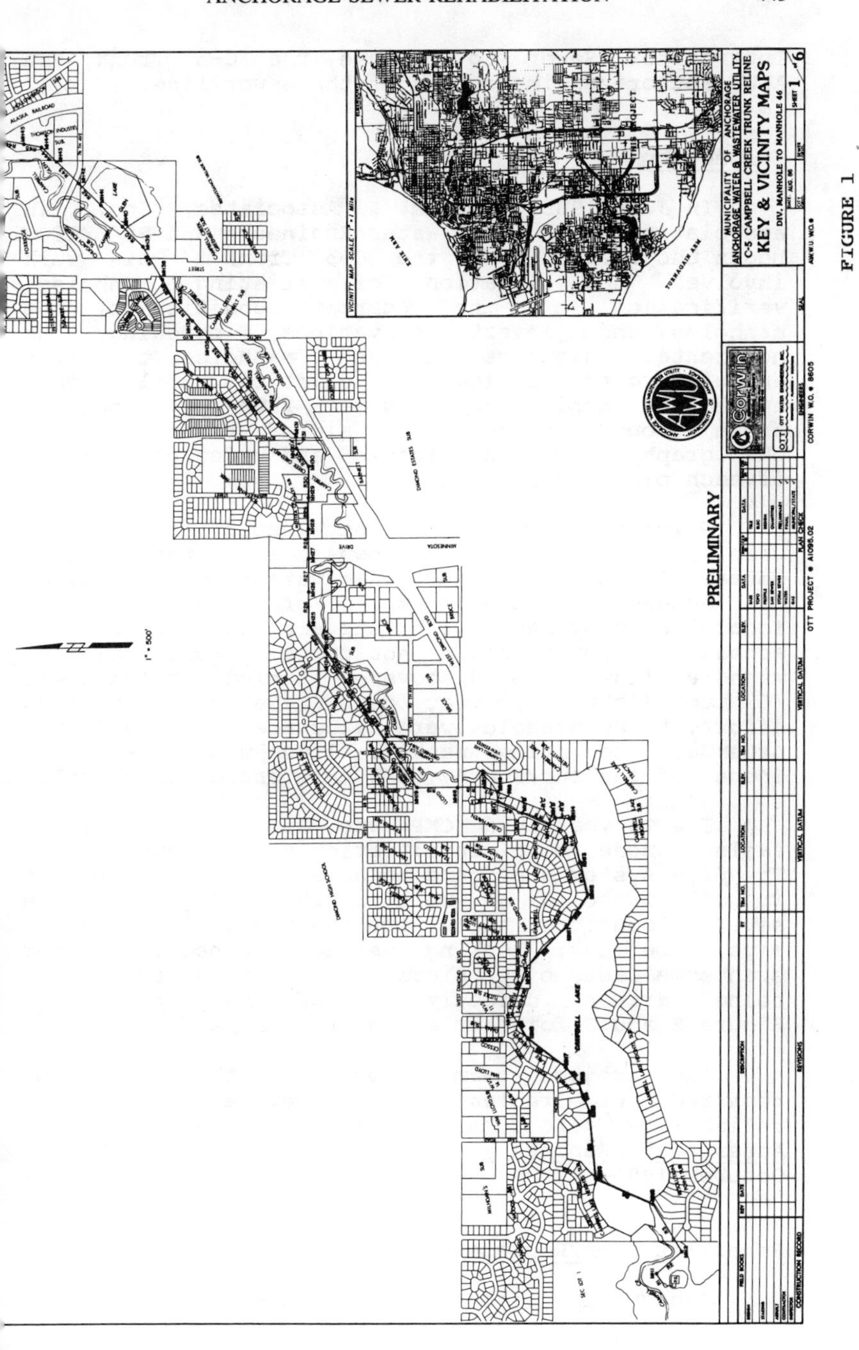
MUNICIPALITY OF ANCHORAGE
ANCHORAGE WATER & WASTEWATER UTILITY
C-5 CAMPBELL CREEK TRUNK RELINE
KEY & VICINITY MAPS
DIV. MANHOLE TO MANHOLE 46
SHEET 1 OF 6
PRELIMINARY
OTT PROJECT # A1095.02
CORWIN W.O. # 8605
VICINITY MAP SCALE: 1" = 1 Mile
1" = 500'
CAMPBELL LAKE
MINNESOTA DRIVE
C STREET
ALASKA RAILROAD

FIGURE 1

Interceptor Sewer to relieve the C-5 Trunk, the AWWU authorized inspection of the sewer line.

Inspection

In June 1985, Corwin & Associates, Inc., in association with Ott Water Engineers and SEAL/TEC, began the inspection of the C-5 Trunk. This work involved the location of existing manholes, verification of the current condition of the manholes and invert elevations, checking the horizontal distance between manholes to match as-built data, review of existing soils data available, monitoring flow at five key manholes, above ground photography along the pipeline, photography of each manhole and video photography of each pipeline segment.

Prior to accomplishing the inspection, notices were sent to all property owners along the pipeline route. At locations where excavation was required to access manholes, most of the work was accomplished by hand with shovels due to limited access by construction equipment. Some existing manhole lids were discovered buried at least six (6) to eight (8) feet deep. During inspection effort, these manholes were excavated by hand and extended to bring manhole lids within six (6) inches of the surface by AWWU maintenance personnel.

The 20 year old CMP was found to be in a failure mode with infiltration at numerous joints. The pipe system was found to have corrosion evident in all manholes as well as the piping itself. In several locations, there were flattened sections of pipe, bellies, leaking service connections, and even some cases of physical damage such as H-Pile being driven directly through the pipe. (See Figure 2 and 3 for sample condition maps).

The project team recommended the pipe be repaired or replaced to provide long term reliability to the community. A Condition Assessment Report [2] was delivered to AWWU in October 1986.

Rehabilitation/Replacement Report

Recognizing the need to effect repair or

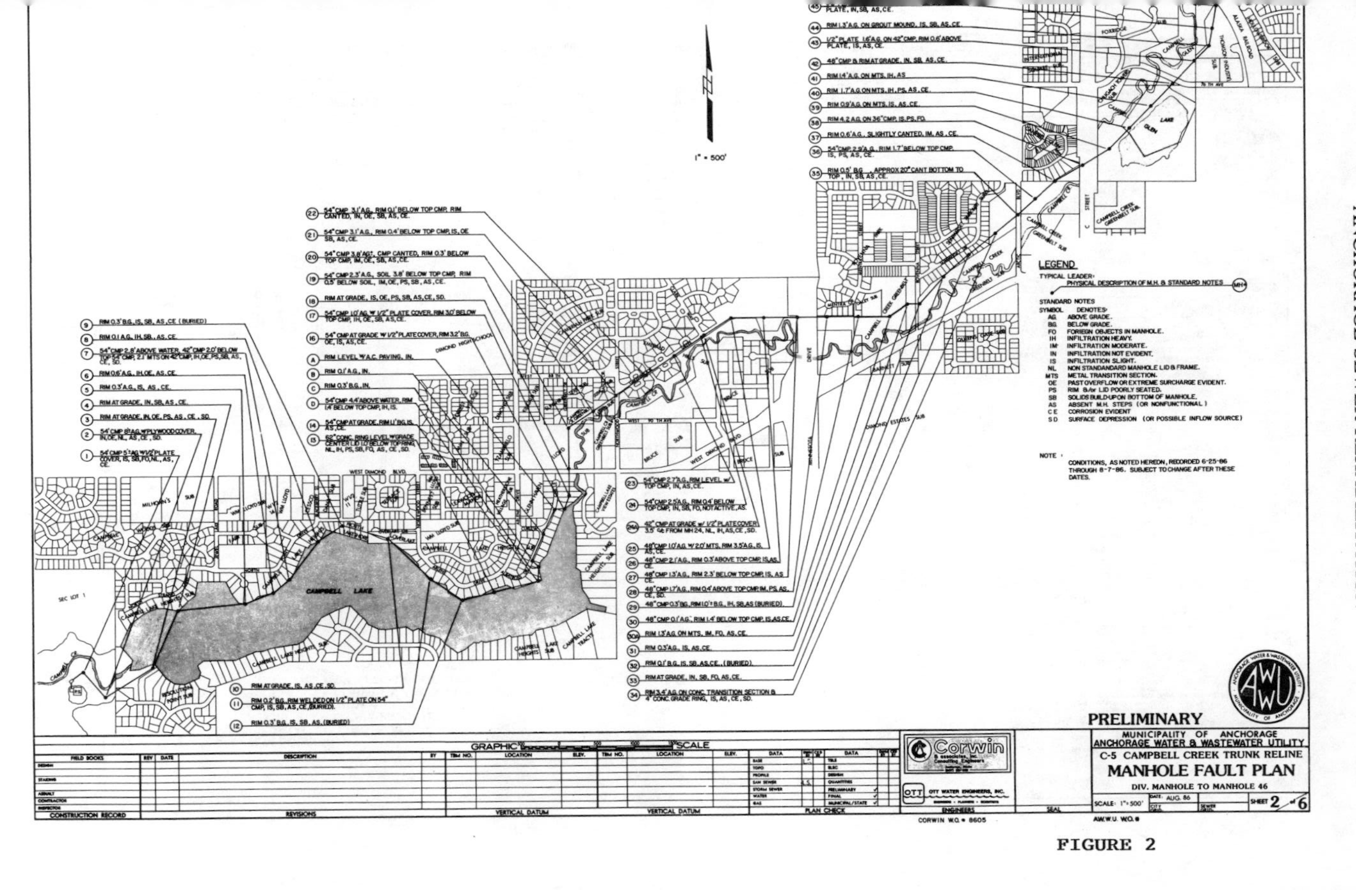
1" = 500'
LEGEND
TYPICAL LEADER:
PHYSICAL DESCRIPTION OF M.H. & STANDARD NOTES
STANDARD NOTES
SYMBOL DENOTES:
AG ABOVE GRADE.
BG BELOW GRADE.
FO FOREIGN OBJECTS IN MANHOLE.
IH INFILTRATION HEAVY.
IM INFILTRATION MODERATE.
IN INFILTRATION NOT EVIDENT.
IS INFILTRATION SLIGHT.
NL NON STANDANDARD MANHOLE LID & FRAME.
MTS METAL TRANSITION SECTION.
OE PAST OVERFLOW OR EXTREME SURCHARGE EVIDENT.
PS RIM &/or LID POORLY SEATED.
SB SOLIDS BUILD-UP ON BOTTOM OF MANHOLE.
AS ABSENT M.H. STEPS (OR NONFUNCTIONAL)
CE CORROSION EVIDENT
SD SURFACE DEPRESSION (OR POSSIBLE INFLOW SOURCE)
NOTE: CONDITIONS, AS NOTED HEREON, RECORDED 6-25-86 THROUGH 8-7-86. SUBJECT TO CHANGE AFTER THESE DATES.
CAMPBELL LAKE
GRAPHIC SCALE
PRELIMINARY
MUNICIPALITY OF ANCHORAGE
ANCHORAGE WATER & WASTEWATER UTILITY
C-5 CAMPBELL CREEK TRUNK RELINE
MANHOLE FAULT PLAN
DIV. MANHOLE TO MANHOLE 46
SCALE: 1" = 500'
DATE: AUG. 86
SHEET 2 of 6
CORWIN W.O. # 8605
OTT WATER ENGINEERS, INC.

FIGURE 2

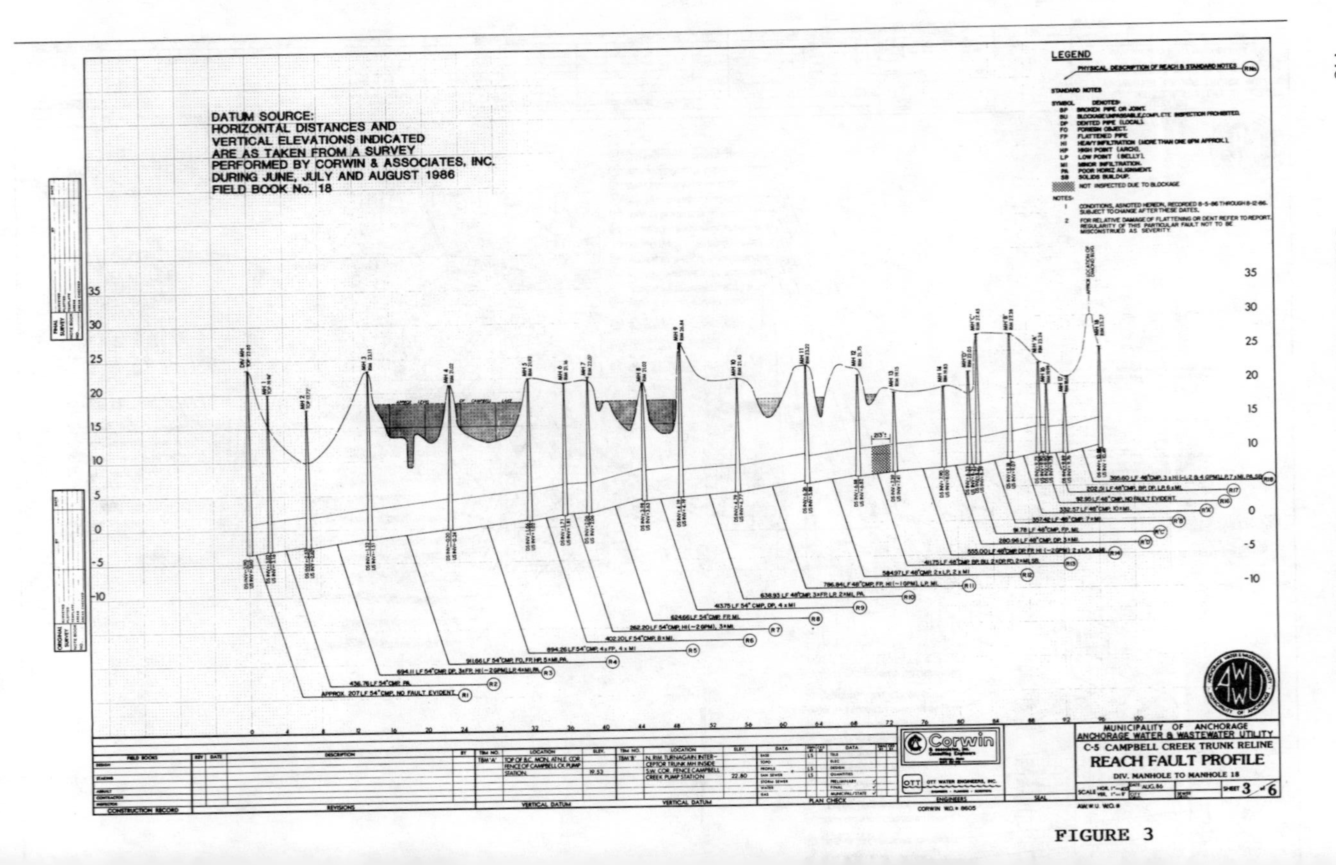
DATUM SOURCE:
HORIZONTAL DISTANCES AND
VERTICAL ELEVATIONS INDICATED
ARE AS TAKEN FROM A SURVEY
PERFORMED BY CORWIN & ASSOCIATES, INC.
DURING JUNE, JULY AND AUGUST 1986
FIELD BOOK No. 18
LEGEND
PHYSICAL DESCRIPTION OF REACH & STANDARD NOTES
STANDARD NOTES
SYMBOL DENOTES
BP BROKEN PIPE OR JOINT.
BU BLOCKAGE/UNPASSABLE/COMPLETE INSPECTION PROHIBITED.
DP DENTED PIPE (LOCAL).
FO FOREIGN OBJECT.
FP FLATTENED PIPE.
HI HEAVY INFILTRATION (MORE THAN ONE GPM APPROX.).
HP HIGH POINT (ARCH).
LP LOW POINT (BELLY).
MI MINOR INFILTRATION.
PA POOR HORIZ ALIGNMENT.
SB SOLIDS BUILDUP.
NOT INSPECTED DUE TO BLOCKAGE
APPROX. 207 LF 54" CMP, NO FAULT EVIDENT.
436.76 LF 54" CMP, PA.
MUNICIPALITY OF ANCHORAGE
ANCHORAGE WATER & WASTEWATER UTILITY
C-5 CAMPBELL CREEK TRUNK RELINE
REACH FAULT PROFILE
DIV. MANHOLE TO MANHOLE 18
Corwin
SHEET 3 of 6
CONSTRUCTION RECORD
REVISIONS
VERTICAL DATUM
PLAN CHECK
ENGINEERS
SEAL

FIGURE 3

replacement of the existing C-5 Trunk, the AWWU authorized Corwin & Associates, Inc., in association with Ott Water Engineers, Inc., Black & Veatch Consulting Engineers, and Westech Engineering, Inc., to complete a report recommending the appropriate solution.

Utilizing data from the inspection effort and the EPA Handbook for Sewer System Evaluation and Rehabilitation [3], the project team developed a listing of possible repair/replacement techniques including the following:

- No Action
- Concrete Invert Lining
- Point Repair
- Resin Impregnated Fabric (Insituform)
- Shotcrete
- Chemical Grouting
- Sliplining
- Replacement/Reroute
- Combination of Above

Upon completion of initial screening, a Value Engineering Team (VE Team) was called upon to refine the available alternatives list for selection of at least three (3) alternatives for each section of the pipeline. This team of five (5) people consisted of pipeline rehabilitation experts, an AWWU design person, and an AWWU maintenance person. The Campbell Lake portion of the C-5 Trunk was deemed to have the following practical solutions:

- Reroute/Realignment
- Sliplining
- Resin Impregnated Fabric (Insituform)
- Combination of Above

The VE Team, upon final review of cost and practicality, selected Reroute/Replacement as the most practical solution for the Campbell Lake portion of the C-5 Trunk. The concept selected consisted of lowering Campbell Lake and installing the replacement sewer off shore in the winter months thereby avoiding conflict with summer use of the lake and the disruption that replacement in the existing location or near docks would cause. This approach would be similar to the original installation in 1965. Further, the VE Team selected the Campbell Lake portion of the C-5 Trunk as being the most critical to repair or replace since a failure would cause such devastating effects to the lake and environment.

Several public meetings were held with the Campbell Lake Owners, Inc. (CLO) to obtain approval of the concept to lower the lake and perform construction in the winter months. The CLO accepted the alternative and agreed to the concept of winter construction.

In October 1987, the final Repair and Replacement (R & R) Report [4] for the C-5 Trunk was completed and delivered to AWWU.

Design Development

In July 1988, Corwin & Associates, Inc., in association with Ott Water Engineers, Inc., was authorized to begin design of the Campbell Lake portion of the C-5 Trunk.

The purposes of the proposed project are to:

- avoid failure of the existing C-5 Trunk Sewer and resulting impacts;
- reduce infiltration and the overall operational expenses of the C-5 Trunk Sewer;
- maintain reliable sewer service to service area residents;
- protect public health;
- protect private and public property; and
- protect water quality and fish and wildlife habitat.

This project has been ranked first on the State of Alaska list for EPA funding in the current Federal Fiscal Year (FFY) 1989 (October 1, 1988 to September 30, 1989).

This project extends from Dimond Boulevard, south and west, to the Campbell Creek Lift Station. It includes replacing and rerouting 8,450 feet of the existing pipe with 48-inch ductile iron pipe, along with repairing a 200-foot section of the 24-inch Campbell Creek C-5-2 sub-trunk sewer and a 1,100-foot section of the 18-inch C-5-7 sub-trunk sewer.

Work involved a detailed survey of the proposed alignment including a bathymetric type survey along the north shore of the lake, soils investigations using a raft for the drilling rig in the lake, public coordination for survey and soils effort and a preliminary design layout of the sewer line plan and profile.

Since the proposed sewer line was to be located such that manholes and pipelines would be continuously under water, the issue of manhole access needed to be addressed. After initial investigation, we discovered similar installations in the State of Washington (Seattle area lakes) where manhole tops were placed near the lake bed surface and caissons were utilized for maintenance access. This concept was accepted by AWWU as the most practical means of dealing with the situation since access to the trunk sewer would be limited.

Existing population served by the C-5 Trunk is approximately 20,000 people and design population to the year 2010 is 22,000 people. Design flows were developed based on the following parameters:

Land Use	Daily Flow Gallons	CFS	I & I Contribution
Residential	80 GPCD	--	105 GPCD
Commercial	6,269 GPAD	0.0097 CFS/AC	500 GPAD
Industrial	5,000 GPAD	0.0077 CFS/AC	500 GPAD
Industrial	6,269 GPAD	0.0097 CFS/AC	500 GPAD

These design flow criteria provided a design average daily flow 4.59 mgd and peak flow of 12.4 mgd. A 48-inch ductile iron pipe was selected as the optimum replacement pipeline. Based on analysis of in-situ soils, the soils possess corrosive capabilities and a ductile iron pipe would need to be encapsulated within a polyethylene encasement to survive . The design trench consists of a Class C bedding material with existing native backfill replaced back up to the surface.

The design includes two inverted siphons where

the pipeline crosses Campbell Creek and sliplining where replacement was not deemed necessary. See other design issues later in this paper.

Environmental Issues on Design

After completion of the preliminary design, a meeting was held with the agencies to determine issues affecting the project. At this initial meeting, it was proposed to lower the lake approximately eight (8) feet leaving only Campbell Creek remaining in its pre-lake form. A great concern was voiced over the effect this would have with fish, primarily coho salmon, surviving in the lake over winter.

In order to fully assess effects on fish due to the project, a complete fish study was performed within Campbell Lake. Initially the study was to begin in October 1988, however, due to an early freeze-up and coordination, the work was not performed until December 1988. Results of the Fish Study [5] indicated that coho salmon fry and smolt (15,000+/-) were in numerous locations within the lake, but most of the fry were concentrated at the Campbell Creek backwater where the creek enters Campbell Lake. Other fish were also found within the lake, including sockeye and chinook salmon, dolly varden, etc.

Upon completion of the fish survey the Alaska Department of Fish and Game (ADF&G) assumed this concept would have 100 percent mortality of salmon resources in the lake thereby creating concern by other agencies as well with the proposed alternative of lowering the lake. The issue then became one of "Avoidance Alternatives." Alternatives which could avoid fish mortality were investigated including:

- °Moving pipeline out of the lake altogether and utilizing pump stations
  - a. Force main in street adjacent to lake
  - b. Force Main in existing pipe
  - c. Force Main at edge of lake
- °Barge installation with lake full
- °Sheetpile along shore to allow pipe installation and lake full
- °Separating the lake into two segments (sheetpile)

°Replacement in existing location

All of these alternatives were more expensive than the original option and created design, construction and easement concerns. None were regarded as practical construction alternatives, but they did avoid fish mortality issues. Another alternative was to collect all fish in the lake prior to drawdown, place them in hatchery type pens at the west end of the lake, and monitor them during construction. No one believed we would be successful with such an approach. A fish biologist retained by our project team believed the best approach would be to just leave the fish in the remaining creek after lake lowering.

After nearly three months of revisions to the Environmental Assessment and several meetings with local, state and federal agencies, a compromise solution was reached which allowed the project to continue. This solution involved lowering the lake only three (3) feet in late September or early October prior to freeze-up and then allow the lake to freeze another three (3) feet (+/-). The concept would allow fish adequate acreage to live in and allow construction to take place either on the nearshore ice or the dry frozen lakebed. Final approval of the Environmental Assessment was obtained in August 1989.

The only other major environmental concerns faced by the project were water quality and wetlands impacts. Both are fairly easy to mitigate for since the project has limited effect to either area based on construction procedures.

Other Design Issues

In 1983, the C-5 Trunk Sewer Segment C-5-2 in Campbell Lake was damaged during dredging operations at the upstream part of the lake. The 24-inch pipeline is shallow (1-3 feet cover) and not sufficiently covered to prevent such an occurrence. Under the proposed construction a new three pipe inverted siphon will be used to cross under the creek channel and hopefully prevent the possibility of future damage. Also, all other work related to the lake will be accomplished under the project including removal and rerouting of old small service lines incorrectly attached to the trunk sewer rather than collectors. An additional three pipe inverted siphon is also required to cross Campbell Creek at the west end of the lake.

Over the last several years the influent or upstream portion of the lake has received heavy sedimentation from Campbell Creek causing loss of float plane access to much of this area. This situation prompted the Campbell Lake Owner's (CLO) to request that a portion of the dredging cost be paid in exchange for an easement to allow the pipeline to be installed in the lake.

Two portions of the replacement project will use sliplining. A 230 foot segment of the existing 54-inch CMP at the low end of the existing trunk sewer will be sliplined with a 48-inch High Density Polyethylene Pipe (HDPE) and another segment of the 54-inch CMP near the connection of a 21-inch collector sewer will be sliplined using a 22-inch HDPE. Sliplining will require excavation of access pits to insert the new pipe into the existing CMP.

Other small collectors will be connected to the new trunk sewer at manholes. In order to ensure future maintenance, a pipe within a pipe utilizing an 8-inch HDPE inside a 12-inch HDPE or a 12-inch HDPE inside an 18-inch HDPE will be used. This would allow maintenance personnel to ensure long term performance and provide for future expansion. If the inner (initial) pipe fails, it can be removed and the larger pipe used.

A storm drain project constructed by the Municipality of Anchorage (MOA) Public Works Department in 1989 created concern by area homeowners over sediment and water level control since it also discharged into Campbell Lake. The final resolution was for Public Works to analyze effects on the existing Campbell Lake Dam and provide an adequate spillway/weir facility to pass a peak storm event. As a result of this project, a drain facility for the proposed (3) foot draw down which was needed for the C-5 project was added to the Public Works contract documents.

All of the foregoing discussion regarding lake drawdown to three (3) feet was changed automatically by Mother Nature on August 25, 1989, when heavy rainfall caused the existing dam spillway to be washed out. Currently, Campbell Lake is approximately six (6) feet below its original elevation.

Construction Planning and Scheduling

As currently planned, the project construction will begin in January 1990. The contractor will have a choice of times for actually beginning construction of the pipeline, but must be complete by April/May 1990. Since the project is a part of the original 1982 Wastewater Facilities Plan and the last project in the plan to be built, it must be complete by October 1990 under the Anchorage Segmentation Package approved by the United States Environmental Protection Agency (USEPA). The project is 75% funded by the USEPA. The contractor will also be allowed to use alternative means of construction with an incentive bonus for cost savings potential or ideas he develops. Permits were applied for and obtained in September/October 1989.

The dry, frozen perimeter of the lakebed will be trenched to a depth of 7 to 22 feet for installation of the pipe atop sand and gravel pipe bedding. Excavated native soils will be used for backfill and site restoration.

State water quality standards will be maintained by establishing a minimum 25-foot construction setback from the creek channel, and through a runoff control plan, whereby possible discharges from any necessary dewatering operations will be made directly into the existing sewerline.

As originally planned, the lake was to be lowered three (3) feet prior to construction which would have necessitated fish handling and protection at selected areas. Now this is not needed due to the spillway washout described above. Most of the mitigation required regarding fish protection was eliminated when the lake was lowered by nature.

Personnel will still be required to periodically sample lake water through the ice during construction and monitor turbidity and dissolved oxygen levels.

The trench for the sewerline will be located

within a 20-foot permanent easement. A temporary construction easement (permit) has also been acquired for use of the lakebed for construction.

There will be two access routes to the project site. The eastern side of the lake will be accessed via a Municipality of Anchorage 150-foot-wide utility (storm drain) easement from Victor Road to the east shore of Campbell Lake. This area was disturbed in the summer of 1988 during installation of a new storm drain as part of an unrelated municipal project previously discussed. Temporary culverted crossings of the creek channel on the drained lakebed will provide equipment access from the utility easement to the project site.

A second route will provide access to the west end of the lake. It follows an AWWU access road along the southwest shore of Campbell Lake. This road already has a culverted creek crossing below the dam.

The lake will be refilled to its original level upon completion of construction by May/June 1990. There will be no permanent change to the lakeshore, lake level, or lakebed except insignificant redistribution of native lake bottom soils around the refilled trench, which will have no effect on water quality or navigation. The construction access site below the dam and tidal wetlands disturbed by construction, will be restored according to ADF&G and MOA specifications.

Summary

This has been an exciting project and should prove to be interesting during construction. It is planned to have a separate presentation on the project when construction is complete. Since it is to be constructed in Alaska during winter, the highlights should prove interesting. The end result should be a valid pipeline to serve the area for at least the next 50 years. The Environmental Assessment (EA) produced for this project was one of the most thorough and complete submitted to EPA Region 10 recently and was more in the realm of an Environmental Impact Statement. Unfortunately or fortunately, Mother Nature took control and all the concerns in the EA were negated due to the Campbell

Lake spillway washout in August 1989.

Bibliography

1. Wastewater Facilities Plan for Anchorage, Alaska. Municipality of Anchorage. June 1982.

2. Campbell Creek C-5 Trunk Condition Assessment Report, Vol. I, Corwin & Associates, Inc. October 1986.

3. Handbook for Sewer System Evaluation and Rehabilitation, EPA - 430/9 - 75 - 021, December 1975.

4. Campbell Creek C-5 Trunk R&R Report, Vol. II, Corwin & Associates, Inc. October 1987.

5. Winter Fish Utilization of Campbell Lake - Minnow Trap Survey, John W. Morsell, Northern Ecological Services. December 1988.

WHITFIELD RESERVOIR INLET OUTLET PIPELINE

Wm. H. Blair,[1] M. ASCE, Patrick Creegan,[2] F. ASCE
Theodore Lynch,[3] and Loren Weinbrenner[4]

ABSTRACT: Alameda County Water District reports on the planning, design and construction of a 48-inch water main installed under modern urban circumstances. The location of the pipe and the construction practices had to meet the approval of five regulatory agencies and satisfy various environmental criteria. Eighteen percent of the alignment features tunneling or boring and jacking in order to pass under an interstate highway, two railroads, a local parkway, and two residences. The provision for crossing a major earthquake fault is also described.

INTRODUCTION

In the spring of 1988 the Alameda County Water District of Alameda County, California, inaugurated the Whitfield Reservoir Inlet-Outlet Pipeline. This is a 48-inch diameter treated water line, constructed in an urban situation where both the design and the construction were strongly influenced by environmental conditions. Although not a major-diameter pipeline, it qualifies as an important modern civil work because it represents the trend of the future for urban underground pipeline construction. The line is but 7,100 feet long and yet it features 750 feet of conventionally excavated soft ground tunnel under Interstate 680; 160 feet of tunnel under a private property line between two homes; 65 feet of bored and jacked casing beneath the Union Pacific (UPRR) railroad line; 158 feet of tunnel under the Southern Pacific (SPRR) railroad line; and 54 feet of special cut and cover tunnel at the crossing of the Hayward Fault. Thus, about 17 percent of the alignment was in these five special tunnels. Most of the remaining length was "shoe-horned" into city streets, with the general contractual provisions that the maximum length of open trench shall never exceed two hundred (200) feet.

---

[1]Design Div. Engr., Alameda County Water District, P.O. Box 5110, Fremont, CA 94537
[2]Struct. Engr., Vice Pres., Engineering Science Inc., 600 Bancroft Way, Berkeley, CA 94710
[3]Engr., Design Div., Alameda County Water District
[4]Engr., Engineering Science, P.O. Box 2447, Monterey, CA 93942

In addition, the contractor had to obtain encroachment permits (whose conditions were published at the time of bidding) from five agencies, and operate under the terms of those permits, which could be even more restricting than the job specifications.

With all these special construction situations, the job, designed by Engineering-Science, Inc., and constructed by Mountain Cascade, Inc., came in at $565 per foot of 48-inch diameter pipe, or about $12 per diameter inch, which included quite a bit of ancillary works, namely 550 feet of 24-inch diameter drain line, 1,800 feet of 8-inch diameter water line, blow off structures and project appurtenances. The final contract price was within 6 percent of the bid price, with 75 percent of the overrun being related to field design changes. The unusual design features, coupled with some innovative field modifications to the specifications, proposed by the contractor and approved by the Owner, make for an interesting project.

THE SETTING

Alameda County Water District is located southeast of San Francisco on the east side of San Francisco Bay, 15 miles north of San Jose and about 25 miles south of Oakland. The District, with a service population of approximately 250,000, serves municipal and industrial water to the cities of Fremont, Newark and Union City. Since 1978 the average daily water demand has increased 60 percent. Whitfield Reservoir is a man-made, buried concrete impoundment, storing about 22 MG (67 AF) of treated water. Whitfield Reservoir Inlet Outlet Pipeline is the major conveyance to and from the reservoir.

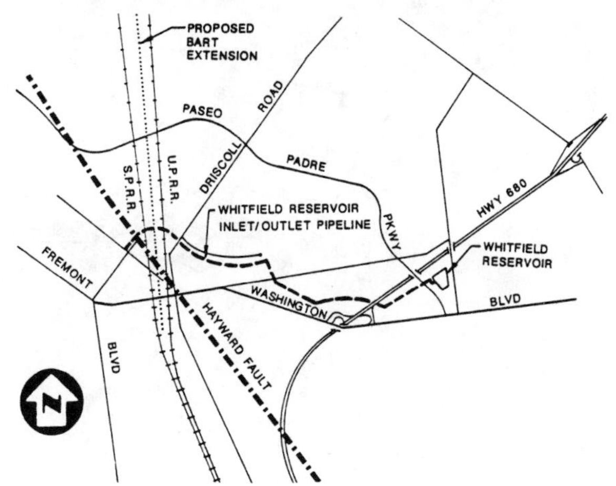

FIGURE 1 - WHITFIELD RESERVOIR INLET/OUTLET PIPELINE ALIGNMENT

In traversing 7,100 feet westerly from the Whitfield Reservoir, the pipeline route (shown on Figure 1) is located mostly in city streets, and has major crossings at the Paseo Padre Parkway, Interstate 680 Highway, Olive Avenue, Driscoll Road, the UPRR and SPRR lines, and the Hayward Fault.

The Hayward Fault is a principal member of the San Andreas related family of faults and lies roughly parallel to and about 19 miles easterly of the main San Andreas (see Figure 2). It trends roughly N35W. Movement along the fault is right lateral aseismic slip (creep), at a rate varying from 3/16 to about 7/16 inches per year. The fault is well defined, having a very narrow rift zone. In some parts of the District it is so well defined that it can be located by curb, fence and building offsets that it has created. Its estimated Maximum Credible Earthquake (MCE) is Richter magnitude 7.5. The Maximum Probable Earthquake (100 year event), used for design of this project, is between 7.0 and 7.25, with 3.6 feet of right lateral movement and 0.7 feet of vertical displacement anticipated. The design accommodates these movements with minimum damage.

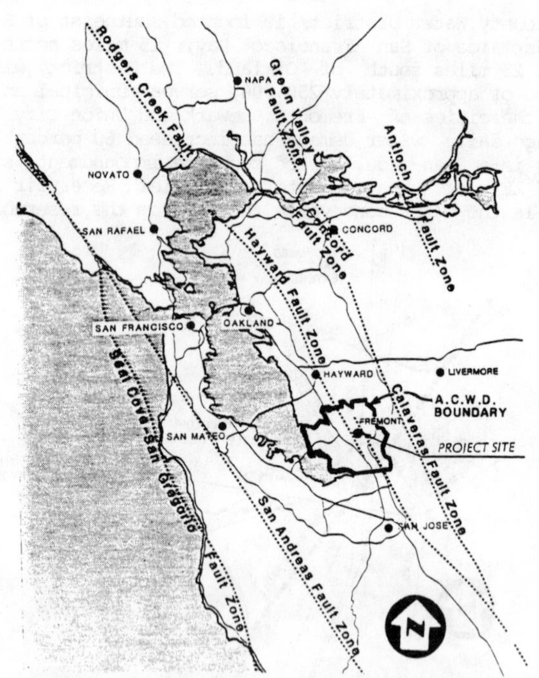

FIGURE 2 - ACTIVE FAULT ZONES IN THE SAN FRANCISCO BAY AREA

PROJECT PLANNING

The Alameda County Water District has the responsibility of serving municipal and industrial water to a fast growing sector of the San Francisco Bay area. When District voters turned down a local water bond proposal in 1971, the Board of Directors of the District adopted the policy of financing all capital improvements on a pay-as-you-go basis, with revenues derived from connection fees. These fees presently stand at $1,570 per single family home and $1,320 for a multiple family dwelling unit.

The District has 3 sources of water supply: the local groundwater basin, imported water from the California State Water Project, and imported water from the City of San Francisco's Hetch Hetchy lines which pass through the District. The Whitfield Reservoir, constructed from 1984 to 1986, and its 48-inch inlet outlet pipeline, constructed in 1987, are a unit serving the most populous pressure zone of the District service area. They provide for meeting peak demands and furnishing emergency storage for three average days of system demand.

Several alternative alignments for the inlet outlet line were studied. A controlling criterion for the alignment was that the reservoir had to be capable of being emptied by gravity flow. To conform to this criterion, the pipeline profile had to be below the elevation of the bottom of the inlet to the reservoir, elevation 175 feet.

The reservoir site is immediately adjacent to a depressed portion of Interstate 680. The pipeline had to cross under this freeway in a casing. The land to be traversed by the pipeline was already developed and a number of existing facilities had to be considered in selecting the alignment. These included a petroleum products line, a 24 inch high pressure gas main, two railroads, a future extension of the Bay Area Rapid Transit (BART), a city park, flood control channels, and numerous existing underground utility lines. In addition, the line would probably have to cross or parallel an intermittently flowing small, heavily wooded creek, and pass through several residential neighborhoods. After a study of alternative alignments, a route was selected and this project was included in a multi-project, major facilities improvement program, for which an Environmental Impact Report (EIR) was prepared in 1980.

In 1984 another facilities improvement program was recommended and an EIR was prepared for it. By this time the supply pipeline route was more definite, and the planned size of the line had been changed from 36 inches to 48 inches. In response to environmental concerns expressed during the previous EIR process about a pipeline alignment in the creek corridor, the District had decided to place the line in a paved street which paralleled the creek for approximately 2,000 feet. In order to reach the street, the pipeline would have to be tunneled between two homes, straddling their property line.

During the preparation of the 1984 EIR the District learned about the plans of the Bay Area Rapid Transit District to construct an extension that would cross the pipeline alignment. The District authorized a focused alternatives study as part of the EIR process to develop a crossing scheme that would meet the needs of both Districts in a cost effective manner. It was not economically feasible to locate the pipeline initially so that it would not be in the way of future BART trackway construction. The adopted scheme permits the initial pipeline construction to be removed for trackway construction, provides outlets for connection of a temporary construction bypass pipeline with a final pipeline location supported by and above the BART trackway.

DESIGN ELEMENTS

Geotechnical Program

Woodward-Clyde Consultants was the member of the design team responsible for carrying out the geotechnical investigation for the project.

Due to their prior experience in the locale, the bulk of their work was conducted after the route studies were complete and the final alignment had been selected. That work included:

1. Trenching and trench mapping at the fault crossing.

2. Reporting on site seismicity from available records.

3. Boring and logging the borings at 10 locations along the profile, to an average depth of about 28 feet, and collecting 59 samples for laboratory testing.

4. Conducting a laboratory and field test program consisting of 3 Atterberg limits, 5 grain size distribution, 25 moisture and dry-density, 23 unconfined compression and 3 corrosivity tests. Additionally, field resistivity measurements were taken in 7 of the 10 borings.

5. Preparing the project geotechnical report.

In all, the geotechnical program accounted for about one percent of the construction costs. There were no serious surprises or changed conditions encountered during construction.

Trenching and Backfilling

Approximately 80 percent of the pipeline length was constructed by conventional open-trench excavation. The selected alignment passes through new and older developed suburban residential neighborhoods which have most of their utilities underground. To minimize utility disruption and relocation, the design was based on installing the pipeline below the existing utilities. Consequently, the average depth of the pipe trench in the residential areas was approximately 10 feet. The deepest excavation was through undeveloped ground in the first 500 feet of the alignment, from the reservoir to where the pipeline crossed Interstate 680. In order to empty the reservoir by gravity, the excavation there reached a depth of 40 feet.

The existing soil conditions were primarily sandy clay and clayey sands, with occasional layers of silty sand. These stood well in vertical trenches when supported by hydraulic trench jacks.

Although there are differences in the stiffness of the steel pipe (AWWA C200) and the pretensioned concrete pipe (AWWA C303) alternatives, it was decided to specify the same pipe bedding and backfill requirements for each alternative thus limiting pipe deflection. Satisfactory performance of each type of pipe depends on adequate support along the sides of the pipe.

Actual construction practices saw the contractor use a sand-cement slurry up to springline in lieu of imported granular material for pipe trenches in the streets. This enabled a much narrower trench to be constructed. This approach is discussed later.

In designing the structural strength of the pipe, a modulus of soil reaction (E') of 1,400 psi was used for the pipe installed in the deep excavation south of Interstate 680, and 1,000 psi when designing the pipe installed in the remainder of the project.

Trenchless Construction

Five sections, about 1,300 feet of the total 7,100 feet length of 48-inch pipeline, required trenchless construction. The surface conditions included street and interstate highway crossings, railroad crossings, and one section where the pipeline alignment in crossing from an open creek area behind a residential area onto the residential street had to pass under and between two adjacent single family dwellings. Two construction methods were specified, a bored and jacked steel casing pipe or mined tunnel using liner plate. Both methods were permitted for the street crossing adjacent to the reservoir. A 750 feet long soft-ground tunnel was required beneath the interstate highway. In order to preclude pavement settlement, the voids outside the liner plate were grouted daily at the end of the shift. Preconstruction and post construction level surveys of the road surface showed that no settlement had occurred. The construction of the pipeline passing beneath the two residences also used the mined tunnel method to prevent surface settlement.

In tunnel situations, contract specifications called for standard bolted 10 ga, galvanized, flanged liner plates for tunnel construction. The tunnel was required to be circular and have a nominal diameter at least 18 inches greater than the outside diameter of the carrier pipe, capable of resisting all earth, erection, and grouting loads.

Crossing the Hayward Fault

The primary design intent for the pipeline crossing of the fault was to accommodate the continual aseismic slip (creep). Special joints, backfill methods, different pipe material design, and other methods used successfully elsewhere were researched and considered for this application.

The fault rift zone is very narrow. Where the fault intersects the pipeline alignment, a zone of intense irregular deformation extends 10 feet out to either side of the well defined fault trace. Because the 48-inch pipe was available in lengths up to 40 feet, it was possible to focus provisions for movement at each end of a single length of pipe.

The first element of the design was to provide a flexible joint on each side of the fault trace. By using a maximum pipe length of 40 feet, and centering the pipe length over the fault trace, the distance of the flexible joints from the fault trace would be almost twice the width of the zone of intense deformation. After researching various flexible joint designs, a nonmetallic expansion joint, molded spherical type, with integral flanges was selected. The flexible joint was specified to provide the following minimum movements: 1 inch elongation; 2 inches compression; 1 inch lateral; 4 degrees angular. Retaining rings and control unit assemblies were installed to control movement in the event of a major seismic event. Based on the maximum projected lateral aseismic creep, these joints should accommodate at least 30 years of creep movement, at the end of which time major maintenance would be required.

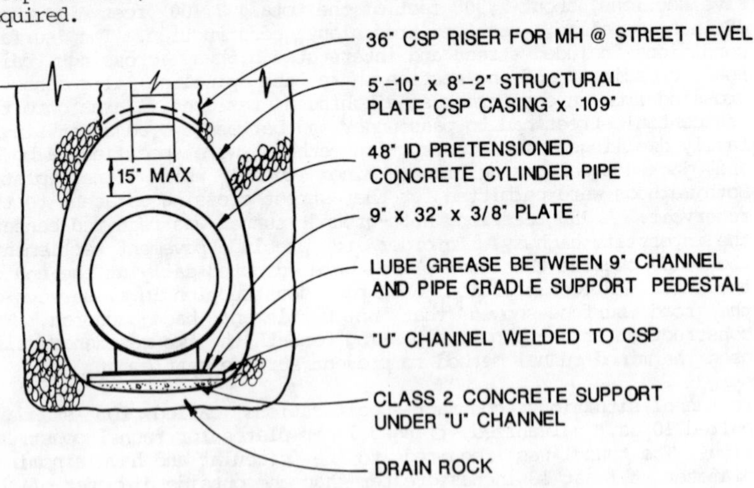

FIGURE 3 - HAYWARD FAULT CROSSING SECTION

The second element of design for the fault crossing was to install the pipe section that included the two flexible joints inside a corrugated metal multi-plate pipe (CMP) casing so as to allow it to deflect with the fault movement (see Figure 3). The bottom section of the CMP casing was installed in an open trench. The 40-foot pipe section was then set into the bottom part of the CMP casing, resting on two pipe saddles, one near each end of the 40-foot section of pipe. The top arch section of the CMP casing was then bolted into place. Then the portals of the CMP casing were closed using concrete-filled burlap bags. The saddles were constructed of steel plate and ride on a "U" channel welded to the CMP. The surface between the channel and saddle was greased to minimize friction during lateral movement. Access manholes were provided at each end of the CMP casing to permit periodic inspection of the flexible joints and to measure actual pipe movement.

Seismic Valve

To prevent significant damages caused by a water line rupture resulting from a major seismic event, two 36-inch diameter isolation butterfly valves were provided in the design, one on each side of the fault zone. Each of these valves is equipped with an automatic seismic shutoff system. The valves are operated automatically by a double acting cylinder operator that is powered by compressed nitrogen gas. The system is triggered by a seismic gas valve that activates a pressure switch when it is subjected to horizontal acceleration of 0.3 g occurring over a period of 0.4 seconds.

Nitrogen was selected for use in the compressed gas system because it is commercially available, is inert, and does not support combustion. Nitrogen is the most commonly used gas for this type of application. An important disadvantage of nitrogen gas is that it is an asphyxiant. Consequently, its leakage could be hazardous to anyone entering the below-grade vault. For this reason an exhaust fan was installed in each structure and portable oxygen sensors are used by personnel entering the vault. Furthermore, all materials and equipment used in the valve operating system were closely reviewed and selected for their low potential for nitrogen gas leakage.

Instrumentation and other provisions were installed to facilitate adding a telemetry system at a future date. This will enable the District to closely monitor valve positions on a continuous basis.

SPECIAL CONTRACT PROVISIONS

The job was bid on a unit price basis. There were 76 payment items, a number of which had alternatives. The principal alternatives related to the several pipe materials allowed. In various locations and combinations the following types of pipe were allowed for bid:

1. Steel, cement-mortar lined pipe, with external dielectric coating (SCML&DC).

2. Pretensioned concrete cylinder pipe (PCCP).

3. Reinforced concrete pressure pipe (RCPP).

For the most part the contractor through his bidding, elected to use the PCCP.

As is conventional, the specifications required that "the minimum width of pipe trenches, measured at the top of the pipe or at the bottom of the trench, shall not be less than eighteen (18) inches greater than the outside diameter of the pipe barrel...exclusive of all trench timbers." This specification was modified during construction as a result of an innovation proposed by the contractor.

CONTRACTOR INNOVATIONS

The first innovation related to the installation of the pipe in the tunnel under the Interstate 680 Highway. The contractor fabricated steel wheeled dollies approximately 2-feet long and curved in cross-section, and installed steel rails on the bottom of the tunnel, at a 30 inch gage. The dollies were placed under each end of each pipe section and were strapped in place with steel bands. Each pipe section was lifted onto the rails and winched forward to position.

The dollies were left in place under the pipe. Later, after all the pipe had been installed and pipe joints had been welded from inside the pipe, the contractor filled the space between the underside of the pipe (bottom 120 degrees) and the invert of the tunnel with grout.

A second innovation made by the contractor was a modification of the pipe trench when in City streets. In variance with the conventional project specifications, the contractor proposed installing the pipe in a "U" shaped trench, placing small mounds of sand under each end of each pipe section, and pouring a lean cement grout along and under the pipe for the full length of each pipe, up to the springline. In order to reduce the volume of grout, the contractor proposed, and the District approved excavating the trench with a round bottomed bucket, sized to provide a mere 3 inches of clearance between the pipe and the lower walls of the trench. The trench section for the pipe installation in the round bottom trench is shown in Figure 4. The pipe was sufficiently heavy to not float when the grout was placed.

This installation method was advantageous to both the District and the contractor. The pipe had uniform bedding in the trench, with no points of concentrated soil reaction. For the contractor, the cost of the grout bedding was more than offset by less excavation, less backfill, less pavement to cut, remove and replace, and a quicker backfill operation.

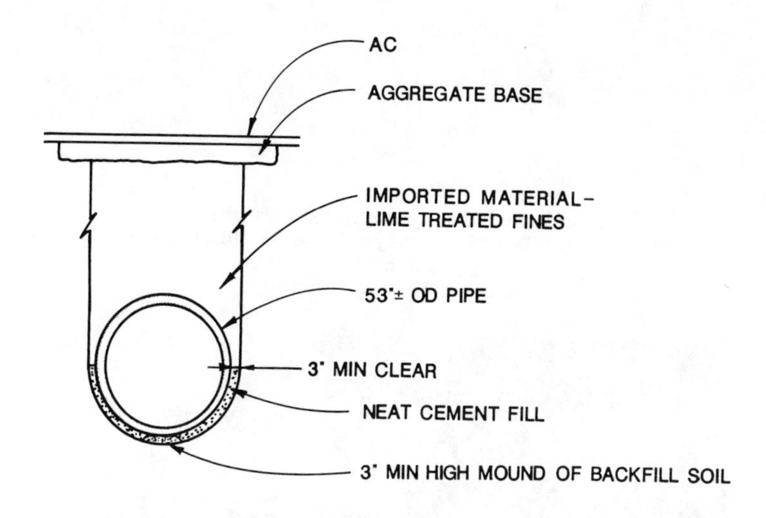

FIGURE 4 - TRENCH SECTION FOR STREETS

CONCLUSION

This was a successful project. Some of the contributing factors were:

- Long range planning to meet District water storage requirements.
- Appropriate consideration of alternative alignments, with particular attention given to geologic hazards, environmental concerns and permitting.
- Qualified consultant and subconsultants.
- Specifications that encouraged price competition between pipe products.
- Specifications that emphasized results rather than methods.
- A contractor with good ideas that could be implemented without compromising the final product.

Pipeline Design and Construction Using High Density Polyethylene (HDPE) Pipe

Stephen P. Tanner[1]
Ronald N. Sickafoose, P.E.[2]

Figure 1. Southerly end of 12" HDPE suspended waterline in McCoy Canyon, Santa Barbara, CA

[1]Project Engineer, Penfield & Smith

[2]Principal Engineer, Penfield & Smith,
111 East Victoria St., Santa Barbara, CA 93101

Characteristics of Conventional vs. Flexible (HDPE) Piping Systems

All pipelines and pipe materials must be designed and constructed to address the limitations of the pipe and fittings in order to insure acceptable performance of the line. Conventional pipeline materials such as PVC, DIP, and cement lined steel are all essentially rigid discrete pipe sections joined by semi-flexible or non-flexible joints which are subject to separation by pulling or deflection. The table below summarizes these characteristics:

Pipe Material:	PVC Pipe	Ductile Iron Pipe	Cement Lined Steel
Pipe Joining Methods	Bell Joint	MJ or Bell Joint	Welded or Flanged
Fitting Joining Methods	Bell Joint or MJ	Bell Joint or Flanged	Flanged or Welded
Ability to Deflect	2°-3° at Joint. Bending Radius = 300D	2°-3° at Joint	2°-3° for MJ Only
Need for Restraint of Thrust Blocking	At all Bends and Tees	At all Bends and Tees	At all Non-welded Bends and Tees
Need for Support	360° Around Pipes, Bends and Fittings	360° for all MJ or Bell Joint Installations	360° for all MJ Installations

When project constraints such as aboveground installation, possible shifting and settlement of soils, or lack of support or bedding are present, conventional pipeline materials and construction techniques may be infeasible or cost prohibitive. The characteristics of HDPE pipe and its installation often prove to be the most cost effective and durable alternative in these adverse pipeline environments. Beneficial properties of HDPE pipe include:

<u>Pipe Joining Method</u>

- ♦ Thermal fusion welding (Creates a single homogeneous line rather than discrete pipe sections)

<u>Fitting Joining Method</u>

- ♦ Thermal fusion welding or flanged

<u>Ability to Deflect</u>

- ♦ Deflection radius up to 20 times pipe diameter (Greatly minimizes the need for fittings)

<u>Need for Support</u>

- ♦ Spacings up to 8-10 times pipe diameter (Allows a wide range aboveground and suspended applications)

<u>Installation Methods</u>

- ♦ Can be pulled (or pushed) into place in lengths of 500 to 1000 feet or more (A solution to many limited access problems)

HDPE pipe should be considered as an alternative to conventional pipe whenever the need for pipe flexibility, aboveground routing, or limited right-of-way access is present. The following case studies illustrate the applications, design, and construction methods in which HDPE pipelines were the superior (and sometimes the only) alternative.

Environmentally Sensitive Areas

12" Gravity Fed Waterline in Santa Barbara, CA

Design

In an effort to acquire new water sources in arid Santa Barbara County, a private developer and public water purveyor teamed up to capture storm runoff water from McCoy Canyon Creek in the nearby coastal mountains and transport it to the water district's treatment facility. During the Environmental Impact Review process the diversion site and pipeline route were identified as riparian and sensitive habitat areas. The constraints which were imposed on the project as a condition of approval were so restrictive it appeared the project would be completely infeasible from a design and construction cost standpoint. Among the project conditions were the following:

- No vehicular access shall be allowed within sensitive habitat.
- All equipment and materials shall be hand carried.
- No trees may be removed, including willows.
- A 5 foot wide path may be cleared by hand for pipeline construction where undergrowth is too thick to permit foot traffic.
- Pipeline construction shall be by hand and aboveground on pylon anchors except at road crossings.

To comply with the project constraints a lightweight, flexible pipe would be needed which could be assembled and placed without heavy equipment within a 5 foot irregular pathway. In addition, the 12" pipeline would need general resistance to environmental conditions such as thermal cracking, sunlight, rain, snow loading, freezing, falling rocks and potential brush fires.

After evaluating the performance specifications of PVC, DIP, steel, Fiberglass Reinforced Pipe, HDPE and others, it was apparent that HDPE was the most durable and cost effective piping material for the project. To handle the estimated 1500 gpm flow of diverted water, a 12" SDR 26 pipe was selected. The pipe material specified was PE3408 grade resin (referred to as extra-high molecular weight polyethylene) with 2% carbon black addition for UV stabilization. Once the piping material and general method of construction was determined, it became necessary to carefully walk the proposed pipeline route to identify any natural features which would interfere with the pulling of the line into place or damage the pipe during installation. The route was essentially a pristine environment with dense underbrush, vines, poison oak, and trees. The geography of the site was highly irregular in profile due to scattered sandstone bedrock outcroppings set between zones of 10-25% slopes of unstable soils. To manage the irregular topography within the capabilities of the HDPE pipe, the site was evaluated in 20' increments in regard to the need for pipe restraint, support, or protection from jagged rocks or abrasive conditions resulting from pipe deflection. In performing this evaluation an inventory of three basic support types, two restraints and a single protective liner were designed to meet the project needs.

Each type of support was designed to allow axial movement of the pipeline from thermal expansion, but to limit lateral deflection of the line. Since supports were necessary only in areas of irregular topography, the line had ample space for lateral deflection (or stress relieving) to occur. All supports were constructed of galvanized steel but were lined with 3/4" thick HDPE material which was fastened to the inner support faces. The liner allows free movement of the pipe, but more importantly prevents grooving of the outer pipe wall. Figure 2 shows two of the H-type supports utilized near one of the primitive wire rope foot bridges constructed to access the diversion site.

Restraints were utilized to limit both axial and lateral movement of the HDPE line. These restraints were used only at sharp bends in the line or at the upper and lower limits of severe downhill sections of the line to limit downhill creep of the pipe. Figure 3 shows a typical C-clamp restraint anchored to a rock outcrop. To prevent gouging of the pipe, 3/8 inch thick neoprene rubber was wrapped around the line at all restraint points.

Sections of the 12" pipeline which ran along sandstone outcroppings were either fitted with a half section of 14 inch HDPE as a liner or restrained from lateral or axial movement.

Figure 2. Adjustable steel pipe supports for 12" HDPE pipeline. Note HDPE liner on steel undersupport.

Figure 3. C-clamp restraint anchored to rock face holding 12" HDPE line. Note neoprene wrap around pipe to prevent cutting of HDPE.

Construction

Due to the severe restriction on the project's construction methods slow progress was made on the clearing of an access trail into the diversion site and pipeline route. To provide compressed air to the site for jack hammering and rock splitting, a 3" SDR 13.5 HDPE line was preassembled into two 450 ft. lengths and dragged up the canyon bottom to serve as a compressed air header for the pneumatic equipment. Workers installed clamp-on service saddles wherever air was required along the route and ran 1" hoses to their equipment. The header was pressurized by two air compressors located outside the canyon. This method worked exceptionally well for the duration of the project even under simultaneous use from four 90 lb. jackhammers and two pneumatic rock drills.

After site preparation was complete the 900 feet of 12 inch pipe, prefabricated supports and fusion welding equipment was air dropped by helicopter (a concession granted by the County Environmental Resources Department) to a 40 by 80 foot staging area in the upper canyon. The 20 ft. pipe sections were preassembled into 60 ft. lengths at this site to expedite the actual pulling operation. A "pulling head" was fused onto the front end of the line which consisted of a heavy wall HDPE nose cone with a steel ring insert to attach to the wire rope pull line. Figure 4 shows the 60 ft. sections being fused together as the line was pulled down the canyon via hand operated winches.

As the line was pulled into place, a crew of workers walked along with the pulling head to guide the front end of the line into the preferred location and to assure temporary supports such as roller fitted pipestands were placed where needed. In general, the line tended to flex and settle along the route of least friction and stress. After the line was in place, both ends were loosely tied off and the line allowed to contract and settle for two days before tieing in to permanent fittings and placing permanent supports and restraints.

Aboveground Pipeline through Unstable Soils Area
8" Pressure Water Line, Santa Barbara, CA

An initial phase of work for the construction of a major oil and gas processing facility was to relocate an 8" water well production line. A 1400' portion of the line

Figure 4. Preassembled 60 ft. lengths of 12" HDPE being placed in fusion welding machine. Note roller fitted pipe stands to allow smooth pulling of pipeline.

was to be routed through a "viewshed" zone in which the view of the hillside could not be visually impacted by any construction activities, according to conditions imposed on the project. Because of the 2:1 and greater slopes on the hillside, a conventional underground line could not be installed without "benching" the pipeline route, creating a minimum 12' wide shelf in the hillside. In addition, the geologist's report identified the native soils material as expansive and unstable. They concluded that the benching of slopes would cause severe sliding and erosion when the winter rains arrived. To avoid any disturbance of the steep slopes and minimize any visual impacts a 2,200' long 8" HDPE line was designed which could be installed on top of the ground in the viewshed area and where slopes were over 20%. The remaining 800 feet of line would be installed underground in a conventional sand bedded trench. When pulling long lengths of HDPE pipe, it is necessary to estimate the pulling force required and the maximum allowable pulling force (sometimes called the Safe Pulling Force) for the particular pipe used. Our in-house approximations for these forces are as follows:

1. <u>Safe Pulling Force</u>

$F_s = 3780\ D^2\ (1/DR)$ For PE 3408 resins only
For PE 2306 of 2407 resins multiply by 0.87

2. <u>Required Pulling Force</u>

$F_A = WL\ (f \cos S \pm \sin S)$

Where:
Fs = Safe pulling force, lbs.
D = Pipe outside diameter, ins.
DR = Standard dimensional ratio i.e. pipe diameter/wall thickness
F_A = Actual required pulling force, lbs.
W = Pipe weight, lbs./ft.
L = Pipe length, ft.
f = coefficient of friction (usually 0.1 to 0.80
s = slope of ground, degrees (if slope is uphill, Sin s is +; if slope is downhill Sin S is -)

For this application a 8" SDR 15.5 pipe manufactured by Polaris Pipe was selected. To calculate the pulling forces the additional information required is as follows:

Maximum slope:	5%
Pipe weight:	6.13 lbs/ft. for PE3408 resin
Maximum pull length:	1400 ft.
Coefficient of friction:	0.3 for grassy slopes
Pipe diameter:	8.65 in. for 8 in. nominal pipe size

The calculated forces are:

F_S = 18,250 lbs.
F_A = 3,350 lbs. (to nearest 50 lbs.)

Since $F_A < F_S$ the line can safely be pulled into place. A winch or other device capable of approximately 3500 lbs. pulling force would be necessary to place the line. Figure 5 shows the 8" line being pulled into place.

After the routing of the line was established, both temporary and permanent supports for the line were designed. For aboveground HDPE waterlines lying on even ground, the major design considerations for supports are thermal expansion of the pipe, potential extreme weather effects such as snow loading or freezing, and creeping of the line due to gravity or shifting slopes. The only concerns for this project were thermal expansion and creep on the 1:1 side hill slopes. Using the manufacturers data for thermal expansion (0.8 x 10^{4} in./in./°F) the maximum seasonal movement was 60 in. for the 1400' aboveground portion of the line. To allow for this movement while preventing pipeline creep down the hill, a series of redwood headers anchored by 2" steel pipe posts were placed every 40 ft. on the downhill side of the pipe (see Figure 6). This minimal restraint allowed the pipe to deflect laterally and stress relieve itself while limiting the downhill movement of the line. Because the line remains full of water, the actual measured expansion is approximately 40% of the calculated value.

In preparation for the pulling operation, a staging area was designated at one end of the pipeline where the pipe would be offloaded and preassembled into 120 ft. long segments. This allowed a more efficient placing of the line once pulling had commenced, and provided manageable sized segments of line to be moved into the fusion equipment as each segment was added on to the pipeline.

After the line was pulled into place, it was allowed to contract and settle while the permanent sidehill supports were installed. The line was then blind flanged at both ends and pressure tested before being tied into the existing well and transmission line.

Figure 5. Pulling 8" HDPE line into place via "pulling head" and cable.

Figure 6. 8" line installed on unstable sidehill slope. Note redwood support headers.

<u>Suspended HDPE Pipeline</u>
12" Suspended Pipeline, Santa Barbara, CA

As part of the previously mentioned 12" pipeline project at McCoy canyon in Santa Barbara, a severely inclined bedrock region was encountered which was bordered by near vertical rock faces at both ends of a 250 ft. wide culvert which dropped down from 25 to 50 ft. to the creekbed below. Upon the geologist's verification that the bedrock was intact and unfractured, it was determined that a suspended section of HDPE pipe would be the most economical and durable material for the crossing. The pipe manufacturer's recommendation for support spacing for the 12" SDR 26 line was 12-14 ft. for an anticipated 1 inch sag in the line.

In consideration of the pipeline requirement for a non-abrasive support bedding and the general project requirements for lightweight, flexible, weather resistent materials, a galvanized steel stirrup was designed with a heavy walled HDPE inner liner which would be hung from a single structural steel cable (see Figure 7).

Figure 7. Pipe suspended by HDPE lined saddle and cables.

Once suspended the HDPE line would be able to deflect laterally to relieve thermal expansion and contraction, and flex under wind conditions while having limiting restraints for sag, creep, and excessive wind or earthquake forces.

Since the line was being pulled into place, a temporary means of support to span the crossing during installation was required. At the same time the permanent main 1-1/4" structural steel cable was placed, a second 3/4" cable was installed which closely profiled the anticipated flowline of the pipe. As the line was pulled into the suspension area, HDPE lined supports were placed on the pipe and attached to the secondary "construction" cable by means of a pulley block. As the pipe continued to be pulled down the canyon, supports on pulleys were added in 20 ft. intervals to ferry the pipeline over the crossing. Figure 8 shows the line supported by this temporary means.

Figure 8. Pulley supported line traversing canyon crossing. Note permanent cable ready for pipe hangers on right side.

After the line had been dragged into place, the pulley-suspended section of the line (approximately 250') was transferred over to the adjacent main cable by attaching the pulley suspended support saddles to the permanent 1/2" steel suspender cables which were already mounted on the main cable. Turnbuckles were installed on each suspender line to adjust the flowline of the pipe after installation (see Figure 1).

Although this line is in service only from November through March, it has been maintained full of water throughout the year to reduce thermal expansion and wind deflection. The "construction cable" running parallel to the pipeline has recently been fitted with a cable car to facilitate inspection and adjustment of the flowline.

Summary

Pipeline projects in recent years have seen a dramatic increase in restrictions on right of ways, environmental preservation, and construction methods, while at the same time the availability of "manageable" pipeline routes and economical land acquisition have all but disappeared. HDPE pipe gives both the design engineer and contractor one more option in their repertoire which lends itself extremely well to adverse piping environments. The pipe manufacturer or distributing representative should always be consulted in regard to specific applications and physical property data of the various grades of and sizes of HDPE.

APPENDIX

Abbreviations Used

DIP - Ductile Iron Pipe
HDPE - High Density Polyethylene
MJ - Mechanical Joint
PVC - Polyvinyl Chloride

Conversion Factors

1 in. = 2.54 cm
1 ft. = 0.31 meter

REFERENCES

Envicom Corp., 84-EIR-13, West Devereux Specific Plan Environmental Impact Report, July 1984.

Penfield & Smith Engineers, Inc., McCoy Diversion Project, Contract # 3 Specifications, March 1987.

Polaris Pipe Co. Inc., Engineering and Systems Design Information for Polaris High Grade Polyethylene Pipelines, 1985.

Plexco, Application Note #10: Safe Pull Strength for Plexco Pipe, 1986.

CHEYENNE STAGE II WATER DELIVERY PIPELINE AND SHOSHONE MUNICIPAL WATER SUPPLY PIPELINE

by Jeff B. Fuller, M. ASCE[1]
Edward A. Katana[2]

ABSTRACT: This paper deals with the design and construction aspects of two pipeline projects in Wyoming.

Located in southeastern Wyoming, the Cheyenne Stage II Water Delivery Pipeline consists of approximately 60.7 miles of transmission pipeline ranging from 33-inch diameter to 20-inch diameter steel pipe and approximately 1.8 miles of PVC pipe ranging from 24-inch diameter to 12-inch diameter. The Stage II Water Delivery Pipeline constructed during 1986 through 1988 parallels an existing pipeline constructed 25 years earlier. Pipeline design and construction includes mountainous terrain, six interceptor structures, twenty-two side hill collectors, six railroad and highway borings, 650 psi maximum working pressure, cathodic protection, instrumentation and telemetry controls, extensive erosion control and water quality program, and evaluation of the performance of the existing pipeline.

Located in northwestern Wyoming, the Shoshone Municipal Water supply Pipeline Project consists of approximately 3.7 miles of 36-inch diameter steel Raw Water Supply Pipeline, and approximately 51.7 miles of Treated Water Pipeline ranging from 36-inch diameter to 16-inch diameter steel pipe and approximately 16.1 miles of 10-inch and 8-inch diameter PVC pipe. The Raw Water Supply Pipeline delivers water from Buffalo Bill Reservoir to a 22 MGD water treatment plant, and the Treated Water Pipeline delivers potable water from the water treatment plant to six cities and several rural water use districts. Pipeline design and construction includes eight river crossings, twelve railroad and highway borings, numerous canal crossings, six pressure control stations, six service

[1]Banner Associates, Inc., Laramie, Wyoming 82070.
[2]Banner Associates, Inc., Laramie, Wyoming 82070.

connections, two pump stations, an aerial river crossing, impressed current and galvanic anode cathodic protection systems, instrumentation and telemetry controls, as well as appurtenant valves. The Shoshone Municipal Water Supply Pipeline Project began construction in 1988 and is scheduled for completion in 1990.

CHEYENNE STAGE II WATER DELIVERY PIPELINE

DESCRIPTION: The Cheyenne Board of Public Utilities, in an effort to maintain an adequate water supply for Cheyenne's future growth proceeded with the development of the Stage II Water Project. The Stage II Water Project consists of:

- o Collection facilities and pipelines to collect runoff from the west side of the Continental Divide;
- o A tunnel through the Continental Divide to divert this runoff to the east side of the Continental Divide;
- o Enlargement of Hog Park Reservoir;
- o Enlargement of Rob Roy Reservoir; and
- o The Cheyenne Stage II Water Delivery Pipeline from Rob Roy Reservoir to Crystal and Granite Reservoirs.

The Cheyenne Stage II Water Project is an enlargement of the Stage I Water Project completed approximately 25 years ago. The enlargement of Rob Roy Reservoir and the diversion and collection facilities associated with the Cheyenne Stage II Water Delivery Pipeline reduce flows into the North Platte River drainage. The waters collected from the west side of the Continental Divide and diverted into the enlarged Hog Park Reservoir are then directed into the North Platte River drainage to replace the flows depleted by the enlargement of Rob Roy Reservoir and subsequent discharging of flows into the Cheyenne Stage II Water Delivery Pipeline.

This paper will discuss only the water delivery pipeline portion of the Stage II Project. The Cheyenne Stage II Water Delivery Pipeline is located in southeastern Wyoming as shown on Figure 1, and consists of four segments as shown on Figure 2. These segments, beginning at the upstream end of the pipeline, are:

- o Segment 3 begins at Douglas Creek below Rob Roy Reservoir and flows east to Lake Owen (10.7 miles);
- o Segment 2 begins at the Lake Owen outlet works and flows east to where it crosses State Highway 287 (29.1 miles);
- o Segment 1 begins as the pipe crosses State Highway 287 and flows east to where the pipe crosses Interstate Highway 80 (9.6 miles); and
- o Segment 1A begins at the Interstate Highway 80 crossing and flows east to discharge into Crystal and Granite Reservoirs (11.3 miles) located approximately 20 miles east of Cheyenne.

The pipeline alignment traverses a wide range of topographical features. The pipeline starts in the Snowy Range Mountains west of Laramie, Wyoming, in pine forests, rock outcrops, peat bogs, marshes, and steep hills. The pipeline then proceeds down the face of Sheep Mountain into the Laramie Plains with rolling hills, bentonite clays, blow-sand, sandstones, and marshes. Finally, the pipeline climbs into the Snowy Range Mountains east of Laramie and a stretch of decomposed granite and solid rock.

The Cheyenne Stage II Water Delivery Pipeline follows the Stage I Pipeline alignment along the majority of its length. The Stage II Pipeline centerline was designed to be approximately twenty-five (25) feet from the Stage I Pipeline centerline. This combining of permanent and construction easements allowed for some easement acquisition cost reduction.

DESIGN: The design of the Cheyenne Stage II Water Delivery Pipeline is based on a pipeline capacity of 18 million gallons per day (MGD). The Cheyenne Stage II Water Delivery Pipeline is a gravity system with the internal pressures ranging from atmospheric (open channel) to a maximum internal working pressure of 640 pounds per square inch (psi). There are no pressure reducing stations as the pipeline extends from an approximate elevation above Lake Owen of 9240 Mean Sea Level (MSL), down to the Laramie plains with a low elevation of about 7280 MSL, and back up to an approximate elevation of 8280 MSL in the vicinity of Crystal and Granite Reservoirs. Any pressure reduction would reduce flows at Crystal and Granite Reservoirs, prevent flows at Crystal and Granite Reservoirs without pumping, or result in larger pipe sizes to maintain the design flow capacity.

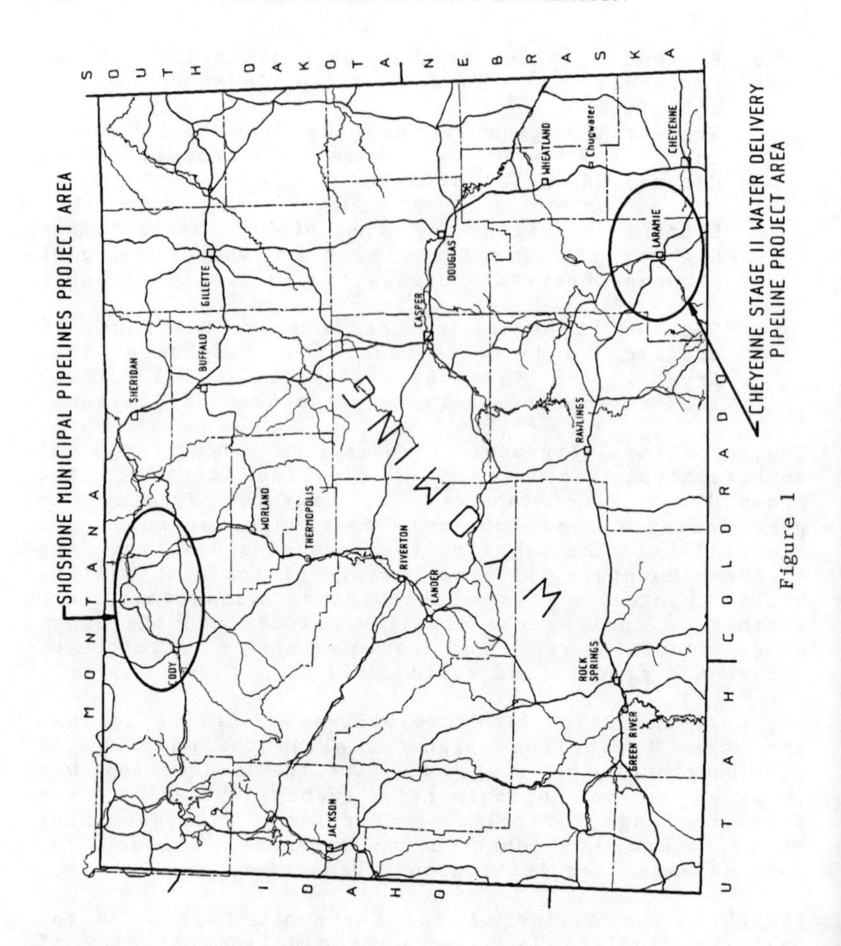
SHOSHONE MUNICIPAL PIPELINES PROJECT AREA
CHEYENNE STAGE II WATER DELIVERY
PIPELINE PROJECT AREA
MONTANA
SOUTH DAKOTA
NEBRASKA
COLORADO
UTAH
IDAHO
WYOMING
SHERIDAN
BUFFALO
GILLETTE
CODY
WORLAND
THERMOPOLIS
RIVERTON
LANDER
CASPER
DOUGLAS
WHEATLAND
Chugwater
CHEYENNE
LARAMIE
RAWLINS
ROCK SPRINGS
GREEN RIVER
JACKSON

Figure 1

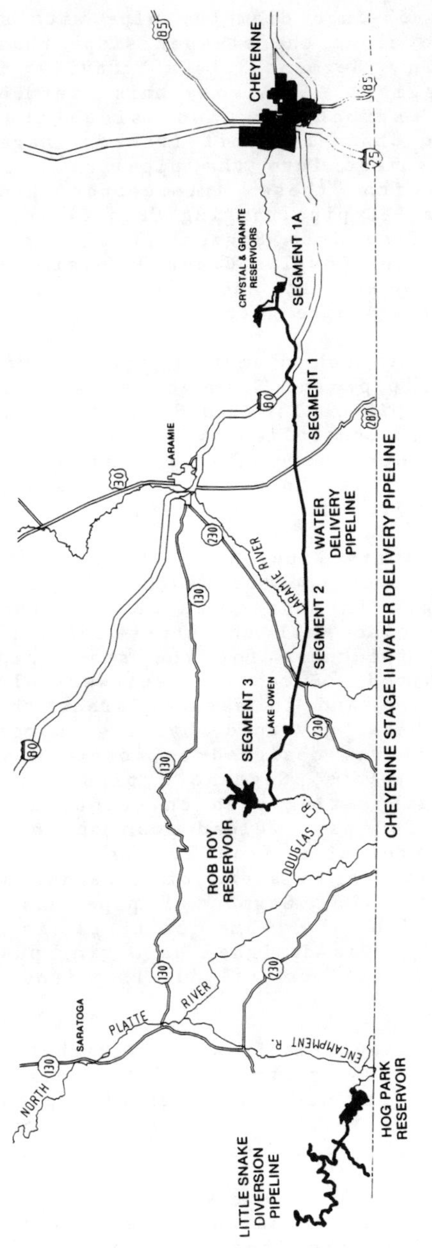

CHEYENNE STAGE II WATER DELIVERY PIPELINE

Figure 2

Segment 3 is 33-inch diameter pipe with some 20-inch diameter pipe along the steeper slope that discharges into Lake Owen. Segment 3 is a "gravity" flow or non-pressure segment. Also, this segment contains interceptor structures and sidehill collection structures to channel runoff from drainages along the pipeline alignment into the pipeline. The pipeline associated with these interceptor and collector structures is PVC pipe ranging from 24-inch to 12-inch diameters. Flow into Segment 3 of the pipeline is controlled at the Douglas Creek Diversion Structure by a sluice gate and at each of the six interceptor structures by a sluice gate.

Segment 2 is 30-inch diameter pipe, Segment 1 is 30-inch diameter pipe and Segment 1A is 24-inch and 20-inch diameter pipe. Flow in Segments 2, 1, and 1A of the pipeline is controlled by a plug valve at the Lake Owen outlet works meter house, and by a plug valve located in the meter house at Crystal Reservoir and at Granite Reservoir.

The Water Delivery Pipeline is buried with four feet minimum cover over the top of the pipe for protection from freezing. In most cases the internal pressures governed the pipe wall thickness design; however, a minimum wall thickness for the steel pipe of 0.188 inch was required. Surge analyses were also performed on the pipeline and in several areas the pipe wall thicknesses were governed by the surge analyses. Segment 3 pipe was designed as concrete pipe with o-ring push on joints, Segment 2 pipe was designed as steel pipe with o-ring push on joints up to internal pressures of 200 psi, welded push on joints in areas of internal pressures from 200 psi to 400 psi, and butt welded joints in areas of internal pressures in excess of 400 psi. Segment 1 pipe was designed as steel pipe with the same criteria as Segment 2. Segment 1A pipe was designed as o-ring push on joints with material options of ductile iron, steel, or concrete.

The project's specifications require that the pipeline be hydrostatically tested to 1.5 times the working pressure which produces internal pipe pressures up to 960 psi during testing that had to be considered during thrust restraint design.

Restrained (welded) joints with a factor of safety of 2 as determined from design procedures from the AWWA M 11 Manual [2] were used for internal pressures of 400 psi or less. If this factor of safety was not provided in the laying schedules as submitted by the

Contractor, thrust blocks were required. Restrained joints as well as thrust blocks were required at areas where the internal test pressures would exceed 400 psi.

The bedding and backfill criteria for the steel pipe were based on flexible pipe theory. A granular bedding material was specified with a minimum of 6 inches under the pipe barrel up to the pipe springline. This material could be material excavated from the trench or imported material. Granular bedding was specified as a well-graded crushed stone or rounded gravel with a maximum particle size of 1/2 inch and a maximum of 5% passing the No. 200 sieve. The select backfill material was placed from the springline of the pipe to 12 inches over the top of the pipe. Select backfill material could also be material excavated from the trench or imported material. Select backfill was specified as a uniformly graded material with a maximum particle size of 3/4 inch and a maximum of 30% passing the No. 200 sieve. Ordinary backfill material was specified from 12 inches over the top of the pipe to the finish grade.

The bedding material and the select backfill material required compaction to a density of not less than 95% of maximum density at -2% to +2% of optimum moisture content as determined by ASTM D 698 [1], or for non-cohesive soils where ASTM D 698 [1] is not applicable, the material required compaction to not less than 70% of relative density as determined from ASTM D 2049 [1]. Ordinary backfill compaction requirements varied depending upon the location.

The vast majority of the pipeline is cement mortar lined in accordance with AWWA C 205 which will function under the normal operating velocities. However, during the filling of the pipeline both initially and after routine shutdowns the section from Lake Owen to Sheep Mountain will be subjected to velocities in excess of 40 feet per second (fps) which required that the pipe be epoxy lined in accordance with AWWA C 210 [2] in this section.

Based on ground resistivity tests, it was determined for corrosion protection that the steel pipe should be cathodically protected. Twelve (12) sacrificial anode ground beds were required along the length of the pipeline. The steel pipe with o-ring push on joints required each joint be bonded. The steel pipe is tape wrapped in accordance with AWWA C214 with a total system thickness of 80 mils. The Stage II Pipeline alignment followed the Stage I Pipeline along the majority of its length and in these areas, the two pipelines were cathodically bonded.

<u>CONSTRUCTION</u>: On a project of this length it is not uncommon to run across an item or two that is out of the ordinary, and this project was no exception. A few constant features that had to be dealt with included the protection of the Stage I pipeline.

The Stage II pipeline had to be located a minimum safe distance from the Stage I pipeline while keeping the right-of-way costs within reason, the possible removal of horizontal thrust block area had to be considered, and the effects of blasting upon the 25 year old pipe was a concern. Also, construction loading on the Stage I pipeline had to be considered.

The pipeline went through two residential areas; Pine Grove Estates and Mountain Meadows Subdivision. The home owners of each area were concerned about the possible effects of blasting on their houses, wells, and septic systems. Consequently, no blasting zones were established within the two areas that extended 500 feet each way from the limits of these developments. A unit price bid item was established to remove rock in these areas by means other than blasting.

The pipeline also went through a grove of aspen trees on the east end of the Project that a few concerned citizens believed were of great aesthetic value. The alignment of the pipeline was altered to minimize the impact upon the aspen grove; the Contractor voluntarily reduced his normal working width in the area to further reduce the impact; and aspen trees within the reduced working width were removed before construction, stored in a nursery, and replanted with a one year watering program.

The Environmental Impact Statement (EIS) for the entire Stage II Project set the parameters under which the pipeline could be constructed; minimum interceptor stream flows were set and mitigation guidelines were established. The Corps of Engineers required that both a Nationwide permit and a 404 Permit be obtained; the Wyoming Department of Environmental Quality required that state water quality standards be maintained; the Wyoming Game and Fish established time windows for stream diversions; and the United States Forest Service established stringent requirements for all work in National Forests.

The work in the National Forest to protect the environment consisted of a comprehensive package of temporary and permanent erosion control measures. The permanent erosion control measures were incorporated into unit price bid items while the temporary measures were incorporated into a $1 million force account bid item. Temporary erosion control was used over the entire project but the majority was used on the 10 miles of the pipeline within the National Forest.

Bids on the Project were opened on November 6, 1986 with Guernsey Stone and Construction of Sheridan, Wyoming submitting the low bid of $41,895,495.10. Guernsey Stone and Construction Company subcontracted with The Industrial Company for installation of the pipe on Segments 1 and 2. The Cheyenne Board of Public Utilities expressed concerns of insufficient contingent funds available for construction with the acceptance of this bid. Negotiations resulted in cost reductions of approximately $800,000. These cost reductions included allowing o-ring push on joints with steel pipe on Segment 3 as an alternate to the concrete pipe specified.

Design of the steel pipe included wall thickness changes in 1/8-inch increments as the pressures fluctuated, additional savings were realized by adjusting the steel pipe wall thickness changes in 1/16-inch increments. Additional miscellaneous items were included to achieve the $800,000 cost reduction.

Construction of the pipeline began on March 10, 1987 and continued until shut down due to winter weather-the first week of November 1987. Approximately 50% of the pipe was installed during the 1987 construction season. Construction began again in March of 1988, and all pipe was installed by October 20, 1988.

Pipe installation operations were completed with three crews; a trenching crew, a pipe laying crew, and a backfill and compaction crew. Normally, the pipe laying crew had the highest production. On maximum production days the pipe laying crew would install up to a mile of pipe a day. The majority of the time a "Henry Pipe Laying Machine" was used to lay the pipe in the trench. A picture of the Henry Pipe Laying Machine is shown on Figure 3. The trenching crew, when using the wheel trenching machine, would excavate up to one-half mile of trench per day. A track excavator was used to dig the pipe trench in areas which contained materials that the wheel trencher was unable to dig or in rocky areas. The track excavator would dig from 1000 feet to 1500 feet of trench per day on a good day and in some difficult areas would dig as little as 200 feet per day. The backfill and compaction crew would complete 1000 feet to 1500 feet per day of trench on good days. The majority of the time the backfill and compaction crew had the least productivity and at times in certain areas would lag three to four miles behind the pipe laying crew. Attempts to reduce the lengths of open trench resulted in the slower production crews working more hours per day and on weekends.

Figure 3

Overall pipe installation; including trenching, laying and backfilling operations, for this project produced production rates of approximately 400 to 500 feet per day.

The construction contract included a substantial completion date of October 31, 1989. With the pipe installed by October 20, 1988, minor clean up operations and some cathodic protection installation operations continued sporadically throughout the winter and into the summer of 1989.

Construction of the Cheyenne Stage II Water Delivery Pipeline was completed ahead of schedule and approximately $2 million under budget.

SHOSHONE MUNICIPAL WATER SUPPLY PIPELINE PROJECT

DESCRIPTION: The Shoshone Municipal Water Supply Project is located in northwestern Wyoming as shown on Figure 1. The Shoshone Municipal Water Supply Project is a regional water development project proposed to meet the water needs of municipalities and rural water users in northwest Wyoming through the year 2030. Communities participating in this project include Cody, Powell, Byron, Lovell, Deaver, and Frannie. Potential rural water users are located in the unincorporated areas of Park and Big Horn Counties adjacent to the proposed pipeline route. A Joint Powers Board comprised of representatives from each of the communities as well as the rural water users was formed to administer the Shoshone Project. A map of the project location is shown on Figure 4.

The major components of the Shoshone Municipal Water Supply Project include:

- The Shoshone Municipal Raw Water Supply Pipeline to deliver water from Buffalo Bill Reservoir and the Shoshone River to a water treatment plant;

- The Shoshone Municipal Water Treatment Plant located west of Cody; and

- The Shoshone Municipal Treated Water Delivery Pipeline to convey water from the water treatment plant to participating communities and rural water users situated along the pipeline route.

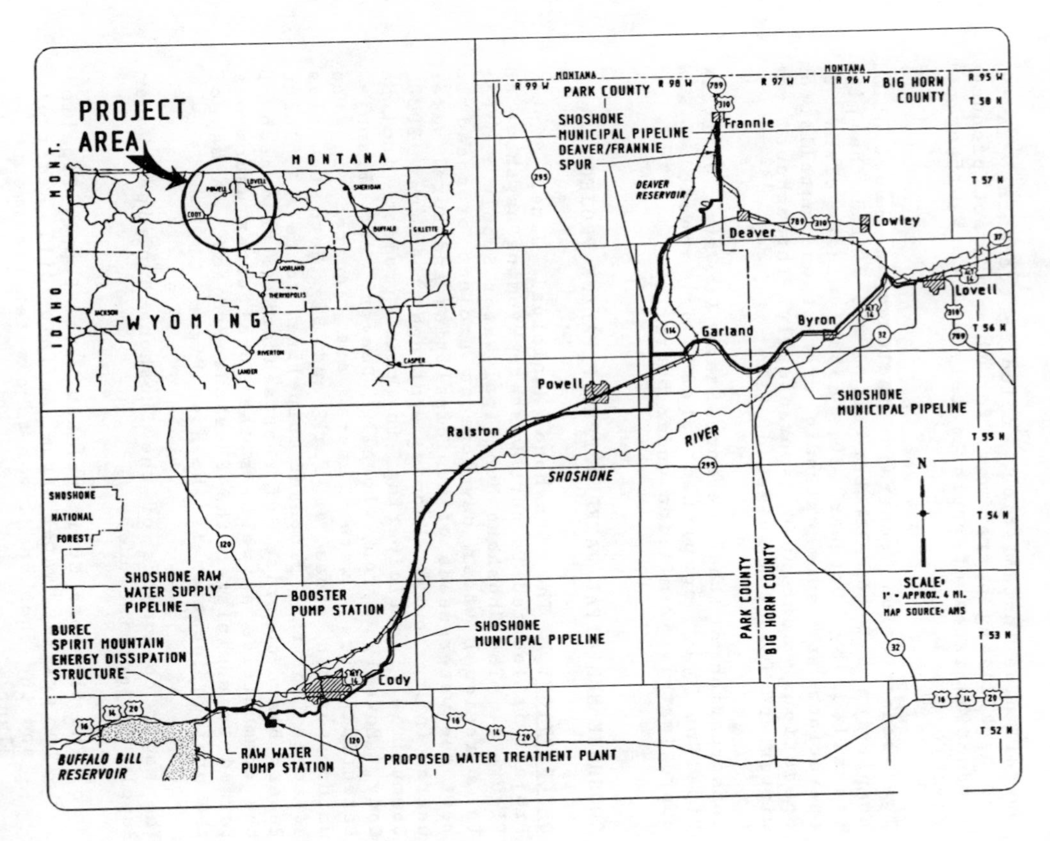
PROJECT AREA
MONT.
MONTANA
IDAHO
WYOMING
POWELL
LOVELL
CODY
SHERIDAN
BUFFALO
GILLETTE
WORLAND
THERMOPOLIS
JACKSON
RIVERTON
LANDER
CASPER
R 99 W
R 98 W
R 97 W
R 96 W
R 95 W
PARK COUNTY
BIG HORN COUNTY
T 58 N
T 57 N
T 56 N
T 55 N
T 54 N
T 53 N
T 52 N
SHOSHONE MUNICIPAL PIPELINE DEAVER/FRANNIE SPUR
Frannie
DEAVER RESERVOIR
Deaver
Cowley
Lovell
Garland
Byron
Powell
SHOSHONE MUNICIPAL PIPELINE
Ralston
RIVER
SHOSHONE
N
SCALE:
1" = APPROX. 4 MI.
MAP SOURCE: AMS
SHOSHONE NATIONAL FOREST
SHOSHONE RAW WATER SUPPLY PIPELINE
BOOSTER PUMP STATION
SHOSHONE MUNICIPAL PIPELINE
BUREC SPIRIT MOUNTAIN ENERGY DISSIPATION STRUCTURE
Cody
BUFFALO BILL RESERVOIR
RAW WATER PUMP STATION
PROPOSED WATER TREATMENT PLANT

Figure 4

The population to be served by the project was projected to the year 2030. Population projections provide design criteria for determination of pipeline and water treatment plant capacity. Each community's historic water use patterns were evaluated where this information was available. Based upon this evaluation, design water need factors were developed for each entity. Water consumption varies significantly with the seasons, day of the week, and time of day. Therefore, allowance was made to consider these variations. The design factor used to size project facilities was the peak day demand. The peak day demand is a per capita figure used to estimate future water needs based upon population projections.

The Shoshone Municipal Water Supply Project is sized to deliver and treat water at a rate that satisfies the peak day demand for all Joint Powers Board entities simultaneously. The raw water and treated water delivery pipelines are sized to meet the peak day demands estimated for the year 2030.

This paper will discuss only the raw water supply pipeline and the treated water pipeline components of the Shoshone Municipal Water Supply Project.

DESIGN:

Raw Water Supply Pipeline: The raw water supply pipeline is a 36-inch diameter pipeline extending from the outlet works facilities at Buffalo Bill Dam, through the Shoshone Canyon, to a water treatment plant located west of Cody. The raw water supply pipeline was designed to deliver the cumulative peak day demand of 21.41 million gallons per day (MGD) or 33.1 cubic feet per second (cfs). The raw water supply pipeline is approximately 3.7 miles long.

The raw water supply pipeline will normally flow under gravity conditions when the Buffalo Bill Reservoir is above average pool level, regardless of demand. Discharge by gravity is still possible at lower pool levels, but at rates less than the design flow. During periods of high water demand, low Buffalo Bill Reservoir levels, or a combination of both conditions, booster pumping will be required to deliver water to the water treatment plant, consequently a booster pump station was provided.

The raw water supply pipeline contains an emergency raw water pump station which will provide raw water from the Shoshone River to the water treatment plant in the event of a scheduled or emergency interruption of raw water supply from the Buffalo Bill Reservoir outlet works facilities.

The raw water supply pipeline also includes an aerial river crossing, a highway bore, and a bore beneath the U.S. Bureau of Reclamation facilities.

The raw water supply pipeline was designed for a minimum five feet of cover over the top of the pipe for protection from freezing; a maximum internal working pressure of 160 psi; and an internal test pressure of 1.5 times the maximum working pressure. The majority of the time the pipe will flow under gravity conditions, and the maximum working pressure occurs with full pipe and no flow (static conditions) when Buffalo Bill Reservoir is at the normal high water line. Internal pressures during pumping operations are kept below this maximum working pressure. Surge analyses were performed on the pipeline with both the emergency pumps and the booster pumps in various scenarios of on and off sequences as well as failure modes. Surge pressures were addressed by limiting the speeds involved with pump turn on and shut off in connection with valve openings and closings. Since the internal pressures are relatively low, pipe wall thicknesses were governed by handling constraints and external loads applied with various depths of cover.

Two pipe materials were allowed, ductile iron and steel, and the contract documents included tables of the various wall thicknesses of steel pipe and thickness classes of ductile iron pipe required along the length of the pipeline. Pipe wall thicknesses for the steel pipe were determined from procedures as outlined in the AWWA M11 Manual [2], and for the ductile iron pipe thickness classes were determined using criteria outlined in the Ductile Iron Pipe Research Association (DIPRA) Handbook of Ductile Iron Pipe [3] and AWWA C150 [2]. Both the ductile iron pipe and the steel pipe options required o-ring push on joints and cement mortar lining.

Compaction and backfill requirements for the different pipe materials were developed; however, all bidders submitted the steel pipe option and these requirements were similar to those as discussed for the Cheyenne Stage II Water Delivery Pipeline and will not be reiterated here.

From soils resistivity tests, cathodic protection systems were developed for the two pipe options. The steel pipe option includes: an external tape wrap coating in accordance with AWWA C214 [2] with a minimum total system thickness of 80 miles; galvanic ribbon anodes installed within the trench on each side of the pipe at pipe springline for the length of the pipeline; all pipe joints bonded for continuity; and test stations installed along the length of the pipeline.

For ductile iron pipe, the corrosion protection system consists of: a polyethylene encasement in accordance with AWWA C105 [2]; bonded joints; and the installation of test stations. No anode system was required. However, the joint bonding and test stations will allow monitoring of the pipeline to determine if anodes would be required at some future date.

<u>Treated Water Pipeline</u>: The treated water pipeline will deliver treated water from the water treatment plant to the Joint Powers Board entities under gravity flow conditions. The pipeline extends east from the water treatment plant to the communities of Cody, Powell, Byron, and Lovell. A spur pipeline delivers water to Deaver and Frannie. Rural water users along the pipeline route may also be served by the pipeline. The majority of the pipeline alignment is located within existing Wyoming State Highway Department rights-of-way or easements. Pipeline sizes range from 36-inch to 8-inch diameter pipe. In all, the treated water pipeline is approximately 66.3 miles in length. The length and size of the segments of the Treated Water Pipeline are:

	Diameter (inches)	Length (miles)
Water Treatment Plant to Cody	36	3.4
Cody to Powell	24	23.8
Powell to Garland	18	6.5
Garland to Byron	16	11.5
Byron to Lovell	14	6.1
Garland to Pressure Control Station #6	10	4.1
Pressure Control Station #6 to Frannie	8	10.9

Pipe sizes were calculated assuming each community, or rural water district, would require the peak day demand simultaneously. Therefore, capacity to meet cumulative downstream peak day demands is provided at all points along the pipeline route. The designed capacity of the pipeline leaving the water treatment plant is 21.41 MGD (33.1 cfs) and decreases downstream to account for projected deliveries to each Joint Powers Board entity.

Other appurtenant facilities and design features included in the Shoshone Municipal Treated Water Pipeline are;

- o Six pressure control stations to control excess pipeline pressure;
- o Six service connections to communities to deliver water into distribution systems, storage facilities, or both at existing system pressures;
- o Provisions for connection to rural water districts;
- o Eight river crossings;
- o Twelve railroad and highway borings;
- o Numerous canal crossings;
- o Pipeline corrosion protection measures;
- o Storage facilities consisting of a 400,000 gallon elevated tank at Byron, a 350,000 gallon buried tank at Deaver, and a 250,000 gallon elevated tank at Frannie.
- o Instrumentation and telemetry controls; and
- o Miscellaneous pipeline isolation valves, air release valves, air vacuum valves, and blow off valves.

For purposes of discussion, the treated water pipeline is divided into the main pipeline and the Deaver/Frannie spur pipeline.

Main Pipeline: The design parameters and material options for the treated water main pipeline are similar to those discussed for the raw water supply pipeline and will not be reiterated, except for the following distinctions:

Five pressure control stations are located along the length of the main pipeline such that maximum working pressures are 200 psi or less. The hydrostatic test head elevation was established for the lengths of pipe between pressure control stations. This hydrostatic test head elevation was established to provide a test pressure of 1.5 times the maximum working pressure (300 psi). This hydrostatic test also allows for a total failure of the upstream pressure control station without failure of the pipeline.

The cathodic protection system consists of five deep well anodes installed along the length of the pipeline and is required for both the ductile iron and steel pipe options. The ductile iron pipe material option requires the exterior tape coating in accordance with AWWA C214 [2] rather than the exterior polyethylene encasement allowed for the raw water supply pipeline.

Deaver/Frannie Spur: The Deaver/Frannie spur of the treated water pipeline requires 10-inch and 8-inch diameter pipe to deliver the design flows. AWWA has developed standards for PVC pipe for use with potable water for these pipe sizes. Soils resistivity tests determined that highly corrosive soils exist along a majority of the Deaver/Frannie spur alignment and PVC pipe would not require cathodic protection. This criteria influenced the utilization of PVC pipe for the Deaver/Frannie spur of the treated water pipeline.

One pressure control station was utilized to keep maximum working pressures below 160 psi, allow for hydrostatic test pressures of 1.5 times maximum working pressures, and remain within the acceptable pressure limits of PVC pipe.

CONSTRUCTION: This project required an archeological survey of the entire alignment prior to construction to identify areas of possible archeological finds that would require monitoring during excavation. These areas were identified, monitored, and in several instances archeological findings temporarily shut down construction operations. However, none of these shut down periods were lengthy and operations were not greatly impacted.

The majority of the pipeline alignment is within existing Wyoming Highway Department rights-of-way, city streets, county roads, and Burlington Northern Railroad rights-of-way. These areas also contained other utilities, both crossing and paralleling the Shoshone pipelines. Identifying these utilities and their locations during design and providing for field adjustments during construction had a major impact on this project. The construction of this pipeline included input from numerous agencies as well as the utility companies. These agencies included the cities of Cody, Powell, Byron, Lovell, Deaver, and Frannie; Park and Big Horn Counties; The U.S. Department of the

Interior Bureau of Reclamation, and Western Area Power Administration; The U.S. Army Corps of Engineers; the Wyoming State Historical Preservation Society; the Wyoming State Game and Fish Commission; the Wyoming Department of Environmental Quality, and the Wyoming Highway Department.

Raw Water Supply Pipeline: Approximately 1 mile of the raw water supply pipeline was constructed across Bureau of Reclamation lands and this portion of the raw water supply pipeline was constructed as a Change Order to an existing Bureau of Reclamation construction contract. Bids for the remaining portion of the raw water supply pipeline were opened on October 27, 1988 with The Industrial Company of Casper, Wyoming submitting the low bid of $4,024,841.00. The contract has a substantial completion date of July 1, 1990. The Industrial Company of Wyoming began construction on May 7, 1989 and shut down for the winter at the end of October 1989. During this time period the Contractor worked only on the construction of the pumping stations. A majority of this pipeline lies in the highway right-of-way which is the main east entrance to Yellowstone National Park. No construction is allowed within this right-of-way from May 15 through September 15 to prevent construction delays due to high traffic volume during the tourist season. Therefore, the Contractor has scheduled to restart construction of the raw water supply pipeline during the early spring of 1990.

Treated Water Pipeline: Bids for the treated water pipeline were opened on July 20, 1988 with Barcon Inc. of Sheridan, Wyoming submitting the low bid of $24,176,000.00. The contract has a substantial completion date of October 31, 1990. Barcon Inc. began construction on September 22, 1988 and worked through the winter of 1988/1989. The Contractor was forced to suspend operations due to weather for only a few short periods.

The Contractor completed pipe installation for the project on September 12, 1989, a year ahead of the substantial completion date. However, some reclamation and miscellaneous items remain to be completed. The Contractor's operations utilized from one to three pipe installation crews working simultaneously to meet his schedule. Each pipe crew included trenching operations, pipe laying operations, and backfilling and compaction operations. The contract documents placed limitations on the amount of open trench allowed; therefore, the

Contractor adjusted working hours of the various operations.

The Contractor's pipe installing operations utilized a wheel trencher and a track excavator for trenching operations; and a Henry Pipe Laying Machine and a track excavator for pipe laying operations. Numerous areas of wet soil conditions were encountered and the documents provided for installation of river/marsh anchors in these wet areas. Installation of these river/marsh anchors is shown on Figure 5.

Figure 5

The pipeline alignment crossed the Shoshone River, a Class I fishery, twice and a "Port-a-Dam" was used to divert the river channel to one side while installing pipe on the protected side of the river channel. The Game and Fish Commission approved of the "Port-a-Dam" method of river diversion, and preferred this method to the conventional construction of earth cofferdams to divert the river. The "Port-a-Dam" method of river diversion allows for less water turbidity and has less impact on the fishery. Installation of the "Port-a-Dam" is shown on Figure 6.

Figure 6

SUMMARY

This paper briefly discussed the design and construction aspects of two recent water transmission pipeline projects in Wyoming. From the brief discussion, the reader should become aware of the various design and construction aspects of these projects as well as the similarities and differences associated with these two projects. Particular aspects of these projects could be discussed in greater detail as the subject of separate reports; however, the intent of this paper was to provide a brief overview of these two projects.

REFERENCES

1. American Society for Testing and Materials (ASTM), Annual Book of ASTM Standards, Standard Specification D 698 (1978 Revision) and D 2049 (' Revision).

2. American Water Works Association (AWWA), American National Standard, C 105 (1982 Revision), C 150 (1986 Revision), C 205 (1985 Revision), C 210 (1984 Revision), C 214 (1983 Revision), and M 11 Manual (2nd Edition, 1985).

3. Ductile Iron Pipe Research Association (DIPRA), Handbook of Ductile Iron Pipe, (6th Edition, 1984).

PIPELINE EXPLODES, TWO DEAD, 31 INJURED, 11 HOMES DESTROYED

Kenneth K. Kienow, P.E.[1,2]
Member ASCE

ABSTRACT

A high pressure pipeline in a suburb of San Bernardino, California exploded in a deadly fireball on May 25, 1989. Damaged as a result of a train wreck several weeks earlier, or by the procedures employed in removing the wreckage, the 14-inch steel pipeline was carrying gasoline at 1700 psi pressure at the time of the rupture. The circumstances surrounding the train wreck and subsequent pipeline explosion are discussed. Recommendations regarding hazardous liquid pipelines in urban neighborhoods to lessen the toll in property damage and lost lives by preventing such incidents in the future are made.

INTRODUCTION

On May 12, 1989, a 69-car Southern Pacific train, overloaded by 600 tons, and with a partially inoperative braking system, gained speed as it traveled down the steep grade of Cajon Pass.

Accelerating to an estimated 100 mph, the train was unable to negotiate a curve at the base of the mountain, jumped the tracks, and plowed into a residential neighborhood in San Bernardino, California, killing four people, and destroying seven homes.

---

[1]President, Kienow Associates, Inc., 612 E. Sunset Drive North, Redlands, CA 92373. Ph.(714)792-9629, FAX (714)793-0389.
[2]Chairman, ASCE Pipeline Division Technical Committee on Pipeline Location and Installation.

The wreckage landed directly on top of the alignment of a petroleum product pipeline. Thirteen days later, on May 25, 1989, the petroleum pipeline suddenly exploded in a fireball. Two more people were killed, 31 injured, 11 homes burned, and four others were damaged.

The 14-inch diameter pipeline was constructed in 1970 to convey petroleum products from Colton, California, to Las Vegas, Nevada, a distance of about 250 miles. The pipeline was conveying gasoline at 1,700 psi at the time of the explosion. This incident illustrates an avoidable man-made disaster of catastrophic magnitude. It was the second pipeline explosion to follow on the heels of a train wreck in Southern California in six months. Figure 1 illustrates the accident site and tabulates the disaster toll.

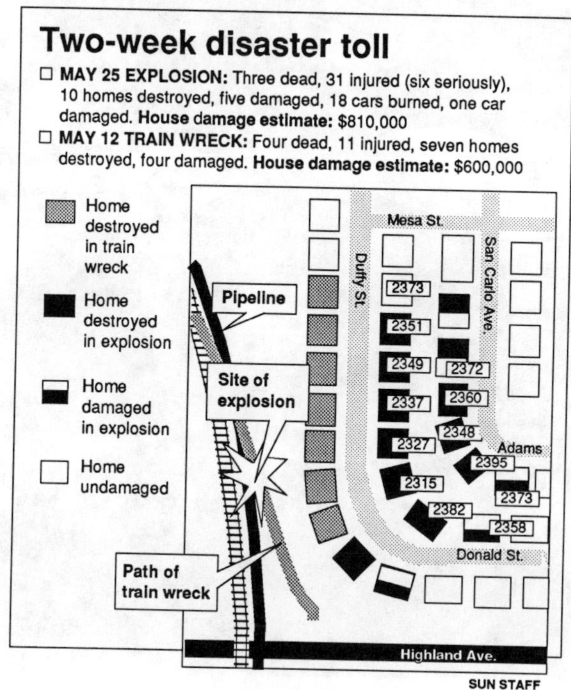

FIGURE 1
(from Ref. 1)

Metalurgist Examines Rupture in Pipe.

FIGURE 2

Were it not for the death toll, (which included two young boys, a mother, and two railroad personnel) and the extensive property loss, this entire episode, from the circumstances leading to the runaway train to the pipeline explosion, would be considered a "comedy of errors". Poor decisions followed worse decisions in a tangle of jurisdictions in which everyone thought someone else was in charge. The pump station operator tried to restart the pumps three times after the modern computerized pipeline control system went into automatic shutoff. The decision not to attempt further restarts was made only when a huge ball of flame and smoke was observed seven miles away.

BACKGROUND

Some contributing factors to the serious nature of the pipeline incident include the extremely high operating pressure; a very shallow burial depth; the pipeline location (along the rear lot line of a row of homes on Duffy Street); and the fact that the rupture was located on the top of the pipe. This combination of circumstances resulted in a high-pressure geyser of gasoline shooting like a fountain hundreds of feet into the air, spraying a gasoline mist on people, pets, houses, streets, fences, trees, and sidewalks. Upon ignition, the entire neighborhood was an instant fireball. Figure 2 is a photograph of the pipe rupture. (Courtesy of the San Bernardino, CA Sun)

Other factors which exacerbated the property damage, injury, and loss of life were the failure of a check valve to operate properly, resulting in a very large discharge of gasoline (300,000+ gallons), and the fact that the failure occurred at the base of a mountain pass, so that pressures of nearly 1000 psi existed even after the pumps were shut off.

Fortunately, there was virtually no wind blowing at the time of the failure. The San Bernardino Valley is known for wind conditions locally called "Santa Ana winds". Santa Ana winds occur frequently and reach and maintain speeds of 40 to 60 miles per hour for days at a time. If a Santa Ana wind condition had prevailed at the time of the rupture, gasoline mist could have been sprayed over a densely populated area for a distance of one-half to one mile downwind. Ignition, under such circumstances, would have produced an unstoppable fire storm which could have consumed hundreds, and perhaps thousands, of homes. A fire of such magnitude generates it's own "weather" and winds, with near hurricane conditions advancing the fire-front at incredible speed.

The pipeline transports petroleum product up a mountain pass, and within twenty-eight miles from the explosion site, climbs several thousand feet to the summit. The 14-inch diameter pipeline contains about eight gallons of fluid per foot of pipeline, or 42,000 gallons of fluid per mile of pipe. The first check valve, located at milepost 6.9 one-tenth mile uphill of Duffy Street, would have limited the spill to 4,000 gallons if it had closed. The distance to the next check valve was just over eight miles. The predictable result was that, in spite of the shutdown of the pumps, approximately 340,000 gallons of gasoline, reversing direction, continued to flow, under very high pressure, back down from the eight miles of pipe in the mountain pass, and was discharged for hours hundreds of feet into the air at the rupture site. It required 14 hours to extinguish the fire.

CONTRIBUTING FACTORS

Some factors regarding check valve operation and placement, and valve inspection and maintenance procedures which need to be reviewed if similar disasters are to be avoided in the future include:

1. According to the National Transportation Safety Board (NTSB), the valve which twice failed to operate was a 14-inch "All Clear Check Valve", a side hinged check valve, Model ACB-976, manufactured by Wheatley Pump and Valve, Inc., of Tulsa, Oklahoma. The NTSB found that over the 20-year life of the pipeline the pipeline operators had performed no maintenance on, or operational tests of, these valves. This valve is completely within the pipe and normally buried and inaccessible.

2. The current regulations require valve position indicators on valves other than check valves. The requirement for a position indicator should be extended to check valves as well. Current pipeline regulations do not clearly state that check valves are included in the requirement to inspect mainline valves twice each year. No citations were issued by either DOT or the State of California for failing to inspect check valves over a 20 year period.

3. A method is needed to induce sufficient pressure differential across a check valve to operate the valve, or otherwise verify proper valve operation. Statements were made that the critical check valve at mile post 6.9 did not close when the line was shut down after the derailment because gasoline had to be withdrawn through a two inch hose at the Colton pump station.

4. A large diameter pump bypass line, or other emergency system, should be required so that pressure in the line can be reduced to a low level in a reasonable period of time. Such a bypass or system should be capable of draining the pipeline at a reasonable rate into existing product storage tanks at the pump station or other emergency storage facilities. Attempting to drain, many miles of fourteen-inch pipeline via a 2-inch hose indicates a serious flaw in the pipeline system. As a result of the inability to rapidly drain the line, the line pressure remained at 800 psi during the derailment cleanup work.

Pipeline location decisions also contributed to the disaster toll. In 1955 the subdivision map, including every Duffy Street lot, was recorded in the County Recorders' Office. In 1957, the area was annexed to the City of San Bernardino. In 1970 the pipeline was constructed, fifteen years <u>after</u> the Duffy Street lots and homesite locations were made a matter of public record, yet the pipeline was placed immediately adjacent to the rear lot line of the subdivision. The west side of the tracks was low-lying vacant land, and a county flood control right-of-way on which homes would never be built, and recorded lots existed on the east side of the tracks. The prudent pipeline location would have been the west side, away from the subdivision lots. "Runaway" trains travel in the downhill direction and jump the tracks on the outside of the curve. There was no other logical least-risk location than the west side of the tracks.

EARTHQUAKE FAULT ZONES

Although not a factor in the Duffy Street accident, the pipeline crosses both the San Jacinto and San Andreas earthquake faults within several miles of the explosion site. These faults have been identified by geologists as two of the most active faults in Southern California. A State of California "earthquake scenario" disaster study, which evaluated the potential effects of an 8.3 Richter

Scale magnitude earthquake on the San Andreas fault, predicts that the pipeline will rupture at the fault crossing.

SUGGESTED RESTRICTIONS AND REGULATORY CHANGES

Some of the factors previously cited could be mitigated by changes in design, operating procedures, and/or pipeline regulations. Some needed changes in pipeline regulations include:

Designer Qualifications: Stringent rules for design and operation of hazardous liquid pipelines in or near densely populated urban areas are needed. California law required that the engineer who designed the curb, sidewalk, sewer, and water lines on Duffy Street be a Registered Civil Engineer, but there are no similar requirements for the designer of a 1700 psi gasoline pipeline along the back fence of residential lots. It should be required that all such pipelines be designed, sealed, and signed by a Registered Professional Engineer.

Pipeline Location: The decision processes, procedures, and regulations governing hazardous material pipeline location, which allowed or resulted in the extremely poor choices made in the Duffy Street area, should be reviewed and changed as necessary. Land use planning, zoning, and subdivision regulations should take into account hazardous pipeline locations. Pipeline location policy should require the approval of local planning and building authorities.

Maximum Pressures: Reasonable limits on maximum allowable pressures are needed for pipelines in urban areas. A pipeline operating at 1700 psi along back yard fences of a neighborhood is a ticking time bomb. In fact, if a citizen reported to authorities he had discovered a 14-inch metal container of gasoline pressurized to 1700 psi, and buried four feet deep in his backyard, the area would be cordoned off and a bomb squad would be dispatched to remove it.

Burial Depths: Deeper burial should be required for pipelines in urban areas. The current Code of Federal Regulations governing hazardous liquid pipelines is 40CFR, Part 195. The Part 195 requirement of 36 inches cover in industrial and residential areas, and 48 inches cover within 50 feet of buildings, is grossly inadequate, particularly in high hazard areas.

Hazard Areas: Location of pipelines should be restricted in natural high risk areas such as known earthquake fault zones; in the vicinity of airport runways; at the base of known landslide/avalanche areas; and at man-made risk areas such as the outside of curves at the base of long steep railroad grades.

Urban Geyser Protection: In densely populated urban areas, a steel reinforced concrete slab or similar barrier should be required over the top of the pipeline extending several feet beyond each side of the pipeline. The barrier would protect the upper part of the pipe from surface impact damage, and prevent a break in the upper part of the pipe from turning into a geyser which, in a high wind condition and delayed ignition, could result in the destruction of an entire city.

Check Valve O & M: Regulations should require the same twice yearly inspection of mainline check valves as are required for mainline valves. Check valve spacing should be governed by a limit on the maximum quantity of product which can be discharged in the event of a rupture. Such a limit would decrease fire danger, reduce damage to the environment, and minimize potential pollution of the groundwater supply. If such a limit were set at 50,000 gallons, check valves on a 14 inch line would be required every 6300 feet, or every 1.2 miles. A 50,000 gallon discharge is equivalent to dumping six full truck and tanker trailer loads of gasoline at a single intersection.

Seismic Zones: Because of the soil and geologic characteristics of fault zones, a hazardous material spill may result in immediate and deep contaminant penetration and serious damage to the local groundwater supply. Seismically activated valves should be required at active earthquake fault crossings, and pipelines articulated to withstand movement at the fault without rupture.

Allowable Stresses: The philosophy of higher safety factors for high risk areas should apply. Because of increased potential for property damage, bodily injury, and loss of life from pipeline ruptures in densely populated urban areas, the allowable pipe wall stress (expressed as a percent of the steel yield strength) should be reduced below that permitted in unpopulated areas. The current 72 percent may be acceptable for unpopulated areas, but allowable steel stress should be reduced to 50 percent of minimum steel yield strength, or less, for densely populated urban areas.

a. AWWA Pressure Water Pipe

The American Water Works Association limits the steel stress used in the design of steel pressure water piping to 50 percent of the yield strength of the steel. The failure of a hazardous liquid pipeline in an urban area is potentially much more serious than failure of a high pressure water line. Why is a steel gasoline pipeline designed to operate at 72 percent of yield stress, 44 percent <u>higher</u> than the standard for pressure steel water piping in our cities?

b. Plastic Pressure Pipe

In plastic pressure piping service, irrigation pipe is designed to operate at 50 percent of the HDB stress, domestic water pipe at 40 percent of HDB stress, and gas service piping at 32 percent of HDB. A similar reduction of steel pipe design stress commensurate with the damage potential of a failure should apply to hazardous liquid steel pipelines.

c. Part 195 Hazardous Liquid Pipelines

The current Code of Federal Regulations, Part 195, Paragraph 106(a), reduces the allowable steel stress at design pressure to 60 percent of steel yield strength when the pipe is on an offshore platform or on a platform in inland navigable waters, presumably to minimize pollution of surface waters due to pipeline rupture and subsequent fuel spill. The residents of a densely populated urban neighborhoods deserve the same consideration.

<u>Check Valve Spacing</u>: If the check valve at mile post 6.9 (one-tenth mile above Duffy Street), had operated as designed, a spill of 4,000 gallons would have occurred instead of over 300,000 gallons. This valve also failed to close when the line was shut down after the train wreck, and some of the gasoline was drawn off at the pump station, leaving the line at 800 psi during wreckage removal operations, a dangerous situation which should not have been allowed.

<u>Authority to Order Relocation</u>: 49 CFR 195.402(C)(6) concerns the requirement of the pipeline operator to "Minimize the potential for hazards identified under

Paragraph (C)(4)...and the possibility of recurrence of accidents analyzed under Paragraph C(5) of this Section."

There is a reasonable possibility of another runaway train leaving the tracks on the east side and damaging the pipeline again. In order to "minimize the potential for hazard", based on the recent runaway train wreck/pipeline damage/explosion scenario, the pipeline should be moved to the west side of the tracks. Extensive study and investigation is not needed to reach this "common sense" conclusion. Yet no one had the authority to order the pipeline operator to relocate the pipeline. The City Attorney was defeated in numerous attempts to obtain court orders to relocate the line away from homes to the other side of the tracks.

On the basis of compliance with Part 195, Section 402(c), the Office of Pipeline Safety, State Fire Marshal's Office, and Department of Transportation would seem to have a valid engineering reason, indeed, an obligation, to order the relocation of the pipeline. If the current regulations do not allow them to issue an order to relocate a pipeline, they should be granted such authority.

POST ACCIDENT DAMAGE ASSESSMENT/INSPECTION

In the 13 day interval between the train wreck and the pipeline explosion, numerous procedural errors and poor judgment calls were made. Some regulation is needed in these areas:

<u>Assume Damage</u>: It does not seem reasonable to assume that a 65-car freight train with five locomotives could hurtle off a 20-foot high embankment at 100 miles per hour and deposit itself directly over a pipeline as a compact 14,000,000 pound pile of wreckage only 600 feet long, and NOT cause damage to a pipeline with only a few feet of soil cover. Yet the 600 feet of pipeline in the train wreck area was only spot checked by "potholing" every 50 feet. It must be assumed the pipeline is damaged, until 100% visual inspection proves otherwise.

<u>Hydro Test</u>: Hydrostatic testing of the pipeline with water at 125 percent of the maximum operating pressure would have detected the damage caused by the wreck or wreckage removal activity that finally resulted in the failure of the pipe line. Since the pipe line was not uncovered and visually inspected for the full length of

the impact area, a 125 percent hydrostatic water test should have been required.

Don't Delegate, Investigate: The State Fire Marshal's (SFM) Office should have required a full length visual inspection, and should have required a water hydrostatic test to 125 percent of the operating pressure. The SFM's chief pipeline inspector, fired six months prior to the accident, was quoted by the newspaper as saying that his replacement was an "engineer" with no pipeline experience. The DOT Office of Pipeline Safety should have overruled the State Fire Marshal's office and required the appropriate tests, but California is one of several states in which the DOT Office of Pipeline Safety delegates it's responsibility to a state agency. This series of accidents suggests we have too few qualified pipeline safety engineers without expecting each state to have their own. This "delegation" policy of responsibility for pipeline safety is a step backward in pipeline safety.

Mandate: A DOT representative has been quoted as saying "we were investigating a train wreck, not a pipeline failure" to explain their activity between the wreck and explosion. Since DOT has jurisdiction over both interstate railroads and pipelines, future DOT policy should be to learn of the existence of, and possible damage to pipelines, and immediately take responsibility for inspection procedures as well as the train wreck, or other investigation. A 125 percent hydrostatic test and 100 percent visual inspection should be a mandatory requirement in future derailments, and any other such incidents which have the potential of damaging a hazardous material pipeline. A statement was made to the press by the State Fire Marshal's Inspector that "we often test pipelines at 50 percent of operating pressure". A hydrostatic test at 50 percent of design pressure does not constitute a "test" in the engineering sense of the word. Canada requires a test to 156 percent of design pressure for HVP pipelines in urban areas.

CONTAMINATED SOIL REMOVAL ACTIVITY

The State Fire Marshal and the pipeline operators have speculated that the pipeline damage might have been the result of wreckage removal operations, rather than the train wreck impact itself, indicating their awareness of the potential dangers of heavy equipment operating in close proximity to a pipeline. However, after the pipeline explosion, pipe replacement, hydrostatic testing of the replaced 600 foot section, and after placing the pipeline back in operation, additional excavation was conducted using a large front-end loader to remove gasoline and diesel fuel-contaminated soil adjacent to the pipeline. This excavation extended to a depth of six to eight feet, within two feet horizontally of the pipeline location centerline markers. The contaminated soil on either side and immediately beneath the pipe is still there.

It is possible that the pipeline has again been damaged, either through direct contact, or by the loader bucket striking a buried piece of train wreckage, rail, or other foreign material and forcing it into the pipeline, unknown to the operator or observers.

Procedures for removing contaminated soil and wreckage, pipeline repair and replacement, and any other post-accident activities must be integrated, well-defined, and enforced. Total contaminant removal requires relaying of the pipeline.

If the replacement pipeline had been relocated on the opposite side of the tracks, the contaminated soil removal could have been done without potential for damage to the new line. This decision would have placed a 20-foot high earth embankment between the residents and the pipeline, affording greater protection to the neighborhood.

CONCLUSIONS

Five 200 ton locomotives and 65 cars, each weighing 100 tons, could not derail at 100 miles per hour, hurtle down a 20 foot embankment, bounce around on top of a pipeline buried three to four feet deep, and not cause damage to the line. Some rail sections were skewered into the ground like spears. One 400 pound, ten foot section of rail was found within a few feet of the ruptured pipe.

Whether the train wreck or the clean-up operations damaged the pipe is academic: the pipe failed 12 days later. The rupture caused a 4 -in. wide by 28-in. long gash in the pipe. Dents were observed in the pipe adjacent to the rupture, and part of a railroad car braking system was found 15-inches to one side of the rupture. A piece of locomotive cowling was several feet from the pipe. A long, deep scrape in the outside of the pipe began on one side about the springline, passed over the top of the pipe on a line nearly coincident with the rupture, and continued down the other side of the pipe. This "scrape" was approximately eight feet long.

Tensile tests on samples cut from a piece of discarded pipeline found on site showed a very high ratio of yield strength to ultimate strength. This may be the result of some form of embrittlement which has occurred over the past 20 years. The high yield to ultimate ratio may result in a line which is more sensitive to damage. The tensile data are given in Table 1 below.

Sample	Yield	Ultimate	Ratio Y/U
1	72,000	78,000	.92
2	69,000	78,500	.88
3	73,500	82,000	.90
4	70,000	84,000	Weld
7	66,000	82,500	.80
8	68,500	82,000	.84
	Average of 5 tests		.87 > .85

TENSILE TEST DATA AND YIELD TO ULTIMATE RATIO

TABLE ONE

There is a requirement in 49CFR Part 195.106 which permits use of steel pipe of unknown mill specification properties (such as used pipe) for hazardous liquid pipelines. In the absence of the manufacturer's mill test data, Part 195 provides that tensile tests may be made on samples cut from the pipe. For design purposes, the yield strength for the pipe may be assumed as either 80 percent of the average of the yield strength in the tests, or the lowest yield strength recorded in the test.

There is a provision in Part 195, however, that states if the average of the yield to tensile strength ratios is .85 or higher, the yield strength for that steel must be assumed to be no more than 24,000 psi.

The tests on the Duffy Street pipe had yield/tensile ratios as given in the right hand column in Table 1. Since the average of the test ratios (.87) is above 0.85, if this 20 year old pipe were currently being put into service, and the allowable yield stress determined by the above test procedure, the allowable steel yield stress would be only 24,000 psi, rather than the current 52,000 psi. Rather than operating at 1700 psi, the pipe could legally be operated at only 770 psi.

This 20 year old pipe, if bought on the used pipe market and tested, could legally be operated at only 773 psi, as compared to the current operating pressure of 1700 psi, because of the loss of ductility that has apparently taken place over the past 20 years. Further study of embrittlement and other aging effects on steel in older high strength steel pipelines in petroleum products service is needed.

LEGISLATIVE ACTION

Some of our cities are laced with buried hazardous liquid pipelines operating at high pressures. Many are old and corroding, and some may be losing their ductility. Many cannot be located by the local fire department, or by the pipeline operators themselves. The existence and location of shut-off valves and their operating condition are often not known. These pipelines run through, or adjacent to neighborhoods, shopping malls, gas stations, and schools. Some are exposed, strapped 16-feet in the air to the underside of the bridges we drive under every day. Surely there are many more check valves which have not been inspected or tested for 20 years, in spite of the fact that the check valves represent the only safety feature between a community and a spill of hundreds of thousands of gallons of gasoline or jet fuel.

Rep. Glenn H. Anderson, D-Long Beach, Chairman of the House Public Works and Transportation Committee's Subcommittee on Oversight, said of the Duffy Street accidents, "This scenario has presented us with a regulatory black hole." Rep. George E. Brown, D-Colton, said "It appears that the end result is the companies are regulating themselves."

Rep. Brown and Representative Jerry Lewis, R-Redlands, have asked the Federal Emergency Management Agency to conduct a 2-year $300,000 study of the dangers of earthquakes and other hazards along crucial energy and utility lines in the Cajon Pass.

In addition, Rep. Brown said that he is proposing changes in the National Pipeline Safety Act, which is up for reauthorization this year, which will give the Office of Pipeline Safety authority to order a pipeline to be relocated when it is deemed a hazard in it's present location.

The problem may be to find qualified inspectors who can recognize a "hazard" when they see one.

In California, Gov. Deukmejian signed a bill in October, 1989 to tighten up state regulation of hazardous liquid pipelines near railroad tracks. Assembly Bill AB381, by Assemblyman Don Elder of Long Beach, requires the State Fire Marshal to conduct a detailed study of all pipelines - interstate and intrastate - located within 1,000 feet of any railroad line. The bill also requires the Fire Marshal to adopt a separate set of regulations for construction, testing, inspection, and operation of intrastate pipelines near railroads.

The governor has asked for an additional $285,000 in the new budget for the State Fire Marshal's office to monitor construction of pipelines, inspect them and oversee their operation. The money will also be used to train the staff in handling pipeline emergencies and informing local fire departments about pipeline locations, design, and construction.

These are all moves in the right direction, but much remains to be done. As an important first step, all hazardous liquid pipelines should be designed by a Registered Professional Engineer licensed to practice by the State. In addition, a thoroughly researched rewrite of CFR Part 195 is in order.

The public is outraged, and rightfully so, at the recent rash of pipeline explosions, leaks, and refinery fires. Unless the engineers in the petroleum products and pipeline industry introduce effective criteria for improving the regulations, we will see a continuing rush of piecemeal legislation. We engineers in the pipeline field can considerably influence the regulatory process and protect the public health and safety.

REFERENCES

1. San Bernardino (CA) Sun Telegram, May 12, 1989; and May 25, 1989; various dates through October 3, 1989.

2. Code of Federal Regulations, 49 CFR, Part 195, Transportation of Hazardous Liquid by Pipeline, October 1988.

3. House Subcommittee on Investigations and Oversight of the House Committee on Public Works, Public Hearing, July 7, 1989, San Bernardino, CA.

4. National Transportation Safety Board, Board of Inquiry, Investigation into the Derailment of Southern Pacific Transportation Company Freight Train on May 12, 1989, and Subsequent CalNev Pipeline Rupture, May 25, 1989, San Bernardino, CA, Public Hearing, August 28 - Sept. 1, 1989.

CORROSION PROTECTION OF CAST IRON AND DUCTILE IRON PIPE SAN DIEGO, CALIFORNIA

MICHAEL S. TUCKER[1]
M. ASCE

ABSTRACT

This paper addresses the use of a loose unbonded plastic covering (polyethylene encasement) for protection of cast and ductile iron pipe in an extremely corrosive soil. The case study in the San Diego area involves six separate inspections of parallel lines - a 16-inch cast iron and a 24-inch ductile iron pipeline. Detailed examination of the exposed pipelines was made on each inspection as well as analysis of the soil and the loose polyethylene encasement. Numerous individuals from outside the Ductile Iron Pipe Research Association (DIPRA) attended and participated in the inspections. Photographs of the pipelines were taken with a report prepared for each inspection.

INTRODUCTION

The use of cast iron pipe for water and sewer pipelines dates back to the 1400's, the earliest installation being in Germany in 1455. In 1664, King Louis XIV of France ordered the construction of a cast iron water main extending fifteen miles from a pumping station at Marly-on-Seine to Versailles to supply water for the fountains and town (DIPRA Handbook, 1984). This cast iron pipe is still functioning after more than 300 years of continuous service.

In the United States, there are at least seven eastern cities that have cast iron pipe in continuous operation in their system after more than 150 years of

---

[1]Senior Regional Engineer, Ductile Iron Pipe Research Association, 1016 East Rosewood Avenue, Orange, California 92666.

service. Also, there are over 300 municipalities and water districts belonging to the Cast Iron Pipe Century Club with over 100 year-old cast iron pipe still in continuous operation. With these impressive statistics, it is apparent that cast iron pipe (and its replacement, ductile iron pipe) holds up remarkably well in long-term service for utilities.

However, there are certain areas of the United States where soil conditions can be corrosive to cast and ductile iron pipe. Bare cast iron or ductile iron pipe installed in these areas could experience wall penetration due to soil conditions in as little as 5 years. The Serrano Valley area of San Diego, California, is one such site and is the subject of this paper.

MATERIALS

Research for corrosion protection of cast iron pipe and bolted mechanical joints dates back to the 1930's. The Cast Iron Pipe Research Association (CIPRA), the predecessor of DIPRA, started investigating various methods of protection for cast iron pipe in corrosive soils in the early 1950's. One of these methods was loose polyethylene encasement (4 mils thick).

In 1952 a test site was initiated in Everglades, Florida, an extremely corrosive soil environment (highly organic muck) exhibited by very low resistivity, neutral pH with negative redox and the presence of sulfate reducing bacteria. A fluctuating water table provided saturated conditions throughout the years. A number of 6-inch mechanical joint pipe sections were originally installed in this test site with the 4-mil loose polyethylene encasement around the joint. After 18 years of regularly scheduled exhumations of pipe sections, the 4-mil loose polyethylene encasement provided excellent protection for the covered pipe and bolts, while the remaining (bare) portions of the pipe sections were extremely corroded (see Figure 1). This method of protection represented a departure from the standard practice of using a protective medium tightly bonded to the pipe surface (Smith, 1968).

The polyethylene serves as an unbonded film preventing direct contact between the pipe and the surrounding soil. The electrolyte available to support the corrosion process is effectively reduced to any moisture that might be present in the small annular space between the pipe and the polyethylene film. Because polyethylene encasement is not a watertight system, some seepage of groundwater beneath the wrap will typically occur (see Figure 2). However, initial corrosion reactions soon deplete the available dissolved oxygen in the annular space, and further corrosion activity is effectively mitigated. Furthermore, the polyethylene

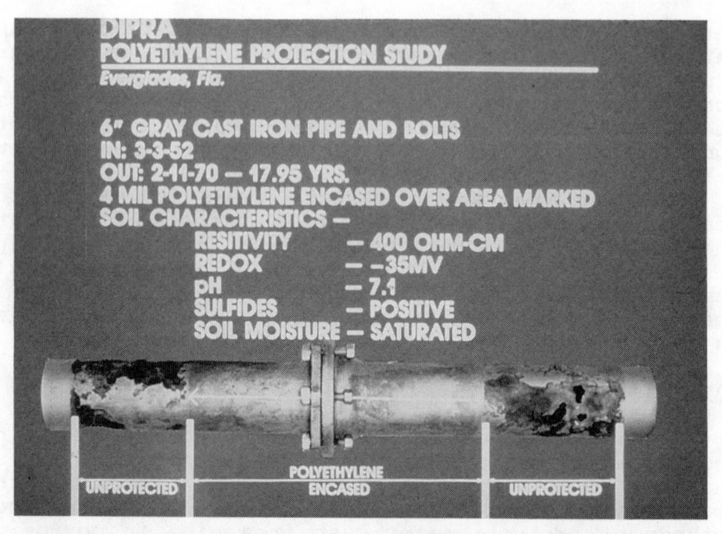

Figure 1. Pipe specimen from Everglades, Florida test site.

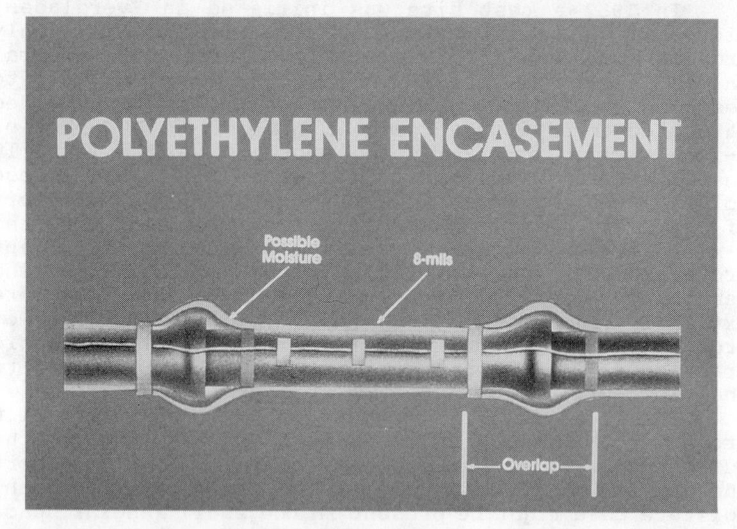

Figure 2. Schematic diagram of a pipeline encased in loose polyethylene for external corrosion protection. Note the overlap at the joints to provide an extra layer of protection against tears caused by bolted joints.

film also provides an essentially impermeable barrier that restricts the renewal of oxygen reaching the pipe surface, and the diffusion of corrosion products away from the pipe surface. Additionally, the pressure of surrounding compacted soil compresses the polyethylene tightly against the pipe, thus preventing any further significant exchange of groundwater between the wrap and the pipe. As a result, the polyethylene provides an essentially uniform environment against the pipe, thereby eliminating local galvanic cells caused by variations in soil composition, pH, aeration, etc. - the principal causes of corrosion on underground pipelines (Stroud, 1989).

Test sites were initially set up in Florida, New Jersey and Alabama and continued thereafter in other sites across the country. The first field installation of polyethylene encasement on cast iron pipe in an operating system was in Philadelphia, Pennsylvania in 1958. Since that time, polyethylene encasement has been used to protect cast and ductile iron pipe in thousands of installations across the United States and in many other countries.

Because of its widespread, successful use and field-proven corrosion protection capabilities, the first Joint American National Standards Institute and American Water Works Association Standard for polyethylene encasement (ANSI/AWWA C105/A21.5) was adopted in 1972 (DIPRA, "Polyethylene Encasement", 1987).

PROCEDURE

To complement CIPRA's own test sites in evaluating the effectiveness of polyethylene encasement to protect cast and ductile iron pipe in corrosive soils, a program was initiated in 1963 with the City of San Diego Department of Utilities to excavate and inspect cast iron pipe protected with polyethylene encasement in their operating system (as opposed to a test site). This joint program has greatly expanded (there are over 70 field inspections to date), and has allowed DIPRA to gain additional research data and information by working with various utilities; and, at the same time, give utilities using polyethylene encasement confidence in their choice of protection for cast and ductile iron pipe. The Serrano Valley area of San Diego has become one of the more prominent of these inspection sites, due to the following:

(1) one of the earliest polyethylene installations
(2) regularly occurring field inspections
(3) extremely corrosive nature of the soil

The usual procedure for performing the inspection

starts with the utility providing a backhoe and crew to uncover the pipe. Typically a pipe section of approximately 10 feet is uncovered using the backhoe and a crew to hand dig around the pipe. This ensures that there is no damage to the polyethylene when the pipe is excavated totally free from the native corrosive soil. At this point DIPRA staff take over by cutting and carefully removing as much polyethylene as is practical, while observing any irregularities in the original installation. A sample of the removed polyethylene is then sent to the DIPRA laboratory for analysis and a sample of the soil adjacent to the pipe is collected for evaluation.

After the polyethylene is removed, the pipe is cleaned with a stiff wire brush and washed. A close inspection of the entire pipe surface is conducted, using a sharp geologist's hammer (see Figure 3). It is DIPRA's practice to invite interested consultants and or utilities to view the dig-up and inspection; at this point, they are invited to get a close-up look at the excavated pipe. Pictures of the dig-up are taken from beginning to end, with particular attention to the field-cleaned pipe. Prior to backfill, the pipe section is re-covered with polyethylene encasement. A detailed report is then prepared by the DIPRA regional engineer.

Figure 3. The pipe surface is carefully examined for possible pitting or graphitization.

BACKGROUND

In 1960, the San Diego Department of Utilities installed 1,500' of 12" and 7,500' of 16" cast iron pipe wrapped in polyethylene encasement in the Serrano Valley area. These new lines replaced an old 18" cast iron water line installed (with no corrosion protection) sometime prior to 1960 that had failed within a few years due to corrosion. A 1975 report indicated that "due to the extensive corrosion difficulty ... and known corrosive soil characteristics - caused somewhat by a direct connect (saline intrusion of groundwater) with the Pacific Ocean near a salt water inlet - the pipe was encased in 8-mil, loose polyethylene tube at the time of its installation." (Smith, 1975)(ANSI/AWWA C105/A21.5 calls for 8-mil thickness). In 1967, a 24" ductile iron pipeline (Soledad Valley Pipeline Phase II) was installed parallel to and approximately 10 feet from the existing 16-inch and 12-inch cast iron water lines. As was the established practice, losse polyethylene encasement was installed for corrosion protection. The San Diego Department of Utilities, along with a CIPRA member company, initiated the first investigation of the polyethylene encased 16" cast iron water line. Working with the Utilities Department, CIPRA performed the initial investigation on this pipeline in November of 1963, with the purpose of evaluating the effectiveness of the loose polyethylene encasement by instigating a plan to repeat inspections over the next few years. Thus began the start of successive inspections (see Table 1) in this area.

INSPECTIONS

#1-1963

At the initial investigation in 1963, three separate sections of pipe were exposed. Representatives from CIPRA, one member company and the San Diego Department of Utilities were present to observe the findings. The first section of pipeline exposed was in a relatively dry area with high resistivity soils (4030 ohm-cm, measured by a single probe).

The second and third locations were very wet, with standing water in the trenches. Soil resistivity readings were very low (see Table 1). At both locations, the polyethylene encasement (originally clear) was discolored with a thin, black scale on the pipe surface. Both the pipe and polyethylene were in excellent condition with only superficial oxidation and discoloration. There were no signs of corrosion pitting (Smith, 1963)(see Figure 4).

Figure 4. Pipe section from 1963 San Diego investigation.

#2-1966

The next investigation was performed in January of 1966, after 5-plus years of service. It was estimated by Mr. Otto Waters, Corrosion Engineer for the San Diego Department of Utilities, that the pipe wall of bare cast iron pipe would be penetrated by these very corrosive soils in 5 years if unprotected (Smith, 1966). Representatives again were present from CIPRA and the Department of Utilities as well as other utilities and one CIPRA member company. Four separate sections of pipe were exposed. The first section revealed a relatively dry area with high resistivity soils (4000 ohm-cm). The next three sections exhibited low resistivities (320-550 ohm-cm), saturated conditions and discoloration of the polyethylene with a thin black superficial scale under the wrap. No corrosion pitting was evident on any exposed pipe after thorough examination using a wire brush and sharp geologist's hammer. At this point, it was apparent the polyethylene encasement was effectively mitigating significant corrosion to the iron pipe; which, presumably, would have experienced severe corrosion if installed unprotected (see Figure 5).

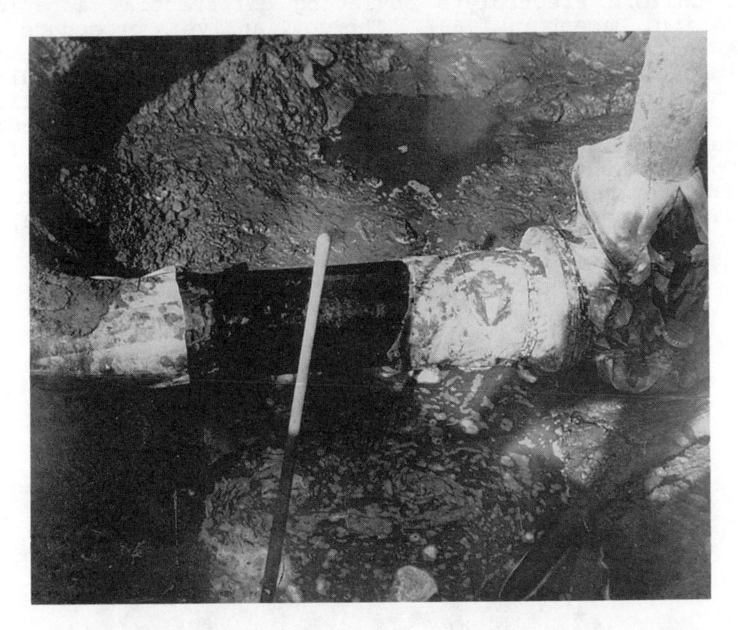

Figure 5. Pipe section from 1966 San Diego investigation.

#3-1968

The third inspection was made in the Fall of 1968 after 8 years of service. CIPRA was represented, along with one CIPRA member company, the San Diego Department of Utilities and Mr. Otto Waters, who now worked as a Consulting Engineer. Two separate sections of 16-inch cast iron pipe were exposed for inspection. The first location included a joint, revealing discoloration of the polyethylene with water inside the tube along with a thin reddish-brown and black scale (3 mils). There was no measurable pitting. The next section of the 16-inch cast iron pipe also included a joint with severe discoloration, the same thin scale of black corrosion products and one pit on the transition section of the bell. A considerable quantity of foreign material including clay was found between the bell and the polyethylene, which was the probable cause of the noted pit.

Analysis of the water under the wrap indicated high levels of chlorides and sulfates, which would seem to be corrosive but with a lack of oxygen (O_2 tested at 0 ppm). Consequently, the absence of the progression of corrosion indicates stabilization through the moisture serving as a uniform electrolyte obviating differential aeration (Smith, December, 1968). The soil analysis procedure in AWWA C105 Standard, Appendix A, was utilized for this inspection (as well as the future dig-ups); results indicate a very corrosive environment for cast or ductile iron pipe. The Appendix A details the various tests performed and the points assigned for each result. The "point-count" from the soil evaluation ranged from 14 to 23.5 (highest available) with the majority in the "worst-case" range of 23.5 points (see Figure 6 and Table 2).

Figure 6. Pipe section from 1969 San Diego investigation.

#4-1975

In 1975, another inspection was made on one section of the 16-inch cast iron main. Several outside utilities were present along with representatives from the U.S.

Navy, and Waters Consultants. Five feet of polyethylene was removed from the 16-inch pipe; pumping was required to lower the groundwater below pipe level. Observations of the exposed pipe revealed considerable discoloration, including rust and a thin black scale-like material over most of the pipe; however, after cleaning and a thorough examination with a geologist's hammer, no measurable pitting was observed. As expected, soil analysis confirmed a corrosive environment with a low resistivity reading of 480 ohm-cm (Smith, 1975)(see Figure 7).

Figure 7. Pipe section from 1975 San Diego investigation.

#5-1981

In October of 1981, an investigation of both the 16-inch cast iron pipe and the newer 24-inch ductile iron pipe was made. A number of interested representatives from various utilities, were present, as well as the U.S. Navy, Waters Consultants and the U.S. Bureau of Reclamation along with several DIPRA staff members.

Soil analysis from the 16-inch cast iron pipe section indicated extremely corrosive conditions (23.5 points). No joints were exposed as 8 to 10 feet of pipe was excavated and 5 feet of polyethylene was removed. Groundwater again had to be pumped out of the excavation to allow for complete inspection, revealing no measurable pitting.

The 24-inch ductile iron pipe was then exposed. Approximately 5 feet of the polyethylene along the top of the pipe was apparently torn during the original installation. Underneath the wrap were signs of soil and decayed organics (caused by the tear) resulting in a reddish-brown discoloration (especially at the tear) and a thin, black, scale-like material. The pipe was cleaned and examined. There were several small, shallow (.04-.05") pits under one deposit of soil which were difficult to distinguish from the original surface roughness. It appeared that these pits were a result of the initial corrosion reaction that subsequently slowed and became dormant under the wrap. No other pitting was observed (see Figure 8).

At this inspection, samples of the removed polyethylene were sent to the DIPRA laboratory for testing and analysis with comparison to the original polyethylene encasement material. Results from the testing indicated no deterioration of physical strength nor signs of brittleness (Higgins, 1981).

Figure 8. Pipe section from 1981 San Diego investigation.

#6-1986

The last inspection to date of these pipelines was performed in June, 1986. Along with DIPRA staff, representatives were present as usual from the San Diego Department of Utilities as well as two corrosion engineering firms. Again the 16-inch cast iron pipe and the paralleling 24-inch ductile iron pipe were uncovered and inspected. Soil analysis from both locations revealed extremely corrosive conditions (23.5 points).

Three feet of polyethylene was removed from each side of a joint on the 24-inch ductile iron. Again pumping of ground water in the excavation was necessary.

The ever-present reddish-brown discoloration, minor rusting and thin black scale-like deposits noted in previous inspections were present; however, the original surface was in excellent condition. A minor pit (approximately 8 mils - due to contact with the corrosive soil) was noted under a small tear in the polyethylene. There were no other signs of pitting (see Figure 9).

Figure 9. Pipe section from 1986 San Diego investigation.

The 16-inch cast iron pipe was excavated with 5 feet of polyethylene removed - no joint was exposed. The pipe was in excellent condition with only mild discoloration and no pitting except again at a small tear in the wrap (less than 5 mils). The polyethylene samples removed from both pipe sections were tested at the DIPRA laboratory. Average values of the test results were greater than the minimum required by the standard (Tucker, 1986). Testing of the original polyethylene encasement has been a standard procedure on all investigations since 1981 (see Table 2).

SUMMARY OF RESULTS

Exposure to the extremely corrosive soil conditions in the Serrano Valley area from 1960 through 1986 has proven the effectiveness of loose polyethylene encasement in preventing major external corrosion on cast and ductile iron pipe. This effectiveness is demonstrated through the series of six separate investigations, which included inspections of the pipe and the polyethylene encasement, as well as over 70 investigations throughout the United States. The testing of the original polyethylene encasement from the later dig-ups has demonstrated its long-term integrity. Methodical and thorough inspections of the pipe surface using a stiff wire brush, sharp geologist's hammer, washing, etc., have clearly demonstrated how the loose polyethylene creates a uniform environment effectively curtailing corrosion activity. Soil samples, at pipe depth, were obtained at each inspection to identify the soil conditions; most of these samples proved to be very corrosive (see Table 2).

Signs of minor galvanic corrosion pitting were present on several inspections. These were attributed to tears in the original polyethylene encasement installation and/or soil underneath the wrap. Generally, at each inspection, signs of superficial discoloration were present (due to initial oxidation), as well as a thin black scale-like material adhering to the pipe exterior. Both were easily removed with a stiff wire brush and washing, and were determined to cause no significant metal loss. Moisture was evident under the wrap at most locations, and was tested at the 1968 inspection, revealing the presence of chlorides and sulfates, but a lack of oxygen, thus obviating corrosion.

CONCLUSIONS AND FINDINGS

Corrosion has been effectively mitigated on cast iron and ductile iron water lines in the very corrosive soils of the Serrano Valley area in San Diego by the installation of loose polyethylene encasement. The six investigations to date of the various sections of the pipelines gives strong support to the fact that properly applied polyethylene encasement provides a relatively impermeable barrier that restricts additional oxygen and creates a uniform environment that is not conducive to galvanic corrosion. The ongoing field investigation program in San Diego, started over 26 years ago, has demonstrated that a utility can use polyethylene encasement with confidence in very corrosive soil conditions, where unprotected pipe could experience exterior corrosion failures in 5 to 10 years.

Table 1. Inspections - Serrano Valley Area
San Diego, California

Number	Date
1	November, 1963
2	January, 1966
3	December, 1968
4	February, 1975
5	October, 1981
6	June, 1986

Soil Conditions
(Test procedures per ANSI/AWWA C105/A21.5 Standard)

Insp. No.	Year	Resistivity ohm-cm.	Redox mv.	pH	Sulfides	Moisture	Pts.
1	1963	4030				Dry	<10
	1963	113.5				Saturated	>12
	1963	144				Saturated	>12
2	1966	4000				Dry	<10
	1966	550	+18	7.4	Negative	Saturated	16
	1966	340	-120	7.3	Positive	Saturated	23.5
	1966	320	-138	7.2	Positive	Saturated	23.5
3	1968	310	+168	7.5	Trace	Saturated	13
	1968	300	+162	7.7	Positive	Saturated	15.5
	1975	480	+80	7.3	Trace	Saturated	16.5
4	1981	420	-158	7.2	Positive	Saturated	23.5
	1981	495	-140	7.1	Positive	Saturated	23.5
5	1986	348	-40	6.9	Positive	Saturated	23.5
	1986	280	-40	6.7	Positive	Saturated	23.5

Table 2. San Diego, California 1981
Clear, Low Density Polyethylene

Parameter	Tested*	Minimum**
Transverse Tensile Strength (psi)	1798.8	1200
Longitudinal Tensile Strength (psi)	1724.8	1200
Transverse Elongation (%)	530.8	300
Longitudinal Elongation (%)	439.5	300

*Tested values are average values of five (5) specimens.
**Minimum values are set forth in AWWA C105.

APPENDIX - REFERENCES

Ductile Iron Pipe Research Association, 1984, Handbook of Ductile Iron Pipe, Sixth Edition

Ductile Iron Pipe Research Association, 1987, "Polyethylene Encasement, Laboratory and Field Investigations Installation Procedures"

Higgins, Michael J., 1981, Report on Inspection of Cast Iron Pipe and Ductile Iron Pipe Protected by Loose Polyethylene Encasement, San Diego, California

Smith, W. Harry, 1963, Report on Observation of Corrosion Protection of Cast Iron pipe by Loose Polyethylene Wrap, San Diego, California

Smith, W. Harry, 1966, Report on Observation of Corrosion Protection of Cast Iron Pipe by Loose Polyethylene Wrap, San Diego, California

Smith, W. Harry, 1968, Report on Corrosion Resistance of Cast Iron and Ductile Iron Pipe, Cast Iron Pipe Research Association

Smith, W. Harry, December, 1968, Report on Observation of Corrosion Protection of Cast Iron Pipe by Loose Polyethylene Wrap, San Diego, California

Smith, W. Harry, 1975, Report on Observation of Cast Iron Pipe Protection by Loose Polyethylene Encasement, San Diego, California

Stroud, T. F., 1989, CORROSION/89, Paper No. 585, National Association of Corrosion Engineers, New Orleans, Louisiana

Tucker, Michael S., 1986, Report on Inspection of Cast Iron Pipe and Ductile Iron Pipe Protected by Loose Polyethylene Encasement, San Diego, California

Measured Performance and Numerical Analysis of Buried Pipe

Yoshiyuki Mohri[1], Yujiro Tsurumaru[2], Isamu Asano[3]

Abstract

This study shows that the behavior of a soil pipe system can be assessed by a nonlinear elasto-plastic finite element analysis. The results of numerical calculations are compared to the measured behavior of a large scale flexible pipe in a field test. The behavior of buried pipe along a pipeline vary considerably and a deflection of about less than 1.0 % measured. The analytical results agree qualitatively with the field investigations and give promise of the support of the current design procedure.

Introduction

Buried pipeline has been constructed over 100 km per year and increased in size of culverts for irrigation in Japan. Especially, large scale culverts are used as a main line in irrigation pipeline system, and these are installed in various field, such as steep mountain side, soft clay field, along a sea shore. In case of soft clay field, these pipes are installed in trench with side wall protected by sheetpiles. The pipe behavior is much affected by the sheetpile extraction work. In actual, failure of pipe is easy to cause during a construction. Most of these buried pipes are empirically designed. The conditions are complicated and engineer can not predict the behavior of the pipe with current design procedures. Finite element analytical design methods are most valuable tool among the current procedures. In recent years, a large number of constitutive models have been proposed to predict the behavior of soil structure. These models must be assessed by the field verification. In this study, non-linear generalized elasto-plastic constitutive model is used for finite element analysis.

[1]Research Engr., Dept. of Const. Engrg., National Resarch Inst. of Agri. Enger.,Kannondai 2-1-1,Tsukuba, Ibaraki, 305 Japan.
[2]Head, Agri. Land Improvement Division, Tottori Pref., Higashi-machi 1-220, Tottori,680, Japan
[3]Research Engr., Dept. of Const. Engrg., National Resarch Inst. of Agri. Enger.,Kannondai 2-1-1,Tsukuba, Ibaraki, 305 Japan.

The validities of this procedure has been established though a comprehensive evaluation procedures. The calculated results are compared with a large scale field test data. Recent surveys has been carried out in a soft clay field. One of these has been used as a model for present numerical analysis.

The specific objectives of this study are ;1) to evaluate the capability of finite element program to represent the soil-pipe interaction ; and 2) to interpret the actual performance of buried flexible pipeline in field investigation.

Computer Program

The analytical tool used to represent the soil-pipe system is a finite element program based on nonlinear elasto-plastic constitutive model. A detailed description of the capabilities of this program is given in Ref.2 & 5. This program allows nonlinear analysis developed by Nelson [3] inside the elastic region and based on the critical state mechanics involving the cap type model in the plastic region. Fig.1 shows the elliptical cap as a yield surface and the Drucker-Prager's failure line in the two dimensional stress axis. In this model the ellipse is expressed by stress invariants I_1 and J_2 as follows.

$$(I_1 - P_0)^2 / a^2 + J_2 / b^2 = 1 \ , \quad a/b = R \tag{1}$$

The critical state line which has a constant slope can be expressed as the following equation.

$$\alpha_{cs} I_1 + J_2^{1/2} = 0 \tag{2}$$

The expansion and contraction of the yield surface is related to the plastic volumetric strain. The size of the yield surface is fixed by specifying that the intersection of the surface Pc with I_1 axis corresponds to the isotropic normal consolidation line. Then Pc is given by the slope of the virgin isotropic consolidation and elastic unloading curves λ , κ .

$$P_c = -\exp((B - e_p)/A),$$
$$A = (\lambda - \kappa)/(1+e)/2.3 \ , \ B = A \ln(-3 p_0) \tag{3}$$

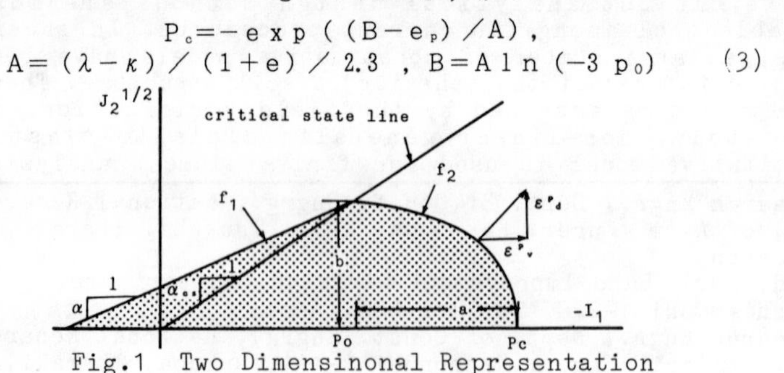

Fig.1 Two Dimensinonal Representation of Generalized Elasto-plastic Model.

The failure surface is defined as the form of the Drucker-Prager's yield criterion.

$$f_1 = \alpha I_1 + J_2^{1/2} - K = 0 \qquad (4)$$

Constants α and K depends on the plastic volumetric strain.

$$K = -(\alpha_{cs} - \alpha) p_0 \qquad (5)$$

And it is assumed that non-associated flow rule applies to the Drucker-Prager's yield surface.
The soil parameters of this model are determined from conventional triaxial compression test and isotropic consolidation test.

Field Observation and Analysis : FIELD 1

Description of Test Site and pipe

The soils in the test field are mostly sandy silts with some areas of clay. The trench geometry is shown in Fig.2. The excavation was carried out in the natural ground using steel sheetpiles for protection from failure of trench wall. Medium sand was used for backfilling around the pipe and compacted in layers of no more than 30 cm thickness.
The pipe was Fiber Reinforced Plastic Mortal(FRPM) pipe of 1350 mm diameter and 6 m long. The cover height being 2.1 m. Test field consist of 70 pieces of pipe. In order to make a detailed investigation in one section along a pipeline 20 two-components load cells were mounted to the outer surface of the pipe. Deflection and settlement of 70 pieces of pipe , total length of which was 420 m, were investigated in detail. Inner horizontal and vertical diameter measurements were made at mid location of pipe. Settlement of pipe was measured at bottom of the joint section of each pipe by level meter.

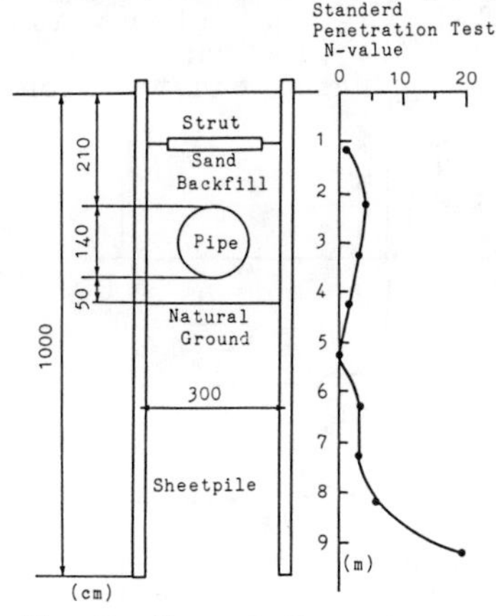

Fig.2 Trench Geometry

Table 1. Soil Property

Backfill Soil

Gs	Uc	Max.grain size (mm)	e_{min}	e
2.69	7.1	9.5	0.707	0.799

Pr(%)	C (kgf/cm^2)	ϕ(°)	Soil type(USCF)
95	0.11	38.0	SW

Natural Soil

Gs	Uc	Max.grain size(mm)	e
2.68	22.5	0.84	0.989

C (kgf/cm^2)	ϕ (°)	Soil type (USCF)
0.22	18	SM

The finite element grid used in the analysis of the field test is shown in Fig.3. The soil-pipe system contains 302 isoparametric quadrilateral elements and 981 nodes. Only one half of the system was represented. The culvert was represented by 12 solid elements. Interface elements were not used. The analysis contains the following step so as to simulate the incremental construction. 1) Excavation of the foundation between sheet piles, 2) installation of the pipe and backfilling, 3) extraction of the sheetpiles. To simulate the sheetpile extraction is a very difficult procedure due to a large deflection of the surrounding soil during the extraction. In this study, sheetpile extraction is simulated as follows; 1) excavate the bottom element of sheetpile, 2) buried the excavated vacant space with natural soil, 3) excavate the next upper sheetpile element, and repeat these steps to the top element of the sheetpile. The width of the excavated space is assumed to be the same as the total thickness of pipe and contacted soil which is 15 cm.

The basic properties of the natural soil and the structural backfill are listed in Table 1. And the soil parameters for the analysis were determined by performing the standard triaxial compression tests and isotropic consolidation test.

Deflection of Pipe

The deflection of pipe depends on the earth pressure on the pipe and soil-pipe stiffness factor. And the soil property around the pipe is much affected by some construction operations. Especially sheetpile extraction much affects the soil movement and the deformation of pipe. Fig.4a and 4b shows the behavior of pipe measured during backfilling and sheetpile extraction. Vertical deflection of the pipe was measured continuously by displacement meter during construction.

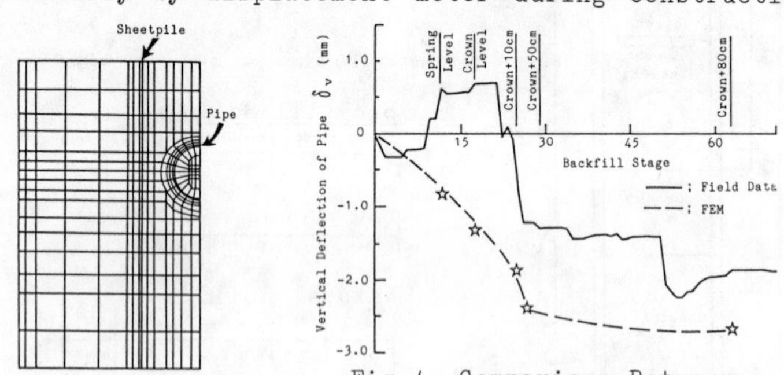

Fig.3 Finite Element Grid Used in Analysis

Fig.4 Comparison Between Measured and Predicted Vertical Deflection

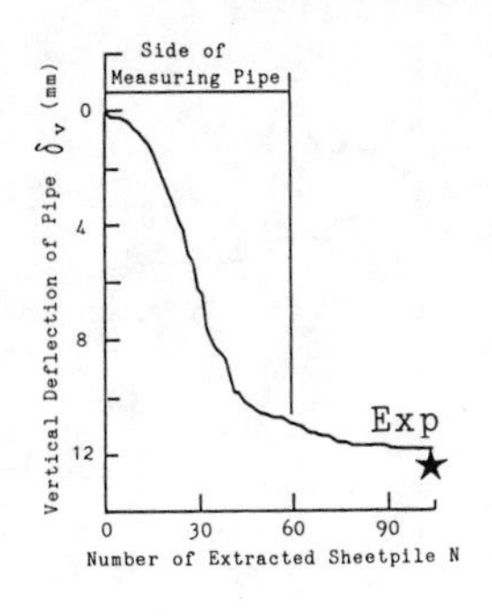

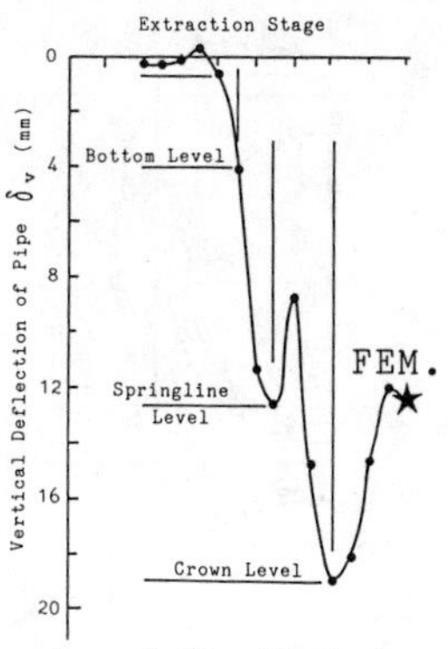

Fig.4 Comparison Between Measured and Predicted Vertical Deflection
a) During Backfilling
b) During Sheetpile Extraction (Measured)
c) During Sheetpile Extraction (FEM)

Vertical diameter increased as the height of backfill was increased to the crown level. And during the filling over the top of pipe, vertical diameter decreased rapidly. After completion of backfilling the pipe deflected 0.2 % of diameter in vertical direction. Although the calculated deflection of the pipe at the fill height of 80 cm, which is before cut of struts, agree approximately with the field measurements, but it can not fully simulate this trend of the deflection of the pipe. Major discrepancies were found in the behavior during the early backfilling stage until the fill height reached the crown level. These discrepancies are believed to be caused by compaction effects. The pipe can easily be displaced laterally inward a significant magnitude by cutting off the struts. This trend is not demonstrated in figure.
Fig.4b shows that, during the extraction of 60 pieces sheetpiles which is laid on both side of pipe, the computed deflections agree quite well with the field measurements. Lateral outward deflection of the pipe increased drastically during the excavation of the sheetpile element over the bottom level. It indicates that the lateral soil resistance of backfill decreased, and vertical load on pipe increased due to elimination of the friction along the backfill-sheetpile interface.

Earth Pressure Distribution on Pipe

The loads measured during the backfilling and sheetpile extraction are shown in Fig.5a and 5b. All the load cells had frictional pressure plate. Arrangement of cells were made symmetric to the vertical center line of

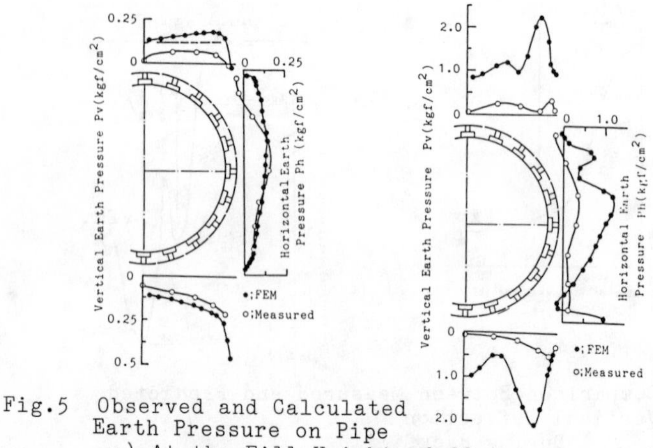

Fig.5 Observed and Calculated Earth Pressure on Pipe
a) At the Fill Height of 80cm
b) After Sheetpile Extraction

the pipe and the plates covered all circumferential area of pipe (see Fig.5a). This type of load cell enables normal and tangential load on the pipe surface to be measured independently. The normal and tangential pressures measured in the field test were projected on the vertical and horizontal pressure distributions in this figure. The distributions were demonstrated by the average values of the earth pressure on both left and right side of the pipe.

At the end of fill height of 80 cm, it appeared that the horizontal earth pressure shows mountain shaped distribution with the peak value at the springline of the pipe. Calculated results also showed this trend and agree quite well with the field measurements. Vertical earth pressure on bottom of pipe were concave distribution and this trend was obtained in both results.
The vertical earth pressure above the crown showed uniform distribution over the width of pipe.

At the end of the sheetpile extraction, FEM solution showed too large earth pressure on pipe compared to the measured results. Major discrepancies between calculated and measured results may be caused by one of the following reasons: 1) only permit a rigid connection at the pipe and backfill materials, 2) lack of proper representation of the fill compaction, 3) incorrect model for the sheetpile extraction. In this study soil-pipe system is treated as the continuous body, so if a large flow of backfill material occurs during the sheetpile extraction, the deflection of pipe and earth pressure distribution on pipe can not be predicted by FEM analysis.

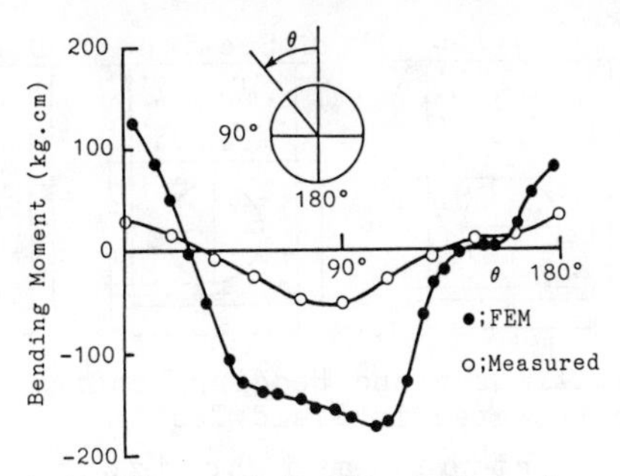

Fig.6 Distribution of Bending Moment along a Circumferential Section : After Sheetpile Extraction

Strain Distribution of Pipe

Fig.6 shows the typical example of the bending moment distribution along the circumference of the pipe after the extraction of sheetpile. The bending moment which generates compressive strain at the outer surface of the pipe is taken positive in this figure. The maximum bending moment occurred at the springline in both results. The calculated distribution of bending moments was not similar pattern to, and larger in magnitude than the field measurements.

Field Investigation ; Field 2

Field was 1800 m section with buried pipe of 2800 mm diameter with 3.1 m of backfill. This pipeline consist of about 400 pieces of FRPM pipes. The actual field conditions are quite uniform in horizontal direction. The pipe was installed in 6.7 meter deep ditch which was excavated with sheet piles. This buried test was conducted accompanying with an actual construction work. One part of the line was installed in a trench with four different conditions of bedding and backfill, in order to examine the effects of various beddings on the soil-pipe system. Each section was prepared for the length of 32 m which consist of 8 pieces of pipe. Fig.7 shows the trench geometry and bedding conditions. The bedding and backfilling conditions are as follows,

1) A section;backfill material was gravel (C-40) and was compacted with pneumatic tampers
2) B section;backfill material was gravel (C-40) and was compacted with vibratory-plate tampers
3) C section;backfill material was gravel (C-20) and was compacted with pneumatic tampers
4) D section;backfill material to depth of 0.5 times the outside diameter of pipe was soil cement and was compacted with pneumatic tampers.

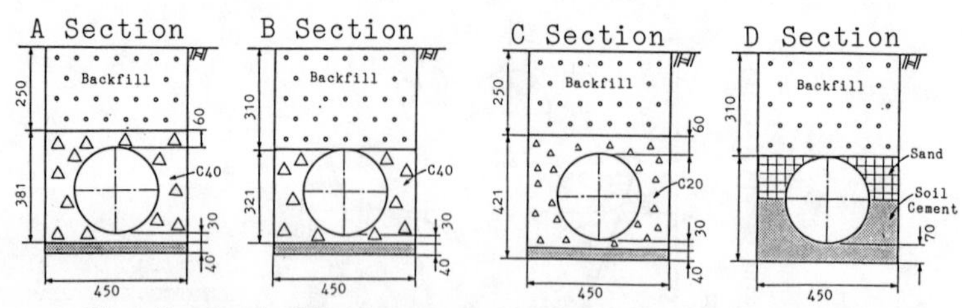

Fig.7 Installation and Bedding Configurations Incorporated in Field Test

Internal vertical and horizontal diameter measurements were made at center location of pipe in each section. The deflections along a pipeline were measured at selected construction steps and to 1.5 years after construction, to make clear the range in deflection of buried flexible pipe.

DEFORMATION OF PIPE

Fig.8 shows the change in average deflection of pipe during the construction. In four section, internal vertical diameter decreased just after buried to ground level. The deflection of flexible pipe increased drastically during the sheet pile extraction which were laid on both sides of the pipe. These change in deflection of four sections was approximately the same, 1.0-1.3% in diameter. Vertical deflections were less than 2.5 % after 1.5 years of construction. Deflection of D section which used soil cement backfill material was about 1.4 % which is 0.6 times of C section with crushed rock material C-20. Deflection range of three section with crushed rock backfill material range between 2.0 and 2.4 %. The deflection much depends on the soil-pipe

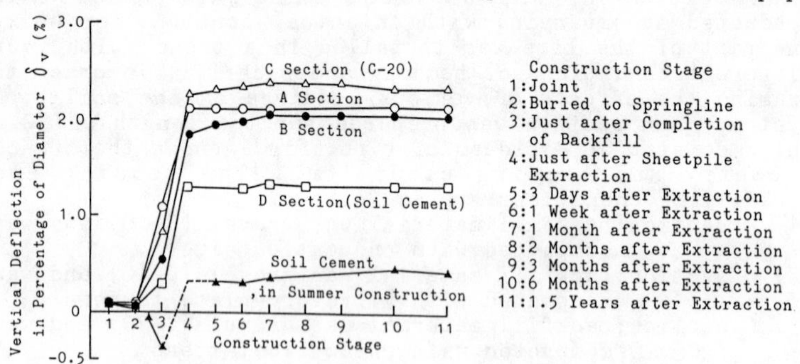

Fig.8 Observed Vertical Deflection during Construction

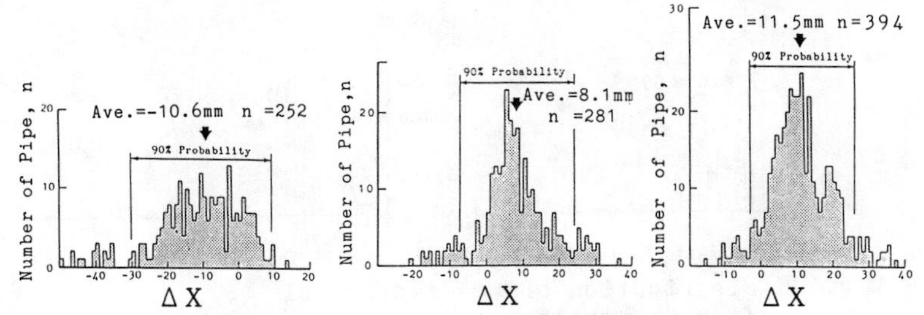

Horizontal Deflection ΔX (mm)

Fig.9 Distribution of Horizontal Deflection along Pipe line : (Soil Cement Backfill)
a) Just after Completion of Backfill ;
b) 1 Week after Sheetpile Extraction ;
c) 1.5 Years after Sheetpile Extraction ;

stiffness factor. In case of the soil cement backfill section, the deflection was much restrained due to high lateral resistance of soil cement. Although the deflection was much less among the four sections, high strength backfill material is not available for the flexible pipe due to the unexpected stress concentration on the pipe.

The deflection along a pipeline of 1800 m with soil cement backfill was measured at center of each pipe. The initial deflection was measured just after connecting to the next pipe without backfilling. Fig.9 shows the deflection range in some construction steps. Horizontal deflection of pipe vary considerably just after completion of backfilling. And as general behavior, the crown was displaced upward due to compaction of the backfill layers. Then horizontal diameter increased drastically and reached 8.14 mm in average in 1 week after sheetpile extraction. And it increased gradually to 11.45 mm in 1.5 years after extraction. This trend of deflection increasing gradually depend on soil type, degree of compaction of backfill, and the pipe stiffness. Fig.10 shows the variation in the ratio of deflection lag factor Dl. This factor converts the deflection of 1 week after sheetpile extraction to the deflection of the pipe after 1.5 years. Average value of Dl was 1.41 and over 90 % of the data was within 2.0. This trend for the deflection lag factor was similar to a value of 1.5 as recommended by Spangler. The deflection lag factor was 1.23 for 1 month after sheetpile extraction. Deflection lag factor depends on the time when the initial deflection was measured and of course on the compressibility of natural ground. So this time deflection lag factor Dl must basically be treated as the

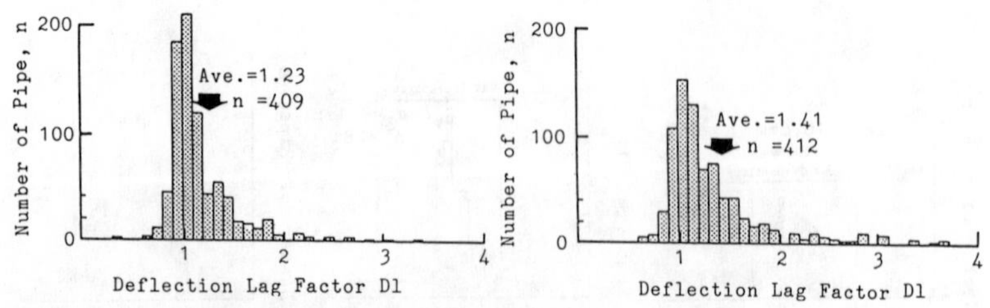

Fig.10 Distribution of Lag Fuctor Dl after Sheetpile Extraction : (Soil Cement Backfill)
a) 1 Month after Sheetpile Extraction ;
b) 1.5 Years after Sheetpile Extraction ;

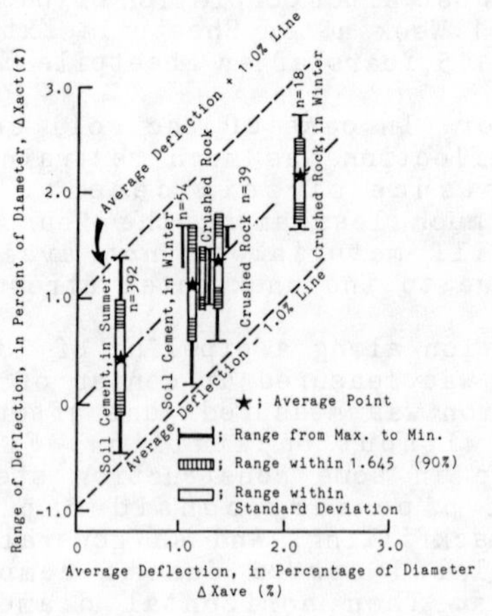

Fig.11 Range of Measured Horizontal Deflections along Pipelines

empirical factor of range of 1.0-2.0 in these conditions.

The deflections along a pipeline vary considerably due to the construction work like as the sheetpile extraction and normal soil variations and inherent differences in compacting backfill material along a pipeline. The range of measured deflections along a section of pipeline are plotted against average deflections for each section in Fig.11. About 100 % of measured deflections were within $\pm$1.0 % deflection of the average value. And over 90% of the data was within $\pm$ 0.5% deflection of the average value.

CONCLUSIONS

This paper presents the results of the behavior of buried flexible pipe in field tests and evaluates the capability of a finite element analysis by comparing with the measured behavior of a buried pipe. The principal conclusions obtained from this study are as follows;

1) A large scale flexible pipe in ditch must be treated as soil-pipe interaction problem, which can be modelled using the finite element method involving the built-up analysis.
2) In the analysis of the buried flexible pipe, the calculated final deflection of pipe agreed well with the measured behavior in the field at the end of the backfilling. However, it can not simulate the trend that vertical diameter increase during the early stage backfilling operation until fill height reached the crown level.
3) During the sheetpile extraction, the deflection of pipe is drastically increased. The predicted deflection of pipe agreed well with the field measurements.
4) The earth pressure distributions calculated by FEM analysis agreed only approximately with the measured results at the completion of the backfilling. But the distributions after the extraction of sheetpile showed considerable discrepancy with the observations.
5) In the field investigation, the deflections along the pipeline vary considerably due to the construction operations, as filling up, cut of struts, and sheetpile extraction. But on 1.5 years after the end of construction, 100 % of measured deflections of the pipelines which were installed in the crushed rock and soil cement backfill materials were within ±1.0 % of the average value.

References

[1] Chang,C.S.,et al., "Computer analysis of Newton creek culvert", Journal of the Geotechnical Engineering Division,ASCE,GT5,May,1980,pp.531-556.
[2] Mohri,Y. "Load study and numerical analysis of buried pipeline", Symp. on underground excavation in soils and rocks,Bangkok,1989.11.
[3] Nelson, I. "Investigation of ground shock effects in nonlinear hysteretic media", report-2 modeling the behavior of real soil,U.S.Army Engineer Waterways Experiment Station. 7,pp.399-417.
[4] Selig,E.T.,et al., "Measured performance of Newton creek culvert", Journal of the Geotechnical Engineering Division, ASCE,GT9,September,1979,pp.1067-1087.
[5] Tanaka,T. "Generalized elasto-plastic model of cohesive soils including strain softening and finite element analysis", Bull.Nati.Res.Inst.Agrec.Engng. Japan,1979,Vol.18,pp.101-122. (In Japanese)
[6] Tohda,J. "Earth pressure on underground concrete pipe in a field test", 1985 ASCE Specialty Conf. on Advances in Underground Pipeline Eng.

Design of Drainage Pipes Below Landfills

Hoch A.,[1], Zanzinger H., [2], Gartung E., M. ASCE [3]

Abstract

Drainage pipes are an essential component of landfills. Even though their embedment conditions differ substantially from those of pipes for municipal utilities near the ground surface, it has been common practice in Germany to apply the ATV - procedures to the design of drainage pipes below some very high landfills. Since a number of structural pipe failures resulted from this design practice, it was considered mandatory to review the ATV - rules, especially their soil mechanics background and their input recommendations for analyses.

Introduction

Landfills for the disposal of solid municipal and industrial or hazardous wastes are complex geotechnical structures. They have to be designed in such a way that a high degree of safety against the emissions of toxic substances can be achieved. In Germany, the multi-barrier concept has been adopted for this purpose. (Stief, 1986). In general, the following barriers can be considered:

- The solid waste material should be in a condition which reduces the solubility and mobility of toxic components.

---

[1] LGA - Stuctural Engineering Dept.

[2,3] LGA - Institute of Geotechnical Engineering, Nuremberg, FR Germany

-Operations of the landfilling process should be executed such, that all emissions are reduced to the unavoidable minimum.

-Final and intermediate covers of low permeability should reduce the access of precipitation to the disposed waste material.

-The leachate has to be collected by means of a reliable leachte collection system, (and it has to be submitted to adequate treatment).

-The landfill has to be isolated from the natural ground and groundwater by an impervious soil layer which in most cases has to be supplemented by a flexible membrane liner.

-The geological conditions of the site should prevent the migration of pollutants into the ground water.

-Supervision and monitoring during the construction-, landfilling- and post - operational phases should secure the safety of the landfill.

In this sense the reliability of leachate collection systems is a very important barrier with respect to the environmental safety and acceptability of landfills. So great attention has to be payed to all members of the leachate collection systems, drainage blankets, collection pipes, conduits, risers, maintenance - and control shafts. This paper summarizes the present design practice of drainage pipes below landfills in Germany.

General Requirements for Leachate Collection Systems

Gravel blankets and perforated pipes have to facilitate rapid drainage near the bases of landfills in order to avoid hydraulic pressure gradients in the impervious soil liners. Failures of the pipes by clogging or rupture can have serious consequences, especially if they occur near the points where pipes penetrate the impervious seals of the landfill. It also has to be kept in mind, that any remedial work for the replacement of damaged drainage pipes at the bottom of a closed or an operating landfill is very difficult, time consuming and costly. Special technologies for relining or trenchless replacement of drainage pipes are still in an experimental stage. The question how to evaluate the structural stability of a pipe installed by pipecracking or burstlining methods below a high cover of landfill has not yet been answered. First research efforts into these problems are merely at the beginning in Germany at the present time.

So it has to be emphasized, that leachate collection systems, and especially all details of the drainage pipes must be designed and executed with great care in order to achieve reliably functioning leachate collection systems.

General requirements that have to be met by all materials used for landfill drainage pipes are:

- -High resistance against chemical and biological attack,
- -Sufficient thermal rigidity, since the temperatures of leachates may well reach 50 ° C,
- -Adequate long-term tensile and flexural strength under the conditions of an adverse chemical milieu for periods up to 100 years or more.

Past Experience

Until 1986 various pipe cross sections and various materials were used for leachate collection pipes in landfills, e.g. Porosit consisting of porous concrete, asbestos-cement, clay pipes with deformation mats above, horse-shoe shaped PVC-pipes and corrugated thinwalled plastic pipes. Many of these experienced serious damage under high loads and had to be replaced. Concrete pipes, asbestos-cement pipes and thinwalled plastic pipes (wall thickness less than 10 millimeters) are considered inadequate for leachate collection systems because their resistance to corrosion is insufficient or because their low bearing capacity would permit only a relatively shallow cover of waste material. Experience with failures of clay pipes indicates, that the deformation mats made of PE-foams which are placed above the crown of the pipes do not serve their purpose. Neither are they sufficiently resistant against chemical attack (probably volatile organics) nor do they exhibit the amount of yield under loading which is required for the development of arching in the soil above the rigid clay pipes. The mechanical properties of the deformation mats are not well known, they are not sufficiently documented and their application resulted in a number of clay-pipe failures.

Up to now, leachate collection pipes made of solid PEHD of wall thicknesses of 20 to 35 millimeters, have performed most reliably. Independent of hydraulic requirements, the minimum internal diameter should be 250 millimeters to facilitate easy television camera ispection, cleaning an maintenance operations.

Present Design Practice

In order to facilitate a systematic structural design of drainage pipes below landfills, the embedment conditions had to be established. So the details shown in Fig. 1 were suggested by a committee and published in a draft of DIN 19 667 (Hoch 1989). The angle of the sand craddle at the base of the pipe has to be 120° to facilitate free entry of the leachate, although a conservative assumption for the angle of 90° is considered in the structural analysis. It is recommended to place a wedge of sand of the gradation 0/4 mm, below the pipe and cover the sand wedge by a nonwoven geotextile in order to separate it from the gravel drainage layer above. The flexible membrane liner is placed below the sand wedge. Required properties of leachate collection pipes for landfills and testing methods for their evaluation will be compiled in DIN 4266, part 1 (1st draft 1989) for pipes made of PVC-U, PEHD and PP, and in part 2 for clay pipes without plastic deformation mats. The items to be treated in DIN 4266 are as follows:

- Range of application
- Mechanical material properties
- Reduction coefficients for long term-, chemical-, and temperature effects
- Geometry of pipes including perforation for the entry of leachate (which amounts to > 100 cm^2 openings per meter of pipe)
- Quality control tests to be done by the manufacturer and by the quality assurance institution.

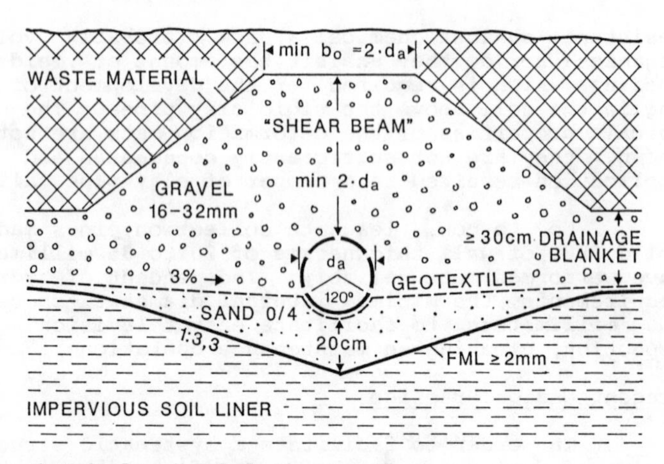

Fig. 1 Embedment of drainage pipes below Landfills

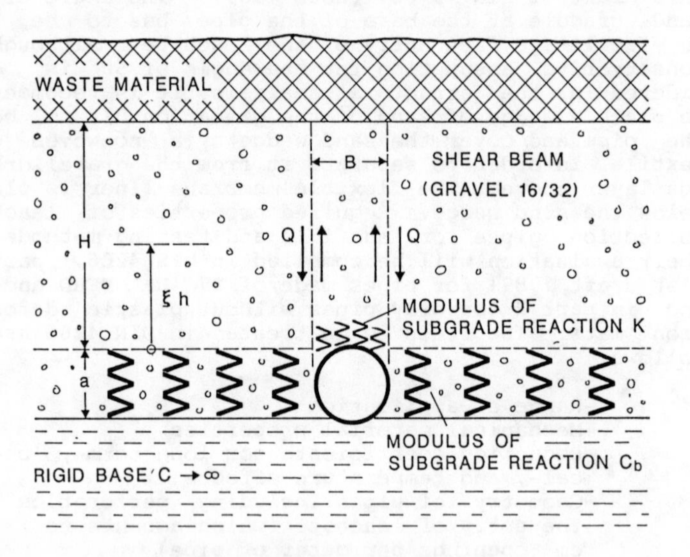

Fig. 2 ATV - Pipe analysis model

In addition to the stability calculations as usually carried out for utility-pipelines under relatively shallow cover, the design of pipes under landfills requires the determination of stresses under long term conditions and their comparison with permissible values, the determination of safety against limiting strains (tensile and compressive) and checking of the degree of safety against buckling failure.

In these calculations it has to be taken into account that the upper sections of drainage pipes are perforated over 240 ° and contain at least 100 cm² of holes or slits per meter.

For the nonperforated pipe under long term loading conditions the pipe stiffness S_R is expressed by

$$S_R = E_R \cdot I/r_m^3$$

where E_R = long term "(creep)" modulus of the pipe material, usually for 50 years

$I = 1/12 \cdot s^3$ with s = wall thickness

r_m = radius of the pipe (middle of wall)

The effect of the perforations is taken into account by a coefficient k_s = 0,7 to 0,9 in the computation of the pipe stiffness $S_{R,s}$ of the perforated pipe

$$S_{R,s} = k_s \cdot S_R$$

The reduction coefficient for the chemical environment is typically A_2 = 0,9 and the reduction coefficient for an elevated temperature of 40 ° C is A_3 = 0,75 for PEHD material. Considering all these, the long term deformation modulus E_{RL} becomes

$$E_{RL} = E \cdot A_2 \cdot A_3$$

for PEHD with an initial modulus E = 150 N/mm²

$$E_{RL} = 150 \cdot 0{,}9 \cdot 0{,}75 = 100 \text{ N/mm}^2$$

for example.

For structural analyses, either one of the following methods is employed:

-Finite element method, which studies the pipe-soil interaction by a numerical continuum model

-Elastic subgrade reaction model which excludes tensile pipe-soil interaction forces

-Computational model of ATV, A 127

The finite element method and the analysis according to the theory of elastic subgrade reaction are internationally well known. However, they are hardly applied in common design practice of pipes in Germany, where the ATV design procedure is most widely used. So this model will be treated in more detail in the present paper.

ATV-Design Model

In Germany the municipal authorities, representatives of the construction industry and public corporations who are dealing with problems of sewage collection, treatment, management and associated construction problems are represented by an association called "Abwassertechnische Vereinigung", abbreviated "ATV". Among the recommendations, standards and regulatory publications of this association, the paper ATV - A 127, (2 nd edition of Dec. 1988), contains the recommended procedure for the structural design analysis of embedded pipelines. The principal assumption of the analytical model consists of the load bearing and distributing action of a ficticious "shear beam" above the pipe supported by elastic springs as schematicly shown on Fig. 2. This model requires that the soil layers above the pipe have sufficient strength to perform as a "shear beam" a prerequisit which cannot be assigned to most of the waste materials of sanitary landfills. So in these cases, only the thickness of the gravel drainage blanket above the pipe can be taken into account for the load distrubution. In order to be structurally effective, the thickness of the drainage layer has to be greater than twice the external diameter of the pipe (Fig. 1.). The material should consist of clean, washed gravel of rounded particles with only a low content of soluble calciumcarbonate. It has to be chemically and physically stable under the adverse landfill conditions.

The weight of the waste material is regarded as a spread loading at the top of the "shear beam". It is generally greater than $q_v = 0,1$ N/mm², the limiting value given in ATV - A 127 for which the deformation moduli E_B of the soil materials of ATV - A 127 are applicable. If deformation moduli for soils are used directly from the tables of ATV - A 127, too large deformations are computed because in reality the soil materials surrounding the pipe exhibit strain hardening behavior. So, higher deformation moduli should be used for elevated stress levels. Fig. 3 proposes a relationship between vertical stresses in the soil and deformation moduli, which could be used for pipe analyses. (Hoch et al. 1989)

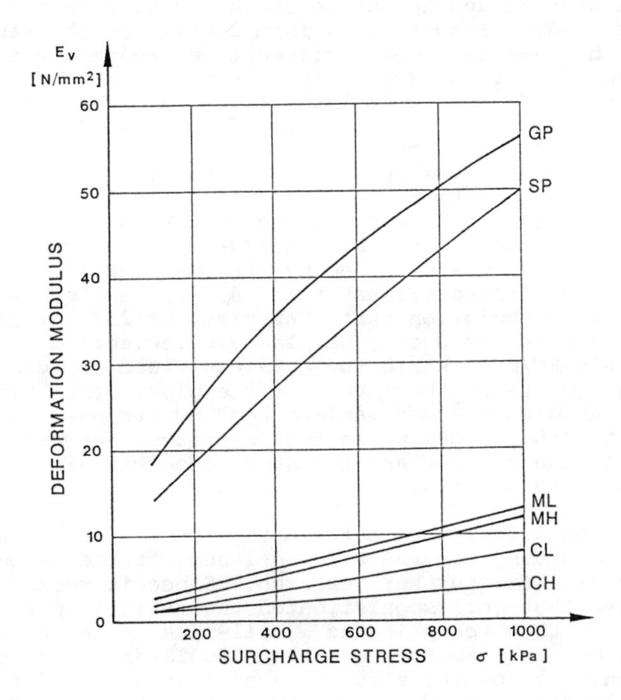

Fig. 3 Deformation modulus of soils for pipe analyses

Since the analytical model of ATV - A 127 was developed mainly for pipelines of municipal utilities under shallow cover and for pipes embedded according to the specifications of DIN 4033 most of its boundary conditions differ from the situation in landfills. This is, why the ATV-rules should not be used without adequate modifications for the design of drainage pipes in landfills. In order to be able to judge the applicability of the ATV-model to the design of pipes below landfills, the results of computations using the ATV-model were compared with finite element studies. Surcharge loads of 360 kN/m², 720 kN/m² and 1000 kN/m² were applied, and the soil parameters were varied systematically. Furthermore, the influences of different pipe diameters and the angle of the sand-base-craddle were studied. Without going into details of the study, it can be said that good agreement was reached between ATV and FE - results in case of rigid pipes such as clay pipes and even for relatively rigid pipes such as PVC-pipes. However, the analyses of flexible PEHD-pipes yielded much larger bending stresses from ATV-computations than from FE-calculations. On the other hand, the calculated deformations under a load of 1000 kN/m² were less than the permissible value of 6 % of the pipe diameter when calculated by the ATV-model, whereas this limiting value was exceeded by the deformations calculated by finite element studies. The latter would lead to the conclusion that PEHD-pipes of 250 millimeters diameter could only be placed below spread loads of < 550 kN/m². While the ATV-model lead to maximum bending moments at the invert of the pipe, the finite element studies yielded maximum bending stresses in the pipes at the points where the interface between the gravel drainage layer and the sand-wedge-craddle intersect the external pipe surface.

The differences between the results of calculations according to the ATV-model and finite element studies call for further research. Since it seems very unlikely that more sophisticated theoretical work can clarify the complicated soil-pipe interaction mechanisms, the authors are planning large scale model tests which allow the simulation of loading conditions below landfills upon well instrumented flexible PEHD-pipe sections.

Summary and Conclusions

Drainage pipes below landfills are exposed to high structural loads, elevated temperatures, corrosive chemicals, and they have to perform over long periods of time. The embedement conditions are characterized by the properties of impervious soil liners below and gravel drainage blankets above the pipes. These conditions are not well modelled by conventionally applied design methods. So some failures of pipes below landfills were caused by inadequate structural design procedures. In order to improve the present situation in Germany, specific standards are being prepared which will approximate actual landfill conditions in a better way. However, a number of unsolved questions in soil-pipe-interaction will remain. For a better understanding of these problems large scale model tests have to be carried out.

Appendix I. References

Abwassertechnische Vereinigung e.V.:
ATV - Arbeitsblatt A 127, Ausgabe Dez. 1988, 2. Ausgabe

DIN 4033, Nov. 1979, Entwässerungskanäle und -leitungen
Richtlinie für die Ausführung
Sewage channals and pipe lines; codes of practice for the construction

DIN 4266, Teil 1, Sickerrohre aus PVC-U, PE-HD und PP für Deponien, Anforderungen Prüfungen
7. Manuskript Entwurf Okt. 89
PVC-U and PE-HD and PP drainage pipes for dumps; requirements and testing

DIN 19667 (1st draft März 1989)
Dränung von Deponien, Techn. Regeln für Bemessung, Bauausführung und Betrieb

DVS-Richtlinie 2205, Teil 1, Stand Juni 1987
Berechnung von Behältern und Apparaten aus Thermoplasten, Kennwerte

Hoch, A., Zanzinger, H., Gartung, E. (1989)
Grundsatzuntersuchung "Rohrleitungen unter Abfalldeponien", Research Report for Bayerisches Landesamt für Umweltschutz, unpublished

Hoch, (1989). Die statische Berechnung von Schächten und Sickerrohren in Deponien, Veröffentlichungen des LGA-Grundbauinstituts, Heft 54

Schmiedel U: Standsicherheitsnachweise von Schächten und Rohren aus HDPE, Fachveranstaltung "Umweltschutz durch sichere Deponien" am 30.-31.01.1989, Essen

Stief K.: Das Multibarrierekonzept als Grundalge von Planung, Bau, Betrieb und Nachsorge von Deponien, Müll und Abfall 18 (1): 15 -20, E. Schmidt Verlag, Berlin 1986

FE ELASTIC ANALYSIS OF MEASURED EARTH PRESSURE ON BURIED RIGID PIPES IN CENTRIFUGED MODELS

J. Tohda[1], M. Mikasa[2], M. Hachiya[3] and S. Nakahashi[4]

ABSTRACT

FE elastic analyses of earth pressure on buried rigid pipes yielded results that agreed well with the earth pressure measured in 42 centrifuged models, confirming both the reliability of the model tests and the applicability of elastic theory to the earth pressure problem. The effect of soil's elastic moduli and three boundary conditions of the ground (at lateral and bottom boundaries and at the pipe surface) on the earth pressure were studied. The difference in the earth pressure among three types of pipe installations, ditch type with sheet-piling, ditch type without sheet-piling and embankment type, was found to be due to the difference in conditions at the lateral boundary of the ground, leading to a new design concept of the earth pressure in actual construction works.

INTRODUCTION

Earth pressure acting on buried pipes is a soil-pipe interaction problem that includes several complex factors such as geometry, material properties and boundary conditions. Since finite element method (FEM) can easily cope with these conditions, this problem has hitherto been analyzed very often by using it. However, the

1 Lecture, Civil Eng. Dept., Osaka City University, 3-3-138, Sugimoto, Sumiyoshi-ku, Osaka, 558, Japan.
2 Professor, Civil Eng. Dept., Setsunan University, 17-8, Ikedanaka-machi, Neyagawa, 572, Japan.
3 Engineer, Chuou Fukken Consultants Co. Ltd., 3-5-26, Higashimikuni, Yodogawa-ku, Osaka, 532, Japan.
4 Engineer, Nihon Koei Co. Ltd., 11-1, Komatsuhara-machi, Higashimatsuyama, 355, Japan.

analyzed results, not being sufficiently backed up by experiments in most cases, have seldom been applied to the actual design and construction of buried pipes.

The authors have investigated the earth pressure acting on buried rigid pipes during this decade by means of centrifuge model tests, and evaluated the effect of most major factors on it [Tohda et al. 1986]. This paper reports FE elastic analyses for 42 centrifuged models to investigate the applicability of the elastic theory to this problem, as well as the reliability of the centrifuge model tests. The calculated earth pressures showed good agreement with the measured ones, leading to a new explanation on earth pressure change due to different types of pipe installations. Since the methods of the tests and analysis have already been detailed by Tohda et al. [1985a and 1988], they will be briefed in this paper.

OUTLINE OF THE CENTRIFUGE MODEL TESTS

In the centrifuge tests, a rigid model pipe (Fig.1) having an outer diameter of 9 cm was buried in either dense or loose dry-sand ground (Table 1 and Table 2), and both normal and tangential components of earth pressure acting on the pipes were measured under a centrifugal acceleration of 30 g. The surface of the model pipe was finished smooth. Fig.2 shows 1/30 scaled models for the following three types of pipe installations: (a) ditch type with sheet-piling (Ditch-S), (b) ditch type without sheet-piling (Ditch-0), and (c) embankment type (Embk.). Ditch-S type is the most common pipe installation method in Japan.

Table 3 shows the conditions of five test series,

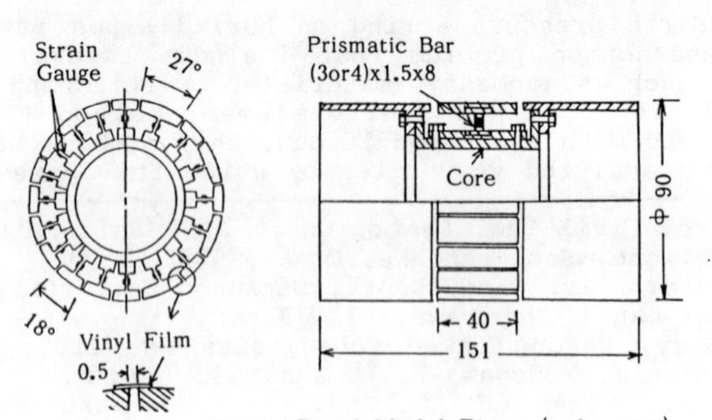

Fig.1 Rigid Model Pipe (unit: mm)

where effects of following five factors, together with those of density of sand ground, were investigated: A) type of pipe installation, B) roughness of pipe surface, C) cover height H, D) thickness of sand bedding H_b, and E) ditch width B_d. The narrowest ditch width of B_d=13 cm was selected as the standard value in Ditch-S and Ditch-0

Table 1 Properties of Silica Sand

G_s	Grain Size	Uc	ρ_{dmax}	ρ_{dmin}
2.65	0.24-1.4mm	1.75	1.58t/m³	1.32t/m³

Table 2 Density and Strength of Sand and Friction Angle ϕ_p against the Pipe Surface

Ground	ρ_d	c_d	ϕ_d	ϕ_p Smooth	ϕ_p Medium	ϕ_p Rough
Dense	1.55 t/m³	0	47°	17°	42°	45.5°
Loose	1.43	0	36°	16°	40°	44.5°

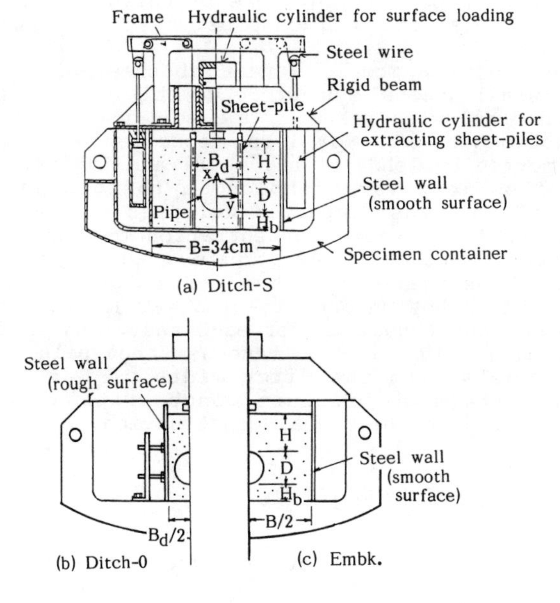

Fig.2 Models for Three Types of Pipe Installations

Table 3 Test Conditions

Series	Pipe Installation	Pipe Surface	H (cm)	H_b (cm)	B_d (cm)	Ground	Number of Tests
A	Ditch-S Ditch-0 Embankment	Smooth	9	4	13 —	Dense Loose	6
B	Ditch-S Embankment	Smooth Medium Rough	9	4	13 —	Dense Loose	12
C	Ditch-S Ditch-0 Embankment	Smooth	4.5 9 12	4	13 —	Dense Loose	18
D	Ditch-S Embankment	Smooth	9	1 2 4	13 —	Dense Loose	12
E	Ditch-S Ditch-0	Smooth	9	4	13 17 21	Dense Loose	12

models, because the ratio of $B_d/D=1.44$ (=13 cm/9 cm) is used in most cases of ordinary construction works in Japan. In test series B, a rough pipe suface and a medium rough pipe surface were prepared by pasting sand papers on the smooth surface pipe.

In Ditch-S model, a pair of sheet-piles were extracted simultaneously under 30 g. This method was adopted by the following reason. The soil used in the actual construction has more or less cohesion. In the field test reported by Tohda et al. [1981 and 1985b], extracted sheet-piles left vacant holes in the ground when the backfilled soil was compacted well; as a result, a significant earth pressure concentration on the pipe, particularly on its top, was induced, and concrete pipe failures took place. In the centrifuge model tests, however, the cohesionless dry-sand was used to reproduce the same ground condition for each test, and in this sand model ground only the simultaneous sheet-pile extraction in the models with the ditch width of B_d=13 cm could actualize the high pressure concentration similar to those observed in the actual construction.

FINITE ELEMENT REPRESENTATION

The test results were analyzed by using a FE program with an isoparametric element [Hinton et al. 1977] under plane strain condition. Fig.3 shows the FE mesh and material properties, where the x and y axes are taken as the vertical and horizontal directions, respectively. The materials (soil and pipe) were assumed as linear elastic bodies. The three boundary conditions are given as follows.

Lateral Boundary Conditions in the Ground

(a) for Embk. model (Fig.4-①):
"K_0-condition" at y=22.5 cm=2.5D.

(b) for Ditch-0 model (Fig.4-②):
"Rough ditch-wall condition" at $y=B_d/2$ with separation, failure and frictional slip on the rough ditch wall.

(c) for Ditch-S model at $y=B_d/2$ (Fig.4-③):
* before the sheet-pile extraction:
"Smooth sheet-pile wall condition".
* during the extraction:
"Vacant hole condition".
* after the extraction:
"Vacant hole condition" below, and "K_0-condition" over, the pipe mid-height.

The components u_i and F_i shown in Fig.4 denote respectively the nodal boundary displacement and force; the subscript i denotes the direction.

Interface Boundary Conditions between Pipe and Soil

(a) for the smooth surface pipe:
"Smooth interface condition" (full slippage).

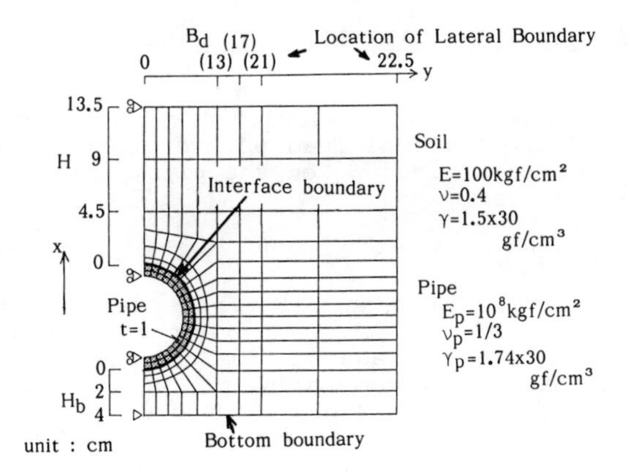

Fig.3 Finite Element Mesh and Material Properties

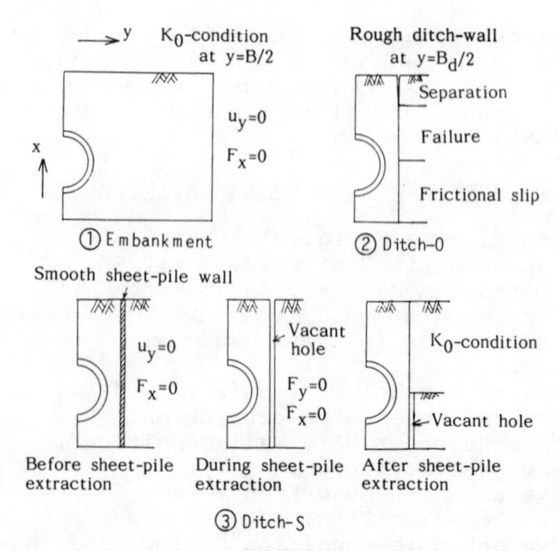

Fig.4 Lateral Boundary Condition in the Ground for Three Pipe Installations

(b) for the rough and medium rough surface pipe: "Fixed interface condition" (no slippage).

Bottom Boundary Conditions in the Ground

"Fixed bottom condition" was employed. "Smooth bottom condition" was also employed in test series D.

COMPARISON BETWEEN MEASURED AND CALCULATED EARTH PRESSURE

The measured and calculated results were compared in terms of the vertical load p_v, vertical reaction load p_r, and horizontal load p_h on the pipe, which are calculated as follows:

$$p_v \text{ and } p_r = \sigma + \tau \tan\theta, \quad p_h = \sigma - \tau \tan\theta$$

where the compressive normal earth pressure σ and the downward tangential earth pressure τ are taken as positive; the angle θ is measured from the pipe top. The distributions of these three loads, p_v, p_r and p_h, normalized by the overburden pressure γH, are shown in Figs.5-10. "Measured loads" in the figures are limited to those of the dense ground condition except for Fig.9, because the tests on the loose ground generated results similar to the results on the dense ground in most cases. Test series D shown in Fig.9 is one of a few cases in which appreciable effects of ground density was observed.

Earth Pressure Change due to Sheet-pile Extraction

Fig.5 shows the load distributions in the standard Ditch-S model for three test stages. The measured loads noted as "During sheet-pile extraction" are the data obtained when the normal earth pressure at the pipe top reached its maximum during the sheet-pile extraction. In the calculated load, tension load appears by not allowing the separation between the pipe and soil. However, another FE calculation allowing the separation generated results close to the calculation not allowing the separation in the region of compression load. The loads calculated under different lateral boundary conditions well explain the change of the loads measured in centrifuged models.

Effect of Pipe Installation Type: Test Series A

Fig.6 illustrates the load distributions for the three types of pipe installations, Ditch-S, Ditch-0 and Embk.. The measured loads in Ditch-S model are the data "during the sheet-pile extraction". The calculated loads are very similar to the measured ones, showing that the marked difference in the measured loads for the three

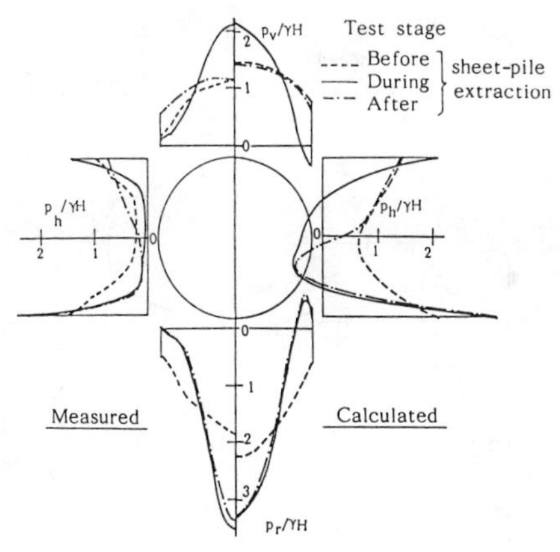

Fig.5 Change of Measured and Calculated Loads during Ditch-S Standard Test

types of pipe installations is exclusively due to the difference in the lateral boundary condition of the ground.

Effect of Roughness of Pipe Surface: Test Series B

Fig.7 illustrates the load distributions for different interface conditions between the pipe and soil

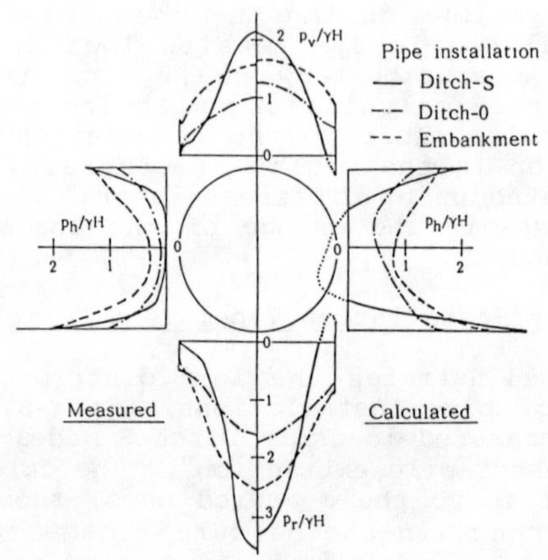

Fig.6 Measured and Calculated Loads for Three Types of Pipe Installations (Series A)

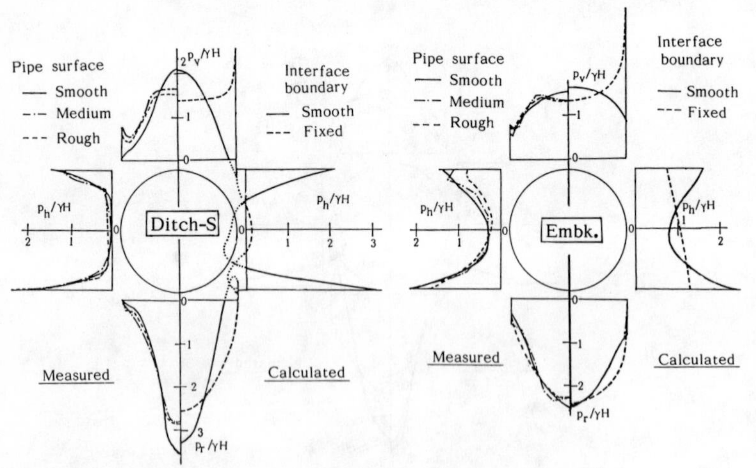

Fig.7 Measured and Calculated Loads for Different Interface Conditions (Series B)

in Ditch-S and Embk. models. The mountain-shaped vertical loads measured for the smooth surface pipe conform well to those calculated under "Smooth interface boundary condition". On the other hand, the trapezoidal vertical loads measured for the two rough surface pipes do not conform to those calculated under "Fixed interface boundary condition", which is explained by the sand slide along the side surface of the two rough surface pipes in the tests [Tohda et al. 1988].

Effect of Geometrical Parameters: Test Series C, D and E

Figs.8-10 show how the changes in the three geometrical parameters, cover height H, thickness of sand bedding H_b and ditch width B_d, affect the loads in a similar way both in the analysis and in the model tests, except for a few Ditch-S models: one with the least

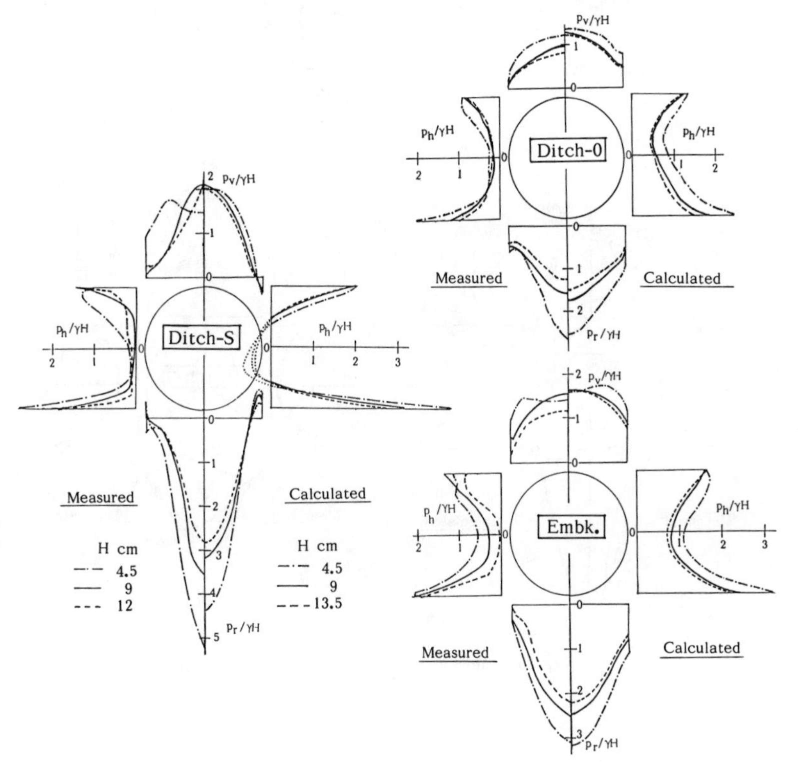

Fig.8 Measured and Calculated Loads for Different Cover Heights (Series C)

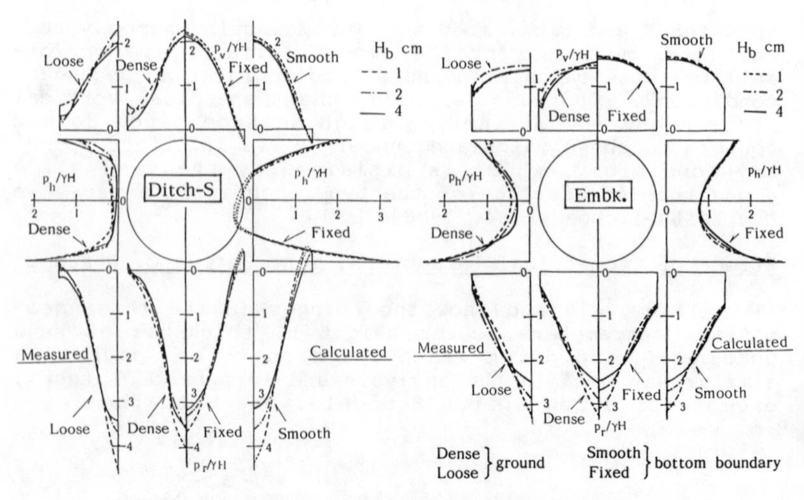

Fig.9 Measured and Calculated Loads for Different Thickness of Sand Bedding (Series D)

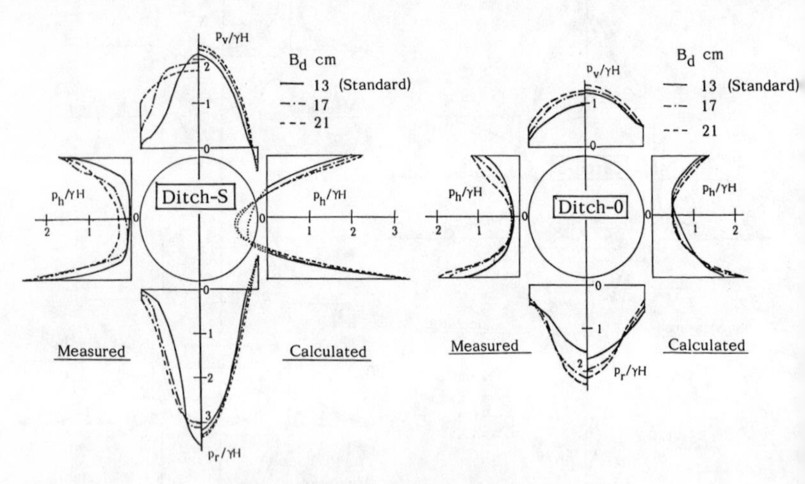

Fig.10 Measured and Calculated Loads for Different Ditch Width (Series E)

cover height of H=4.5 cm in test series C, and two with widest B_d values in test series E. The latter will be discussed in the following section.

RELIABILITY OF THE MEASUREMENT AND ANALYSIS

Ditch-0 and Embk. Models

According to Figs.6-10, the measured loads for both Ditch-0 and Embk. models conformed well to the calculated loads. Thus, the measurement and analysis back up each other sufficiently in these cases.

Ditch-S Model with Wider Ditch Widths

Fig.10 shows that in Ditch-S models with the ditch widths wider than B_d=13 cm the measured loads did not conform to the calculated ones. In these tests, the sand was observed to fill up promptly the spaces vacated by the sheet-pile extraction (cf. Fig.11); the stress condition at $y=B_d/2$ during the sheet-pile extraction, therefore, is naturally different from "Vacant hole condition" employed in the analysis, resulting in less earth pressure concentration on the pipe top and bottom.

In this case, the measured and calculated results are considered to correspond to different field conditions: the former to the field case not producing vacant holes with light-compacted backfill, while the latter to the case producing the vacant holes with well-compacted backfill.

Ditch-S Model with Narrowest Ditch Width

Figs.6-10 show that in the case of B_d=13 cm the measured and calculated loads during the sheet-pile ex-

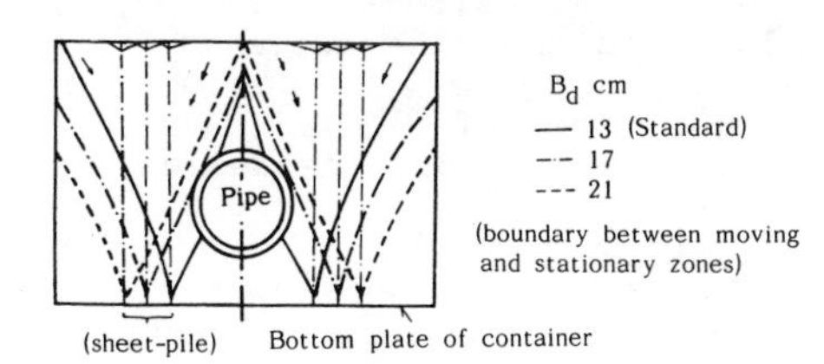

Fig.11 Movement of Sand during Sheet-pile Extraction (Series E, Ditch-S)

traction showed good agreement. As shown in Fig.11, the sand movement for this ditch width is prevented by the pipe till the lower ends of sheet-piles are elevated up to about the level of the pipe top, keeping the loose state at $y=B_d/2$ similar to "Vacant hole condition".

After the sheet-pile extraction, the sand at $y=B_d/2$ still maintain the loose state below the pipe mid-height, leaving the high earth pressure concentration on the pipe bottom just as in the calculation (cf. Fig.5).

Thus, the standard Ditch-S model test of B_d=13 cm luckily yielded the results similar to those of the field test that produced vacant holes after the sheet-pile extraction. These results, measured in the field and in the centrifuge models, are backed up sufficiently by the analysis, and provide a useful data for the buried pipe design.

APPLICABILITY OF ELASTIC THEORY FOR EARTH PRESSURE ON BURIED RIGID PIPES

Traditionally, the earth pressure on buried rigid pipes has been treated as an ultimate-equilibrium problem using soil strength, just as the earth pressure on retaining walls. It was made clear, however, by this study that: 1) the results of centrifuge model tests are well explained by the elastic analysis, and 2) soil density and, therefore, soil strength do not affect the measured loads appreciably. These facts lead us to an important conclusion that the earth pressure on buried rigid pipes, contrary to the traditional concept, is a problem to be treated by the elastic theory.

In the following, the effects of soil's elastic moduli (E and ν) will be discussed to support the above conclusion.

(a) Effect of soil's E value:

Soil's E value of 100 kgf/cm^2 was employed in this analysis. Further FE calculations with different soil's E values clarified that the earth pressure on the rigid pipes having $E_p=10^8$ kgf/cm^2 is not affected by soil's E value for a wide range of 5-500 kgf/cm^2. This must be a very useful information for pipe design.

(b) Effect of soil's ν value:

Fig.12 shows that the loads in both Ditch-0 and Embk. models are considerably dependent on ν, whereas the loads in Ditch-S model are not. In the former two models, different ν value yields different horizontal

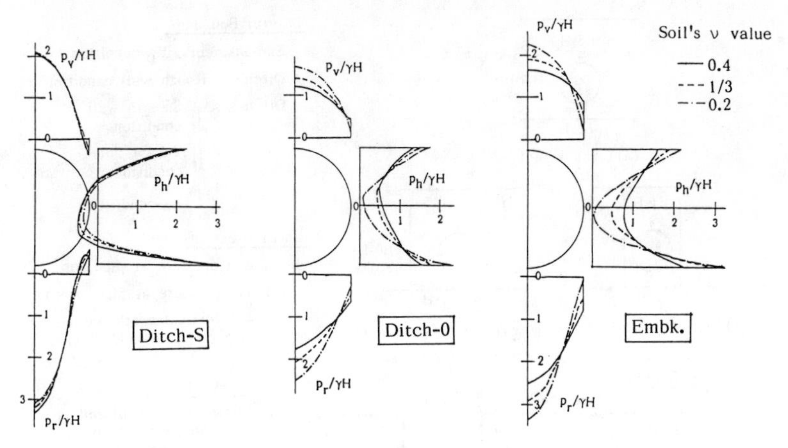

Fig.12 Calculated Loads for Different Poisson's Ratio of Soil

earth pressure on the lateral ground boundary, whereas in Ditch-S model the horizontal earth pressure on this boundary is always null. This difference in the stress condition at the lateral boundary is the cause of the different effect of ν values.

LATERAL BOUNDARY CONDITION IN ACTUAL CONSTRUCTION

In the conventional design methods, different load formulae are being used for different types of pipe installations. This study revealed, however, that the marked difference of earth pressure for different pipe installations is attributed exclusively to the difference in the lateral boundary conditions of the ground.

Fig.13 summarizes the pipe installation type, lateral boundary condition of the ground and the earth pressures for the actual construction works. The earth pressures shown in the figure are for the case of H=D. This figure is a summary of the achievements of ten years' experimental and analytical studies by the authors, and will provide a new view point and a useful criteria for considering the earth pressure problem on buried rigid pipes.

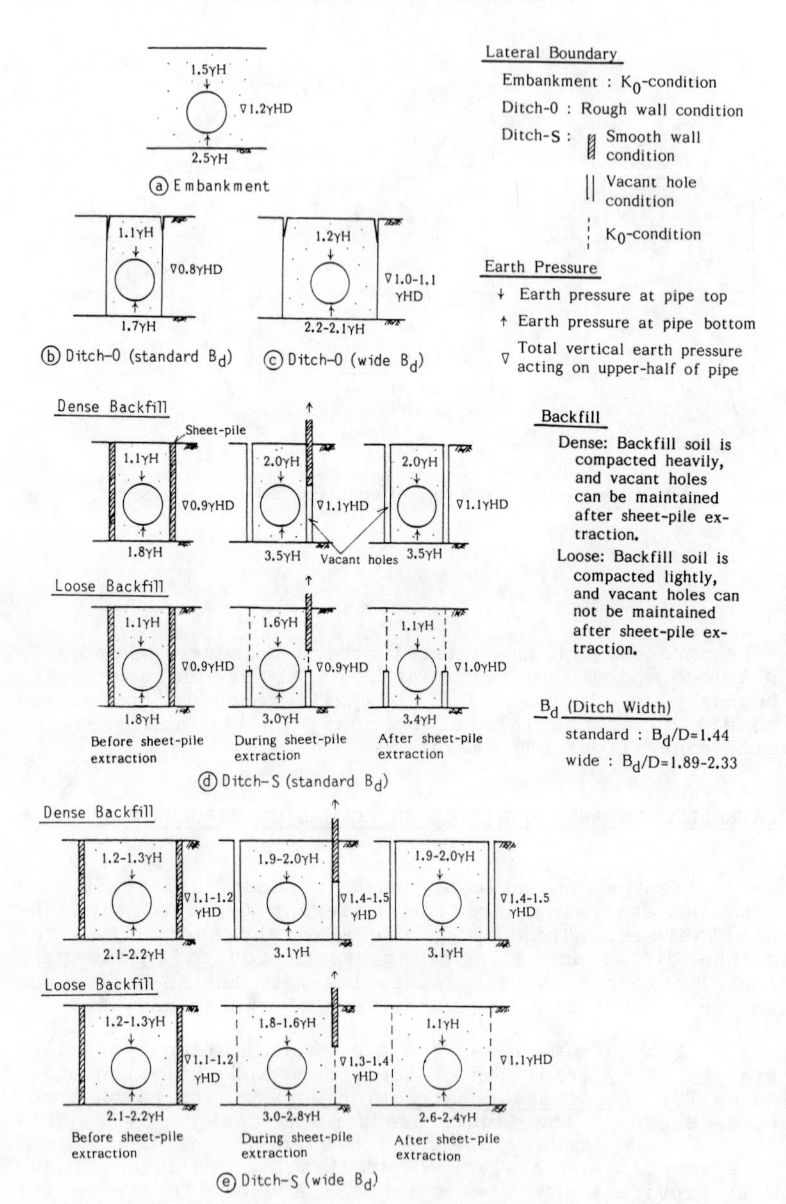

Fig.13 Lateral Boundary Conditions and Earth Pressure for Three Types of Pipe Installations in Actual Construction

CONCLUSIONS

The comparison between the measured earth pressures on buried rigid pipes in 42 centrifuged models and calculated ones by FE elastic analyses for these models leads to the following conclusions:

1) Measured earth pressures in the model tests were sufficiently supported by the elastic analysis.
2) The earth pressure is affected almost exclusively by the following three boundary conditions: the lateral and the bottom boundary conditions of the ground and the interface boundary condition between pipe and soil.
3) The marked difference in earth pressure among three types of pipe installations (ditch type with sheet-piling, ditch type without sheet-piling, and embankment type) was explained quantitatively by the difference in the lateral boundary condition of the ground.
4) The complicated load change by the sheet-pile extraction in Ditch-S models as well as the effect of soil's ν value on the load could also be explained by the concept of lateral boundary condition of the ground.
5) Soil's E value does not affect the earth pressure on buried rigid pipes appreciably, though soil's ν value does.
6) The earth pressure problem of buried rigid pipes has hitherto been considered as an ultimate-equilibrium problem. However, it was concluded that this problem should be treated by the elastic theory.
7) The earth pressure on buried rigid pipes is classified as shown in Fig.13, which will provide a new design concept of this problem.

REFERENCES

Hinton E. et al. 1977. Finite element programming. Academic Press Inc. Ltd.

Tohda J. et al. 1981. Soil Pressure on, and structural behavior of, concrete pipe bedded on concrete cradle. Proc. of JSCE. Vol.310.

Tohda J. et al. 1985a. Earth pressure on underground rigid pipe in a centrifuge. Proc. of ASCE Int. Conf. on Advances in Underground Pipeline Eng.

Tohda J. et al. 1985b. Earth pressure on underground concrete pipe in a field test. Proc. of ASCE Int. Conf. on Advances in Underground Pipeline Eng.

Tohda J. et al. 1986. A study of earth pressure on underground rigid pipes by centrifuged models. Proc. of JSCE. Vol.376/3-6.

Tohda J. et al. 1988. Earth pressure on underground rigid pipes: Centrifuge model tests and FEM analysis. Proc. of Int. Conf. on Geotechnical Centrifuge Modelling.

EVALUATION OF COMBINED LOAD TESTS OF PRESTRESSED CONCRETE CYLINDER PIPE

Mehdi S. Zarghamee*, Member, ASCE

Abstract

Combined load tests of prestressed concrete cylinder pipe, presented by Tremblay in the proceedings of this conference, are analyzed. Data gathered during the tests includes first visible cracking, vertical deflection, strains in concrete, mortar, steel cylinder, and prestressing wire at combined loads. In this paper the test results are analyzed and compared with the corresponding values calculated using an analytical model that accounts for the nonlinear stress-strain relationships of the constituent properties, including tensile softening and cracking of core concrete and coating mortar, and that also accounts for moment redistribution resulting from the changes in the stiffness of sections around the pipe under combined loads. Agreement between the observed and calculated combined loads at first visible core and coating cracks, vertical deflection and wire strain is good to excellent. However, there are some discrepancies between the observed and calculated strains in the concrete core, the mortar coating, and the steel cylinder. These are explained in terms of nonuniformity of the strain field in concrete and mortar after microcracking.

Introduction

Prestressed concrete cylinder pipe (PCCP) consists of a concrete cylinder, referred to as the core, which is compressed by a helical high-strength steel wire, wrapped under high tension around the core. A mortar coating is applied to the exterior of the prestressed pipe to protect the wires against corrosion. The concrete core usually has a thin-gage steel cylinder which is either embedded within the core, or is on the exterior of the core. Concrete cores with an embedded steel-cylinder are vertically cast and the corresponding pipe are referred to as "embedded-cylinder pipe"; those with an external steel cylinder are usually spun cast against the steel cylinder and are referred to as "lined-cylinder pipe."

The purpose of this paper is to present the results obtained in a series of combined-load tests performed under the auspices of the American Concrete Pressure Pipe Association on three embedded- and on three lined-cylinder pipe and to compare test results with the results of an analytical model developed to predict the behavior of prestressed concrete pipe subjected to combined loads

* Principal, Simpson Gumpertz & Heger Inc., 297 Broadway, Arlington, MA 02174

(Zarghamee and Fok 1990). This prediction is based on an assumed state of the prestress in the concrete core, steel cylinder, steel wire, and the mortar coating, which is computed using the step-by-step integration procedure of Zarghamee and Dana (1989). The analytical model accounts for nonlinearities of the stress-strain relationships of the constituent materials, including tensile softening and cracking of the core concrete and coating mortar, and for moment redistribution as the stiffness at sections around the pipe circumference changes with load.

Analytical Approach

To predict the post-cracking behavior of prestressed concrete pipe, the pipe wall is divided into 24 layers, as shown in Fig. 1, and the behavior of each layer is determined by a nonlinear stress-strain relationship and a failure criterion as described below. The stiffness of prestressed-concrete pipe subjected to combined loads and internal pressure varies around the pipe circumference, and such variation of stiffness affects moment distribution. Furthermore, as the external load is increased, tensile softening and cracking of the core and the coating, yielding of steel cylinder, and nonlinearity of the stress-strain relationship for the prestressing wire beyond its elastic limit may cause additional variation of stiffness and moment redistribution. The analytical procedure used accurately accounts for the effect of the change in stiffness on moment distribution around the pipe.

The computation procedure is as follows. Based on the assumed states of stress and strain in a pipe, subjected to the combined effects of internal pressure, external load, and pipe and fluid weights, the bending and the axial stiffnesses at sections around the pipe are computed. Using the computed stiffnesses, the pipe deformations, thrusts and moments, and finally stresses and strains are calculated. The calculated stresses and strains are used with the assumed stress/strain relationships to modify the pipe wall bending and axial stiffnesses at sections around the pipe circumference. The computation is repeated until convergence is achieved.

The model is applied to the prediction of the behavior of embedded- and lined-cylinder pipe subjected to combined-load tests. The predicted behavior of test pipe is compared with actual test results. The comparison is performed in terms of the combined-loads at first visible cracking, pipe deformations, and wire strains.

Properties of Constituent Materials

The prescribed material properties are based on data collected from the American concrete pressure pipe industry and from published data in the field. In this section the properties used for each constituent material, and the bases for their selection are presented.

The stress-strain relationships of concrete and mortar in compression are based on the work of Carreira and Chu (1985). They present a method for representing the stress-strain relationship of concrete in compression which compares favorably with data obtained from carefully conducted tests.

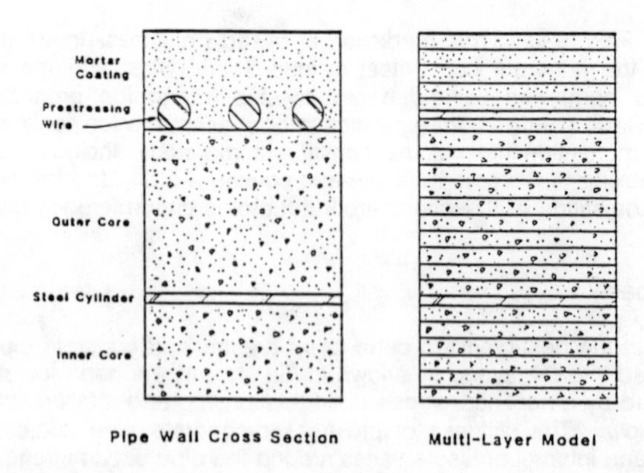

Fig. 1 – Multilayered model of prestressed concrete pipe wall

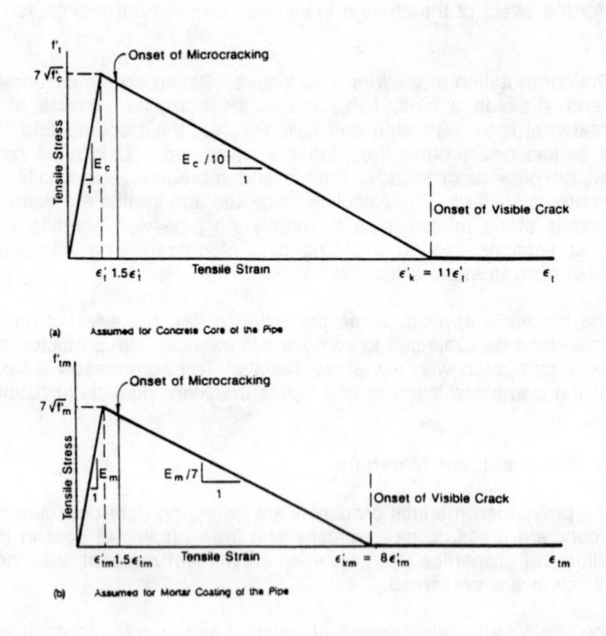

Fig. 2 – Stress-strain relations for concrete and mortar in tension

The behavior of concrete and mortar in tension is modelled with trilinear stress-strain curves, as shown in Fig. 2. The ascending branch of the curve hasa modulus of E_c (or E_m for mortar), which is the same as that for compression, and terminates when f'_t (or f'_{tm} for mortar) is reached. The descending branch of the curve is linear with a slope of $-E_c/10$ (or $-E_m/7$ for mortar) and terminates when the stress drops to zero. The last part of the curve depicts the behavior of visibly cracked concrete where the stress is equal to zero across the crack.

The modulus of elasticity of concrete, E_c, and mortar, E_m, are calculated from the formula by Pauw (1960).

The stress-strain relationships for prestressing wire and the steel cylinder in the model were obtained by best-fitting the experimental data for unpre-stretched wire and cylinder. The stress-strain relationship for prestressing wire is linear up to a gross-wrapping stress $f_{sg} = 0.75\ f_{su}$ and then nonlinear thereafter. The stress-strain relationship for the steel cylinder is bilinear with tensile yielding at f_{yy}. The exact expressions used for the stress-strain relationships of concrete, mortar, prestressing wire and steel cylinder are given in Zarghamee and Fok (1990).

State of Stress in Test Pipe

The three embedded-cylinder test pipe, denoted as E60-100, E60-150 and E60-200, are all 60 inches in diameter, and are designed to operate at working pressure levels of 100 psi, 150 psi and 200 psi, respectively. The three lined-cylinder test pipe, denoted as L48-100, L48-150 and L48-200, are all 48 inches in diameter, and are designed for working pressure levels of 100 psi, 150 psi, and 200 psi, respectively. All six pipe specimens were tested about 90 days after manufacturing. The test procedures are fully described by Tremblay (1990).

The embedded-cylinder pipe were stored in a laboratory environment with 46% average relative humidity and the lined-cylinder pipe were stored outside in 62% average relative humidity environment. The state of stress in the pipe was computed using the step-by-step integration procedure of Zarghamee and Dana (1989), using the average relative humidity of the storage environment. Table 1 shows the state of stress in these pipe segments at the time of testing.

TABLE 1

STATE OF STRESS IN TEST PIPE

	E60-100	E60-150	E60-200	L48-100	L48-150	L48-200
Decompression Pressure (psi)	103	158	224	122	204	298
Inner Core Stress * (psi)	315	582	880	977	1,477	2,208
Outer Core Stress (psi)	839	1,179	1,606	–	–	–
Mortar Prestress (psi)	−250	−124	−17	−313	199	−96
Steel Cylinder Stress (psi)	18,473	23,209	28,244	16,378	22,248	29,320
Prestressing Wire Stress (psi)	−158,791	−155,503	−152,028	−160,414	−156,706	−151,746

* compression is shown as positive.

Note that the exposure condition for the test pipes has resulted in significant differences of prestress between the inner and the outer cores; for a pipe subjected to a buried, moist soil environment, the inner and the outer cores have virtually the same prestress (Zarghamee and Dana 1989).

Analysis

To demonstrate the results of the analysis of pipe behavior under combined loads, we have defined certain stress and strain limit states and have calculated the combined loads corresponding to these limit states. The limit states considered are as follows:

1. Serviceability Limit States

At Invert or Crown

- Onset of visible cracking in the inner fiber of the core when strain reaches ϵ_k.

At Springline

- Onset of visible cracking in the outer fiber of the coating when strain reaches ϵ_{km}.

2. Elastic and Strength Limit Sates

At invert or Crown

- Yielding of cylinder for embedded-cylinder pipe

At Springline

- Stress in wire reaching $f_{sg} = 0.75\ f_{su}$.
- Stress in wire reaching yield, defined as 0.85 f_{su}, corresponding to 0.2 percent offset in a virgin (unprestretched) wire.

3. Ultimate Strength

At Springline

- Compressive strain in the inner fiber of the core reaching 0.3 percent.

The analysis for each of the combined loads was performed as if the pipe were virgin; the tests performed after the onset of cracking were not, of course, conducted on virgin pipe. Since the combined-loads were applied with increasing magnitudes, the effect of previous loading history is expected to be small.

The details of the analysis are presented in Zarghamee and Fok (1990).

Results

In the following paragraphs, significant experimental and theoretical data are presented for the six pipe that were tested. The results are in the form of combined-load diagrams that show the theoretically predicted first-crack lines and the design curves, based on the new proposed design procedure, for the working and working plus transient combined loads (Heger, Zarghamee and Dana 1990 and Zarghamee, Fok and Sikiotis 1990) and the test combined loads at which either no visible crack appeared or a core or a coating crack was observed along with the measured crack width. Additionally, representative experimental and theoretical plots are provided for changes in pipe deflection, prestressing wire strain, steel cylinder strain, core and coating strains with pressure applied simultaneously with three-edge bearing load. In these tests, the pipes were subjected to a constant three-edge-bearing load while the pressure was varied.

The combined-load diagrams, containing the theoretical limit state lines and the test results are shown in Figs. 3 and 4. The magnitude of the core and coating crack widths measured at the combined test loads exceeding the working plus transient design curve and after the removal of the internal pressure are shown also in these diagrams.

During the test, the pipes were tapped with a hammer to detect any sign of delamination of the coating. No delamination was found except for L-48-150 and L-48-200 and then only after they were subjected to a pressure of 370 psi and 510 psi, respectively, about 250% of their pressure class.

The experimentally obtained and the theoretically calculated vertical deflection-pressure diagrams and wire strain-pressure diagrams for Class 200 embedded-and lined-cylinder pipe are shown in Figs. 5 and 6. In these tests, a three-edge bearing load equivalent to 6 ft of earth cover was applied first and then internal pressure was increased to a level that produces f_{sg} limit stress in the wire and then decreased to zero, Internal pressure was again increased cyclically to 110, 125, and 135 percent of the pressure level that produces f_{sg} limit stress in the wire.

Fig. 7 shows the experimentally obtained and the theoretically calculated strains at the inner surface of the core and the steel cylinder at the invert, and the mortar coating at springline for a 60-in.-diameter Class 100 embedded-cylinder pipe, subjected to a three-edge bearing load of 4.4 kip/ft and varying internal pressures.

Discussion

The results, shown in Figs. 3 and 4, demonstrate that the proposed limit states design procedure precludes visible core and coating cracks when the pipe is subjected to the allowable working plus transient combined loads. They also show that the theoretical model is capable of accurately predicting the combined-loads at first visible core and coating cracks.

A significant feature of prestressed concrete pipe is its ability to control coating cracks. The combined-load tests demonstrate that if pipe is cracked

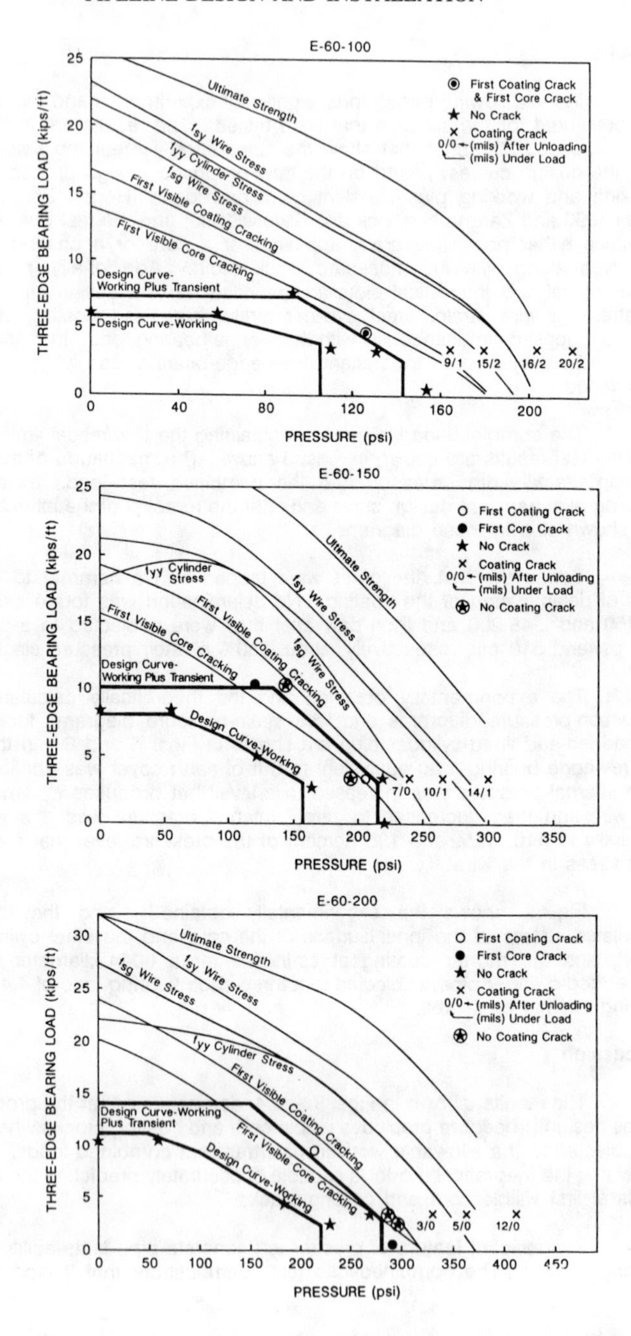

Fig. 3 – Measured and calculated combined loads for Classes 100, 150, and 200, 60-in. embedded-cylinder pipe

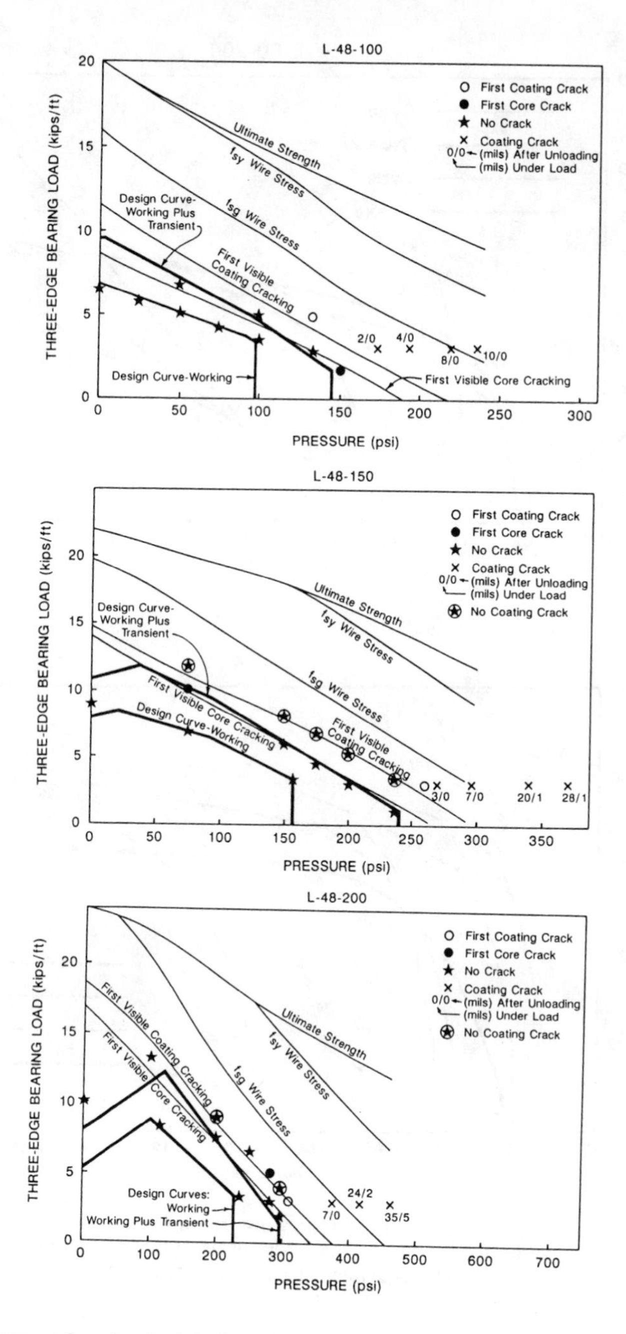

Fig. 4 – Measured and calculated combined loads for Classes 100,150, and 200, 48-in. lined-cylinder pipe

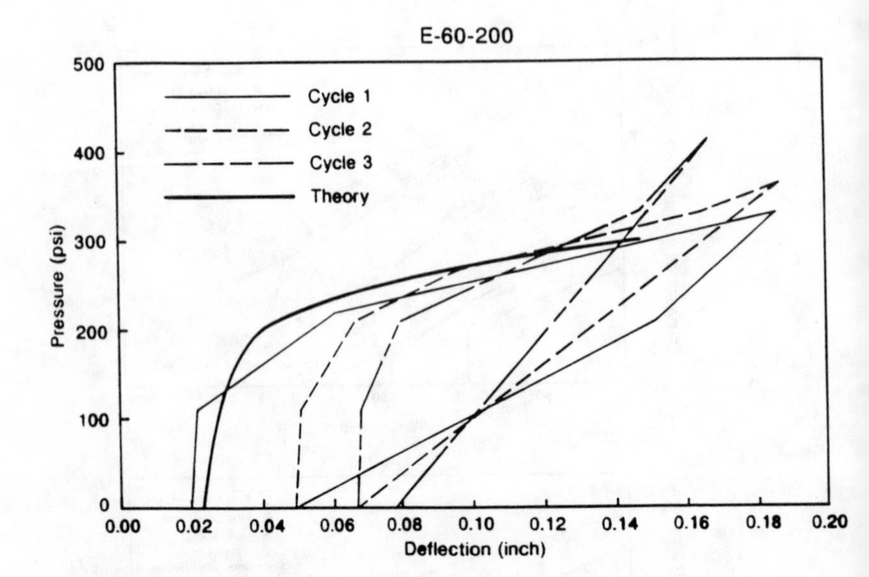

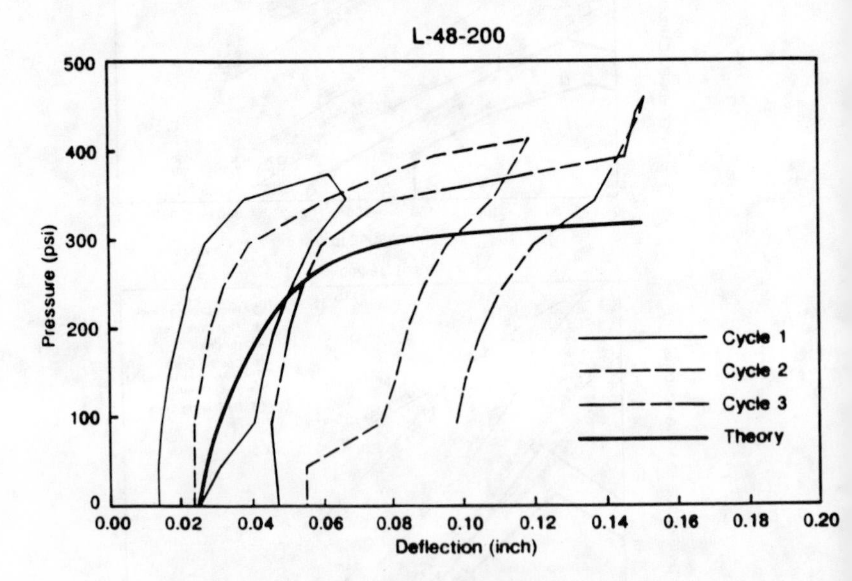

Fig. 5 – Pressure-vertical deflection diagrams for Class 200 60-in.-diameter embedded-cylinder and 48-in. lined-cylinder pipe with a three-edge bearing load equivalent to 6 ft of earth cover

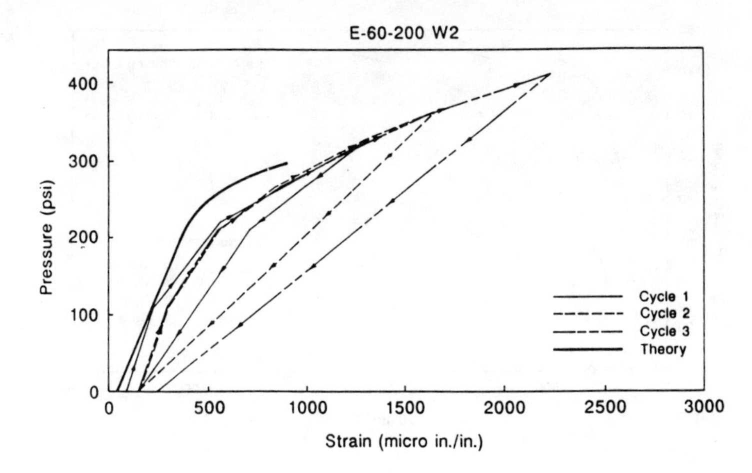

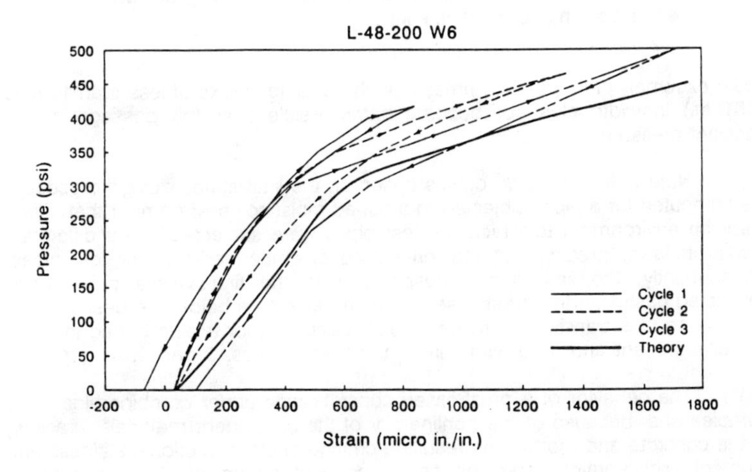

Fig. 6 – Pressure-wire strain diagrams at springline for Class 200, 60-in.-diameter embedded-cylinder and 48-in. lined-cylinder pipe with a three-edge-bearing load equivalent to 6 ft of earth cover

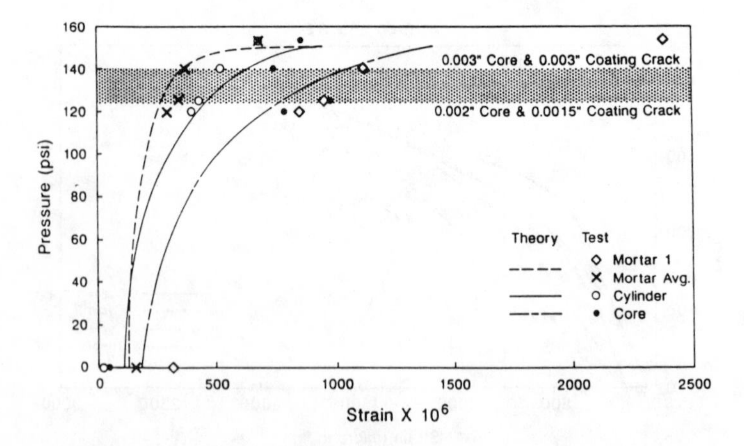

Fig. 7 – Measured and calculated strains in the inner fiber of core and in the cylinder at the invert and of a single gage and the average of four gages on the outer fiber of coating at the springline of a Class 100, 60-in.-diameter embedded-cylinder pipe with a three-edge bearing load of 4.4 kips/ft

under extremely high transient pressures, the coating cracks of less than 10 mils (0.01 in.) in width close or become barely visible after the passage of the transient pressure.

Note that the design curves based on the limit states design procedure are computed for a pipe subjected to a buried moist soil environment that differs from the environment to which the test pipes were subjected. The difference results in lower prestress in the inner core of embedded-cylinder test pipes. Consequently, the limit line corresponding to the first visible core crack, calculated based on the prestresses given in Table 1, is below the design curve for working plus transient combined loads, calculated for a pipe buried in moist soil environment and filled with water, as shown in Figs. 3 and 4.

The behavior of a prestressed concrete pipe under combined-load is a complex one, because of the nonlinearity of the constituent materials, cracking of the concrete and mortar, the resulting change in cross sectional stiffness and moment redistribution, and the time-dependent nature of the response of concrete to load.

Concrete and mortar do not behave as homogeneous materials under combined loads. The concrete strain in the microcracked state consists of an elastic component that is proportional to the stress and an inelastic component. Microcracking does not in general occur uniformly; in fact, cracking occurs in discrete fracture process zones, spaced appreciable distances apart. Between the fracture process zones, the concrete is unaffected by the microcracking and therefore is free of inelastic strains. In combined-load tests of prestressed concrete pipe, the concrete and mortar strain gages in the fracture process zones may detach and those away from the fracture process zones will show strains that are equal to the elastic strain component, significantly smaller than the average strains that are used in our theoretical development. The gages attached to the steel cylinder and prestressing wire may show strains that are smaller or larger than the average strains, depending on whether or not they are near a fracture process zone. Of course, there is a higher probability of being away from a fracture process zone than of being near one. Therefore the majority of wire and cylinder strain gages are expected to show strains that are smaller than the average calculated strain, and only infrequently will strain gages show measured strains larger than the calculated average strain.

The pipe under load is not a statically determinant structure and the bending moment redistributes when the cross sectional stiffness changes around the pipe. In typical prestressed concrete cylinder pipe subjected to combined external load and internal pressure, the bending stiffness of the pipe wall at the invert and crown is larger than that at the springline because of the participation of the coating. As the external load or internal pressure is increased and the core cracking load at the invert and crown is approached, the bending stiffness of the pipe at the invert and crown will decrease and bending moment will redistribute to the springline. Subsequent cracking of the core at the springline may cause an additional redistribution of the bending moment from the springline to the invert and crown. Although the theoretical model used for calculating strains accounts for changes in the stiffnesses of the constituent materials resulting from material nonlinearities in general and strain softening, cracking, and yielding in particular, the idealized stress-strain relationships of the materials contain approximations that may contribute to the discrepancies between theoretical and experimental results, especially when the material undergoes unloading and reloading.

The concrete and mortar do not respond to load changes instantaneously. In fact the response of the pipe to the instantaneous changes in combined loads is somewhat gradual and may take several minutes. Although time, of the order of one to several minutes, was allowed after each load cycle for the pipe to reach equilibrium, it is not known whether the allotted time was enough.

Fig. 5 shows changes in the deflection of pipe with the application of external load and with the change of internal pressure. An interesting phenomenon supported by both the theoretical and experimental results is that the downward vertical deflection of the pipe increases with internal pressure; both the calculated and the measured deflections increase as the axial and bending stiffnesses of the pipe decrease with added internal pressure. The same phenomenon is observed for L-48-200. Such an increase of the vertical deflection of the pipe with added internal pressure is apparently a result of the tensile softening and cracking of concrete and mortar. The agreement between the changes of the calculated and measured deflections is excellent, although there appears to be a residual deflection after each cycle of loading. This phenomenon suggests that when prestressed concrete pipe is strained by internal pressure well beyond the onset of visible core cracking, the unloading process may be inelastic and time-dependent. Such an inelastic and time-dependent behavior can affect pipe deflection in the subsequent loading cycles.

The experimentally measured and calculated wire strains, Fig. 6, show close agreement for embedded-cylinder pipe, although the measured values are slightly higher than the calculated ones, possibly due to previous history of the pipe. For the lined-cylinder pipe, the fact that the measured strains are smaller than the calculated strains should be viewed as a manifestation of the nonuniformity of the behavior of the cracked pipe.

The calculated tensile strain of the core at the invert and of the coating at the springline, as shown in Fig. 7, increase slowly with internal pressure until the pressure approaches the cracking value. Near the first visible core and coating cracks, tensile strains increase rapidly with small changes in combined loads.

Furthermore, the nonuniformity of tensile strain in the coating is evident from Fig. 7 where measured strains are plotted for a strain gage that show a rapidly increasing strain with pressures that even exceeds 0.2% strain while other gages hardly reach one-tenth of this strain level. This gage is obviously in the fracture process zone while others are not. Note that the calculated mortar strain agrees quite well with the average strain measured by all four strain gages, including the one in the fracture process zone.

Strain gages at the invert and crown on the inside face of the core have a tendency to detach. Of the four core strain gages at the invert and crown only one showed a gradual increase with internal pressure.

The experimentally measured and calculated steel cylinder strains are in good agreement up to 125 psi internal pressure, but for very high pressures at which the core is cracked or nearly cracked, the measured steel cylinder strains lag the calculated average strain.

Conclusions

Based on the forgoing discussion the following conclusions can be made.

1. Agreement between the calculated and the test combined loads at the first visible core and coating cracks is good.

2. The strain magnitude measured in the core, coating, and the steel cylinder have a significant variable component most probably caused by the nonuniform distribution of the inelastic deformation of concrete and mortar in discrete fracture process zones.

3. Prestressed concrete pipe subjected to self weight, water weight, external load and internal pressures substantially in excess of the pressure that produces the f_{sg} stress limit in the wire develop coating cracks that close or become less than 0.002 in. wide when the pressure is reduced to zero.

4. In prestressed concrete pipe designed for the combined loads according to the proposed limit states design procedure, visible core or coating cracks do not develop under the working combined loads and are not expected to develop under the working plus transient combined loads.

References

Carreira, D. J., Chu, K. (1985). "Stress-Strain Relationship for Plain Concrete in Compression," ACI Journal Proceedings, Vol 82, No. 6, November-December, 797-804.

Heger, F. J., Zarghamee, M. S., and Dana, W. R. (1990). "A Limit States Design Concept for Prestressed Concrete Pipe, Part A: Criteria," Journal of Structural Engineering, ASCE, to appear in August issue.

Pauw, A. (1960). "Static Modulus of Elasticity of Concrete as Affected by Density," ACI Journal Proceedings, Vol. 57, No. 6, December, 679-688.

Tremblay, A.W. (1990), "Combined-Load Testings of Prestressed Concrete Cylinder Pipe," Proceedings, International Conference on Pipeline Design and Installation, ASCE Pipeline Division, Las Vegas, NV, March 1990.

Zarghamee, M. S., and Dana, W. R. (1989). "A Step-by-Step Integration Procedure for Computing State of Stress in Prestressed Concrete Pipe," presented at ACI National Convention, Atlanta, to be published.

Zarghamee, M. S., and Fok, K. L. (1990), "Analysis of Prestressed Concrete Pipe Under Combined Loads," Journal of Structural Engineering, ASCE, to appear in July issue.

Zarghamee, M. S., Fok, K. L., and Sikiotis, E. E. (1990), "A Limit States Design Concept for Prestressed Concrete PIpe, Part B: Procedure," Journal of Structural Engineering, ASCE, to appear in August issue.

MSZ38-89.mw

THE EFFECT OF COMPACTION ON BURIED FLEXIBLE PIPES

N.F. Zorn[1], P. van den Berg[2]

Abstract

This contribution presents an analysis of the installation of flexible pipes in granular soil. The deformation of the pipe cross section is measured in field and laboratory tests. It is shown that the effect of soil compaction forms a major load on the flexible structure. The behaviour of the coupled system is simulated with numerical, finite element method, calculations.

Introduction

In order to determine realistic values for stresses and strains in buried flexible pipelines a research project commissioned by the N.V. Nederlandse Gasunie and VEG Gasinstituut N.V. was performed. Current design methods which are based on stress distributions in terms of a loading and bedding angle or predescribed deformation patterns of the pipe are sufficient for safe (conservative) design but proved to be inadequate for prediction of deformations and corresponding stresses and strains in a layered soil system and during installation.

The study started with field tests and was continued with laboratory tests and numerical, finite element, analysis. It was aimed at the interaction of the flexible structure with the soil during all phases of it's service life, i.e. from installation to replacement after years of service [Zorn, 1990].

The pipes analysed were made of PVC/CPE and had diameters varying between 110 and 200 mm with a wall thickness of 2.7 and 3.4 mm. They were buried at ± 0.8 m depth. Two types of soil, granular sand and clay were considered. This paper however is restricted to behaviour of pipes embedded in sand and specially focussed on the analysis of the coupled system pipe-soil during the installation phase.

---

[1] Head Fundamental Research, [2] Research engineer, both DELFT GEOTECHNICS, P.O. Box 69, NL-2600 AB Delft
The Netherlands

Field tests

The field tests consisted of measurements of the horizontal and vertical dimensions of the pipes under different loading conditions. As first reference the 'in situ' condition of a pipe that had been in service for 15 to 20 years was documented. Then the pipe was excavated and the unloaded configuration measured. The deformation measurements were completed by soil investigations, consisting of cone penetration tests (CPT's) and the analysis of soil samples in the laboratory to determine the sieve curve and the density of the sand.

The deformation measurements were performed with the LGM-Cruise [Bezuijen, 1985], a pipe mounted on sledges that has displacement pickups mounted in horizontal and vertical direction. The Cruise is pushed through the pipe with sounding rods and the vertical and horizontal pipe dimensions are digitally displayed. Figure 1 presents an example of 'in situ' measurement results that were performed on a 110 mm diameter pipe embedded 0.7 m in sand. The location was beneath a sidewalk in a suburban area. The deformations are noted in %, referring to the nominal diameter of the pipe. It can be seen that the deformations show considerable scatter, very probably depending on local variations of the installation quality. The scatter was noted in statistical parameters, but only the mean values were used as reference.

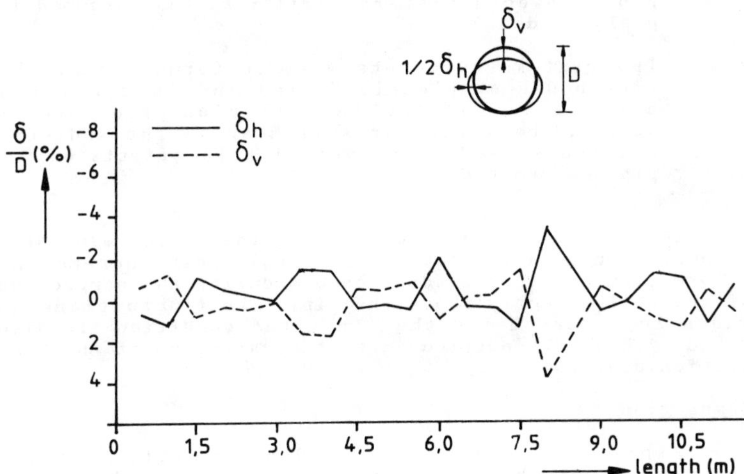

Figure 1: Deformation measurement results, existing pipeline in sand

Monitoring of the deformation of a replacement pipe section was performed with a series of measurements all performed at 0.5 m intervalls. The different steps of this procedure are explained including the qualitative (averaged) results for a good installation under optimal field conditions in the following:

- First the initial dimensions are documented for reference after the pipe section is laid in the trench. All deformations refer to this configuration.

- Then the soil is filled into the trench next to the pipe, until the top of the pipe is reached and compacted. The second measurement is performed after this step. For good soil compaction it shows an ovalisation of the pipe with increasing vertical and decreasing horizontal dimensions.

- The third measurement was performed after the trench was filled with the soil amounting to a cover of 2 to 3 pipe diameters. The effect of this dead weight loading on the pipe was marginal.

- The next step in a good installation procedure is to compact this soil fill. The subsequent measurements showed considerable pipe deformations, with an ovalization orientated in horizontal direction. The horizontal dimensions thus increased from less than reference to more than reference, and the pipe showed a pattern similar to that assumed in design codes.

- The next measurements were performed after the trench was completely filled and in the final situation. For the installation procedure as outlined here further soil dead weight and compaction showed relatively little effects on the pipe deformation.

The field measurements during the replacement of a pipe section clearly showed that the largest incremental deformation and thus loading occurs during the compaction of the soil directly above the pipe. The fluctuations as shown in figure 1 for the 'in situ' condition are also found - they are contributed to the variation of the compaction process.

Compaction

Since the effect of soil compaction on the deformation of the flexible pipe is such an important parameter when compared to the direct effect of the dead weight loading, it is of interest to briefly review what takes place during this process. A more elaborate review can be found in [Zorn et al, 1986]. However as first step to introduce this process into the design procedures of flexible structures interacting with soil basic information can be sufficient.

Compaction of sand refers to a process during which the granular assembly of sand grains is changed from an initial configuration into a more dense configuration. The void volume decreases during this process. In geotechnical engineering the density of sand can be described with two parameters: the porosity n, defined as ratio of the void volume V_v and the total volume V_t, and the relative density D_r that refers to the configuration of a sand volume with respect to two extreme conditions - the most dense n_{min} and the most loose n_{max} configuration. These two extreme values are functions of the sieve curve and thus refer to each type of sand individually. Both can be determined in laboratory tests. The definitions are:

$$n = V_v / V_t$$

$$D_r = (n_{max} - n_x) / (n_{max} - n_{min})$$

The index x referring to the actual configuration. The relative density of a sand combined with the 'in situ' state of stress define the mechanical properties of the sand, its stiffness and strength. Since the state of stress of the sand does not vary significantly in a trench the relative density is used as governing parameter when modelling sand for the conditions considered. This counts for normal compaction conditions, in which the granular assembly has a lower porosity after the compaction operation.

Under some circumstances, especially when the effect of compaction is measured as increase of the CPT results, prestressing or arching the sand between the undisturbed soil and the pipe in horizontal direction can be wrongly interpreted as compaction. In this case the state of stress in the trench does influence the soil behaviour, often causing creep in the surrounding, sometimes cohesive soils.

One optical indication of compaction is the settlement of the free surface. The grains are reassembled to a more dense configuration and the "soil volume" in the trench thus decreased. The settlement of the free sand surface due to an increase of the relative density is presented in table 1, in which the resulting settlement of the surface is related to the increase of the relative density. The settlement calculations are performed starting from an initial relative density of 30% which roughly represents uncompacted soil in the trench.

ΔD_r	Δu
%	cm/m
10	2.0
20	3.9
30	5.8
40	7.6
50	9.9

Table 1: Free surface settlement due to increase in relative density

Next to direct measurement of the actual density, achieved by analysing samples taken with a known volume for instance, this parameter presents an important and independent check of the compaction operation.
As already mentioned compaction influences the mechanical properties of the soil. Higher stiffness of the sand next to the pipe makes it possible to activate more horizontal soil reaction at small deformations and results in higher load bearing capacities of the coupled system soil and pipe. The importance of this effect will be presented in the numerical calculations.

Laboratory tests

In order to analyse extreme conditions and to provide reference calibration values under carefully controlled conditions for numerical analysis, laboratory tests simulating the installation procedure were performed. An additional aim of which also was to verify the conclusion resulting from the field measurements concerning the local fluctuations of the deformations.

These tests made it possible to perform soil testing following each step during the installation under controlled conditions. This particularly included the measurement of the sand density prior to and after compaction and the measurement of the surface settlement of the sand fill due to compaction.

The test facility used had dimensions of 2.5 x 1.5 x 6 m. It initially was filled with well graded riversand that had been compacted per layer of 15 cm, achieving an average relative density of 75%. This sand volume represented undisturbed homogeneous conditions rather well and was kept at a moisture content of 5%. A trench was excavated and pipes were installed simulating the 'in situ' procedure under controlled laboratory conditions.

The tests performed reproduced the field results qualitatively when comparing the avarage values. An example of final deformation measurements is presented in figure 2. It clearly can be seen that the degree of scatter is reduced. Even under laboratory conditions the variations in the behaviour of the coupled system, soil-pipe, cannot be completely omitted when dealing with such sensitive structural elements as plastic pipes.

Phase		v %
1	compaction next to pipe	+ 2.1
2	dead weight cover 2 - 3 diameters	+ 1.6
3	2 compacted	± 0.0
4	trench filled	± 0.0
5	final configuration	0.0

Table 2: Average horizontal and vertical pipe deformations in sand during installation

Table 2 shows the average deformations in the different stages in vertical direction for good installation in the laboratory. The test results showed that the deformation of the pipe during the installation procedure can be assigned to the effect of compaction for 75%, only 25% of the deformation results from the dead weight of the soil. The effect of compaction, the settlement of the free surface in the trench was also measured and related to the increase in relative density increase predicted in table 2. The agreement of settlement measurements with the achieved and measured increase in relative density was within a margin of 5%.

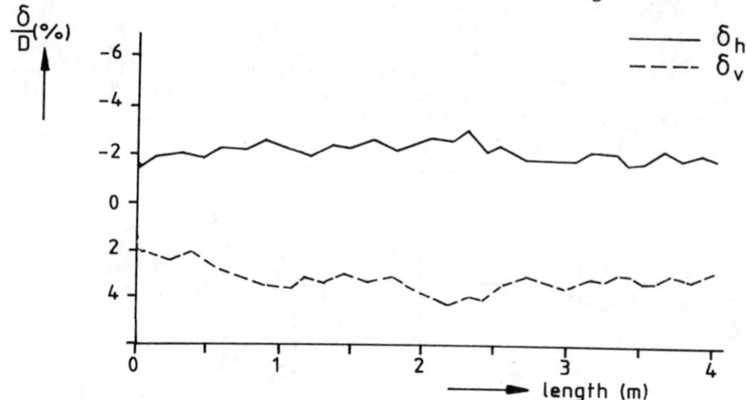

Figure 2: Deformation measurements laboratory tests in sand

Laboratory conditions made it possible to also simulate a bad installation as reference case. This installation, not admissible when monitoring a real pipe replacement, consisted basically of two steps. The the trench was filled in one step and compacted only once at the surface. In this case the pipe deformed due to the dead weight of the soil, it showed little increase due to the compaction operation. It should be remarked that this type of installation will cause uncontrolled compaction due to traffic vibrations and moisture during the service life of the pipe.

Numerical analysis

As already mentioned standard design procedures are not capable of correctly modelling the installation process of a flexible structure in sand since they are generally based on assumptions with respect to how the pipe will be loaded [Spangler, 1941; Leonhardt, 1979]. They do not model the installation phase and try to include the loads that deform the pipe in this stage by overestimating the effect of the soil dead weight by predescribing the stress distribution on the pipe cross section. This generally leads to conservative design [Zorn, 1987]. If however more exact values of the deformation or strain of the flexible structure is required more sophisticated tools and more advanced models are required.

The application of the finite element technique allows to take the features into account that are required and together with numerical modelling and engineering judgement is one of the sophisticated tools available.

Ignoring the non symmetric load and support conditions - a good approximation for the avarage installation - and neglecting axial effects the problem of soil structure interaction for the installation of flexible pipes can be modelled with plain strain finite elements as presented in figure 3.

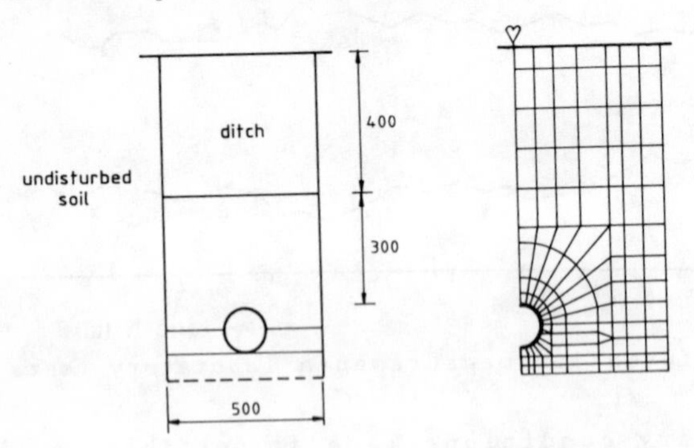

Figure 3: Trench with pipe and finite element model

The discretisation of the soil allows to distinguish different regions within the model, the undisturbed soil around the trench, the different fill levels and the loosened soil underneath the pipe. The pipe itself is also modelled with plane strain continuum elements that prove to be sufficient for the problem. A test calculation in which the deformation was compared with analytical results showed that the error was well below 1% for the deformation modes that were checked. The finite element code DIANA [Kusters, 1983] was used for the calculations, which were performed with a reduced integration scheme.

The different stages of the installation can be modelled by activating the stiffness and the dead weight of the elements that exist in the installation phase of interest. The dead weight loading is included by providing the correct value for the soil density as input, the finite element code then calculates the resulting forces.

In order to model the compaction process it is necessary to apply an additional load. The mechanically correct dynamic loading would require very sophisticated material modells capable of simulating the physical compaction process due to alternating loading and dynamic calculations to model the wave propagation process. A rather long way to go to design a pipe. If however only the effect of the compaction, the deformation of the flexible structure and the therein induced stresses and strains are of interest, a different, more simple model can be used.

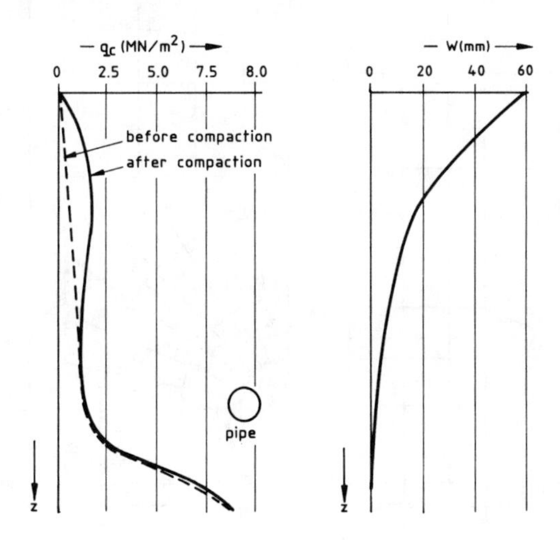

Figure 4: Sand stiffness in the trench, CPT, and resulting compaction displacement field

The loading due to compaction can be reduced to the calculation of the effect of a predescribed soil surface displacement that is applied to the model at every phase of the installation. The material behaviour of the soil may be assumed elastic for every layer and must be updated to take the effect of the previous compaction step on the stiffness into account. The use of this material model for soil restricts the applicability of the results to the interpretation of the deformation, stresses and strains of the pipe. The soil stresses are not correct, the soil displacement field however is realistic.

The importance of the use of the correct input for the soil stiffness distribution is documented in figure 4, in which the displacement field of the soil for bad installation is shown. The figure presents the results for the assumption of a realistic stiffness distribution of the sand in the trench, a linear increase with depth. This stiffness distribution results in a hyperbolic displacement field, which resembles the soil movement during compaction quite realistically. The calculated pipe deformations agree well with the measurements for this installation procedure.

The same procedure can be followed for good installation, in this case the effect of compaction must be updated after every calculation step. The final result is presented in figure 5 in which the complete installation process has been calculated for good and bad installation.

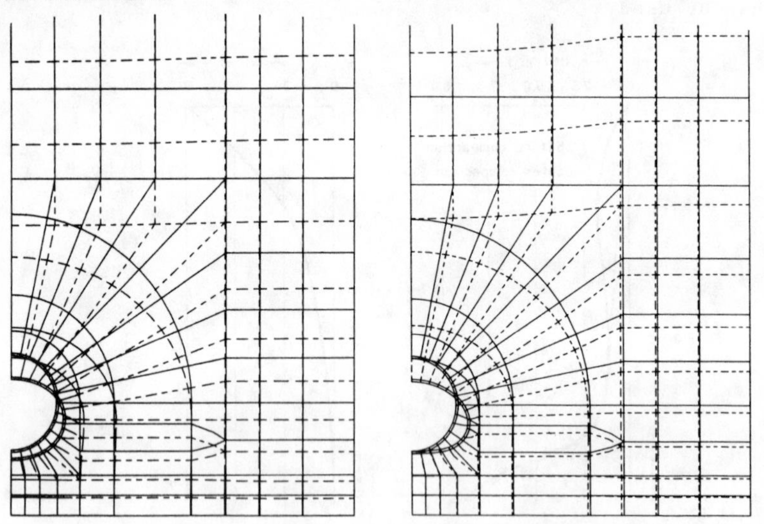

Figure 5: Comparison pipe deformation for good and bad installation

The good installation can be recognized by the settlement of the pipe into the bottom of the trench, the sand was loosed up prior to the installation, and by the fact that the deformation is rather small. In case of a bad installation the pipe has to deform considerably to activate the passive soil resistance that increases the bearing capacity. This deformation includes considerable bending modes which make less effective use of the pipe stiffness than the membrane deformation modes in case of good installation.

In both cases the loads applied are the combination of dead weight and the effect of soil compaction. The resulting deformations, 2% for good installtion and 6 to 8% for bad installation are in good agreement with the average values of the laboratory tests.

Conclusions

The results of the field and laboratory deformation measurements show that the compaction operation of granular fills in the surrounding of flexible buried structural elements presents a significant deformation generating load. The effect of this load, that is responsible for approximately 75% of the deformation of a flexible plastic gas pipe during the installation procedure can be simulated using finite element techniques to calculate the behaviour of the coupled system soil - flexible structure.

The application of numerical simulation techniques allows to perform sensitivity analyses to determine an optimal design for given criteria and installation qualities.

The numerical model derived can also be applied to analyse the behaviour during installation of other structures embedded in soil as for instance flexible culverts. The results when calibrated with measurements in addition present good boundary conditions in terms of the system stiffness for the analysis of the service load conditions, including extreme loads.

References

Bezuijen,A., Zorn, N.F., Measurement of gas pipe deformations, Gas 12, 1985 (in Dutch)

Kusters, G.M.A., et al, DIANA - a comprehensive but flexible finite element system, 4th Int. Symp. on Finite Element Systems, Southhampton, July 1983

Leonhardt, G., Die Erdlasten bei überschütteten Durchlässen, Bautechnik Heft 11, Nov. 1979, in German

Spangler, M.G., The structural design of flexible pipe culverts, IOWA Engineering Experiment Station, Bulletin no. 153, 1941

Zorn, N.F., Berg, P.vd, Postma, T.R., Mechanics of Dynamic Compaction Phase II, Delft Geotechnics Report CO-279431/15, July 1986, Delft

Zorn, N.F., Berg, P.vd., Deformation of flexible pipes and culverts, mechanics, measurements and calculations, Delft Geotechnics Symposium: Pipelines geotechnically discussed, The Hague Oct. 1987. (in Dutch)

Zorn, N.F., Berg, P.vd, Buried Flexible Pipelines - Deformation and Geotechnical Loading Conditions, Technical Paper, ETCE - Pipeline Engineering Symposium ASME, New Orleans, La. 1990

Designing Vitrified Clay Pipe Systems with "EASE"

E. J. Sikora[1] and E. Lamb[2] ASCE

Abstract

NCPI has developed a computerized trench load design program which provides a broad and complete spectrum of design options which will enable engineers and contractors to safely design and install permanent clay pipe systems with the least cost.

Based upon the Marston Equations and ASTM requirements, program "EASE" has been developed with the user in mind. It is completely self contained and user friendly. "EASE" provides three different design options depending on the needs of the user. Where applicable, live loads and impact factors are also included.

This paper also summarizes the clay pipe industry's laboratory, field and university sponsored research which contributed to the development of "EASE". The question of the applicability of Marston's traditional trench load design equations to today's construction techniques will be addressed.

Introduction

A steel cable strung through a buried pipe and extending to the ground surface where it was attached to a wood beam with one end supported on a platform scale does not appear to be a very scientific approach to trench load measurement. On the contrary, that test by

----

[1]Vice President-Director of Technical Services, National Clay Pipe Institute, 253-80 Center Street, Lake Geneva, WI 53147

[2]District Manager, National Clay Pipe Institute, 3031 Palmaire, Phoenix, AZ 85051

Anson Marston(1) and others which followed at the Engineering Experiment Station of the Iowa State College of Agriculture, became the foundational basis for all succeeding analysis and design of the loads on buried pipe. Strangely, the impetus for the Marston study was the need to produce more Iowa corn by draining the wet fields with clay drain tile. By 1911, enough drain tile had already been installed in Iowa alone to extend five times around the world.

With the assignment of developing a design procedure for buried pipe, Marston, and the other researchers that followed, produced a design method that has continued from its introduction in 1913 until this day. The success of Marston is best demonstrated by the worldwide use of the trench load equation that bears his name.

Figure 1. Ames Senior Testing Machine

Despite the tremendous track record of the Marston design, it became apparent that modern pipe installation methods differed in several ways from the basic premises upon which the original equation was based. Most notedly were the deeper and wider trenches, the use of mechanical backfill compaction, the practice of using imported bedding materials and the development of clay pipe with higher manufactured strength, flexible compression joints and longer lengths. As a consequence, the National Clay Pipe Institute began to evaluate the validity of the trench load equation as it applies to current practice. This evaluation resulted in a series of laboratory and field programs which covered test methods, installation practices, bedding material evaluations and the evaluation of bedding factors. These studies included a review of European design practices and an extensive finite element computer analysis conducted at the University of Wisconsin. This NCPI research and review is summarized in Part I of this report. The cumulative result of this research led to the development and implementation of a trench load design computer program called "EASE".

It was John H. Walton(2), the noted English author of the "Structural Design of the Cross Section of Buried Vitrified Clay Pipelines" who stated, "...somehow I have a horror of trying to mechanize or routinize a design job of this kind. I am always afraid that if structural design of pipelines is made "fast and easy" it will be undertaken by people who do not really know what they are doing and sooner or later disaster will ensue."

Notwithstanding Mr. Walton's caution, the clay pipe industry believes that the time is right for a trench load design program which is both "fast and easy". Program "EASE" is the subject of Part II of this paper.

PART I - NCPI RESEARCH SUMMARY

- **Field Test to Measure Trench Loads**(3) - The backfill load on an 8-inch VCP supported on electronic load cells was measured over a 10 year period. The backfill in the 12 foot deep trench was unconsolidated sandy clay. The wide trench width allowed maximum load development.

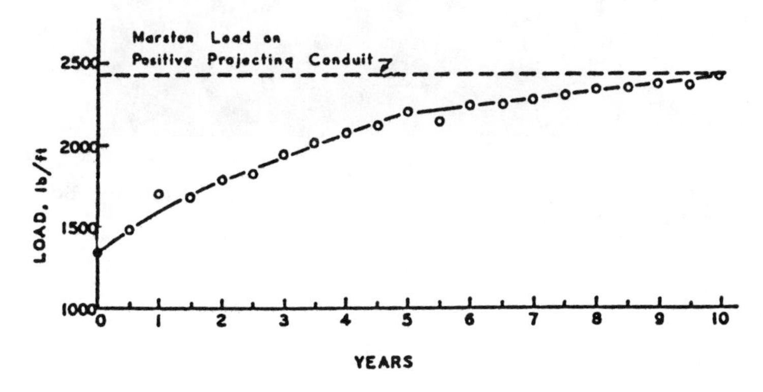

Figure 2. Measured Loads on 8-inch Vitrified Clay Pipe

Main Conclusion - The load on the pipe gradually increased to closely approximate the load predicted by the Marston equation.

- **Field Test to Measure Differential Movement at Compression Joints** - The magnitude of the differential movement between adjacent pipe was measured under 12-foot of uncompacted backfill.

  Main Conclusion - There was a wide variation in the ability of commonly used bedding materials to provide uniform support of the pipe.

- **Laboratory Test to Measure the Stability of Bedding Materials**(4) - The size, shape and gradation of bedding materials were evaluated in terms of ability to resist the downward movement of an 8-inch pipe loaded to 2000 plf and subjected to flooding.

  Main Conclusion - The stability of a bedding material increases with increased particle size and angular materials are more stable than rounded materials of the same gradation.

Figure 3. NCPI Test Equipment for Evaluating Bedding Material Stability

- **Laboratory Test to Measure the Effect of Bedding Material Selection on Load Factors**(5) - The distribution and magnitude of tensile strain was measured with strain gages bonded to a vitrified clay pipe. The strains obtained in different classes of bedding with different bedding materials were compared to the strains obtained in a 3-Edge bearing apparatus.

  Main Conclusion - The current load factors for the design of vitrified clay pipe sewer systems are valid. Also, shovel slicing the bedding material into the haunch areas was shown to decrease the movement of the pipe into the bedding and to increase the resulting load factor. ASTM size 6, 67 and 7 yielded comparable results.

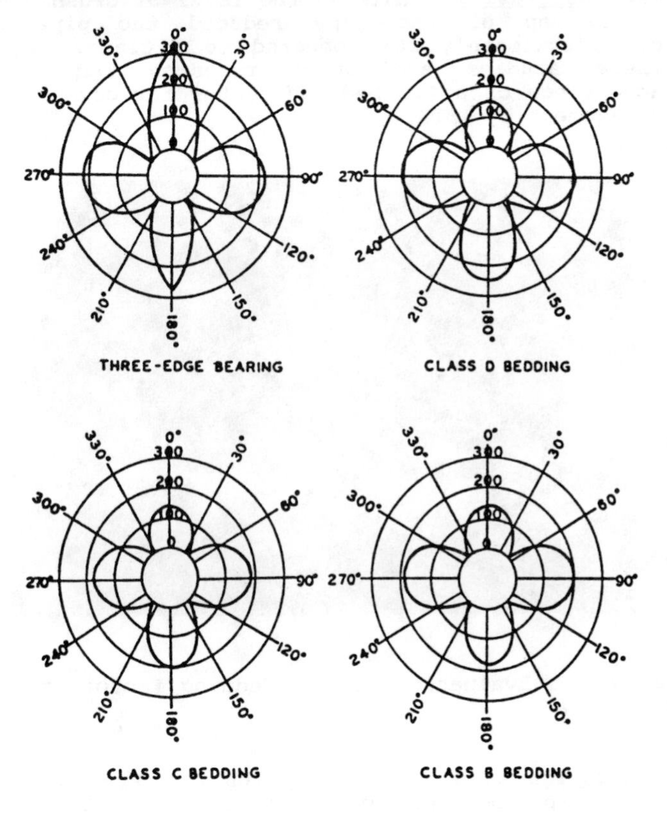

Figure 4. Strain Patterns in Three-Edge Bearing and Different Classes of Bedding

- **Load Tower Tests** - A 12 foot high load tower was constructed to measure the loads imposed by a natural sand fill on similarly installed rigid and flexible pipe.

  Main Conclusion - Sidefill consolidation decreases the load on both rigid and flexible pipe. The loads on a flexible pipe were approximately 80% of the loads on a similarly installed rigid pipe.

- **Orange County, CA Field Test** - This test measured the increase in load factor as a result of bringing crushed stone up to a level equal with the top of the clay pipe.

  Main Conclusion - Raising the level of crushed stone to the top of the pipe reduced the pipe stress by approximately 20% compared to a Class B system. These studies and other research led to the acceptance of a 2.2 bedding factor for crushed stone encasement by ASTM.

Figure 5. Evaluation of 2.2 Bedding Factor at Orange County, CA

- **Evaluation of European Research and Practice** - The Clay Pipe Development Association in England had conducted extensive load factor testing for a number of years and concluded that the commonly used design bedding factors were too low. German engineers used a design approach based upon lower pipe stresses than would be expected with the Marston equation. Both approaches are less conservative than the current practice in the United States.

Main Conclusion - The Marston approach yields a design which is very conservative using today's installation procedures.

- **Finite Element Analysis**(6) - **University of Wisconsin** -This study computed trench loads and bedding factors in several bedding material types and densities, trench widths and depths.

  Main Conclusion - This research essentially confirmed that the Marston design was very conservative under common methods of installing vitrified clay pipe.

- **Uhrichsville, Ohio Field Test**(7) - This test line measured the loads on clay pipe in several trench widths and bedding conditions with a controlled backfill density in an attempt to simulate the Wisconsin computer analysis.

  Main Conclusion -The Uhrichsville loads approximated the prism load on the pipe in a well compacted trench. The loads were higher than the finite element predictions but lower than calculated by the Marston equation.

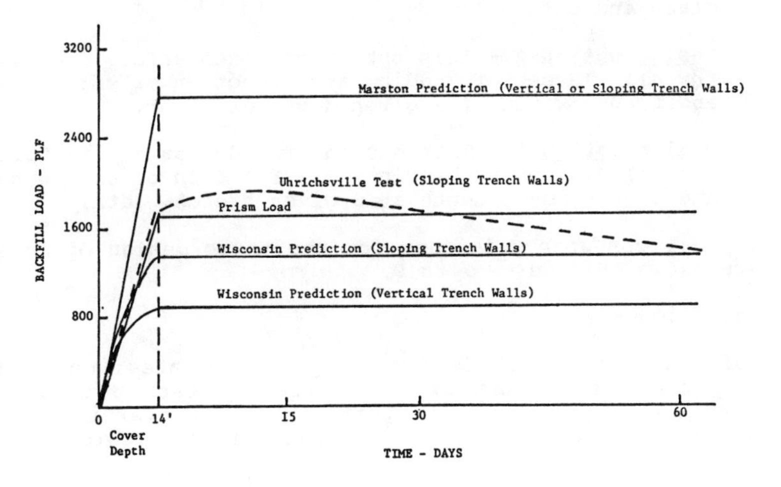

Figure 6. Measured and Predicted Loads on 12-inch Vitrified Clay Pipe at Uhrichsville, OH

COMMENT: The clay pipe industry recognized the potential benefits of adopting a less conservative design as indicated by the NCPI sponsored University research, the NCPI laboratory and field studies and the European practice. However, to retain a conservative design, while encouraging a more cost effective and practical approach to trench load computations, the clay pipe industry developed a computer program called "EASE" which is the subject of Part II of this report.

PART II - "EASE" A TRENCH LOAD DESIGN PROGRAM

Program Features

"EASE" will compute Marston's trench load equations, including live loads and impact factors in virtually any type of soil. All reasonable combinations of trench width and cover depths are included as are ASTM pipe strengths and bedding classes. The computer presents all computations as a function of resulting safety factor.

Available Design Options

The program presents three separate design options.

- **Design Option 1** - This option can be used to compute trench load for a given trench cover depth, width and soil type and weight in terms of bedding class and safety factor.

- **Design Option 2** - This option provides safety factors for all classes of bedding as a function of soil type and trench width at a given depth of cover.

- **Design Option 3** - This option provides safety factors for all classes of bedding as a function of soil type and trench cover depth at a given trench width.

Typical computer printouts for the three design options are shown on Figures 7 to 9.

Operation

This trench load design program operates on IBM compatible equipment and is extremely user friendly. The operator has many opportunities to override the selected default values if other data is available.

```
NATIONAL CLAY PIPE INSTITUTE
TRENCH DESIGN CALCULATIONS 4.2ER

Date of Calculations: 3/26/90

PROJECT NAME: ASCE CONFERENCE

Engineer:
Prepared by:

Manhole Station (From): 1
                  (To): 2

LOAD CALCULATIONS (Using ASTM C-12 Bedding Factors)
        Using ASTM C-700 (Extra Strength)

            Pipe Size:      8 in.
        Trench Width:      32 in.
        Trench Depth:     8.0 ft. over top of pipe.
         Soil Weight:     115 lbs./cu.ft.
       Ku'Value Used:   0.150
      rsdp Value Used:   0.75
           Live Load:   16000 lbs.
       Impact Factor:    1.50
    Transmitted Load:     120 lbs./ft.
       Backfill Load:    1180 lbs./ft.
   Total Load on pipe:   1300 lbs./ft.
    Transition Width:     2.2 ft.
  Load at Transition:    1300 lbs./ft.
```

BEDDING FACTOR		SAFETY FACTOR	SF @ TRANSITION
CLASS D	(1.1)	1.86	1.86
CLASS C	(1.5)	2.54	2.54
CLASS B	(1.9)	3.22	3.22
CR STONE	(2.2)	3.72	3.72
CLASS A	(2.8)	4.74	4.74
CLASS A	(3.4)	5.75	5.75
CLASS A	(4.8)	8.12	8.12

Figure 7. Trench Load Program - Design Option 1

```
NATIONAL CLAY PIPE INSTITUTE
TRENCH DESIGN CALCULATIONS 4.2ER

Date of Calculations: 3/26/90

PROJECT NAME: ASCE CONFERENCE

Engineer:
Prepared by:

Manhole Station (From): 2
                  (To): 3

LOAD CALCULATIONS (Using ASTM C-12 Bedding Factors)
        Using ASTM C-700 (Extra Strength)

            Pipe Size:     18 in.
        Trench Depth:    20.0 ft. over top of pipe.
         Soil Weight:     120 lbs./cu.ft.
       Ku'Value Used:   0.130
      rsdp Value Used:   0.75
    Transition Width:     4.8 ft.
```

	LOAD AT TRENCH WIDTHS							
TR.WIDTH (in):	36	42	48	54	60	72	0	Trans
LOAD (lbs.):	3420	4374	5372	6403	6845	6845	0	6845

BEDDING FACTORS		SAFETY FACTORS							
CLASS D	(1.1)	1.06	0.83	0.68	0.57	0.53	0.53	0.00	0.53
CLASS C	(1.5)	1.45	1.13	0.92	0.77	0.72	0.72	0.00	0.72
CLASS B	(1.9)	1.83	1.43	1.17	0.98	0.92	0.92	0.00	0.92
CR STONE	(2.2)	2.12	1.66	1.35	1.13	1.06	1.06	0.00	1.06
CLASS A	(2.8)	2.70	2.11	1.72	1.44	1.35	1.35	0.00	1.35
CLASS A	(3.4)	3.28	2.57	2.09	1.75	1.64	1.64	0.00	1.64
CLASS A	(4.8)	4.63	3.62	2.95	2.47	2.31	2.31	0.00	2.31

Figure 8. Trench Load Program - Design Option 2

NATIONAL CLAY PIPE INSTITUTE
TRENCH DESIGN CALCULATIONS 4.2ER

Date of Calculations: 3/26/90

PROJECT NAME: ASCE CONFERENCE

Engineer:
Prepared by:

Manhole Station (From): 3
(To): 4

LOAD CALCULATIONS (Using ASTM C-12 Bedding Factors)
Using ASTM C-700 (Extra Strength)

Pipe Size: 30 in.
Trench Width: 60 in.
Soil Weight: 120 lbs./cu.ft.
Ku'Value Used: 0.110
rsdp Value Used: 0.75
Live Load: 16000 lbs.
Impact Factor: 1.50

DEPTH OVER TOP OF PIPE (ft)	PIPE LOAD (lbs)	SAFETY FACTORS CL D (1.1)	CL C (1.5)	CL B (1.9)	CR ST (2.2)	CL A (2.8)	CL A (3.4)	CL A (4.8)
4	3919	1.40	1.91	2.42	2.81	3.57	4.34	6.12
5	3869	1.42	1.94	2.46	2.84	3.62	4.39	6.20
6	4004	1.37	1.87	2.37	2.75	3.50	4.25	5.99
7	4239	1.30	1.77	2.24	2.60	3.30	4.01	5.66
8	4526	1.22	1.66	2.10	2.43	3.09	3.76	5.30
9	4459	1.23	1.68	2.13	2.47	3.14	3.81	5.38
10	4854	1.13	1.55	1.96	2.27	2.88	3.50	4.94
11	5232	1.05	1.43	1.82	2.10	2.68	3.25	4.59
12	5594	0.98	1.34	1.70	1.97	2.50	3.04	4.29
13	5940	0.93	1.26	1.60	1.85	2.36	2.86	4.04
14	6271	0.88	1.20	1.51	1.75	2.23	2.71	3.83
15	6588	0.83	1.14	1.44	1.67	2.12	2.58	3.64
16	6892	0.80	1.09	1.38	1.60	2.03	2.47	3.48
17	7182	0.77	1.04	1.32	1.53	1.95	2.37	3.34
18	7460	0.74	1.01	1.27	1.47	1.88	2.28	3.22
19	7726	0.71	0.97	1.23	1.42	1.81	2.20	3.11
20	7980	0.69	0.94	1.19	1.38	1.75	2.13	3.01
21	8224	0.67	0.91	1.16	1.34	1.70	2.07	2.92
22	8457	0.65	0.89	1.12	1.30	1.66	2.01	2.84
23	8680	0.63	0.86	1.09	1.27	1.61	1.96	2.77
24	8893	0.62	0.84	1.07	1.24	1.57	1.91	2.70
25	9097	0.60	0.82	1.04	1.21	1.54	1.87	2.64
26	9293	0.59	0.81	1.02	1.18	1.51	1.83	2.58
27	9480	0.58	0.79	1.00	1.16	1.48	1.79	2.53
28	9659	0.57	0.78	0.98	1.14	1.45	1.76	2.48
29	9830	0.56	0.76	0.97	1.12	1.42	1.73	2.44
30	9994	0.55	0.75	0.95	1.10	1.40	1.70	2.40

Figure 9. Trench Load Program - Design Option 3

Availability

The program is available to design engineers by contacting any NCPI member manufacturer or District Manager.

Acknowledgment

The National Clay Pipe Institute gratefully acknowledges the contribution to this project by Larry Tolby, NCPI District Manager, Northern California, whose early vision led to the development of this program.

APPENDIX

References

1. A. Marston and A. O. Anderson "The Theory Of Loads on Pipes in Ditches, and Tests of Cement and Clay Drain Tile and Sewer Pipe", Bulletin No. 31 Engineering Experiment Station, Iowa State College of Agriculture and Mechanic Arts, 1913.

2. J. H. Walton "The Structural Design of the Cross Section of Buried Vitrified Clay Pipelines", 1970.

3. R. F. Havell and C. Keeney "Loads on Buried Rigid Conduit - a Ten Year Study", 1976.

4. J. S. Griffith and C. Keeney "Load-Bearing Characteristics of Bedding Materials for Sewer Pipe", 1967.

5. E. J. Sikora "Load Factors and Non-destructive Testing of Clay Pipe", 1980.

6. J. K. Jeyapalan, N. Jiang and W. S. Saleira "Finite Element Analyses of Loads and Bedding Factors for Buried Vitrified Clay Pipes", 1987.

7. E. J. Sikora "NCPI Field Test - Uhrichsville, Ohio" Sewer Sense No. 23, 1988.

OPTIMAL DESIGN OF WATER TRANSMISSION MAINS

Jobaid Kabir[1], M ASCE

ABSTRACT

Capital cost for constructing a water transmission main includes cost of the pipeline and its associated installation costs. Also included in the capital cost is the cost of a pump station. Operations and maintenance costs include power and labor. Selection of a smaller diameter pipeline for transmission of water results in smaller capital cost but higher annual power cost due to high friction loss. On the other hand, selection of a larger diameter pipeline requires higher capital cost and resultant lower power cost.

A Lotus 1-2-3 spreadsheet has been developed to provide the most cost effective design of the transmission facility for the design criteria established by the user. User defined parameters include pipe roughness, flow rate, static head, pipe length, unit labor cost, unit power cost, and ENR construction cost index. Other input parameters to be selected by the user include pump efficiency, annual pump operation time, interest rate, and period of bond issue. The spreadsheet can be used in the optimal design of transmission mains. This spreadsheet can also be used for sensitivity analysis of input variable.

---

1. Engineer, Lower Colorado River Authority, P.O.Box 220, Austin, Texas 78767

INTRODUCTION

A water transmission system is generally composed of two major components: pipeline and pump station. The pipeline for a water transmission system requires large initial capital expenditures and nominal annual operations and maintenance costs. Pump Stations also require a capital expenditure, but the major part of the pump station cost is in annual operations and maintenance expenses, including the cost of electricity.

In designing a transmission system, the use of a smaller pipe size will require a smaller initial capital cost, but the annual cost of electricity will increase with the increase in friction loss due to higher flow velocities. To determine what transmission system is most economical for a given set of conditions, it is necessary to evaluate the trade-off between initial capital costs and annual operations and maintenance costs over the design life of the project. This paper presents a Lotus 1-2-3 spreadsheet capable of designing an optimal conveyance system.

PROCEDURE

A Lotus 1-2-3 spreadsheet was developed for design and financial evaluation of pipe lines and pump stations. This spreadsheet designs the pipeline and pump stations using the parameters defined by the user. This spreadsheet also performs financial evaluation of the system to optimize the capital cost and annual operations and maintenance costs. Unit costs for pipelines from Ref.1 and the relationship for estimating cost of pump stations from Ref. 2 were used in this spreadsheet.

Friction loss in a transmission main is computed by using the Hazen-William formula for which a roughness is assigned by the user to reflect the type of the material of the pipe. If old pipe is used for the project, the age of the pipe will also dictate the roughness coefficient. User defined values of flow rate, static head, length of pipeline, and pump efficiency are used in conjunction with estimated friction loss for designing pump stations.

Financial analysis of the initial capital cost and annual operations and maintenance costs was performed by converting all costs to equivalent annual costs. User defined values of unit power cost, labor cost, annual pump operation time, and length of pipe are used for estimating construction and annual operations and maintenance costs. The Engineering News Record (ENR) construction cost index is used to reflect the geographic location of the project. Assigned values of interest rate and period of bond issue are used to compute amortized capital costs. Table-1 shows a set of example input data used for the spreadsheet.

Table-1. Example Input Data

Input Variables	Example Input Data
Hazen-Williams Roughness Coefficient	100
Labor Cost ($/Hour)	15
Pump Efficiency (Percent)	67
Power Cost ($/Kw-Hr)	0.07
Annual Pump Operation (Hrs/Yr)	3,650
Interest Rate (Percent)	10
Period of Bond Issue (Years)	20
Design Flow Rate (GPM)	1,000
Static Head (Feet)	20
ENR Construction Cost Index	3,184
Pipe Length (Feet)	10,000

This spreadsheet designs pipelines and pump stations for different velocities. These designs are used to estimate initial capital costs for pipeline and pump stations and annual operations and maintenance costs for the pump stations. Amortized annual capital costs are computed by using the financial parameters assigned by the user.

This spreadsheet selects the most cost effective design to reflect minimum annual cost including both amortized capital cost and annual operations and maintenance costs. Figure 1 shows total annual costs for various pipe sizes. This figure shows that with the increase in the pipe diameter, amortized annual capital cost for the pipe line increases. Consequently, both amortized annual capital cost for the pump station and annual power cost decrease with the increase in pipe diameter. This figure shows that total annual cost initially decreases with the increase in pipe diameter and

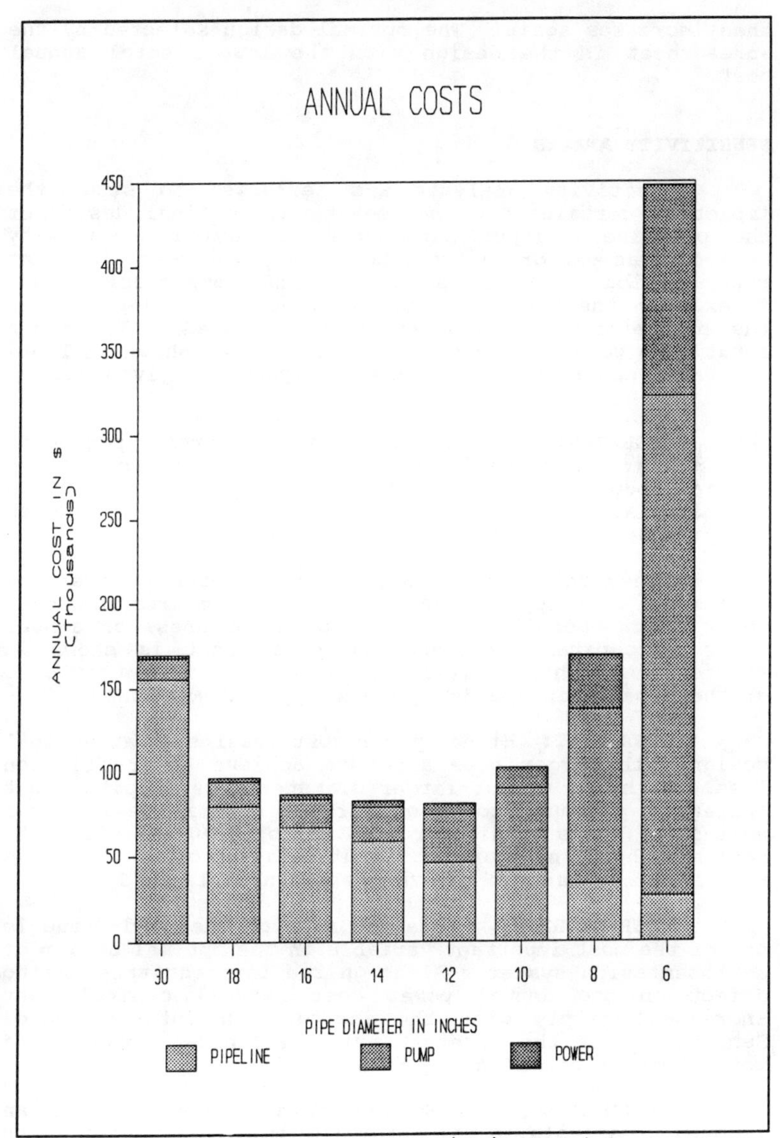

Figure 1 Annual costs of transmission system

then increases again. The optimal design selected by the spreadsheet is the design with the lowest total annual cost.

SENSITIVITY ANALYSIS

Sensitivity analysis was performed to study the impact of certain input parameter on the final design of the pipeline. Input parameters for which sensitivity analysis was performed are roughness, power cost, interest rate, period of bond issue, and ENR construction cost index. In these sensitivity analyses, only the value of the parameter in consideration was changed. All other input data were kept constant to the values shown in Table 1. Brief descriptions of these analyses are given below.

ROUGHNESS: Pipe roughness is an important parameter in designing a transmission main. Value of the Hazen-William roughness coefficient for a given pipe depends on the material of the pipe. Due to deposition of solids on the pipe wall, roughness increases with time.

Higher roughness requires larger pumps to overcome additional friction loss and consequently results in higher power cost. Sensitivity of the pipe roughness on annual costs of capital ivestment and power cost is shown in Figure 2. In this analysis it was assumed that unit cost of the pipe is independent of the type of pipe used.

POWER COST: Higher power cost results in an optimal design with larger pipe sizes to achieve lower friction loss. With the use of larger diameter pipe, capital cost increases. Annual costs of various optimal designs for unit power costs ranging from $0.06 to $0.30 per Kilowatt-Hour are shown in Figure 3. Input data for other variables were kept constant to the values shown in Table 1.

INTEREST RATE: Interest rate of the bond issue is one of the most important variable in the optimal design of a transmission system. Although the interest rate has no effect on the annual power cost, annual capital cost increases sharply with the increase in interest rate. Sensitivity of the interest rate on capital, power, and total cost is shown in Figure 4.

PERIOD OF BOND ISSUE: Period of the bond issue is an important variable in the optimal design of a transmission

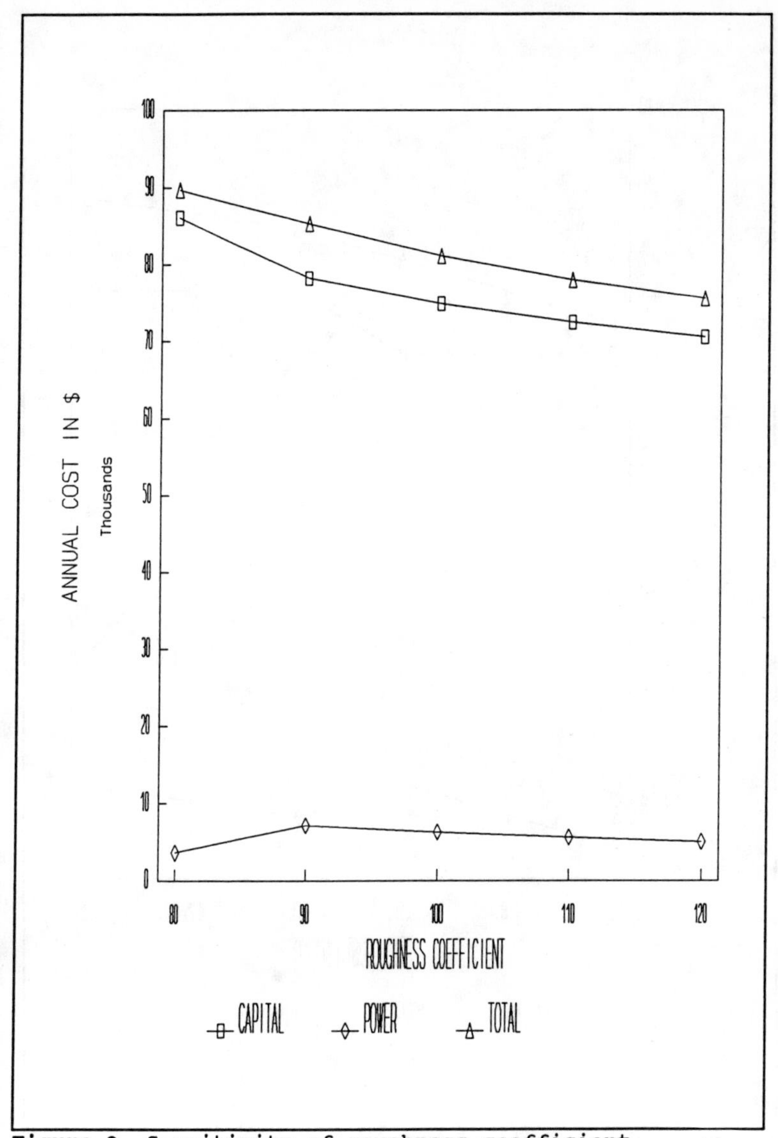

Figure 2 Sensitivity of roughness coefficient

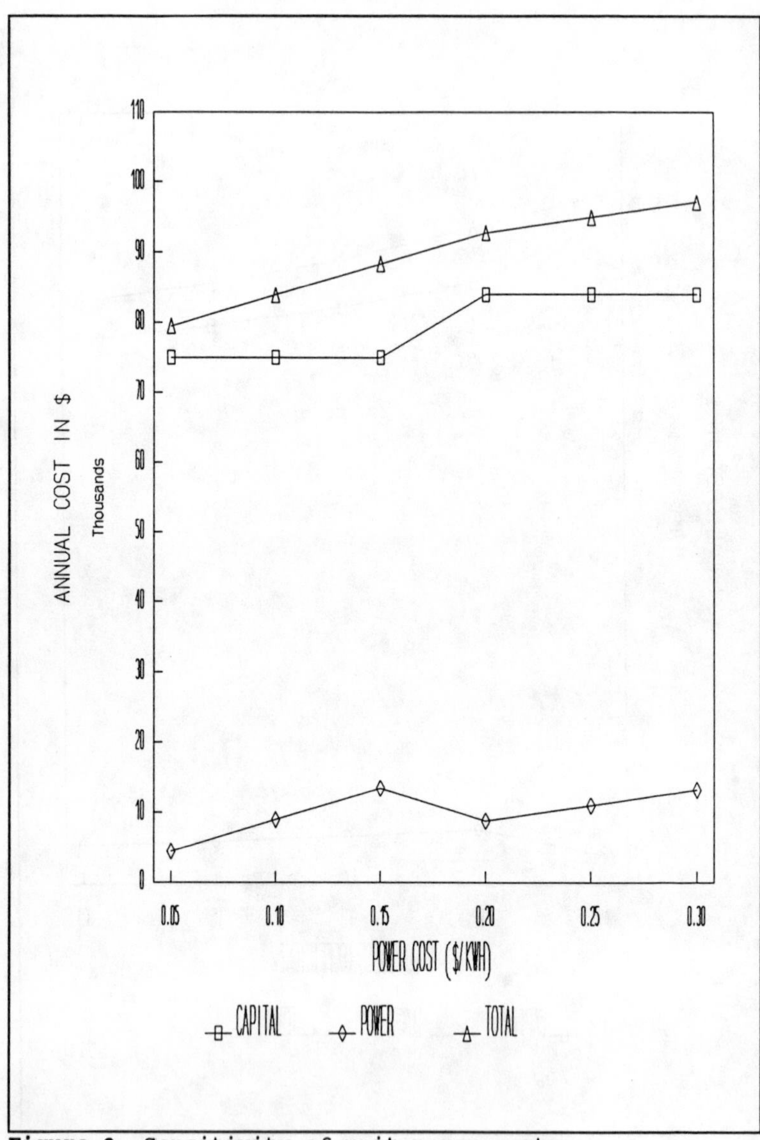

Figure 3 Sensitivity of unit power costs

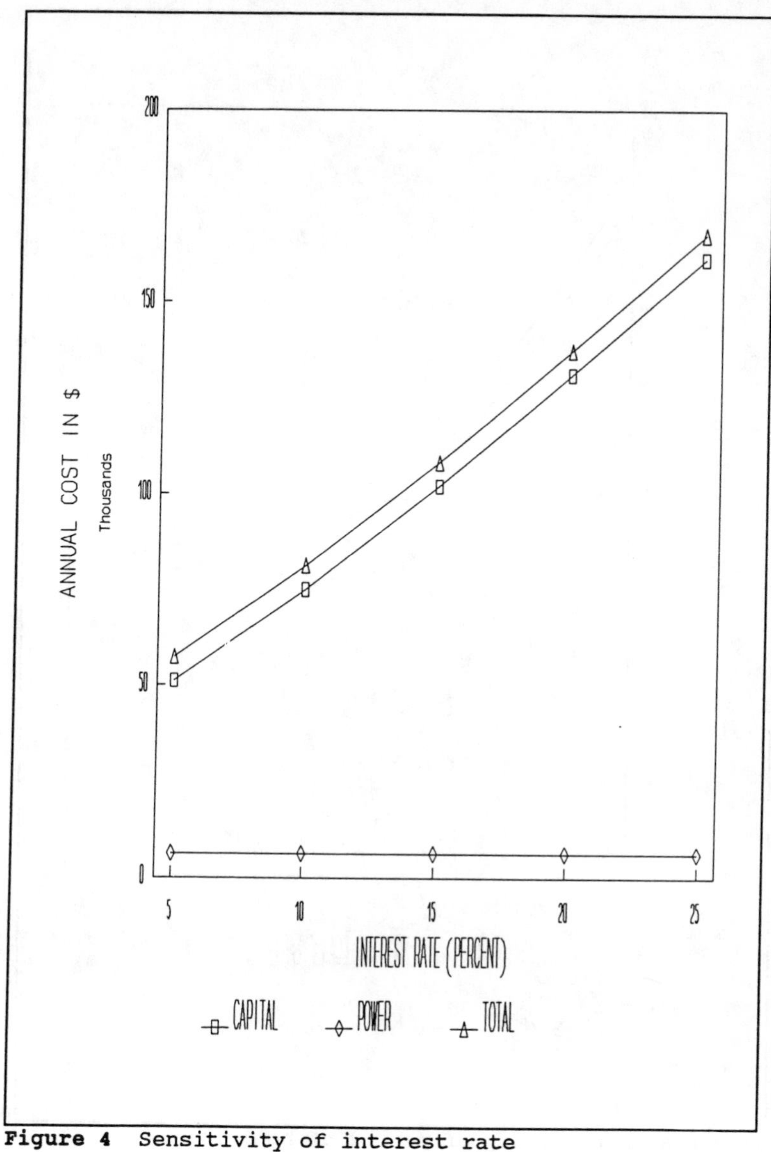

Figure 4 Sensitivity of interest rate

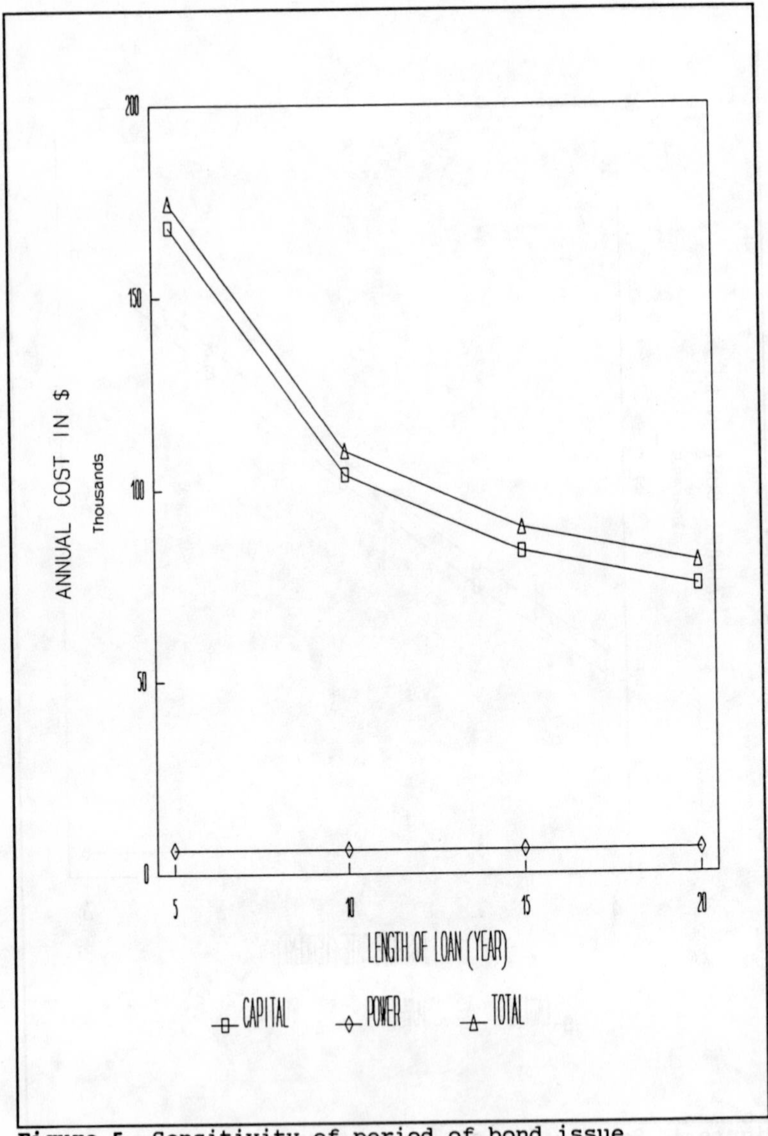

Figure 5 Sensitivity of period of bond issue

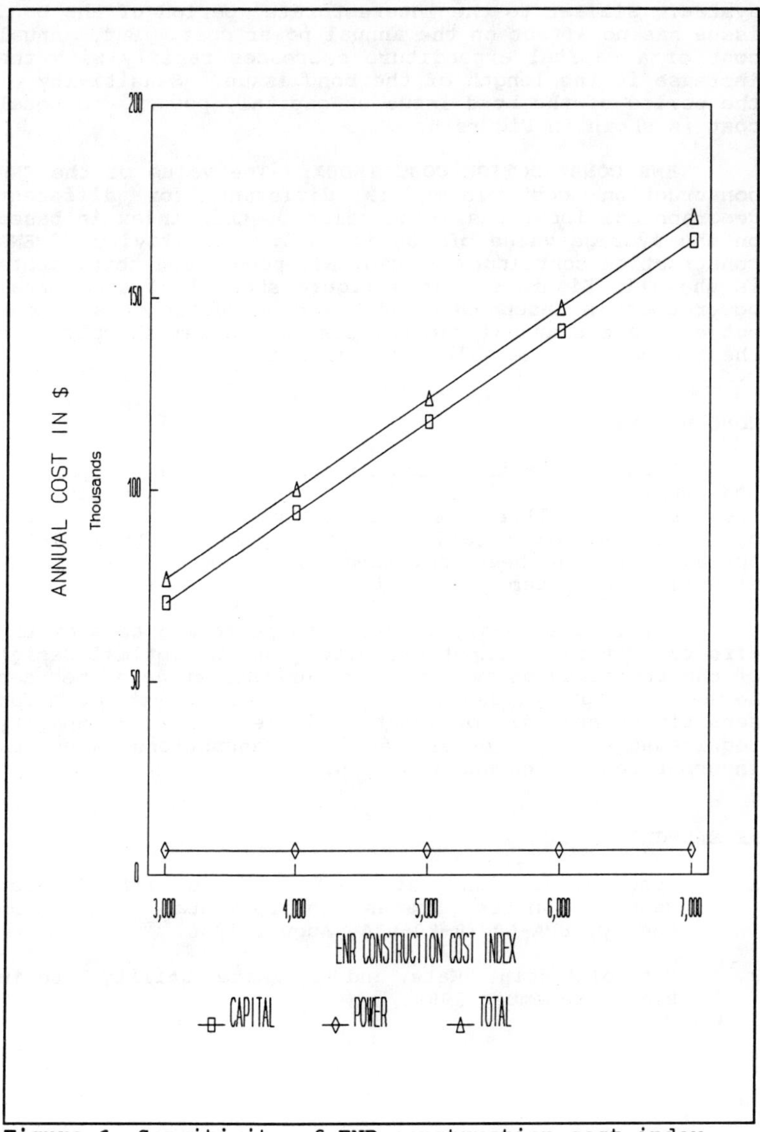

Figure 6 Sensitivity of ENR construction cost index

system. Similar to the interest rate, period of the bond issue has no affect on the annual power cost. But, annual cost of a capital expenditure decreases rapidly with the increase in the length of the bond issue. Sensitivity of the period of the bond issue on capital, power, and total cost is shown in Figure 5.

ENR CONSTRUCTION COST INDEX: The value of the ENR construction cost index is different for different geographical locations. The value of this index in based on the average value of 100 in 1913. Sensitivity of ENR construction cost index on capital, power, and total costs is shown in Figure 6. This figure shows that the annual power cost is independent of ENR construction cost index but equivalent annual capital cost increases sharply with the increase in the value of this index.

CONCLUSIONS

Design of a transmission system includes pipeline, and pump station. Use of a larger diameter pipeline results in a smaller power cost due to lower friction loss but requires larger capital expenditures. A Lotus 1-2-3 spreadsheet has been developed for optimal design of transmission system.

Sensitivity analysis has been performed to show the effects of various input parameters on the optimal design of the transmission system. The spreadsheet developed can be used for optimal design of a transmission system and for sensitivity analysis of input variables. Future capacity requirements and other design assumptions can be incorporated at the users option.

REFERENCES

1. Donovan, J.F. and Bates J.E., "Guidelines of Water Reuse", Unites States Environmental Protection Agency, EPA-600/8-80-036, August 1980.

2. City of Austin, "Water and Wastewater Utility Interim Plan", December 1986.

Analysis of Alternative Pipe Materials for a Major Public Water Agency

Michael T. Stift, P.E.[1], ASCE - Associate Member ASCE
and
George F. Ruchti[2]

Abstract

A major southwestern public water agency has historically specified the exclusive use of prestressed concrete cylinder pipe (PCCP) for large diameter water transmission mains. Due to the large diameter and footage of the pipe required for a new pipeline and the logistics of the PCCP suppliers, it was felt that competition for this project could be severely curtailed. A general overview of alternative pipe materials, capable of meeting or exceeding the project parameters, were therefore evaluated.

In the 66-inch through 90-inch diameter range required of this project, there are only a limited number of products that have been developed for pressure conditions. It was decided to evaluate one product, steel pipe, for use as an alternative material.

This paper presents a product review of engineering and construction considerations for PCCP and steel pipe. Bedding and backfilling, installation characteristics, pipe design criteria, lining, coating and corrosion protection requirements are evaluated, and a recommendation that steel pipe be permitted as an alternate pipe material is made.

----

[1]Civil Engineering Consultant, 610 W. Fifth Avenue, Escondido, CA 92025.

[2]Associate Principal, Rocky Mountain Consultants, 8301 E. Prentice Avenue, #101, Englewood, Co. 80111.

Project Parameters

The project consisted of 5,200 linear feet of 90-inch, 13,800 linear feet of 78-inch and 6,050 linear feet of 66-inch diameter pipe with internal pressures and design flows for each reach shown in Table 1.

Table 1

Diameter (In.)	Ultimate Flow (MGD)	Design Test Pressure PSI
90	186	140
78	108	202
66	73	161

Construction of the proposed pipeline was generally to be within the existing street right-of-way and required installation of jacked casings beneath several existing railroad tracks and a large diversion channel.

Cover conditions over the top of the pipe were to be as shown in Table 2. Typical pipe zone trench widths were to be limited to the outside diameter of the pipe plus three feet clearance (1-1/2 feet each side of the pipe) to allow adequate placement and compaction of the bedding material. Native soils were expected to be silty sand with a maximum unit weight of 110 pounds per cubic foot.

Table 2

Diameter (In.)	Minimum Cover (Ft.)	Maximum Cover (Ft.)
90	5	16
78	5	15
66	5	30

The corrosion potential of the soil was evaluated by a qualified corrosion engineering firm acting as subconsultant to the engineer. In general, the soil was found to be moderately corrosive for most of the 78-inch and 66-inch pipeline, with one stretch of 6,100 feet in the moderate to high category. The soil corrosivity potential along the 90-inch pipe alignment was found to be low.

Material Comparison

PCCP and steel pipe function in different manners, and, as such, do not permit direct comparison. These differences should not be evaluated in the context of strengths and weaknesses, but rather as individual characteristics of materials which have been used successfully for past pipeline installations. An outline review of the primary design criterion and individual characteristics can be found in Table 3. Explanations of these material characteristics are as follows:

Table 3

Design Criteria	Material Characteristics: PCCP	Material Characteristics: Steel Pipe
Structural Components	Heterogeneous composite of cement mortar, concrete, high tensile wire and steel cylinder	Steel Cylinder
Design Analysis	Rigid conduit	Flexible Conduit
Internal Pressure	Resisted by compression in the concrete	Resisted by tension in the steel wall
External Load	Primarily pipe material resistance	Primarily soil interface
Lining	Concrete	Cement mortar (although other material available)
Coating	Cement mortar or concrete	Optional-cement mortar, tape, and coal tar investigated
Watertight Membrane	Embedded steel cylinder	Steel wall
Joints	Rubber gasket	Rubber gasket and welded
Corrosion Resistance		
Low soil corrosivity potential	Bonding joints and alkaline environment with monitoring system	Bonding joints and monitoring system
Moderate soil corrosivity potential	Bonding joints and possible cathodic protection system	Bonding joints and possible cathodic protection system
High soil corrosivity potential	Bonding joints and cathodic protection system	Bonding joints and cathodic protection system

Structural Components

PCCP is composed of a steel cylinder, lined with or embedded in concrete. This concrete and steel cylinder core is prestressed (post-tensioned) by helically wrapping a high strength spring steel wire under tension. The wire is then covered with a brush mortar or cast concrete coating. Minimum standards for production of PCCP are generally covered by AWWA Specifications C301.

Steel pipe is produced by forming a steel cylinder either spirally from coils or circumferentially from plates (with circumferential and staggered longitudinal welds). Steel pipe fabrication for water works applications is generally covered by AWWA Specification C200.

Both products have a long history of water transmission main usage.

Design Analysis

PCCP is analyzed as a rigid structure, capable of supporting itself when loaded externally. Although the pipe does undergo some deflection under external loading conditions, this deflection is relatively minor and does not develop substantial side support from the outer soil envelopes. Efforts in the bedding of the pipe, however, especially under the pipe haunches (up to approximately one-third of the pipe outside diameter), can considerably assist the pipe's load carrying capacity.

Rigid pipe structural analysis requires that the stresses developed by the internal pressure conditions be added to those determined by the external load analysis to create a combination stress design. The governing minimum standard for the design and manufacture of PCCP is AWWA C301.

The appendix to C301 permits structural analysis by either of two procedures: Appendix A which is an empirical analysis comparing internal pressure ratios to external pressure ratios graphically for various operating and transient conditions; and Appendix B, which compares the same conditions through stress analysis. The Appendix B graph evaluates various internal and external loading conditions for specific pipe wall design, whereas the Appendix A graph allows for a rapid evaluation of pipe wall configurations for a fixed set of design conditions. This latter procedure presupposes that three-edge bearing testing for a statistically valid determination of values has been performed.

Pipe strength characteristics are determined by the amount of compression in the concrete and steel cylinder core. Compression is determined by the diameter, spacing and stress level of the helically wrapped prestress wire and the wall thickness of the core. Prestress wire stress levels usually range from 201,000 to 293,000 pounds per square inch (psi), and wire diameters from No. 8 gage (0.162 inches) to 5/16 inches. Core thicknesses are determined by available equipment with minimum dimensions of D/16, where D equals the inside diameter of the pipe in inches. The watertight membrane, the steel cylinder, tends to resist the prestressing action. Dual economy of steel cylinder and prestress wire cost is realized by providing as thin a cylinder as practicable. As the steel cylinder is the primary component to resist thrust forces, as well as to provide beam strength, the thickness must, at times, be increased or steel bars be cast in the core and attached to the steel joint rings.

Steel pipe, on the other hand, is designed as a flexible conduit. Internal pressure, external load, buckling loads, and pipe handling criteria are analyzed independently. The latter two design criteria, buckling and handling, are not normally analyzed for PCCP due to this pipe's inherent ability to resist commonly applied forces. Each of these design criteria are presented later in this paper. The important thing to recognize in the design of a flexible conduit is that the pipe completely resists the internal pressure conditions by placing the wall in tension, but that the primary structural resistance to external loading is provided by the surrounding soil envelope.

Internal Pressure

Internal pressure consists of essentially three conditions: a working or design pressure (Pw) of sustained application; a test, or semi-transient pressure (Pt) of some limited time interval; a surge (water hammer) or transient pressure (Pwh) that is instantaneous in duration. Each of these conditions is handled somewhat differently.

For PCCP, a specific wire force and core dimension is considered and a pressure to obtain zero compression in the core (Po) determined. By trial and error, various combinations of loadings are analyzed to determine the acceptability of the design. Commonly, these combinations include:

A. Pw versus earth (dead) load (Wd) on the design graph;
B. Pw versus Wd plus live load (Wl) on the transient graph;
C. Pt versus Wd+Wl on the transient graph;
D. Pw+Pwh versus Wd on the transient graph.

Once all conditions are satisfied and each condition falls under the respective graph, the design is acceptable.

For steel pipe, the pressures are handled somewhat consistently to the approach for PCCP. As previously discussed, combination loadings for flexible pipe analysis are not considered. The formulas for determining the required steel pipe wall thickness is t = PD/2S, where P equals pressure in psi; D equals pipe diameter in inches (mean diameter although outside diameter is frequently used for convenience); S equals the allowable stress in the steel cylinder wall. S varies with the pressure condition and the yield strength of the material. Commonly, yield strengths of 36,000 and 42,000 psi are specified. Allowable maximum stress levels of 50 percent of yield for Pw, 60 percent for Pt and 75 percent for Pw+Pwh are frequently used.

External Load

As stated for internal pressures, external loads are of the sustained and transient variety. Dead loads (Wd) due to the earth cover on PCCP are determined by the Marston equations for trench and embankment conditions. This analysis considers that the pipe retains shape while the soil on each side settles differentially. In the case of steel pipe, the pipe deflects with the soil and is, therefore, considered to be sustaining the load created by the prism of earth directly over the conduit. The formula is Wd=HBcW, where H equals height of cover in feet, Bc equals pipe outside diameter in feet, and W equals the unit weight of the soil in pounds per cubic foot. Live loading conditions to resist vehicle (truck), railroad or special transport or construction type loadings can be determined using AASHTO (American Association of State

Highway and Transportation Officials) or AREA (American Railway Engineers Association) developed formulas, or by the application of Boussinesg equations.

The resistance of PCCP to these loading conditions were briefly discussed previously. A complete dissertation of the applicable formulas for a stress analysis of PCCP would prove too cumbersome for this paper. This analysis can be found in AWWA C301, Appendix B.

Steel pipe design centers on the ability of the pipe and soil interface to resist deflection. In order to prevent damage to the various lining and coating components of a steel pipe, as well as to the steel shell itself, deflections should be limited to:

Mortar lined and coated steel pipe - 2 percent of outside pipe diameter

Mortar lined and flexible coated steel pipe - 3 percent of outside pipe diameter

The modified Iowa deflection formula, used to determine flexible pipe deflection, is:

$$\text{Delta } x = Dl\ KWr^3/(EI+0.061\ E'r^3)$$

Where Dl is the deflection lag factor (1.0 to 1.5 for steel pipe), K is the bedding constant (generally 0.1 for a well bedded pipe), W is the load per unit pipe length in pounds per linear inch of pipe and r is the pipe radius in inches. The external loading is resisted by two structural members: the pipe itself (EI) and the supporting soil ($0.061E'r^3$). Therefore, the analysis and development of the soil support is the component most relied upon to provide an economical installation by reducing the required steel cylinder wall thickness. A thorough discussion of this formula can be found in AWWA Manual of Water Supply Practice M-11.

Lining

The concrete lining of PCCP is actually an integral structural component of the pipe. Additional protective measures (such as epoxy paints, PVC sheets, etc.) can be applied, but these measures are rarely required for a water pipeline.

There is considerable flexibility in the choice of an interior lining for steel pipe. Cement mortar lining, either shop or field applied in accordance with AWWA C205 or C602 respectively, are generally satisfactory for the vast majority of buried water pipelines. Other options include a liquid epoxy in accordance with AWWA C210, as well as a wide array of paints for special applications and exposed conditions. Due to EPA (Environmental Protection Agency) and OSHA (Occupational Safety and Health Act) regulations, as well as the advent of technologically advanced coatings and linings, coal tar applications in accordance with AWWA C203 have become infrequently specified as environmental clearance has been difficult to obtain.

Coating

Although not part of the concrete core, the PCCP brushed mortar or cast concrete coating provides some structural support and is the only practical method of providing physical and chemical protection for the prestressing wire.

Steel pipe with a brush mortar coating (AWWA C205) provides protection similar to that of PCCP. A tape coating conforming to AWWA C214 provides an excellent dielectric barrier between the soil and pipe interface.

Joints of any coating system should be completed in kind.

Watertight Membrane

For either PCCP or steel pipe, the welded steel cylinder provides watertightness. Every completed cylinder with joint rings attached (where applicable) is hydrostatically tested to assure zero leakage. In the case of steel pipe, this test also assures the structural integrity of the completed pipe with regard to internal pressure conditions.

Joints

PCCP utilizes a Carnegie shaped spigot ring and formed bell ring welded to the steel cylinder. An O-ring gasket provides the seal between adjacent sections of pipe.

Steel pipe may be manufactured using joint rings similar to those of PCCP, bell and spigot ends formed by shaping the ends of the steel cylinder, or by providing joints prepared for field welding. However, in the larger pipe diameters, because of the flexible nature of steel pipe, joints become more difficult to stull to maintain roundness. Therefore, the authors generally recommended field welding of the steel pipe joints for this project.

All of the above joints have a long history of providing an excellent seal throughout the normal range of waterworks pressures. The gasketed joints formed in the ends of a steel pipe provide good continuity of structural characteristics, but are limited in the wall thickness in which they can be formed. Field welded joints provide resistance to longitudinal forces such as thrust and are recommended where internal pressures are very high. Lap field welded joints allow for some misalignment and joint deflection, while butt welded joints provide a somewhat greater degree of strength. Field welded joints are more expensive to install than gasketed joints.

Corrosion Resistance

Coatings providing various levels of corrosion protection were discussed in "Coating" above.

Pipe corrosion at the soil and pipe interface is the main reason for pipeline failures. Soil corrosivity potential can change with time as land use changes. The cost of replacing a pipeline or waiting for catastrophic pipe failure to occur can be overcome by providing a corrosion monitoring system as part of the pipeline design. For PCCP or steel pipelines, electrical continuity by welding of joints or bonding jumpers connected across gasketed joints with a corrosion monitoring system should be inherent to the pipeline design regardless of soil corrosivity potential.

In soils of low corrosivity potential, the cement mortar coating creates an alkaline environment which passivates the steel members, thereby inhibiting corrosion. In more aggressive soils, the application of a cement mortar sealing compound, and/or cathodic protection may be required.

Cathodic protection consists of connecting sacrificial anodic material to the piping system. Sacrificial anodes may be electrically connected at various intervals along the pipe, or may be positioned in one or more "beds" using impressed DC current to encourage electrical activity to protect larger sections of pipe. For longer pipelines, the impressed current system is generally more economical.

The quantity of sacrificial anodic material and corresponding current flow is directly proportional to the surface area of unprotected metal. Cement mortar coating provides a relatively poor dielectric barrier compared to tape or coal tar coatings and, therefore, has higher costs for those pipeline sections requiring the application of a cathodic protection system. Tape and coal tar coatings provide an excellent dielectric barrier, and as a result, long term costs for impressed current systems are generally lower.

The services of a qualified corrosion engineer should be employed to determine the adequacy of the coating system with regard to the surrounding soil corrosivity and electrical interference potential and to make recommendations for and provide design of a corrosion monitoring and possible cathodic protection system.

Competition

Proximity of manufacturing facilities, shipping costs, material costs, and more importantly, specified pipe design parameters, are factors in determining the competitiveness of each product. In the waterworks industry, it has been found that legitimate competitive bidding generally results in overall project cost savings to the owner.

Conclusions and Recommendations

Both large diameter PCCP and steel pipe have proven service records and are reliable products. In general, the manufacturing of steel pipe is less expensive than PCCP. This advantage is offset by increased installation costs for welding steel pipe joints and improving the pipe bedding and soil compaction requirements over that required for PCCP. Generally, the increased construction requirements for steel pipe offset the pipe cost savings in a competitive bid. The selected pipe material is a matter of the optional pipe material cost in conjuction with the contractor's pipe installation proficiency resulting in economy for the owner, without relaxing project design criteria or sacrificing product design life.

For this project, the recommended specifications were as follows:

PCCP - Pipe is to be bedded in sand for 6-inches below the pipe to a point 3/10 of the pipe outside diameter high. The bedding material is to be compacted to 90 percent standard proctor density. A corrosion monitoring system is to be employed for the entire pipeline, entailing electrical bonding jumpers across gasketed joints. In addition, 6100 feet of 78-inch pipe required an exterior epoxy coating and cathodic protection system in known areas of high soil corrosivity. The application of a corrosion monitoring system over the entire pipeline with testing for corrosivity potential and assessment for the need for additional cathodic protection measures is recommended after six months.

Steel pipe - Pipe is to be bedded in sand from 6 inches below the pipe to a point 7/10 of the pipe outside diameter for cement mortar coated pipe and 12 inches over the top of the pipe for tape coated pipe. All bedding material is to be compacted to 90 percent standard proctor density which is expected to develop a minimum E′ of 1,000. Joints for the steel pipe shall be field welded. The 66-inch diameter pipe may have an in-plant cement mortar lining, but larger diameters must be cement mortar lined in-place. The corrosion monitoring and cathodic protection system recommended for cement mortar coated steel pipe is similar to that of PCCP. The entire pipeline is to be placed under cathodic protection for tape or coal tar coated steel pipe.

Based on the design and installation requirements, steel pipe or PCCP will provide a satisfactory level of performance over the life of this project. The addition of steel pipe to the construction documents will increase competition and provide a more economical project for the owner.

A LOOK BACK IN TIME TO VERIFY LIFE CYCLE COST ANALYSES

John J. Meyer, P.E., Member, ASCE[1]

Abstract: The proper engineering design of any hydraulic structure requires consideration of different but inter-related fields of: 1) Planning, 2) Hydrology, 3) Hydraulics, 4) Structural, 5) Installation, 6) Durability, and 7) Economics.

The first 5 fields of pipe design are well established. However, the durability and economic aspects are generally not given proper consideration with pipe materials being selected on a first cost basis for many projects. Life cycle cost analyses have been developed to assist the engineer in his decision making process.

The author will use examples of past installations, actual construction costs, actual inflation rates based on Consumer Price Index and actual historical interest rates based on Federal Reserve System data to verify life cycle equations.

INTRODUCTION

"THE LOWEST BIDDER"

"It's unwise to pay too much, but it's worse to pay too little.

When you pay too much, you lose a little money -- that is all. When you pay too little, you sometimes lose everything, because the thing you bought was incapable of doing the thing it was bought to do. The common law of business balance prohibits paying a little and getting a lot -- it can't be done. If you deal with the lowest bidder, it is well to add something for the risk you run. And, if you do that, you will have enough to pay for something better."

John Ruskin

---

[1]Manager, Marketing Service, The Cretex Companies, Inc., 311 Lowell Avenue, Elk River, MN 55330.

During the latter part of this decade a particular emphasis has been placed on evaluating materials from a life cycle cost analysis viewpoint. The equations that have been developed for this are based upon accepted principles of economics, i.e. the "Present Value Equation." Much debate has been generated, not about the validity of these equations, but how the equations are applied and the factors that are chosen to perform the necessary calculations.

The present value equation provides the user with a measure of all costs, present and future, associated with a project over the project design life. The major factors which generate the greatest amount of discussion are the project design life, the service life of the material and the future interest and inflation rates. While certain engineering studies and policy decisions can define project design life and service life, predicting future interest and inflation rates is a less exacting science. Case studies of past interest and inflation rates show approximately a differential of 2% between the two rates.

The future value of costs can be defined as follows:

$$F = P(1+I)^n \quad \text{EQ. 1}$$

where: F = Future value
P = Amount of current costs
I = Inflation rate
n = Number of years after completion that costs are incurred

An illustration of this concept can be demonstrated by utilizing a chart developed by the U.S. Chamber of Commerce.

Using an inflation factor geared to the consumer price index complied by the Bureau of Labor Statistics, the chart yields the amount needed to match the buying power of a dollar in previous years.

Year	Inflation Factor
1960	3.78
1961	3.71
1962	3.70
1963	3.65
1964	3.61
1965	3.55
1966	3.45
1967	3.35
1968	3.22

1969	3.05
1970	2.88
1971	2.76
1972	2.67
1973	2.52
1974	2.27
1975	2.08
1976	1.97
1977	1.85
1978	1.71
1979	1.54
1980	1.36
1981	1.23
1982	1.16
1983	1.12
1984	1.08
1985	1.04
1986	1.00

The index for each year appears at the right. The figures show that for the 1986 year, it would take $3.78 to match the purchasing power of a 1960 dollar, $2.88 to equal a 1970 dollar and $1.36 for a 1980 dollar.

If one is dealing with larger numbers, say a contract of $300,000, it is just a matter of adding zeros. If the construction cost $300,000 in 1970, then the same construction would cost $864,000 in 1986 (2.88 x 300,000).

The present value factor converts future replacement costs to today's dollars:

$$\mathbf{PV} = \frac{\mathbf{F}}{\mathbf{(1+i)^n}} \qquad \text{EQ. 2}$$

where: PV = Present value
i = Interest or discount rate

An illustration of this concept can be provided by examining the banking system. Interest is the charge for borrowed money with the banking system concerned about the real rate of return. The real rate of return is the amount to be earned after accounting for inflation. For example if for one year the inflation rate is 4% and the interest rate is 6%, the real rate of return for the banking system is 2% (6% - 4%). Therefore, in the life cycle cost analysis one must address the issue of the inflation/interest differential.

Substituting EQ. 1 into EQ. 2 one obtains

$$PV = P\left(\frac{1 + I}{1 + i}\right)^n \qquad \text{EQ. 3}$$

The above equations were provided in order to illustrate the concepts that need to be understood for an effective life cycle cost analysis. The writer will apply these concepts to develop two additional and final equations to be used in the case analyses.

The first case is direct and straightforward: Material design life is equal to project design life.

$$EC = P \qquad \text{EQ. 4}$$

where EC = Effective Cost
P = Amount of Current Costs

The second case to be addressed is the material life being less than the design life of the project. The effective cost is the amount of current costs plus the present value total of all replacement costs adjusted for inflation.

$$EC = P + \Sigma PV \qquad \text{EQ. 5}$$

Substituting EQ. 3 into EQ. 5 one obtains

$$EC = P + \Sigma P\left(\frac{1 + I}{1 + i}\right)^n$$

or

$$EC = P\left[1 + \Sigma\left(\frac{1 + I}{1 + i}\right)^n\right] \qquad \text{EQ. 6}$$

EQ. 6 has appeared in print in a variety of formats other than that that is published here. Another form of EQ. 6 is provided for your reference:

$$EC = P\left[(1 - F^L) \div (1 - F^N)\right]$$

where EC = Total effective cost
P = Bid price (current dollars)
F = Inflation/interest factors
L = Project design life (years)
N = Usable material life (years)

Case Studies

As stated previously, the single most area of debate in using life cycle equations is to predict the inflation and interest rates over the course of the next 25 to 50 years. General rules of thumb can be provided by examining historical data relative to these rates. The author's intent here is to look back in time with the benefit of this historical data to verify documented cases. It should be noted also that the analyses that will be studied do not include such costs as maintenance, rehabilitation, detour, etc. The case studies will deal with actual bid costs, actual inflation rates, actual interest rates and actual replacement costs.

Case Number 1 involves a Wisconsin community. A bituminous coated corrugated metal pipe installed in 1956 was found to be in an advanced state of deterioration in 1984, with the invert corroded to the point where the pipe had to be replaced.

The bid prices in 1956 were $56,713.35 for concrete pipe and $51,994.29 for corrugated metal pipe, a difference of $4,719.06.

This difference was the apparent savings when corrugated metal was selected for this particular project. The service life of the metal pipe can be liberally assumed to be 25 years, based on the advanced state of deterioration found after 28 years of burial. The inflation rate between 1956 and 1984 averaged 6.1 percent per year based on the U.S. Department of Commerce records. The average interest rate (prime) during the same period averaged 8.7 percent per year based on Federal Reserve System data.

To find the effective cost of corrugated metal pipe that is equivalent to the cost of concrete pipe with a 100 year service life Equation 6 may be used with three replacements at 25, 50 and 75 years.

$$\text{EC METAL} = \$51{,}994.29 \left[1 + \left(\frac{1+0.061}{1+0.087}\right)^{25} + \left(\frac{1+0.061}{1+0.087}\right)^{50} + \left(\frac{1+0.061}{1+0.087}\right)^{75} \right]$$

$$\text{EC METAL} = \$104{,}337.38$$

The total cost of reinforced concrete pipe with a material life of 100 years may be found by Equation 4.

$$\text{EC CONCRETE} = \$56{,}713.35$$

If, in 1956, the owners had performed a least cost analysis, they would have found that the corrugated metal pipe had an effective cost of nearly twice as much as the reinforced concrete pipe for a 100 year project design life. The ensuing bid results for this project confirm the results of these calculations.

In addition, certain external factors played havoc on this project. The original pipeline was put in a storm sewer easement in a vacant field. Twenty eight years later, garages, apartment houses and offices are located in such a manner that there is not enough room for a backhoe to swing in order to install replacement pipe. Portions of the street must be removed and replaced for replacement pipe to be installed. Therefore, in this case the true effective cost of the project is not $56,713.35 or $104,337.58, but in excess of $200,000 as the community chose to install steel plates to the existing pipe as the solution to the drainage problem.

Case Number 2 involves highway culverts constructed in 1957. Corrugated metal culvert pipe, 18-inch and 24-inch in diameter, was installed under a state highway. Approximately 1,300 linear feet of the culvert was scheduled to be replaced by 18-inch and 24-inch reinforced concrete culvert pipe. Using the Federal Highway Administration Culvert Inspection Manual, the culverts would be rated two or failure.

The bid price on the original contract for the 18-inch corrugated metal pipe was $3.25 per foot installed; the price for the 24-inch Corrugated Metal Pipe was $4.90 per foot installed. Reinforced concrete pipe was not specified in the original contract.

Twenty-seven years later, with conditions such as bedding, depth and footage the same, bids were taken on replacing the failed culvert. The bid prices were $21.60 per foot installed for the 18-inch corrugated metal pipe and $23.40 per foot installed for the 24-inch corrugated metal pipe. Pavement replacement, traffic barricades and other related items were separate bid items. The alternate of reinforced concrete pipe in 18 and 24 inch sizes were bid at $25.50 per lineal foot and $29.00 per lineal foot respectively.

These numbers suggest a number of alternatives to be examined. If the author uses the 1957 prices as a base and also uses the corresponding Consumer Price Index and Producer Price Index for that period, the 1984 bid prices are approximately 2.5 to 3.0 times too high. This hypothesis does not appear to be correct when the author surveyed similar type work and bid documents for the 1984 construction year. It appears from additional surveying by the author of 1957 records that the base prices used in this case may be low or have been offset by other bid items in the original contract. Unfortunately Department records for doing this type of historical work had not attained the level of sophistication that is presently enjoyed.

This example could be reworked in two ways the first being to project twenty-five year cycle price increases based on the first twenty-five year cycle, i.e. 18-inch corrugated metal pipe \$3.25 (1957) to \$21.60 (1984) and 24-inch corrugated metal pipe \$4.90 (1957) to \$23.40 (1984) for three additional cycles. It should be quite obvious that this exercise will produce results which would be drastically in favor of concrete pipe.

The second alternative would be to use the 1984 bid prices assigned over a 50 year project life and compute one 25 year cycle based on the historical inflation/interest cycle from 1959 to 1984. The average inflation rate was 6.31; the average interest rate was 8.13. Using Equation 6 for the 18-inch corrugated metal pipe would yield $\$21.60 \left[1 + \left(\frac{1+0.06.31}{1+0.0813}\right)^{25}\right] =$ 21.60 + 14.13 = \$35.73 per lineal foot for an effective cost. Similarly, the effective cost for the 24-inch corrugated metal pipe would be $\$23.40 \left[1 + \left(\frac{1+0.0631}{1+0.0813}\right)^{25}\right]$ = 23.48 + 15.31 = \$38.71 per lineal foot.

The effective cost of the 18-inch reinforced concrete pipe remains \$25.50 per lineal foot; the effective cost of the 24-inch reinforced concrete pipe remains at \$29.00 per lineal foot.

Case Number 3 addresses a two lane primary highway that is 14 feet over twin 132-inch structural plate pipe. The invert corroded causing loss of fill and structural strength. The 1966 price of this installation was \$34,201.00.

In 1980, the replacement cost was \$99,481.00 for twin 96-inch reinforced concrete pipe. The inflation rate for the 14 year period from 1966 to 1980 was 7.46percent per year. Using Equation 1, discount \$99,481.00 to the year 1966 at 7.46% or $\$99,481.00 = P\,(1+.0746)^{14}$. Therefore, P = \$99,481.00 x .3652 or P = \$36,330.46 which is the effective cost of the reinforced concrete pipe. The difference between the effective cost of the reinforced concrete pipe at \$36,330.45 and the original cost of installation at \$34,201.00 is \$2,129.46.

The effective cost of the structural plate pipe can be computed over the varying life expectancies of the product. The average interest rate for the stated time period is 8.29 percent per year. The actual, historical life cost to date can be computed by using Equation 6 where $EC = \$34,201 \left[1 + \left(\frac{1+0.0746}{1+0.0829}\right)^{14}\right]$ or EC = 34,201 + 30,708.08 = $64,910.88. Note that this cost of the structural plate pipe is for a 28 year period versus the $36,330.45 of the reinforced concrete pipe for its life expectancy. In fact if a 70 year life expectancy were chosen for both products, the effective cost for reinforced concrete pipe remains at $36,330.45 while the effective cost for the structural plate pipe is $139,358.82.

YEAR	EFFECTIVE COST AT 7.46% INFLATION, 8.29% INTEREST	
	REINFORCED CONCRETE	STRUCTURAL PLATE
0-14	36,330.45	34,201.00
15-28	36,330.45	64,910.88
29-42	36,330.45	92,455.57
43-56	36,330.45	117,176.05
56-70	36,330.45	139,358.82

Several recent publications have suggested that owners of projects fund a project to the level of the highest initial cost, install the project with lowest initial costs, and invest the difference in order to fund the second installation and/or any further installation. The author will apply that strategy to this case study in two ways.

As stated previously, the difference in costs of the two materials in 1966 is $2,129.46. Under the first consideration, the owner would invest this amount to fund only one replacement again using the structural plate material. Equation 2 would be used to solve for the interest rate or $2,129.46 = \frac{64,910.88}{(1+i)^{14}}$. Solving this equation for i, the required investment rate is 27.6%. This rate is 19.31% (27.60 - 8.29) over and above the average interest rate. Again this proposal allowed for no additional replacements.

The second consideration is to invest the initial difference and fund the 1980 replacement using the concrete pipe material. This is a permanent replacement cost. Again using Equation 2, $2129.46 = \frac{99,481.00}{(1+i)^{14}}$ or i = 31.6% which is approximately 3.8 times the average rate of 8.29%. In fact, the owner should have invested $30,621.00 $\left(PV = \frac{99,481.00}{(1+0.829)^{14}}\right)$ in order to fund the second replacement.

SUMMARY

Three historical case examples have been presented in an effort to give the reader a better understanding of certain drainage pipe installations economic impacts by using the principles of least cost analysis. These case studies have the benefit of a "hindsight" perspective and are not meant to be interpreted as a criticism of those who were required to make the original installation decisions. The purpose of this writing to provide future decision makers with historical evidence that product longevity is, in many cases, more valuable than lowest initial cost. There have been a great number of factors which have contributed to the infrastructure crisis facing all project owners; perhaps a portion of this crisis could have been avoided by selecting products based on longevity.

In only one example the author did consider investing funds to pay for required future installations. While this type of philosophy is not particularly applicable to publicly funded projects, it would be beneficial for all public agencies to carefully consider in what manner future projects will be funded. Public and private money supplies simply cannot grow continuously year after year at levels 10, 15 or 20 percent above the interest rate. Product longevity, by all rights, requires careful consideration.

The author wishes to thank the owners of the projects which were used for the case studies. Their patience and understanding have been greatly appreciated.

Life Cycle Cost Analysis
Discount Rates and Inflation

Thomas J. Wonsiewicz[1]

Abstract

Life Cycle Cost Analysis techniques are useful for comparing alternatives that have differing cash flows over the expected life of a project. The selection of an appropriate discount rate and the method of dealing with inflation can have a significant influence on the outcome of the analysis. In general, approaches that place a relatively low value on the money being invested, tend to favor alternatives with larger initial costs and lower future costs. Conversely, approaches that place a higher value on money, favor alternatives with lower initial costs and higher future expenditures.

This paper deals with the considerations that need to be made in selecting an appropriate discount rate. Opportunity costs, costs of capital, minimum attractive rates and costs of borrowing concepts are discussed along with techniques for dealing with inflation. The tendency to use borrowing costs for discount rates is shown to be incorrect. It is recommended that all public agencies follow the policy and guidelines of the Federal Government and adopt a 10% discount rate.

Background

Life Cycle Cost (LCC) or as it is also known, Least Cost Analysis, is the commonly used method to compare alternatives that are characterized by differing expenditures during the design life of a project. The technique is based on the common sense

---

[1]President, Lane Enterprises, Inc., 3705 Trindle Road, Camp Hill, PA 17011.

concept that "time is money." A dollar spent or earned in the future has a different value than a dollar spent or earned today. By placing a time value on money, future expenditures are discounted and brought back to the present period. A direct comparison of the total present values, or present worths, reveals which alternate is lower in cost.

Table 1 shows a comparison between two alternate methods of satisfying a 50-year design life requirement. Alternate A requires no future expenditure beyond the initial investment. Therefore, its present value is equal to the initial cost. Alternate B is expected to require a rehabilitative expenditure in year 40. It is assumed that the value of money is 9%.

Year	Expenditure	Present Value at 9% Factor*	Present Value at 9% Amount
Alternate A			
Year 0	$230,000	1.0000	$230,000
Alternate B			
Year 0	$195,000	1.0000	$195,000
Year 40	50,000	.0318	1,590
Total	$245,000		$196,590

*Present Value Factor = $1/(1+i)^n$ where i is the discount rate and n is the years after the initial investment.

Table 1. Least Cost Comparison

The above illustrates that even though the total expenditures of $245,000 for Alternate B is somewhat larger than for A, its present value is lower by about 15% and is the better buy.

Key Assumptions

Many assumptions and choices are required in establishing the costs associated with each alternative method of satisfying the design requirements for a drainage product. The Least Cost Analysis pamphlet published by the National Corrugated Steel Pipe Association (Leason and Standley, March 1986) provides a straightforward discussion of project design life, material service life, rehabilitation and residual value assumptions. Of the above, material service life can have the greatest influence on the LCC analysis. The designer must carefully and objectively assign a service life which takes into account the pipe material under consideration, as well as the environment, effluent and slope. Care should also be taken to consider rehabilitation techniques as a method to extend material service life. Rehab costs are often far less expensive than outright replacement.

The most influential assumptions involved in LCC analysis are linked with the treatment and selection of discount and inflation rates. It is important to have a firm grasp on how these assumptions affect results. The analyst must also understand the conceptual basis that underly the selection of appropriate rates. If not, confusion can result, particularly when an analyst applies methods supplied by industry that lead to conflicting results.

The problem lies in the functioning of the present value equation shown in Table 1. Most people do not have a good "feel" for how variations in either **i** or **n** affect results. In general, greater significance is given to future spending at low discount rates and less significance at higher discount rates, as shown in the following table and figure.

	Discount Rate		
Year	3%	6%	9%
0	$1.00	$1.00	$1.00
25	.48	.23	.12
50	.23	.05	.01
75	.11	.01	.01

Table 2. Present Value of $1 Expended at Various Intervals and Discount Rates

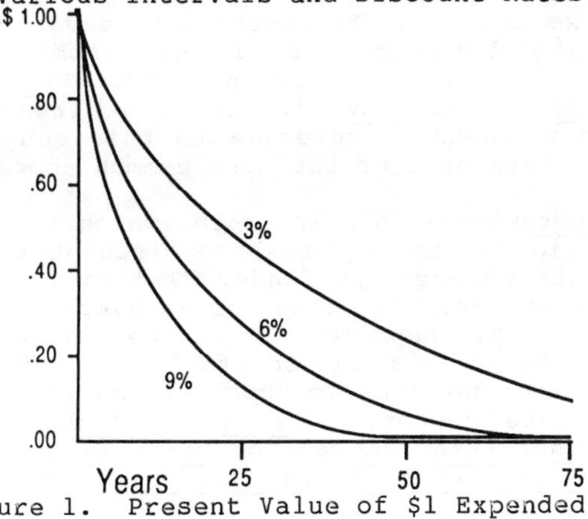

Figure 1. Present Value of $1 Expended at Various Intervals and Discount Rates

Since discount rates are treated exponentially, the three times increase from 3% to 9% at year 50, results in a 23-time decrease in the significance of present value (.23 vs .01). Further, this same aspect accounts for the fact that a 3 percentage point difference at higher rates (9% vs 6%) has much less of a present value significance than the same 3 percentage point difference at low rates (6% vs 3%).

If the analyst can internalize Figure 1, it will go a long way in helping to develop an intuitive feel for how changes in assumptions can affect results. For example, for projects with sizeable future expenditures, a change in discount rate assumptions from 3% to 9% will have a significant affect on results. However, the shift in timing of expenditures assumed in year 50, by ten years, will only have a small influence. It should also be noted that above a 6% rate, the effect of significant expenditures or residuals out beyond year 40, is small. The booklet Life Cycle Cost Analysis: Key Assumptions and Sensitivity of Results (Wonsiewicz, January 1988) provides further detail on sensitivity.

Discount Rates

Although their terminology differs, most experts agree that the discount rate should reflect the investors time value of money. It should represent a value that makes the investor indifferent between paying or receiving a dollar now or in the future. The problems arise in attempting to establish a rate that meets this criteria. It is not possible to establish one value to satisfy all situations. Accordingly, the analyst must become sufficiently involved and informed to adequately discharge his or her responsibility in selecting an appropriate discount rate. The Principles of Engineering Economy (Grant and Ireson, 1970) provides comprehensive coverage to this subject. The following treats some of the more common approaches.

°Cost of Borrowing - This approach, which is often used in the public sector, equates the discount rate to the interest rate on borrowed funds. The rate is based on either current interest rates, or an average over time. Conceptually, this approach forces the "value" of money being invested to be equal to its "cost." The cost of borrowing would never be appropriate and is rarely used in the private sector. The value of money invested must be higher than the cost of borrowing. Otherwise, no gain would ever accrue to the owners of an enterprise.

°Cost of Capital - This approach is more commonly used in the private sector. Conceptually, it sets the discount rate at a value that yields a return high enough to recover the cost of borrowed funds and provide the stockholders with a satisfactory return on their equity investment. Simple to grasp, tough to apply. The determination of a cost of capital rate depends on debt to equity ratios, the cost of borrowing and stockholders' return on equity requirements. Although this value will vary from company to company, most would likely fall in the 10% to 15% range (after-taxes).

The introduction of providing a return on stockholders' equity, makes value determination far less precise. Unlike banks who charge for the use of their funds, stockholders take their funds to the companies where they perceive they can earn an attractive return. Accordingly, a company on hard times may not be able to increase its equity, or even borrow additional funds. If returns are not adequate, the source of funds dries up.

°Opportunity Value - Theoretically, the investor's opportunity value is the most correct value to use for a discount rate. Practically, it is very difficult to determine. Opportunity value is determined principally by the availability and attractiveness of other investment alternatives. When funds are invested in one project, the opportunity to obtain a return by investing the funds elsewhere is foregone. The return on the forgone investment represents the opportunity value. Little more needs be said as to why this concept is difficult to implement. The analyst performing an LCC analysis is generally not in the position to make the judgments required to establish this value. Where it is used, the rate is established in a fairly subjective manner.

The opportunity value concept provides an interesting perspective for those who are inflexible in their position that the discount rate for public sector investment should be tied solely to the cost of borrowing. Since public borrowing is tax advantaged, it carries low interest rates. If these borrowed funds were prudently invested, they can yield a return greater than the cost of borrowing. Why, at the very least, shouldn't the higher earnings rate be used as the discount rate? In the same vein, since taxes are used to repay public debt, then the opportunity value should reflect the taxpayer's lost opportunity to invest the taxes collected.

°Minimum Attractive Rate - This concept builds on the essential aspects of the cost of capital and opportunity value concepts, but implicitly eliminates the need to tie the value to some specific, quantifiable measure.

The use of a minimum attractive rate acknowledges that there are generally more useful projects or good investment opportunities than there is money to fund them. This holds for both the public and private sector. The decision maker's task is to select those projects that represent the best use of the resources available.

When a minimum attractive rate is applied to individual investment decisions, or employed as a discount rate to compare alternatives, it acts as a screen. The limited capital resources only flow to those investments that meet or exceed the hurdle. Rate selection is very subjective, since it is part of a resource allocation process. The rate must be greater than the cost of capital. In the private sector, rates in the 15%-20% range (after-tax) are common.

Inflation

Up to this point, the entire focus has been on discount rates. The treatment given to inflation can dramatically affect the outcome of LCC analysis and make the interpretation of the end result more difficult.

Although everyone recognizes inflation as a reality, we tend to be more comfortable in working with constant dollars (not adjusted for inflation) as we look to the future. On the other hand, we are accustomed to dealing with nominal as opposed to real (adjusted for inflation) discount rates. Simply stated, most individuals would consider a quaranteed 10% nominal rate of return as a good deal, but would not be so sure that a guaranteed 5% "real" return was as good.

Projecting inflation can be complex and, possibly, misleading. Consider how the relative attractiveness of an investment to reduce oil consumption would have changed from the mid-1970's to either the mid-1960's or 1980's. The complexity of the calculations increase dramatically as the desire for accuracy increases. One need only scan the Least Cost Energy Guide published by the National Bureau of Standards (Ruegg and Stevenson, January 1987) to see how complex assessments can become. Typically, there are three general approaches to dealing with inflation.

°<u>General Inflation</u> - Future expenditures are estimated in current dollars, then projected to the future. The inflation factor (I.F.) is equal to $(1+I)^n$ where **I** is the assumed inflation rate and **n** is the number of years after the initial investment. Using the data from Table 1 and assuming a 4% inflation rate, Table 3 shows the development of an inflation-adjusted present value for Alternative B.

Year	Expenditure (Current $)	Inflation Factor (4%)	Present Value at 9% Factor	Present Value at 9% Amount
0	$195,000	1.0000	1.0000	$195,000
40	50,000	4.8010	.0318	7,634
	$245,000			$202,634

Table 3. Inflation Adjusted Present Value

The simplicity of the above calculation quickly fades when one attempts to apply different inflation factors to different cost components (e.g. labor, materials, energy).

°<u>Differential Inflation</u> - The logic here is that the general inflation affects both the cost and benefits of a project over time. However, specific items may not follow the general pattern. As an example, if one believed that the cost of electricity will increase at 5% annually, but that general inflation only will be 4%, then future energy costs would be inflated at a rate of 1% per year. This approach is used by the Army in its manual <u>Economic Studies for Military Construction and Design Applications</u> (Department of the Army, December 1986).

°<u>No Inflation</u> - Again, the underlying logic is that, over time, inflation will affect both future costs and benefits. Inflation is sometimes left out just to simplify calculations and to avoid being unduly influenced for the wrong reason. For example, if a project aimed reducing oil costs looked attractive and contained an inflation assumption of 10% per year for oil, how does it look in today's dollars only? It could be that inflation is providing the bulk of the savings. A prudent investor may think twice and should at least look at the sensitivity of the result over a range of price assumptions.

Policy and Practices

The preceeding sections on discount rates and inflation should help the user understand why it is fruitless to search for clear guidelines that will produce precise values for discount and inflation rates. Policies and practices have evolved that, for many, must be relied upon.

In the late 1960's, the Joint Economic Committee of Congress held extensive hearings on the Economic Analysis of Public Investment Decisions (Joint Economic Committee of Congress, July-August 1968) where notables such as Otto Eckstein had a great deal of input. In essence, it was concluded that for the public sector, the opportunity cost of displaced private spending should serve to define the discount rate. This same conclusion is reached in Measuring the Opportunity Cost of Government Investment (Stockfisch, March 1968) and in The Rate of Discount for Evaluating Public Projects (Mikesell, 1977).

The above concept was then embodied in OMB Circular A-94 Discount Rates to be used in Evaluating Time-Distributed Costs and Benefits (OMB, March 1972) which is still in effect as of October 1989. This policy states that a discount rate of 10% shall be used and that the rate represents an estimate of the average rate of return on private investment, before taxes and after inflation." Inflation is only permitted on a differential basis. That is, general inflation should not be included. These concepts and guidelines have been further embodied in analytical procedures such as Economic and Environmental Guidelines for Water and Related Land Resources Implementation Studies (U.S. Water Resources Council, March 1983) and the previously mentioned Department of the Army economic studies criteria.

States have been slow to follow this lead and few have clearly developed policies and procedures. The private sector tends to structure LCC analysis to suit internal needs, although the use of minimum acceptable rates of return as discount rates tends to eliminate many of the conflicts that can arise in the public sector. Sound LCC analysis principles are embodied in the Standard Practice for Measuring Life-Cycle Costs of Buildings and Building Systems (ASTM, 1983). However, even this standard practice stops short of supplying discount and interest rates. It does state, however, that when selecting discount rates, the earning rate should take precedence over borrowing rates.

Conclusions and Recommendations

The difficulties in choosing discount rates and dealing with inflation are not new. There is less of a problem in the private sector where hurdle rates (minimum acceptable rates) are used to allocate limited capital resources.

The public sector is not unified in its approach. Federal guidelines, in most cases, require that a 10% discount rate be used and that only differential inflation be considered. This reflects the logic that money invested in the public sector must come out of the private sector. The 10% rate is nominated as the opportunity value of taxes in the private sector.

States lean toward using the cost of borrowed money for the discount rate. Unfortunately, this reflects a poor grasp of opportunity value concepts and leads to inefficient use of limited capital resources. Borrowing costs (interest rates) are not opportunity values and should not be used as discount rates. Borrowing does not pay for public projects, taxes pay for the projects. Borrowing is simply a financing vehicle.

The quickest remedy available for any state or governmental agency is to adopt the principles outlined by the Office of Management and Budget in Circular A-94. These guidelines are clear, concise and fairly respect the value of the public's tax dollar and will lead to an equitable allocation of limited financial resources.

References

1. American Society for Testing and Materials, "Standard Practice for Measuring Life-Cycle Costs of Buildings and Building Systems," E917-83, Philadelphia, PA, 1983.

2. Department of the Army, "Economic Studies for Military Construction Design-Applications," Washington, DC, December 1986.

3. Grant, E. L. and Ireson, W. G., "Principles of Engineering Economy," Fifth Edition, The Ronald Press Company, New York, 1970.

4. Leason, J. K. and Standley, R., "Least Cost Analysis," National Corrugated Steel Pipe Association, Washington, DC, March 1986.

5. Mikesell, R. F., "The Rate of Discounting for Evaluating Public Projects," American Enterprise Institute for Public Policy Research, Washington, DC, 1977.

6. Office of Management and Budget, "Discount Rates to be used in Evaluating Time-distributed Costs and Benefits," Circular No. A-94, Washington, DC, March 1972.

7. Stockfisch, J. A., "Measuring the Opportunity Cost of Government Investment," Research Paper P-490, Washington, DC: Institute for Defense Analyses, 1969.

8. U.S. Congress, Joint Economic Committee, Subcommittee on Economy in Government, "Hearings on Economic Analysis of Public Investment Decisions: Interest Rate Policy and Discounting Analysis," 84th Congress, 2nd session, 1968.

9. U.S. Water Resources Council, "Economic and Environmental Guidelines for Water and Related Land Resources Implementation Studies," Washington, DC, March 1983.

10. Wonsiewicz, T. J., "Least Cost Analysis: Key Assumptions and Sensitivity of Results," National Corrugated Steel Pipe Association, Washington, DC, January 1988.

Landfill Gas
Pipe Selection and Installation

William M. Held, Associate Member ASCE[1]
Eric R. Peterson, Member ASCE[2]

Abstract

Sanitary landfills present a unique array of pipe selection and installation criteria when designing landfill gas (LFG) control and recovery systems. Factors that must be considered include chemical corrosion resistance, temperature characteristics, ability to withstand differential settlement, and weatherability. In addition, the ease of pipe installation and repairs must be considered due to the safety precautions required when trenching and working in buried refuse, which is actively producing methane.

Pipe materials historically used in LFG systems include polyethylene (PE), polyvinyl chloride (PVC) and fiberglass. Polyethylene has often proved to be best suited for landfill environments, although varying conditions between landfills (such as LFG temperatures of 160 degrees F) sometimes dictate the use of other materials. Advantages and disadvantages are discussed in the paper with concluding recommendations.

Introduction

The purpose of this paper is to present an overview and examples of pipeline installations in sanitary landfills for conveyance of landfill gas. The paper will present a brief background on landfilling and landfill gas, then discuss pipe materials and pipeline installation in landfills.

----

[1]Project Manager, SCS Field Services, 211 Grandview Drive, Covington, KY 41017.
[2]Senior Project Engineer, SCS Engineers, 11260 Roger Bacon Drive, Reston, VA 22090.

Sanitary Landfilling and Landfill Gas

Sanitary landfilling is a method of disposing of solid wastes on land which protects the environment by spreading the waste in layers, compacting it, and covering it at the end of each day. Most landfills are constructed by excavating below grade some depth, filling to grade, then proceeding some distance above grade. Most new landfills are lined and capped with impervious materials to reduce the infiltration of precipitation and the contamination of the environment with leachate.

Landfill gas, which is comprised mainly of methane and carbon dioxide, is produced in a landfill by the process of anaerobic digestion of organic material in municipal solid waste. The organic, biodegradable fraction of solid wastes begins to undergo decomposition soon after landfilling. Initially, the decomposition is aerobic, due to the air trapped in the landfill.

As the air is depleted, anerobic decomposition occurs. Anaerobic conditions are promoted and maintained by the landfilling techniques mentioned above. After all oxygen has been consumed, methane forming micro-organisms become dominant. These micro-organisms form methane, carbon dioxide, and water. The percent of methane increases as the carbon dioxide level decreases. A typical composition of landfill gas is shown in Table 1.

TABLE 1
TYPICAL LFG CONSTITUENTS (Vogt, 1987)

Component	Component Percent
Methane	45-50
Carbon Dioxide	40-55
Trace Constituents:	
Methylcyclohexane, Acetone, n-Hexane, 1,2-Dichlorethene, Methylene chloride, Xylenes, Ethylbenzene, Vinyl Chloride, Tetrachloroethylene, Toluene	

LFG COLLECTION SYSTEMS

The typical LFG collection system consists of vertical extraction wells drilled into the waste, a pipe network that connects the wells, a blower system that creates a vacuum throughout the network to extract the gas, and either flaring or processing for use on the pressure side of the blowers. Spaced at intervals throughout the pipeline are condensate traps to remove and collect the moisture entrained in the landfill gas. A typical LFG collection system is shown schematically in Figure 1.

DESIGN CONSIDERATION

Materials used to handle landfill gas will be subjected to certain conditions that must be considered in the design of a LFG collection system. The most important are discussed below.

Chemical and Corrosion Resistance

Due to the high moisture content of the gas and some of the components of it (acids, hydrogen sulfide), materials that come in contact with the gas must be corrosion resistant. These materials may also be in contact with leachate.

Extensive research has been done on the chemical resistance of pipe materials and numerous charts are available that give the relative resistance of materials to a specific chemical. Not as clearly understood, however, is the resistance of plastic materials to the mixtures of chemicals that may present themselves to a pipe in actual service conditions in the landfill environment. Research done by the U.S. EPA on plastic materials used for linings has shown a wide variety of changes in physical properties can occur after exposure simulating service conditions. Among these are large weight gains (swelling) and loss of strength. In general, however, plastic piping materials have shown good chemical resistance in landfill gas applications. This is partly due to clean backfill protecting the pipe from direct contact with refuse, and shallow burial depths above the leachate saturation zone.

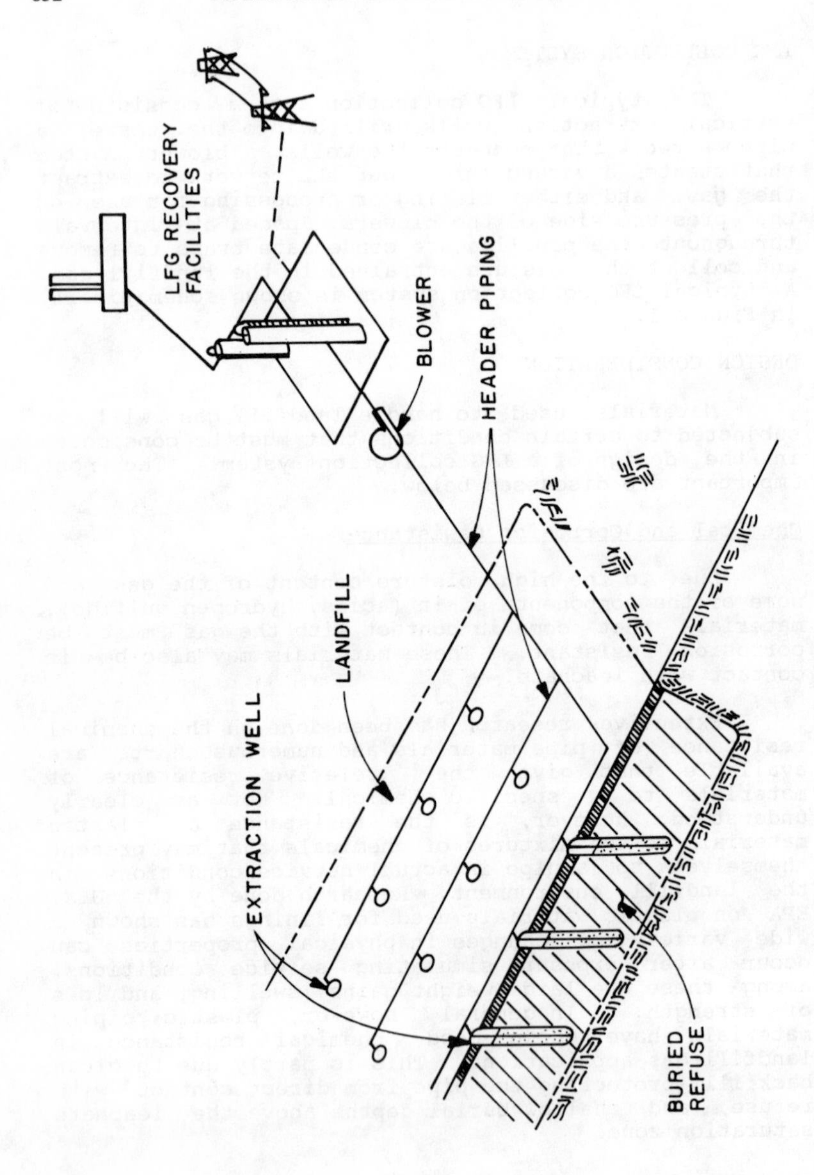

Figure 1. Typical LFG Collection System

Condensate Removal

Another consideration resulting from the high moisture content of landfill gas is the need to remove condensate to prevent blocking of the collector pipe. Low points are intentionally and strategically designed into the collection system to drain or collect condensate. A typical example of one style of condensate trap in shown in Figure 2.

Strength

As a landfill decomposes and ages, significant differential settlement can occur that can affect horizontal collector pipes. Flexible piping or connectors help reduce the effects of differential settlement. To maintain condensate flow in the pipe, slopes on the order of 2 to 4 percent or more are designed into the system. In most landfill gas collection systems, the collector pipes are not buried very deeply (typically less than 10 ft). However, since the collector pipes are operating under a vacuum, one must consider both the vacuum in the pipe, which is tending to pull it in from the inside, as well as the dead load on top of the pipe, which is tending to crush it from the outside. Additionally, any live loadings (such as at a road crossing) must be considered. In active landfill situations, heavy equipment likely will run over portions of the buried pipeline. Pipe encasements and routing of equipment traffic must be considered during design to prevent frequent damage.

Temperature Resistance

Temperature within a landfill can vary greatly, but typically range from 65 degrees F to 120 degrees F (18 C to 49 C). Generally, temperature increases with depth and is also dependent on climate and types of wastes received. Temperatures as high as 180 degrees F (82 C) have been reported in landfills. Persistent underground fires are not uncommon in landfills.

While generally not a concern for the header pipes, which are buried in shallow trenches, temperature resistance is a consideration in well construction. Typical extraction wells average 50 to 75 ft deep and may be subjected to elevated temperatures.

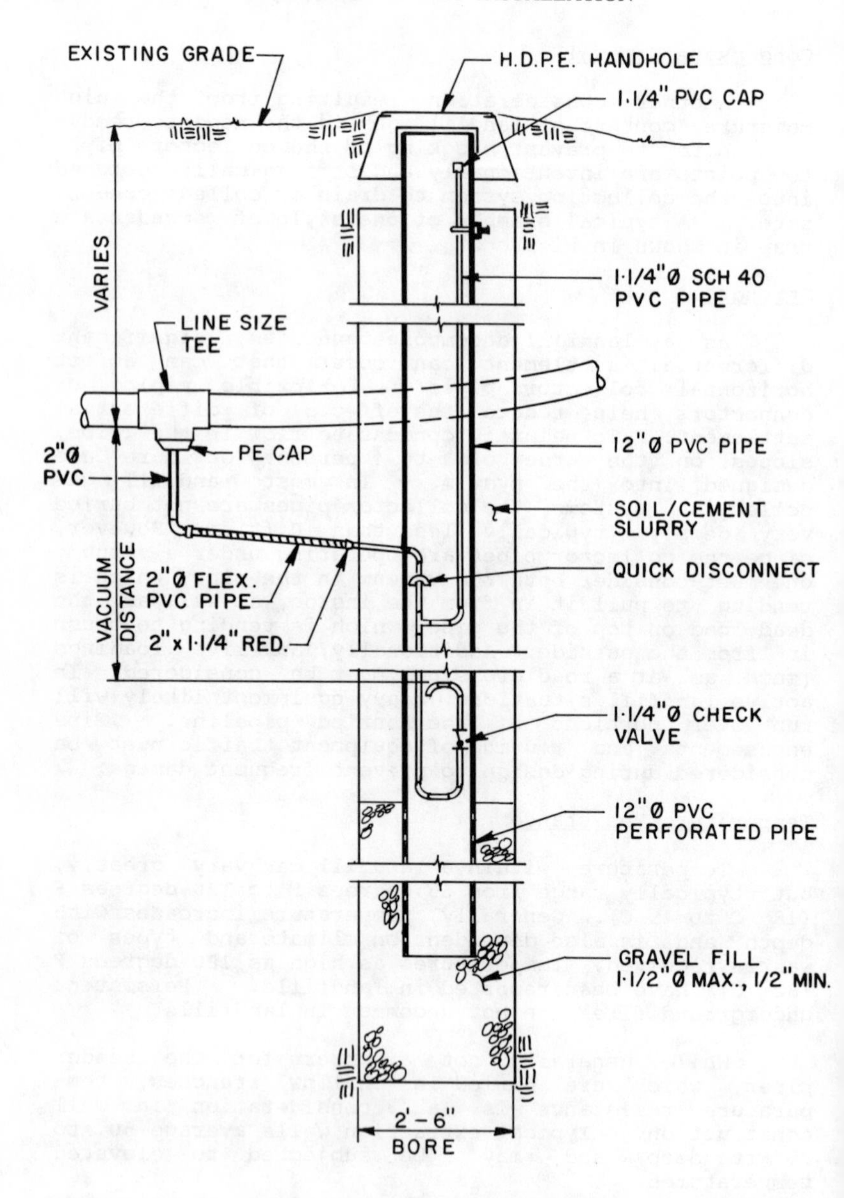

Figure 2. Typical Condensate Trap

Thermal Expansion and Contraction

In a buried environment, where the temperature fluctuations should be minimal and pipe is supported on all sides by soil, thermal expansion is not a major concern. However, in systems where the collector pipes are above ground, thermal expansion and contraction must be accounted for in the design. For example, PE and PVC differ greatly in their respective changes in size and temperature changes. PVC expands 0.000030 inches per inch length per degree F of temperature change (.000054 cm/cm/deg C). PE pipe is three times that at 0.00009 in./in./degree F (.00016 cm/cm/deg C). To accomodate thermal expansion and contraction in abovegrade PVC pipelines, flexible coupling are installed between fixed points such as well head connections. PE pipe is "snaked" along the ground between fixed points to provide slack for expansion and contraction.

Weather Resistance

Changes in the physical properties of pipe can be caused by various kinds of exposure to the outdoor environment. Weather effects can be minimized or eliminated by the proper storage and installation of the pipe. Materials not protected from ultraviolet radiation with the addition of carbon black (e.g., PVC) should be protected both during storage and in service to prevent degradation from UV radiation. PVC pipe used above grade can be painted or covered with a nominal layer of soil to protect it from UV radiation.

MATERIALS

Given the above design considerations, there have been only a few types of material that have been used in LFG collection systems. Each is discussed briefly below.

Polyethylene

Polyethylene pipe used in LFG systems should meet the requirements of ASTM D2513 and ASTM D3350. The polyethylene used is generally a PE 3408 and an SDR (Standard Dimensional Ratio) of 17 or less is recommended.

The primary advantage of PE pipe is its flexible characteristics that allow it to absorb the stresses induced by landfill settlement without cracking. Flexibility along with the hot fusion method for joining PE pipe facilitates installation. Long lengths of pipe (up to 400 feet or more) may be fused above grade and lowered into the trench with one end extending out of the trench for fusing additional lengths. This reduces time spent in the trenches where exposed refuse and elevated methane and reduced oxygen concentrations present obvious safety hazards. Repairs to buried PE pipe, however, are made more difficult by the fusion method since a large excavation is required to bring the fusion machine to the buried pipe.

PVC

Historically, either Schedule 40 or Schedule 80 PVC has been used in LFG systems. Recently, the use of PVC in header lines has been replaced almost entirely by PE. This is due to the tendency of PVC to crack from differential settlement. Years of operation and maintenance of LFG systems indicate that PVC pipelines require more frequent repairs. PVC, however, is still the predominant material used in vertical extraction wells.

Fiberglass Pipe

Although this pipe has been used in both wells and header pipes, it has been limited due to much higher cost than PE or PVC. Its primary advantage over PE and PVC is its ability to handle elevated temperatures. As mentioned previously, elevated temperatures may occur in deep well piping, and fiberglass has been successfully used in such applications. Piping on the pressure side of a blower may also be exposed to high temperatures resulting from the compression of the gas. Hence, fiberglass is sometimes specified for blower-station piping based on the anticipated temperature rise across the blower equipment.

CONCLUSIONS AND RECOMMENDATIONS

About two years ago, the Governmental Refuse Collection and Disposal Association, Inc., (GRCDA) did a survey of members of the LFG community about LFG collection systems and found that PVC and PE pipe accounted for 97.7 percent (72.7 percent PVC, 25 percent PE) of the material used in the horizontal collector pipes and 95.4 percent (88.6 percent PVC, 6.8 percent PE) of the materials used in the vertical well pipes (GRCDA, 1987).

Most likely, if the survey were conducted again in five years, the percentage of use for PE in horizontal collection pipes would greatly increase. The recent surge in PE use is due to its competitive price, and the benefits of flexibility revealed through operation and maintenance experience in the LFG industry.

PVC likely will continue as the favored material for well piping because of its ease of installation. The rigidity of PVC facilitates installation in the well bore, and the ease of joining PVC (through solvent welding) facilitates installation of valves, fittings, and monitoring ports in the LFG well head.

APPENDIX.-REFERENCES

GRCDA, "Results of the GRCDA Migration Control Survey," Governmental Refuse Collection and Disposal Association, Silver Spring, Maryland, February 1987.

Vogt, W.G., and Conrad, E.T., "VOC Emission Rates from Solid Waste Landfills", Proceedings from NSWMA Waste Tech '87 Conference, October 1987.

Twin Pipeline Crossing of the Fraser River

Jack H. Lee [1], David Swanson [2]

Abstract

A new creek water supply was developed for the communities of Mission and Matsqui, British Columbia, Canada, in the early eighties. Over 60 kilometres of 450 mm to 900 mm diameter ductile iron pipe, welded steel pipe and concrete cylinder pipe were installed between 1981 to 1989 for gravity conveyance of potable water to an ultimate service population of 200,000 people. Part of the supply system includes about 1,000 metres of underwater crossing of the Fraser River with twin pipelines. The crossing is unique because one pipe is for potable water and the adjacent pipe is for conveyance of sewage.

Introduction

The District of Mission and the District of Matsqui are communities located approximately 75 kilometres east of the City of Vancouver, British Columbia. The District of Matsqui's water supply was a series of groundwater wells and the District of Mission was a gravity supply from a small lake. Both sources were under capacity and major upgrading was required. A Water Commission was established in 1980 to interconnect the two water systems together with the development of a new water supply. One of the main obstacles to the plan was that these two municipalities are located on the opposite sides of the Fraser River and therefore a connecting watermain across the river was necessitated.

----

[1] Jack H. Lee, P. Eng., Senior Engineer and Partner, Dayton & Knight Ltd., P.O. Box 91247, 626 Clyde Avenue, West Vancouver, B.C. Canada V7V 3N9

[2] David Swanson, Water Supply Coordinator, Dewdney-Alouette Regional District, 32386 Fletcher Avenue, Mission, B.C. V2V 5T1

Supply System

The recommended new water source was from Norrish Creek located about 20 kilometres to the north-east of the District of Mission. The existing lake source and ground wells were retained as backups. Usage of the existing supply mains were maximized in order to reduce the capital cost of construction. The new supply system is illustrated schematically in Figure 1.

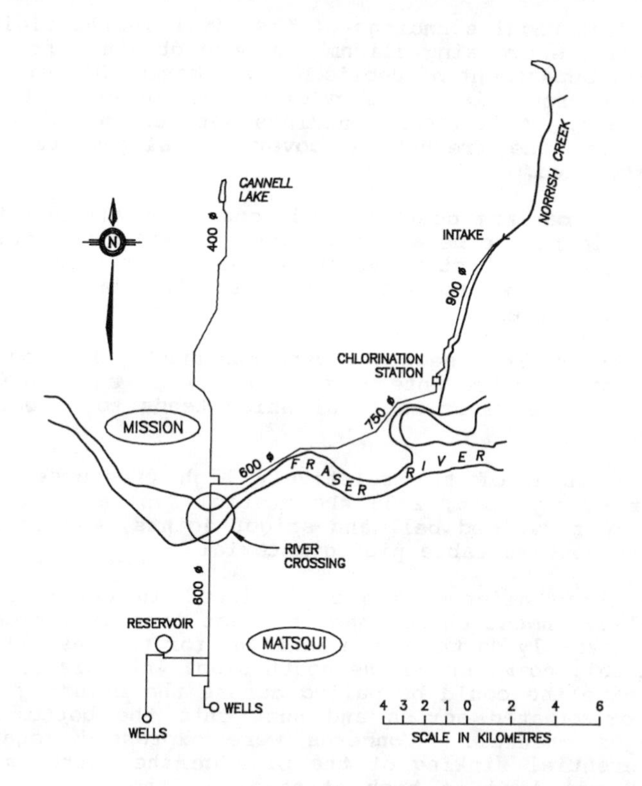

FIGURE 1 SUPPLY SYSTEM SCHEMATIC

The primary source of water is from Norrish Creek. During periods of high turbidity (over 5 NTU), the creek is shutdown. Cannell Lake and the Matsqui wells are activated as supply source. The principal component of the new supply system is the design and construction of the Fraser River crossing which connects the two water systems (Cannell Lake and Norrish Creek with the Matsqui wells).

Design Considerations

During the initial stages of pre-design work it was decided to include a second pipe attached to the watermain which would convey sewage from the District of Mission southward across the Fraser River. This would provide Mission with a connection in close proximity to the proposed watermain crossing on the south banks of the Fraser River.

Historical soundings of the river in the vicinity of possible crossing alignments were obtained from the Federal Department of Public Works - Marine Division and from the Provincial Ministry of Lands, Forests and Water Resources. Additional soundings were conducted before and after the freshet to cover the alignments being investigated.

The maximum depth of bed scour from the soundings was estimated to be about one metre. Maximum variation between the highest recorded to the lowest recorded bed level is approximately three metres. The average is less than two metres.

River cross sections were measured and velocities taken at 30 metre intervals at 0.2 metre and 0.8 metre depths. The river is tidal which tends to affect the velocities.

Because of the relatively high pressures for a water supply main, 2350 kPa static, 9.5 mm wall steel pipe, with welded bell and spigot joints, was selected as the most suitable piping material.

Three alignments were reviewed in detail. The final alignment chosen has the best launching area for pipe assembly on the north bank and for the installation of a pulling winch on the south side. At this location the pipeline could be pulled across the existing river bed or floated across and sunk onto the bottom of a dredged channel. Concerns were expressed regarding differential sinking of the pipe on the latter method which may lead to high stresses in the steel due to excessive curvature of the pipeline. It was decided to pull the pipeline submerged across the river along a dredged channel.

A minimum of 3 metres of cover was designed over the pipe on the main channel to prevent flotation of the pipe and to prevent damage from anchors and debris carried by river bed movement. The variation in river bed levels must also be considered to protect undermining. A minimum of 3 metre cover below the lowest recorded bed elevation was to be provided for dredged river sand material. The area of the proposed alignment is an active borrow pit area. The penetration depths of "spuds" from suction dredges and from boat anchors were also major design concerns with the maximum penetration being about 3 to 4.5 metres for the spuds and 1.2 to 1.8 metres for anchors.

Figure 2 illustrates the river and pipe profile. The profile shown allows for future dredging directly over the pipeline to a depth of 4.6 metres with another 3 metres allowed for pipe protection as well as any possible channel erosion or shift with the final pipeline invert established at 15 metre intervals.

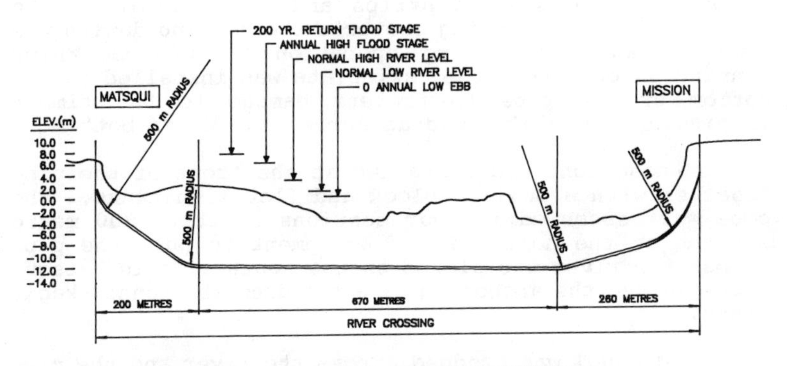

FIGURE 2 RIVER CROSSING PROFILE

Installation

Figure 3 illustrates a section of the twin pipe crossing. The 600 mm diameter watermain and sanitary sewer siphon are connected with a 12 mm thick steel spacer plate at 12 metre intervals. The pipe is coal tar enamel lined and coated to AWWA C203 requirements.

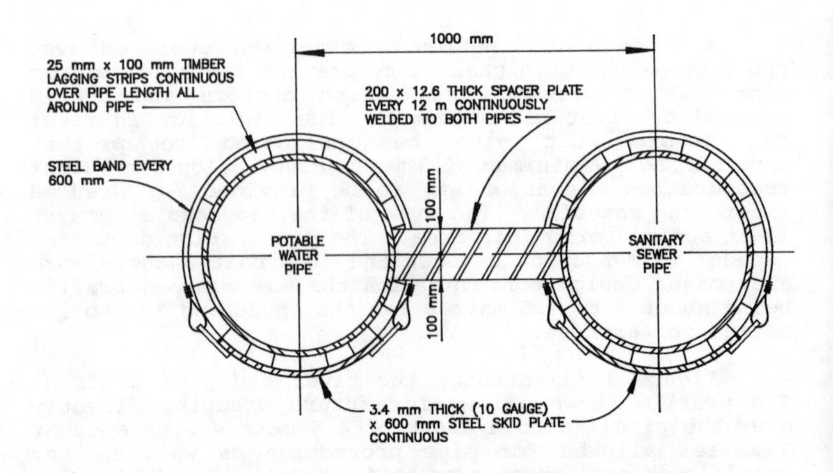

FIGURE 3 RIVER CROSSING PIPE LAGGING, STRAPPING, SKID PLATE AND CONNECTION DETAIL

The external coating was covered with a protective timber lagging laid in strips and held in place with steel bands. The lagging protects the coating during the dragging and backfilling operations. A 3 mm thick continuous uncoated steel skidplate was installed on the bottom of the pipe to prevent damage to the timber lagging and to minimize drag across the river bottom.

A nose cone was installed at the front of the pipe together with a pulling block and flotation tanks. The pipe was assembled in four sections of about 300 metre lengths on the north side. Equipment to pull the pipe across the river was placed on the south side in Matsqui and involved the anchoring of a two drum 150 tonne Skagit winch.

A channel was dredged across the river and the pipe installed in four pulls over a 29 hour period. Two field welds were made for each pull for a total of six welds. Each weld consists of four internal passes and four external passes for the bell and spigot joint. The joints were tested with nitrogen at 2100 kPa for leakage before coating and lining with coal tar enamel. The protective coating was holiday tested before placement of the timber lagging.

A 62.5 mm diameter steel cable was used for the pull and due to its overall length, field splicing by a factory representative was necessaary.

The pipe was backfilled with 3 metres of cover after installation. The remaining trench backfill was completed using the natural action of the river.

Soundings completed the following year showed that the pipe trench had in fact been completely restored to the original river bed profile.

References

1. AWWA Manual M11, Steel Pipe Design and Installation, Manual of Water Supply Practices, 1964.

2. Dayton & Knight Ltd., "Engineering Survey - Watermain Crossing of Fraser River" for Dewdney-Alouette Regional District, 1980.

Relining of Jordan Aqueduct, Reach 3

Douglas H. Wegener[1]

Abstract

As pipelines approach their life expectancy, they may need to be replaced. Or in some instances, replacement is needed much sooner due to defective pipe. Normal replacement operations require reexcavating the existing pipeline, removing it, and installing new pipe. In some instances this open trench excavation may not be desirable. Thus, the need for alternative procedures.

One such procedure is the relining of existing pipelines in place, thus rejuvinating it to new pipeline status. This procedure was done on the Jordan Aqueduct, Reach 3.

Jordan Aqueduct, Reach 3 consists of 66- and 48-inch-inside-diameter pipe. This pipe is embedded cylinder prestressed concrete pipe. In 1984 a section of 66-inch-inside-diameter pipe failed resulting in property damage near Salt Lake City, Utah. The cause of the failure was determined to be defective materials in the pipe. Therefore, the entire pipeline needed to be replaced. The decision was made by the Bureau of Reclamation to reline the existing pipeline instead of replacing it, thus reducing environmental disturbances in a highly urbanized area.

A steel liner was inserted inside the existing pipeline. The annular space between the two pipes was grouted to form a composite section pipeline. The inside of the new steel liner was then cement-mortar lined in place. The pipeline was placed back in service and has performed satisfactory.

Introduction

This paper presents a case study on a method of rehabilitating a structurally unsound buried pipeline without removing and replacing the existing pipe.

In July of 1984, the construction of Jordan Aqueduct, Reach 3 was substantially complete. Reach 3 consists of approximately 2.1 miles of high pressure

[1]Civil Engineer/Design Manager, Bureau of Reclamation, PO Box 25007, Denver, Colorado 80225

pipe which is comprised of approximately 11000 linear feet of 66-inch-diameter pipe and 360 linear feet of 48-inch-diameter pipe. This entire reach of pipeline consists of embedded cylinder prestressed concrete pipe, and was constructed in Salt Lake City, Utah (see Figure 1). This pipeline provides the Salt Lake Valley treated water for municipal and industrial use.

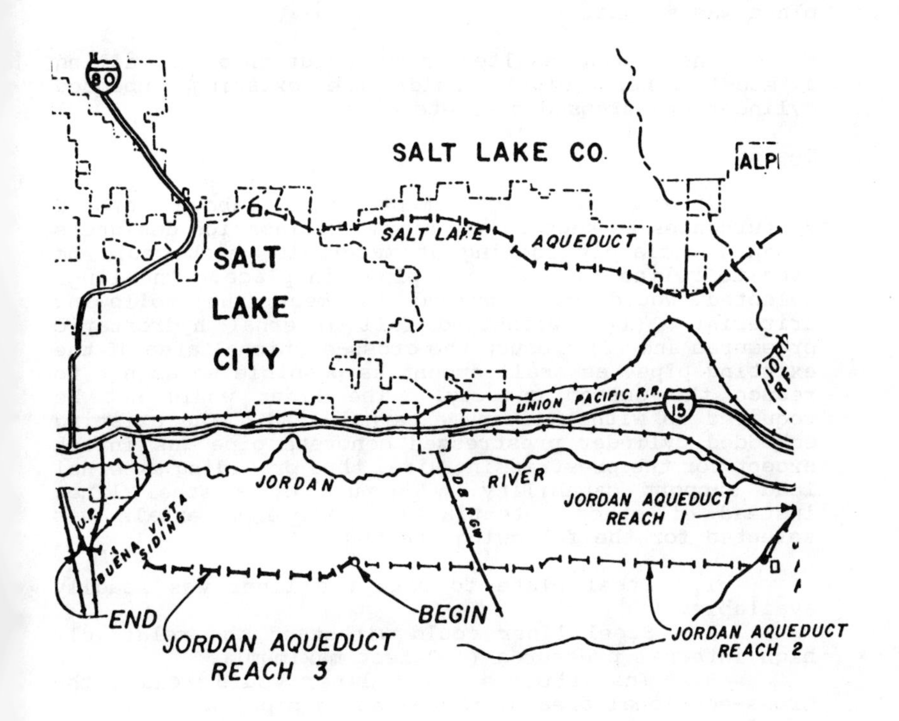

FIGURE 1

For approximately one month, the pipeline was in service then a pipe failure occurred. After an extensive investigation, it was determined that the failure was caused by defective prestressing wire. With concerns that additional failures would occur in other pipe sections which also contained the defective wire, the decision was made by the Bureau of Reclamation to replace the entire pipeline.

However, due to environmental concerns (urbanization of the surrounding area), Reclamation did not want to reexcavate the entire pipeline. Therefore, a method of rehabilitating the pipeline while left in place was sought.

The search resulted in the solution of installing a steel liner (pipe) inside the existing embedded cylinder prestressed concrete pipe.

Design

Given the requirement that above ground disturbances be kept to a minimum, Reclamation designers began immediately looking at materials which could be used to reline the existing pipe, in place. The liner selected would be required to meet the following criteria: (1) withstand full internal hydrostatic pressures and (2) reduce the cross-sectional area of the existing pipe, as small amount as possible so as not to reduce the design cpacity. The liner would not be required to withstand any external load. The existing embedded cylinder prestressed concrete pipe was intack except for the prestressing wire, thus providing external load support capability. The use of a steel liner instead of other potentially suitable materials was selected for the following reasons:

1. Steel plate to make the liner was readily available.
2. Steel liner could withstand the relatively high internal pressures (450 feet maximum).
3. Installing a steel liner would reduce the cross-sectional area of the existing pipe, a relatively small amount, thus not affecting the capacity of the pipeline.
4. Specially trained technicians would not be required to install the steel liner.

After steel had been selected as the lining material, the actual liner design was accomplished. Allowable steel plate options for the liner were

ASTM Designation: A 283, grades C and D, and ASTM Designation: A 36. Welded spiral-seam steel pipe could also be used and made of ASTM Designation: A 139, grades B, C, and D or E. Using these options resulted in a liner plate thickness of three-eighths inch for the 66-inch pipe and one-fourth inch for the 48-inch pipe. The steel liner was designed in accordance with standard Bureau of Reclamation procedures for designing steel line pipe for internal hydrostatic pressure.

The existing 66-inch-diameter pipe was to have a finished inside diameter of 62 inches after lining and the existing 48-inch-diameter pipe was to have a finished diameter of 44 inches.

The annual space between the existing pipe and the steel liner would be stage grouted. Stage grouting would be required to limit the hydrostatic uplift forces caused by the fluid grout which would tend to deflect and/or buckle the steel liner. The grouting would be accomplished in five stages. The first four stages would grout approximately one-fourth the height of the annular space. The fifth stage would grout the crown. The grout would consist of sand, cement, and water, or cement and water.

After the annular space had been grouted, the steel liner would be cement-mortar lined, in place. The cement-mortar lining thickness would be one-half inch, and conform to AWWA Standard C 205.

Designing the steel liner was only half the design problem. The second half of the problem was to develop a viable method of installing the steel liner. The alignment and profile of the existing pipeline contains some minor bends and several structures. The structures are combination manholes and air valves, or manholes and blowoffs. These structures consist of concrete encased steel tees and are cement-mortar lined. Whenever such a structure is encountered, the steel liner would be interrupted and welded to the existing steel tee. An adaptor would be required to taper between the two different diameters.

As far as change in alignments, or deflection angles are concerned, only two existed in the horizontal direction which would not allow a 20-foot-long section of steel liner to pass through. In the vertical direction, seven such locations existed. At or near each of these locations a manhole structure existed. Also, the designers determined that a practical maximum distance to haul steel liner inside the existing pipe would be approximately 800 feet. Therefore, a "break-in" location was specified approximately every 1600 feet. These spacings and the break-in locations were checked to ensure that no steel liner sections would be required to be pulled through a structure or through one of the bends in the pipeline.

The steel liners were designed as full circle cans, 20 feet long. An annular space of 1 1/8 inches was left to allow for ease of installation. As each steel can was moved into place, it was welded to the previously installed can. This process was continued throughout the entire relining operation.

At the break-in locations, new steel pipe sections were installed. These sections were designed to withstand both internal pressures and external loads, in accordance with standard Reclamation procedures for the design of steel line pipe. These pipe sections were 24 feet long and cement-mortar coated. With the installation of these pipe sections, the Jordan Aqueduct, Reach 3 became a continuous welded steel pipeline.

After installing the 1/2-inch-thick cement-mortar lining, the repaired pipeline was required to be hydrotested.

<u>Construction</u>

After reviewing Reclamation's designs, the contractor, who was selected to reline Jordan Aqueduct, Reach 3 made requests for changes in design procedures. The first request was to use larger diameter steel liner cans. The contractor felt he would have no trouble installing the larger cans and that the backfill grouting of the annular space would be substantially reduced. His proposal was to use 64- and 46-inch diameter cans. In conjunction with this proposal, he also requested that he be allowed to install the cans with a lapped longitudinal seam thus effectively reducing the can diameter. Once in place, the cans would be expanded and the longitudinal seam welded. Reclamation approved these changes with requirements that the overlapping seam not make the steel cans exceed their allowable yield strength. As a design note, spiral-welded steel liner cans could not be used with this type of installation procedure. The reason being, that once the longitudinal seam is cut in the full circle can, the spiral-welded seam would not let the longitudinal seam match back up.

Another request by the contractor was to full circle grout in one stage the annular space between the liner and existing pipe. He stated that he had done this before on another pipeline successfully. Due to the relative thin thickness of the steel liner this request required an extensive design analysis. Both overall deflections of the pipe and localized buckling needed to be considered. To reach an acceptable solution, the actual yield strength of the steel supplied was used in the analysis instead of allowable stresses. Blocking points to keep the steel liner centered in the existing pipe were carefully sized and located. Grouting pressures were to be closely monitored. Even with all this extensive analysis and strict construction requirements, the contractor still buckled a section of liner, which had to be replaced.

To install the steel liner cans, the contractor obtained a motorized sled which transported the steel cans inside the existing pipeline and placed them in the correct position. Once in place, the overlapped steel cans were unloaded and allowed to expand. Some hand work with jacks was required to expand the steel cans to their original diameter. After the steel liner cans were expanded, the longitudinal seams were tack welded to hold them in place. A major problem occurred during this process. In places, the inside diameter of the existing pipe was smaller than the minimum allowable inside diameter. This resulted in the loss of annular space which made the expanding of some liner cans extremely difficult.

Once several cans were in place, the longitudinal seam was machine welded. After the longitudinal seam was welded, the circumferential joints between steel cans were hand welded. Another critical problem occurred at the intersection of the longitudinal welds and circumferential welds. Several of the welds at these intersections were defective. These defective welds were discovered when bleed water from the grouting of the annular space leaked into the inside of the steel liner. As a result of these leaks, grouting was stopped and a leak test programs was developed. The leak test consisted of adding water in the annular space, under pressure, and visually searching for leaks inside the steel liner. After all the defective welds were repaired, the grouting was completed.

The remaining three construction activities went smoothly. The steel pipe sections located at the break-in locations were installed, these pipe sections were backfilled, and the inside of the steel liners and pipe sections were cement-mortar lined in place.

This repair work was substantially complete in April of 1988 and the pipeline filled. It has been in operation since that date without additional failures.

Conclusions

From the experiences gained on the relining of Jordan Aqueduct, Reach 3, Reclamation designers concluded lining, in place, large diameter pipelines is a viable solution to rehabilitation of pipelines. In this case, the cost per linear foot of relining the pipeline was approximately $350.00, while the cost of constructing a new pipeline per linear foot would be approximately $560.00. These costs do not include costs for environmental factors which in some instances may be quite large.

Conclusions reached regarding technical concerns are:

1. A pipeline may not be suitable for in-place relining depending upon its alignment and profile. If numerous large angle bends occur along the pipeline or numerous in-line structures exist, the number of "break-in" locations would be greatly increased.

2. Ample annular space should be left between the existing pipe and liner to allow for ease of installing the liner.

3. With proper equipment to transport the lining, haul distances inside the pipe is not a concern.

4. To limit uplift forces, grouting of the annular space should be accomplished by stage grouting. The risk is too great of buckling or deflecting the liner if full circle grouting is allowed. Any small rise in grouting pressures can cause immediate failure of the liner.

5. The longitudinal cutting and overlapping of steel liner cans is a method of obtaining more clearance for installing the liner. However, this method also generates the problem of having to weld the longitudinal seam inside the pipe. This tends to slow production considerably. Therefore, depending on the number and magnitude of bends that the steel liner cans need to pass through, the preferred method would be not to make longitudinal cuts.

Evaluation and Selection Methodology for Determination of a Preferred Pipeline Alignment

Michael T. Stift, P.E.[1] - ASCE Associate Member
Larry Hobson, P.E.[2]

Abstract

The San Diego County Water Authority (SDCWA) had need to construct a regional 108 to 72 inch (2.7 to 1.8 meter) diameter pipeline through San Diego County to increase water supply to its member agencies to meet current and future water demands as a result of urban growth and development throughout the county. Determination of a preferred pipeline alignment must weigh environmental, community, institutional, economic and classical engineering issues. The nature of these concerns was examined and guidelines used to rank or reject alternative alignments were developed. A case history provides insight into actual selection analysis relative to tunnel construction, rock blasting, utility conflicts, as well as community concerns over impacts to businesses, residences, traffic, visual impacts from disturbing natural vegetation in park areas and other factors. Consistent evaluation criteria were developed to assess the main issues and a selection matrix was used to evaluate weighted impacts relative to each alternative alignment. To demonstrate an understanding of environmental, engineering and other concerns in selecting a preferred pipeline alignment, the evaluation criteria and selection matrix were presented in the project's Environmental Impact Report summary.

---

[1]Senior Civil Engineer, San Diego County Water Authority, 610 W. Fifth Avenue, Escondido, CA 92025.

[2]Engineering Manager, San Diego County Water Authority, 610 W. Fifth Avenue, Escondido, CA 92025.

Background

San Diego County is located in a semi-arid region where natural occurrence from rainfall and groundwater can provide a secure supply of water for about 10 percent of the present population of 2.25 million people. Most of the local reservoirs were constructed between 1890 and 1943 to impound rainfall runoff, which along with pumped groundwater, usually met the demands of the following summer. With the increase in population and industry during World War II, it was evident by the mid-1940's that the local water supply would be inadequate for the urban communities on the coastal plain and foothills of San Diego County.

The San Diego County Water Authority (SDCWA) was organized on June 9, 1944, to provide the special governmental powers to design, construct and operate a system of water facilities to convey water through the county. Traditionally the mandate of SDCWA has been to supply high quality water to its member agencies at the lowest possible cost and to meet the demands of its member agencies without restriction.

SDCWA supplies 90 percent of the water currently consumed within San Diego County. Most of this water supply is imported from northern California and the Colorado River and then distributed within the county through the First and Second San Diego Aqueducts as shown in Figure 1. The aqueducts are the regional water supply "backbone" system. Five existing large diameter gravity flow pipelines convey treated and raw water throughout the county. Rapid growth within SDCWA's service area has resulted in increased water demand. The capacities of the existing SDCWA pipelines are rapidly being approached as water demand increases. Peak demand in some areas currently exceeds pipeline capacities, which has required restrictions on the use of water by some member agencies. Therefore, the SDCWA Board of Directors adopted the Water Distribution Plan (WDP). The WDP objectives are:

1. Increase the capacity of the existing aqueduct system.
2. Increase reliability and operational flexibility of the aqueduct distribution system.
3. Obtain additional imported water supplies.
4. Increase yields from existing water filtration plants.

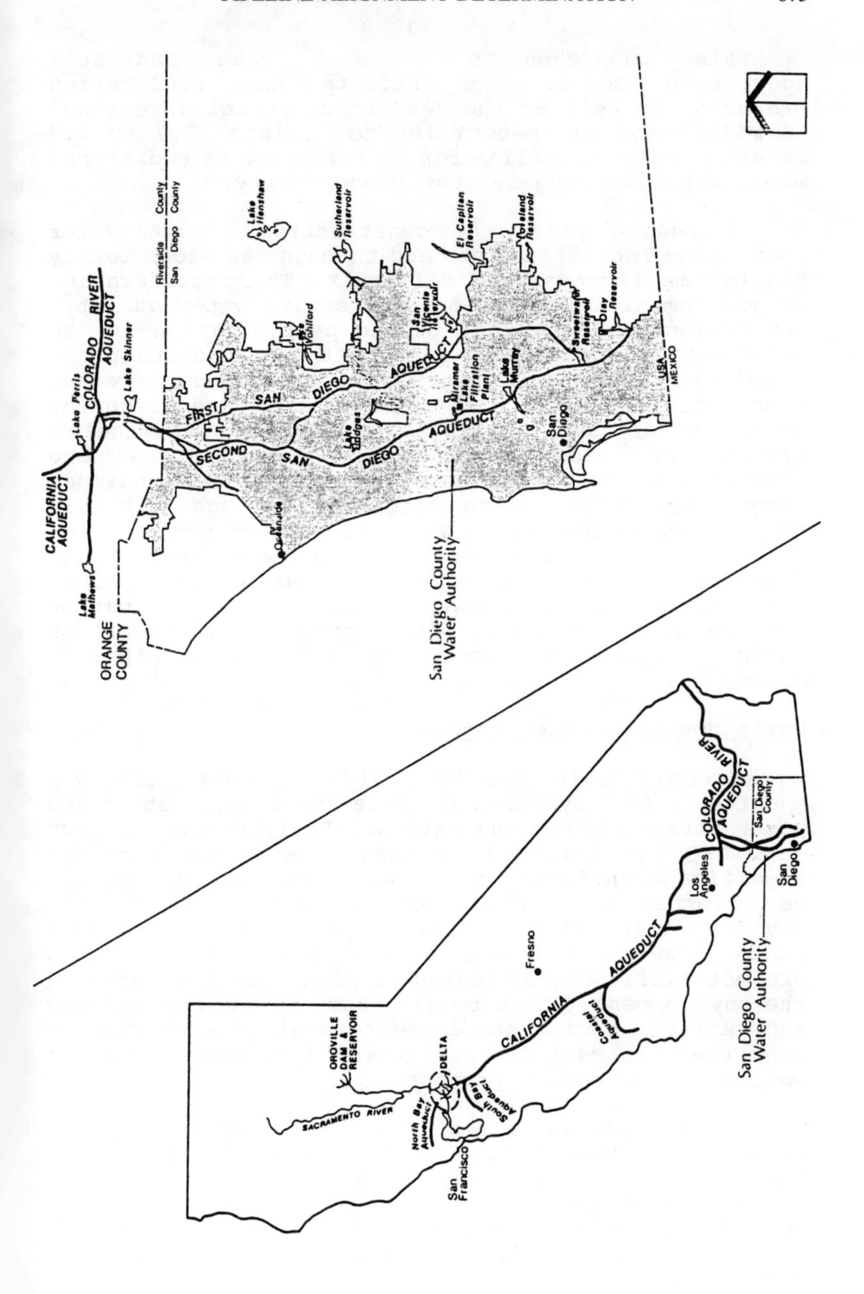

Figure 1 Aqueducts Serving SDCWA

This comprehensive regional plan addresses aggressive conservation efforts and reclamation projects, as well as the need to construct a regional 60 mile (96.6 kilometer) 108 to 72 inch (2.7 to 1.8 meter) diameter pipeline for distribution of additional water supply throughout San Diego County.

In recent years the construction of new water conveyance facilities into and through San Diego County has become increasingly difficult. The proliferation of governmental regulations addressing water quality, environmental issues and public participation in the decision making process, along with urban development crowding existing and future pipeline routes, have had substantial impacts on the timing, cost and alignment selection of new facilities. This paper examines a methodology used to present environmental and engineering factors in choosing a preferred regional large diameter pipeline alignment through a highly developed area along an 8 mile (12.9 kilometer) section of the overall pipeline between Lake Murray and Sweetwater Reservoir as shown in Figure 1. The process was a means for public and institutional participation and provided an on-going public record of the decision making process in selecting the final pipeline alignment.

Environmental Review Process

The California Environmental Quality Act (CEQA) was enacted in 1970 by the California State Legislature and was patterned after the National Environmental Policy Act. CEQA is intended to ensure decisions consider significant environmental impacts and that the public be informed of and allowed to participate in the environmental documentation process. CEQA also states that "economic and social changes resulting from a project shall not be treated as significant effects on the environment." As a result, some of the engineering and social decisions in determining a preferred pipeline alignment are not intended as part of CEQA's environmental review process.

Under CEQA, an Initial Study was prepared for the project to identify possible adverse environmental impacts of a regional large diameter pipeline. If no significant adverse impacts had been found, a Negative Declaration would have been prepared. However, due to the project's magnitude, especially in traversing highly developed and urbanized areas, significant environmental impacts were identified and an Environmental Impact Report (EIR) was prepared. Public input was solicited and the Draft EIR then circulated for public review and comment prior to completion of the Final EIR. The EIR process provided and documented the available information for the decision makers consideration.

A review of initial public meeting comments indicated that the decision process to be applied by project planners and engineers in determining a preferred pipeline alignment should be well documented. A method had to be found to summarize and present the information, as well as allow others to clearly understand the reasoning in choosing one alignment over another. It was also determined that part of the decision in constructing a regional pipeline to meet existing, as well as future water demands, and affecting the region and population it was to serve, involved consideration of economic, social and engineering issues as well as environmental issues. Finally, the public review and EIR process were not just a presentation of information for review and comment. For the planner and engineer, it was also an educational forum that required relating technical data and construction techniques to non-technical parties to facilitate a working public environment so that informed comments could be received.

Determination of Alternative Pipeline Alignments

The location of existing SDCWA and member agency water distribution facilities provided the basis in defining the study area. The beginning and ending locations for a pipeline project are fixed as water is located in one area and is to be supplied to another. The purpose of the pipeline between Lake Murray and Sweetwater Reservoir was to provide additional water for emergency storage and subsequent treatment at existing water filtration plants, as well as to serve existing member agency turnouts and metering stations located between the reservoirs.

Selection of alignment alternatives involved exhaustive investigation of various pipeline routes. Restrictions due to pipeline system hydraulics eliminated some alternatives immediately. Alignments that could be considered growth inducing were abandoned. Finally, SDCWA is a public agency and has a fiduciary responsibility to its member agencies and other constituents to spend money wisely. Therefore, pipeline alignments that added considerable length and cost to the project, but had equivalent environmental and engineering impacts, were eliminated. This process of elimination defined the study area.

Environmental and engineering information was then identified and evaluated within the study area and possible pipeline corridors were developed. From a progressive review of technical information and public comment, various focused pipeline alignments emerged as possible alternatives. For the purpose of this paper, only three of the alternative pipeline alignments that were studied are presented. The alternatives have been labeled the Eastern, Central and Western pipeline alignments.

Environmental Considerations

A team of planners and engineers developed a summary list of environmental evaluation criteria that would incorporate the essential environmental factors for each alternative pipeline alignment. The following summary environmental evaluation criteria were used:

- Community Disruption
- Traffic Impacts
- Environmental Resources
- Utility Conflicts
- Geotechnical

Community disruption factors included impacts that preclude or effect existing and future land use for residential, commercial and public institutions such as schools, fire and police departments, hospitals and convalescent homes. Impacts that were considered included length of time to construct the pipeline through an area, relative residential density along each alignment, business and home frontage access, disruption of public transportation systems such as bus service and public safety issues.

Evaluation of traffic impacts is closely tied to community disruption and public safety issues, but was assessed separately due to the specific significance in traveling to and from work, as well as maintaining emergency vehicle access. Other factors that were considered included street closures to all but local traffic, impacts to pedestrian traffic, truck traffic to and from the construction area, parking of equipment and worker's vehicles, major intersection impacts and traffic circulation and safety assessment due to traffic control in relation to level of service for a particular street.

Environmental resources evaluated aesthetic impacts to the community such as visual scarring across an area by disturbing existing vegetation from open trench excavation. Other considerations included dust, noise and vibration from construction equipment and effects on sensitive receptors such as hospitals and convalescent homes, and disturbance of cultural resources such as archeological sites and sensitive biological species.

Evaluation of utility conflicts and geotechnical concerns involved both environmental and engineering factors. The primary environmental consideration for utility conflicts as a result of pipeline construction was interruption of utility service to areas for extended time periods. Environmental geotechnical concerns included impacts from erosion, landslides, blasting for rock and boulder excavation and tunnel construction.

Engineering Considerations

Classical engineering considerations were also evaluated. The following summary engineering evaluation criteria were used:

- Capital Costs
- Pipeline Right-of-Way and Permits
- Hydraulics
- Utility Conflicts
- Geotechnical

Capital costs were estimated for each pipeline alternative. These included costs for pipeline construction, easement acquisition, permits, traffic control, and mitigation measures. Assumptions relative to land values for easement acquisition, permit fees and pipeline maintenance costs were determined from past projects. Preliminary construction costs were estimated by calculating cost for each main segment of pipeline. Contingency costs were then determined by multiplying by construction difficulty factors (CDF) to reflect degree of difficulty and constructibility issues along each alignment. For example, as open country (non-paved areas) trenching was the least costly type of pipeline construction, construction in a 4-lane road might have a CDF of 1.2 while a 2-lane road might have a CDF of 1.4 because of traffic control considerations.

Acquisition of right-of-way and special permits were important parameters in evaluating the alternative pipeline alignments. Encroachment permits from public agencies were required prior to construction where the pipeline was to be installed on lands in their jurisdiction. Evaluation included the number of permits required and consequential construction requirements. Also, when routes were not within existing public right-of-ways, negotiating and acquiring easements, as well as project schedules were considered.

Initial and future pipeline hydraulics for each alternative were also evaluated. In some cases, there were specific benefits due to reduction of energy costs relative to pumping requirements. In addition, construction of connection piping by member agencies to the main pipeline required consideration of impacts associated with appurtenant pipe installation and existing pressure zone requirements.

Engineering considerations for utility conflicts included the number of utilities to be crossed or relocated and the proximity of parallel utilities to the pipeline alignment. Geotechnical concerns included the type of soil materials encountered in trench and tunnel construction, the need for intermediate air shafts in tunnel construction, contamination from hazardous materials and pipe corrosion due to soils and stray electrical currents. Generally, these factors affected the length of time to construct a segment of pipeline or were work related safety issues.

Selection Matrix

A weighted selection matrix was used to integrate environmental and classical engineering decisions to determine the preferred pipeline alignment. Alternatives for each alignment were analyzed using the list of evaluation criteria that incorporated essential environmental and engineering factors. Table 1 shows an example evaluation criteria summary that was used in the alternative alignment analysis.

Evaluation criteria was subjectively ranked for relative impact by a team of planners and engineers,

Table 1

ENVIRONMENTAL AND ENGINEERING EVALUATION CRITERIA SUMMARY

	Evaluation Criteria (length)	Eastern (56,850 Ft.)	Central (45,550 Ft.)	Western (53,275 Ft.)
1.	COMMUNITY DISRUPTION - land use, public safety and convenience, residential, commercial, restricted public access, public institutions (school, church, hospital, fire/police, convalescent homes, etc.), development density, frontage access to homes/business, public transportation	- effects redevelopment project on Fletcher Parkway - landlocks Mt. Helix and Bancroft areas (non-mitigable) - slow construction along Bancroft due to narrow, winding ROW. - greatest length/impacts longer - significant disruption of Navajo Road per San Carlos Town Council & Navajo Community Planning Group	- open cut through Mission Trails Golf Course and park area - pipe installed in wide roads only. More work area. - 24 hour/day operation at La Mesa tunnel portal - restricted access to Lake Murray Park - eliminates 42 parcels residential and commercial land use for waterline purposes	- open cut through Mission Trails Golf Course and park area - business impact at 70th and University Ave. - restricted access to Lake Murray Park - significant disruption to San Miguel and Massachusetts per City of Lemon Grove - significant disruption of 70th St. per City of La Mesa
2.	TRAFFIC IMPACTS - traffic lane closure and traffic control, street closure and detours, pedestrian traffic, construction truck traffic, signalization, parking equipment and worker's vehicles, effect on cross traffic-intersections, traffic safety assessment.	- close Bancroft Drive to all but local traffic - Bancroft Drive is winding road with limited vehicular sight distance in places - Impacts Navajo Road, major ingress/egress to communities of San Carlos and Navajo - unmitigable traffic impacts on Bancroft Drive	- fewest impacts of all alternatives - traffic congestion at Baltimore, Fletcher Pkwy. and Sweetwater - least length of paved road - intense 24 hr. traffic at the La Mesa tunnel portal	- cross major intersections of Jamacha and Parkway - construction in arterial roads of Skyline, Elkelton, Jamacha, Massachusetts and 70th - less adverse than Eastern - business traffic impacted on 70th Street - close San Miguel to all but local traffic - cross Massachusetts and Broadway intersection
3.	RESOURCES - noise, dust, vibration, archaeological sites, cultural impacts, biological impacts, visual impacts, impacts to sensitive receptors.	- open cut across Dictionary Hill will leave visual cut - coastal sage impacted at Dictionary Hill - noise in residential areas - remove 6000 feet landscaped median along Fletcher Parkway - 8 sensitive noise receptors	- coastal sage impacted at Lake Murray - noise in residential area at Baltimore and Sweetwater Road - 24 hr/day construction for one year at tunnel portal - significant air quality and noise hotspot at La Mesa tunnel portal due to length of time to construct tunnel - 2 sensitive noise receptors	- coastal sage impacted at Lake Murray - noise in residential area of San Miguel, Elkelton and Arran Way - 8 sensitive noise receptors
4.	UTILITY CONFLICTS - number of utilities crossed, proximity of parallel utilities, #/length of utility relocating, utility service disruption, future utility coordination with developers/agencies. special construction permit requirements.	- greatest disruption of existing utilities - relocate 2000 feet of electric conduit in Kempton & Presioca St. - relocate 10,000 feet of electrical, telephone & 10" sewer in Bancroft Dr. - lateral crossings approx 200	- least disruption of existing utilities - lateral crossing approx 130 - relocate approximately 1200 feet of 2" gas in Arran Ave.	- relocate 2,000 feet of 1" gas in San Miguel Ave/ 1200 feet gas in Arran Ave. - utility lateral crossings approx 250 - relocate 2200' of 8" sewer in San Miguel - relocate 800' of 1 1/2" gas in Kiowa Drive

Table 1

ENVIRONMENTAL AND ENGINEERING EVALUATION CRITERIA SUMMARY

	Evaluation Criteria (length)	Eastern (56,850 Ft.)	Central (45,550 Ft.)	Western (53,275 Ft.)
5.	GEOTECHNICAL - type of soil (blasting, unstable trench, amount of spoil and import backfill, etc), groundwater, tunneling, landslide activity.	- blasting along Navajo Road, Grossmont Center & Dictionary Hill - 2,650 feet long tunnel in Grossmont area and under SR-125, short tunnel at Jackson Drive crossing - potential for erosion on Dictionary Hill	- 11,000 feet long La Mesa tunnel is required - groundwater at La Mesa tunnel construction, golf course, Lake Murray area - possible existing groundwater contamination at La Mesa tunnel. Providing safe supply and not spreading contamination may be unmitigable - parallels metro railway for greater distance/possibility of pipe corrosion from stray electrical currents	- groundwater in Lake Murray area - blasting areas required north of I-8, SR-94 and possibly in Massachusetts - 3 short tunnels at I-8, Lemon Grove Ave., and SR-54
6.	CAPITAL COST - construction, constructibility, traffic control, easements and permits, mitigation measures.	- $91,552,000	- $104,111,000	- $82,204,000
7.	RIGHT-OF-WAY/PERMITS - #/area of permanent easement for construction, maintenance and access; drainage easements, temporary construction easements, number of permits, permit construction restrictions.	- 7,650 LF easements required - remainder in public right-of-way (4 encroachment permits	- 19,900 LF easements - CALTRANS policy no encroachment in future SR-125 - condemn 15 homes, 8 multi-family units and 2 businesses - may conflict with SR-125 federal funding requirements La Mesa tunnel - 4 encroachment permits	- 10,000 LF easements - remainder in public right-of-way (4 encroachment permits
8.	HYDRAULICS - initial/ultimate hydraulics, length of connection piping to member agency facilities.	- prevent Alvarado pump back except need 9,000 ft pipeline from Water Filtration Plant to Navajo connection - 500 ft pipeline for Helix WD connection - 1,200 ft pipeline for San Diego #13 connection	- 1,800 ft pipeline for Helix WD connection - 500 ft pipeline for San Diego #13 connection - 800 ft pipeline for Otay WD #5 connection	- 1,800 ft pipeline for Helix WD connection - 500 ft pipeline for San Diego #13 connection - 800 ft pipeline for Otay WD #5 connection

as well as from comments received from the public, public agencies and citizen advisory committees. The eight criteria were assigned a total relative impact of 100. Environmental considerations were given a weight of 60; including 25 for community disruption, 20 for traffic, 10 for environmental resources, 2.5 for utility conflicts and 2.5 for geotechnical considerations. Classical engineering considerations were given a weight of 40; including 25 for capital costs, 5 for right-of-way and permits, 5 for hydraulics, 2.5 for utility conflicts and 2.5 for geotechnical considerations.

The alternatives were given an ordinal ranking for each criteria. The most favorable or highest ranking corresponds to the highest assigned number. In ranking of the environmental evaluation criteria, short term, mitigable impacts were ranked highest. Long term, mitigable impacts were ranked lower. Non-mitigable impacts were always ranked lowest and assigned a value of 1. For example, in Table 2 there are 3 alternative pipeline alignments, each alignment is given a ranking of 1 through 3 for the community disruption evaluation criteria. The Western alignment has the highest ranking of 3 (most preferred) as the impacts to land use are significant, but short term and mitigable. The Central alignment required the condemnation of 42 land parcels which contained 15 single family homes, 8 multi-family units and 2 businesses. Relocation of displaced persons and businesses and permanent loss from the area was a significant, long term land use impact, but was considered mitigable by fair compensation and relocation assistance. Therefore, the Central alignment was given a rank of 2. The Eastern alignment has the lowest ranking of 1 (least preferred) due to the significant impacts resulting from the longest construction time within arterial roads and non-mitigable impacts from landlocking of significant areas from freeway access. In cases where the alignments were essentially the same for a criteria, each alignment was given an equal ranking by averaging the combined total. For example, under hydraulics the connection to existing member agency facilities off the main pipeline was equivalent for the Central and Western alignments. Of the three alternatives, the Eastern alignment is inferior and was assigned a rank of 1, while the Central and Western alignments were considered equivalent and each received a rank of 2.5 (3+2/2).

The relative impact weight assigned to each

evaluation criteria was multiplied by the subjective ranking and then summed for all evaluation criteria to derive a total score for each pipeline alignment. The preferred pipeline alignment was determined by summation of the weighted rankings for each alternative alignment and the highest ranked route, the Western pipeline alignment, was the environmental and engineering preferred pipeline alignment. This analysis was the process utilized to present to the public and decision maker a methodology for selecting a preferred pipeline alignment. An example selection matrix is presented in Table 2.

<u>Summary</u>

The EIR document analyzed the alternative alignments in terms of potential environmental impacts, identifying non-mitigable impacts and required mitigation measures to reduce adverse impacts to less than significant along each alignment. The environmental evaluation criteria included the conclusive information presented in the EIR, as well as input from the concerned public, citizen groups and public agencies. The engineering evaluation criteria included institutional, social, economic and classical engineering concerns. The selection matrix integrated environmental and engineering criteria, and provided a method of subjectively identifying the relative impact of various issues. The summary evaluation criteria and selection matrix were included in the EIR summary, and as such, were not a part of the EIR, due to consideration of non-environmental issues. However, the inclusion of these facts provided the decision maker with comparative environmental and engineering information in a concise format enabling a more informed decision. The result of this evaluation and selection methodology was selection of the Western alternative as the preferred pipeline alignment. General public acceptance and environmental clearance was obtained by SDCWA and the project has proceeded to the final design stages. Construction is scheduled to begin for this portion of the pipeline by late summer of 1990.

Table 2

Pipeline Alignment Selection Matrix[a]

Environmental Evaluation Criteria	Relative Impact	Eastern Rank	Eastern Wt. Rank	Central Rank	Central Wt. Rank	Western Rank	Western Wt. Rank
1. Community Disruption	25	1	25	2	50	3	75
2. Traffic	20	1	20	3	60	2	40
3. Resources	10	1	10	3	30	2	20
4. Utilities[b]	5	1	5	3	15	2	10
5. Geotechnical[b]	5	2	10	1	5	3	15
Engineering Evaluation Criteria							
6. Capital Costs	25	2	50	1	25	3	75
7. ROW/Permits	5	2	10	1	5	3	15
8. Hydraulics	5	1	5	2.5	12.5	2.5	12.5
TOTALS	100		135.0		202.5		262.5[c]

a Ranking 1 through 3, with 3 being the environmental and engineering preferred ranking.

b Half (2.5) of the Relative Impact weight for Utilities and Geotechnical evaluation criteria considers environmental factors and half (2.5) considers engineering factors for a total Relative Impact of 5.

c The Western alignment has the highest weighted rank and is the preferred pipeline alignment.

Field Experiment on Behavior of Continuous Water Main with a Miter Bend

Nobuhisa Suzuki*, Hiroshi Shima* and Yoshiyuki Mohri**

Abstract

A field experiment was conducted to investigate behaviors of a buried continuous steel water main with a 90-degree miter bend subjected to internal pressure and ground settlement. Deformations of the pipe during backfilling are also discussed. A 812.8mm (32-inch) outside diameter, 7.0mm (0.276-inch) wall thickness and 50m (164 feet) long test pipe was used for the experiment. The ground settlement effect was created by dissolving a salt layer compacted beneath the test pipe. Displacements, deformations and strains induced in the test pipe were measured and the data were compared with a nonlinear finite element solution.

Introduction

Segmented buried pipelines require thrust blocks to resist two kinds of forces acting on a pipe bend which are an out-of-balance force induced by internal pressure and a centrifugal force due to momentum of water flow. The size of the thrust block mainly depends on the magnitude of internal pressure, water velocity and modulus of surrounding soil. The total weight of the thrust block can be occasionally heavy enough to cause nonuniform ground settlement which may result in significant damage to the buried segmented pipelines.

On the other hand, continuous pipelines absorb the forces generated at the pipe bend without the massive thrust block because of continuity of the pipeline in the longitudinal direction. Welded continuous pipelines can also endure unexpectedly large ground settlement due to the bending flexibility of the pipeline and the high

* Japan Association of Water Steel Pipe, Tokyo JAPAN
** Ministry of Agriculture, Forestry and Fisheries, Tokyo JAPAN

ductility of the material. The ductility is advantageous to endure large deformation of the pipe and prevent leakage of water.

The purpose in this paper is to demonstrate the safety of the continuous pipeline through investigating the behaviors induced by internal pressure and ground settlement. The field experiment was conducted using a 50m long welded test pipe with a 812.8mm outside diameter and a 7.0mm wall thickness. The experiment was conducted in a test field of the National Research Institute of Agricultural Engineering of Ministry of Agriculture, Forestry and Fisheries. Deformation and strains induced in the test pipe during backfilling, a hydraulic pressure test and a deformation test due to ground settlement were measured. These measurements are compared with the nonlinear finite element solution obtained by applying four node flat shell elements (See Figure 6).

Field Experiment

Set-Up of Experiment

A plan view of set-up of the field experiment is schematically illustrated in Figure 1. The test pipe consisted of two 25m long straight pipes and a four piece 90 degree miter bend. Two massive concrete anchorages were provided to constrain the longitudinal displacements at the both ends of the test pipe.

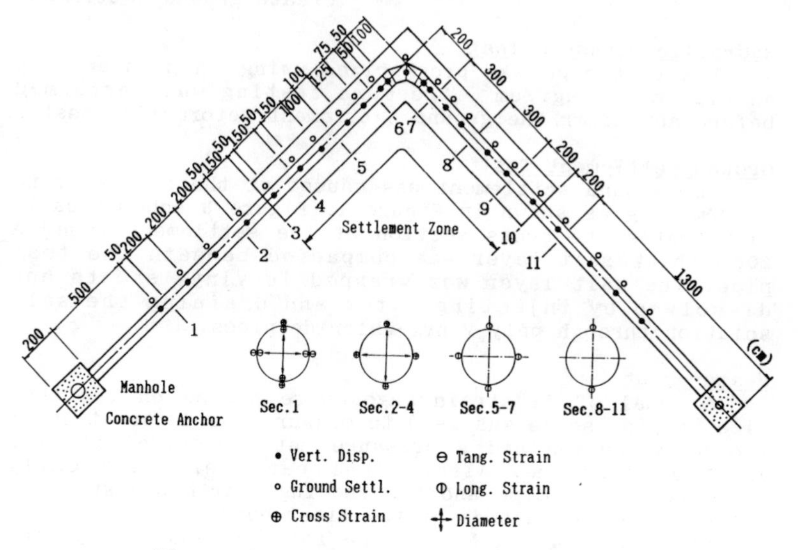

Figure 1. Plan View of Test Set-Up

The test pipe was placed in a trench excavated as shown in Figure 2. The pipe was covered by 1.2m of soil. The bedding of the trench was composed of two layers of sandy soil classified SW in the Unified classification system. Layers were 30cm and 20cm thick. The sides of the pipe and a 30cm thick layer above the pipe crown were filled with sandy soil classified as SM. The upper layers were filled with native material and compacted very carefully every 30cm.

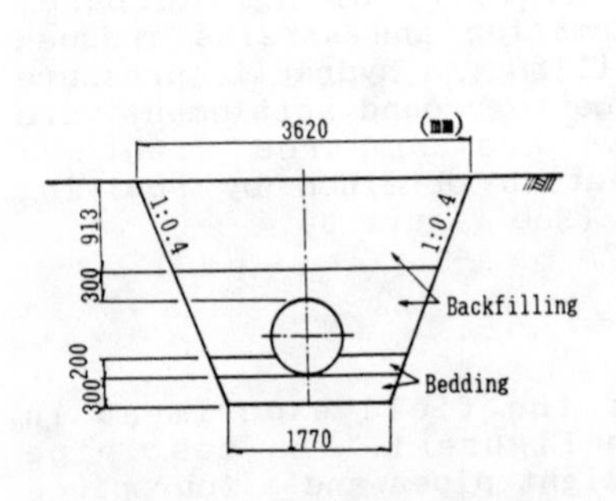

Figure 2. Test pipe and soil layers in trench

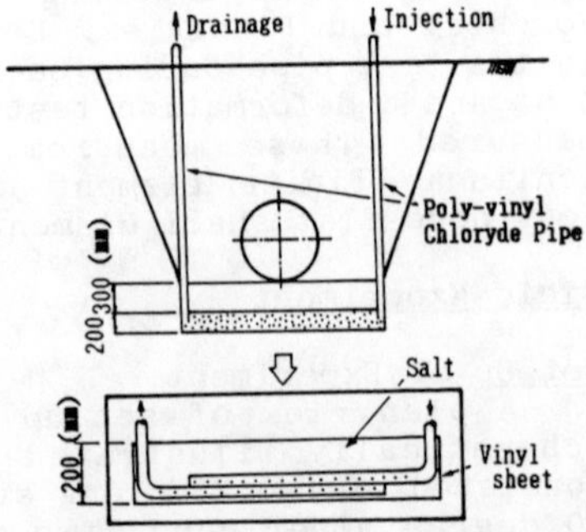

Figure 3. Salt layer to create ground settlement

Hydraulic Pressure Test

The test pipe was pressurized using a plunger pump as high as 24 kgf/cm^2. Pressure testing was performed before and after the ground settlement deformation test.

Ground Settlement Test

The ground settlement was caused at the central zone of 20m long as shown in Figure 1. Figure 3 schematically illustrates the cross section of the settlement zone. A 20cm thick salt layer was compacted beneath the test pipe. The salt layer was wrapped in vinyl sheets and dissolved by injecting water and draining the salt solution through poly-vinyl-chloride pipes.

Measurement

A total of 54 strain gages were placed on the test pipe, a pipe scale was used to measure the deformation of the pipe section and a pressure meter provided for the internal pressure. Also, steel bars were vertically attached to the pipe and the bedding. Earth pressure was measured on the outside of the miter bend.

Figures 1, 4 and 5 show the location and arrangement of the measuring instruments mentioned above.

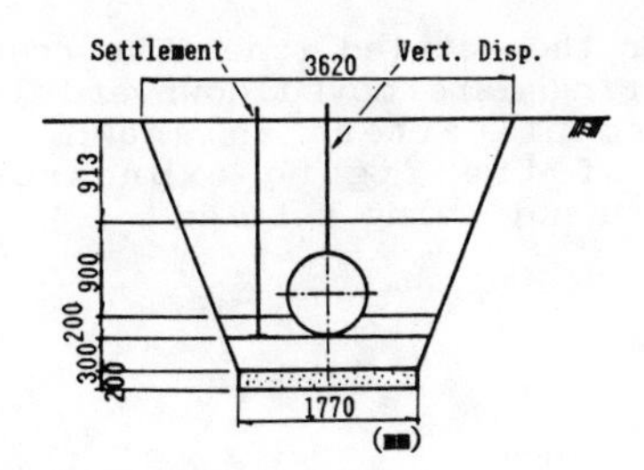

Figure 4. Steel bar for vertical displacement measurement

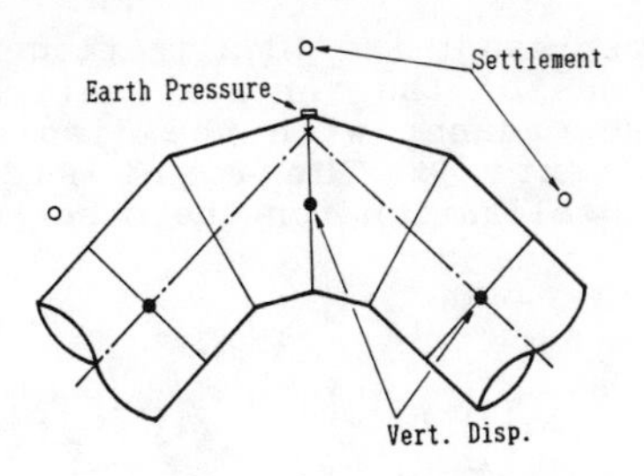

Figure 5. Measurement locations at miter bend

Idealization for Finite Element Analysis

Figure 6 shows mesh idealization applying four node flat shell elements which are connected to nonlinear soil springs to represent soil-pipe interaction. The nonlinear soil springs are defined as Figure 7 referring to the results of the Lateral Load Tester (LLT).

Figure 6. Modeling of pipe

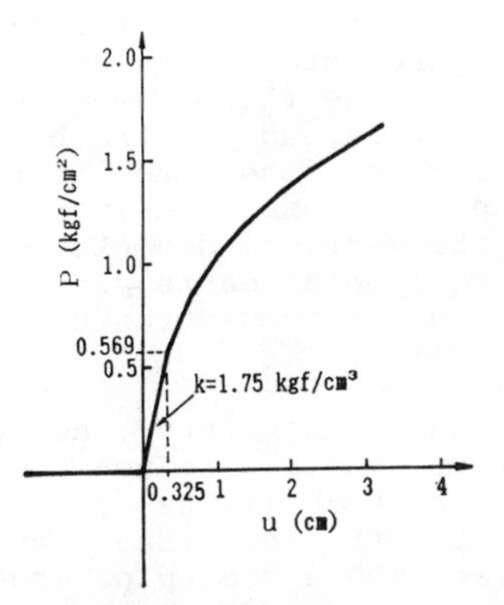

Figure 7. Soil spring

The internal pressure is idealized into concentrated loads to be applied to every node with the same incremental steps as the hydraulic pressure test.

The soil-pipe interaction during the deformation test is modelled as Figure 8. The left side of the figure represent the idealization for the settled zone. The free ends of the vertical soil springs are moved downward in accordance with the displacement pattern as shown in Figure 9. The right side of the figure expresses idealization for the other zone not being settled.

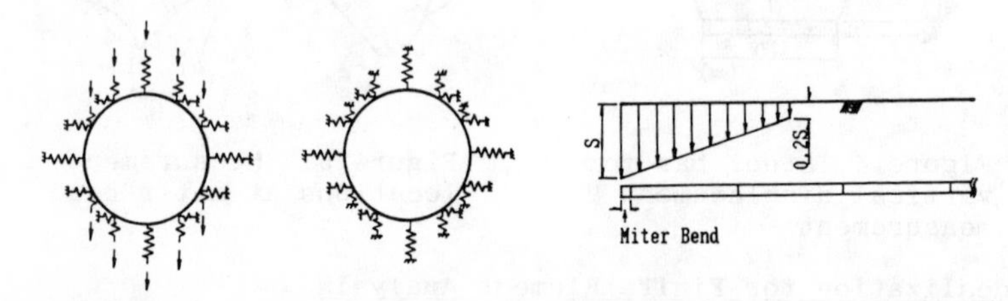

Figure 8. Pipe-soil interaction

Figure 9. Settlement

Deformation of the Pipe during Backfilling

Figure 10 plots changes of inside diameter of the pipe measured after completion of every layer of backfill. The data before completion of the third layer, which corresponds to the backfilling of the sides of the pipe, represent an increase in the vertical diameter due to horizontal earth pressure caused by compaction of the backfill.

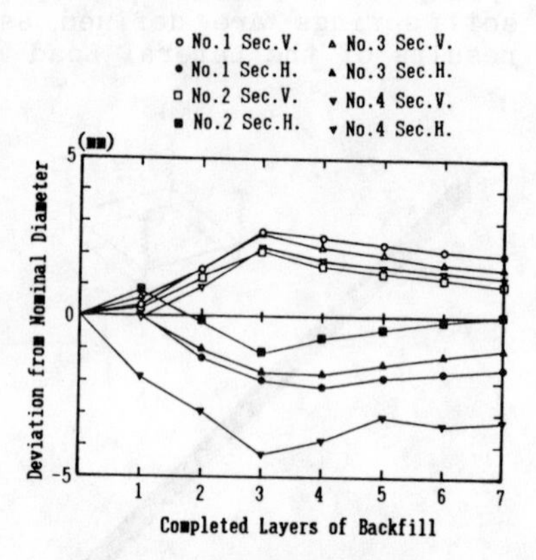

Figure 10. Variation of diameter of pipeline during backfilling

On the contrary, the subsequent data obtained during the backfilling between the fourth and the seventh layers above the pipe crown show increase in the horizontal diameter due to increase in vertical earth pressure. The cross section of the buried pipe, however, remained an ellipsoid with the vertical axis being longer after completion of the backfilling. This is because the compaction of the soil at the sides of the pipe were performed very carefully. Figure

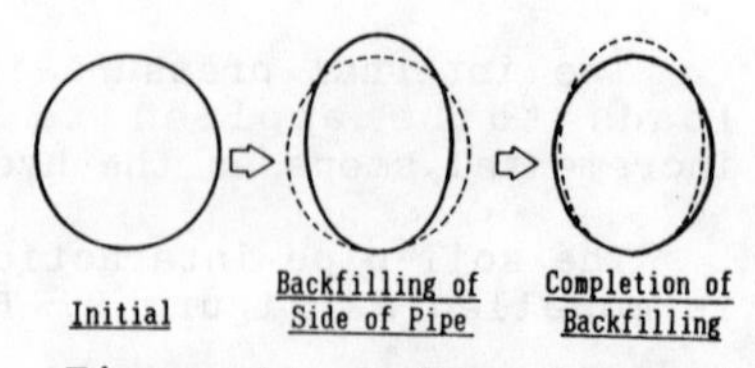

Figure 11. Deformation of pipe section during and after backfilling

11 shows these variations of the deformation of the pipe.

Behaviors during Pressure Test

Strains

Some data of the strains induced in the pipe during the pressure test are illustrated in Figures 12 and 13. All the strains shown in the figures are positive by the effect of the internal pressure. The longitudinal component of strains are mainly generated by the thrust force acting on the miter bend. The hoop strains increase linearly. The longitudinal strains represent a slight nonlinear tendency due to the frictional soil spring characteristic.

Figures 14 and 15 show comparisons of the measured strains with the nonlinear finite element solution at p=24 kgf/cm^2. Furthermore, elastic strains induced in a pressurized thin wall cylindrical vessel in air can be estimated from the elementary theory as 128x10^{-6} for the longitudinal and 652x10^{-6} for the circumferential direction. The average value of the measured longitudinal strains is approximately 80x10^{-6}, which is far small compare with the strains obtained from the elementary theory by the effects of the surrounding soil.

The nonlinear finite element solution with respect to the longitudinal direction does not coincide precisely with the measured strains, however, the patterns of the strain distributions are well predicted. On the other hand, the circumferential strains do not agree with the measured data because the strains induced in the pipe are apt to be affected by the initial imperfection of the cross section. In other words, the ellipsoid tends to deform into a circle.

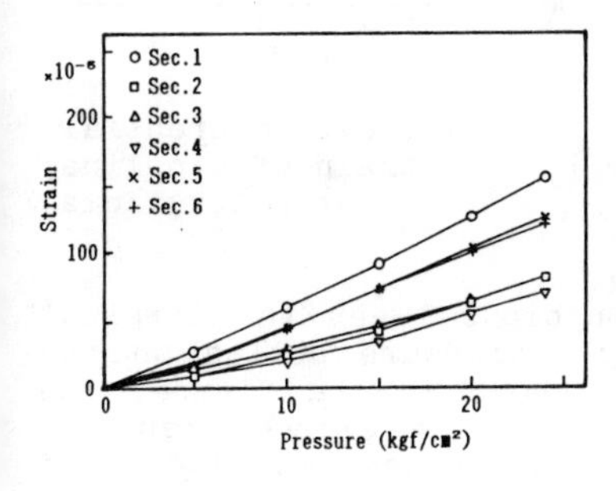

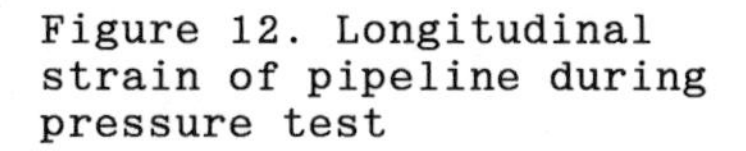
Figure 12. Longitudinal strain of pipeline during pressure test

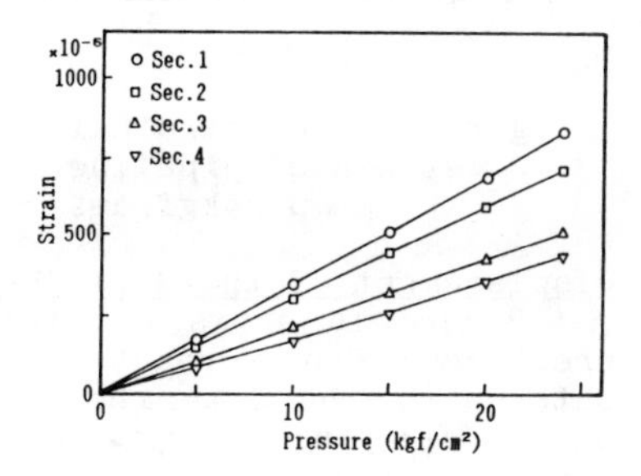

Figure 13. Circumferential strain of pipeline during pressure test

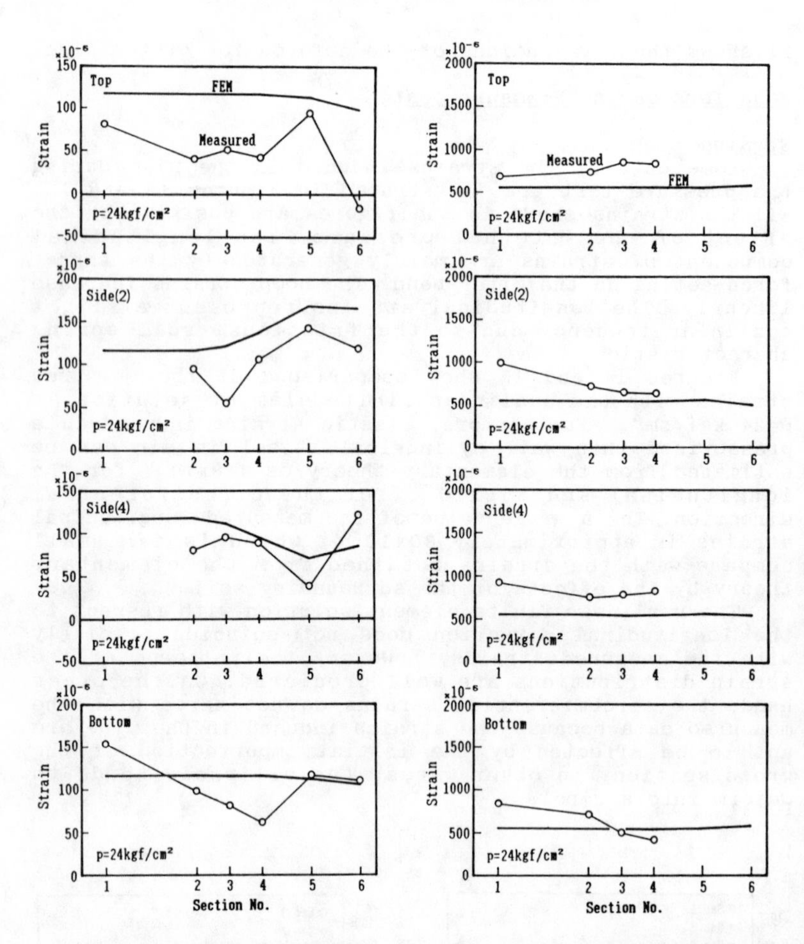

Figure 14. Longitudinal strain of pipeline at p=24kgf/cm2

Figure 15. Circumferential strain of pipeline at p=24kgf/cm2

Displacement of the Miter Bend

Figure 16 shows a relationship between the internal pressure and the reactive earth pressure acting on the outside of the miter bend. As shown in the figure, the earth pressure increases as the internal pressure increases which proves that the miter bend moved outward as illustrated in Figure 17 due to the force induced by the internal pressure.

The maximum displacement of the miter bend can be estimated to be as large as 2.6mm by integrating the measured longitudinal strains at p=24 kgf/cm^2 which are illustrated in Figure 14. The similar estimation can be made considering a horizontal displacement and a tilting angle of the bar scale placed vertically on the miter bend to measure the settlement. This procedure yields a 2.3mm horizontal displacement of the miter bend. Finite element method prediction for the displacement is 3.7mm.

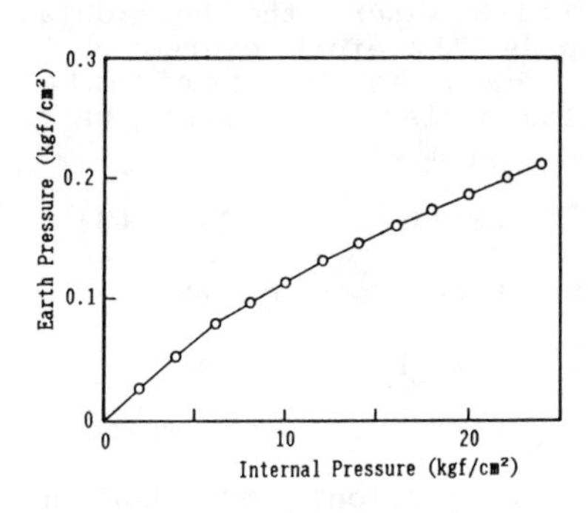

Figure 16. Earth pressure at outside of miterbend

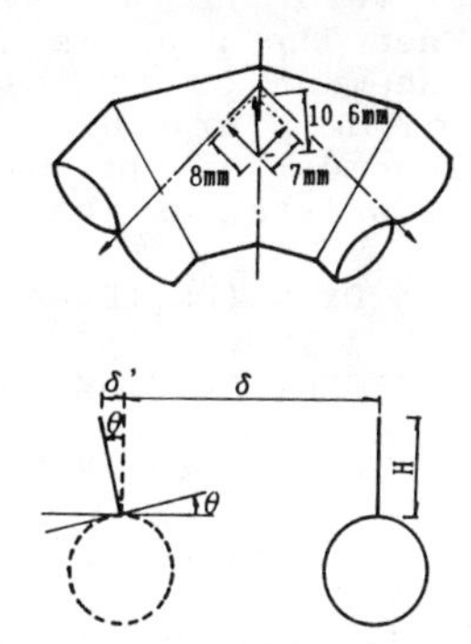

Figure 17. Displacement and deformation of miter bend

<u>Deformation Recovery of Pipe due to Internal Pressure</u>

The elliptical cross section resulting from backfilling has a longer vertical axis. The vertical axis is liable to reduce the diameter due to the internal pressure as shown in Figure 18.

The total external loads acting on the pipe wall can be simply discretized into three loading states as shown in Figure 19. States 1 and 2-1 represent the earth pressure during backfilling up to the pipe crown and completion of backfilling, respectively. The pipe is filled with water at State 2-2. The maximum horizontal

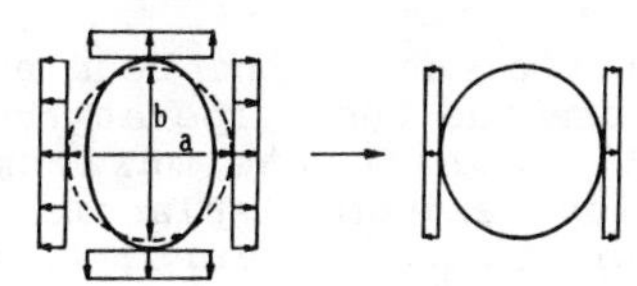

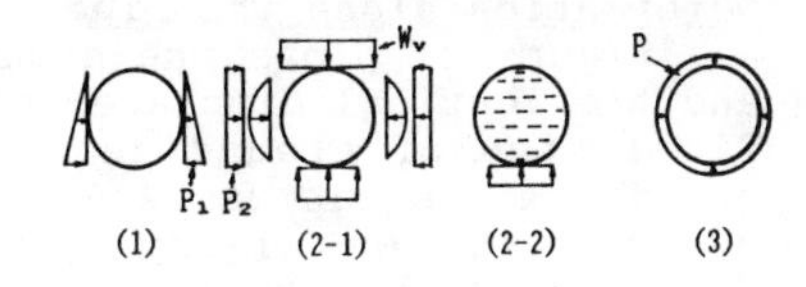

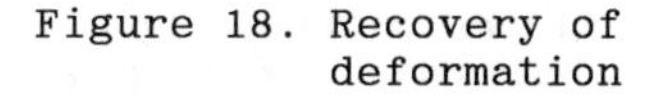

Figure 18. Recovery of deformation

Figure 19. Loading conditions

displacements Dx_1 and Dx_2 of the pipe can be expressed as the following equations [1,2 and 3].

$$Dx_1 = p^2R^4/(12EI) \quad (1)$$

$$Dx_2 = (2KxWvR^2-1/6p_1R^4+2KoWoR^5)/(EI+0.061E'R^3) \quad (2)$$

Where, Kx; a deformation coefficient due to vertical earth pressure (0.089), Wv; vertical earth pressure (γH), EI; bending rigidity of a pipe, R; radius of a pipe, p1; active earth pressure (0.333Wv), p_2; horizontal resistant earth pressure, Wo; unit weight of water, Ko; a weight factor (0.075).

When the pipe is filled with water, the horizontal displacement of the pipe can be therefore expressed by sum of Dx_1 and Dx_2. If we express the horizontal displacement of the pressurized buried pipe as Dx, which can be written as the following equation [4].

$$Dx = (Dx_1+Dx_2)/(1+pR^3/(12EI)) \quad (3)$$

Furthermore, bending moment M can be expressed as,

$$M = C_1p_1R^2+K_1WvR^2+KrE'RDx_2+C_2p_2R^2+C_1pRDx \quad (4)$$

where, $E'=((2KxWvR^4-1/6p_1R^4)/Dx_2-EI)/0.061R^3$, which is the soil modulus in Spangler's equation. And resultant stress s can be written as,

$$s = M/Z + pR/t \quad (5)$$

where, Z; a section modulus expressed by $t^2/6$.

Figure 20 compares the measured and the calculated circumferencial strains which can be defined by equations (1) through (5). The measured data do not agree precisely with the calculated data. However, the calculated data are very effective for prediction of distribution of the circumferential strains including the recovery of the buried circular pipe section due to the internal pressure.

Behaviors during Ground Settlement

Deformation along the Pipe

Figure 21 plots the ground settlement at the bedding and the vertical displacements along the test pipe at the final step of ground settlement in which the maximum ground settlement was 200mm and the maximum displacement of the pipe was 141mm. As shown in the figure, the distribution of the ground settlement represent a little asymmetric pattern and different from the expected configuration because of plugging of the dissolved salt drain pipes. However, the displacement of the pipe appeared symmetrical about the miter bend placed in the center of the test pipe.

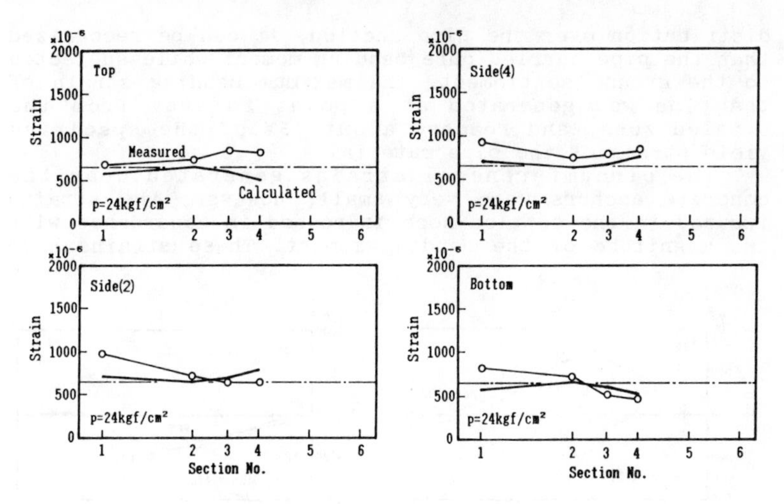

Figure 20. Internal-pressure-induced circ. strains

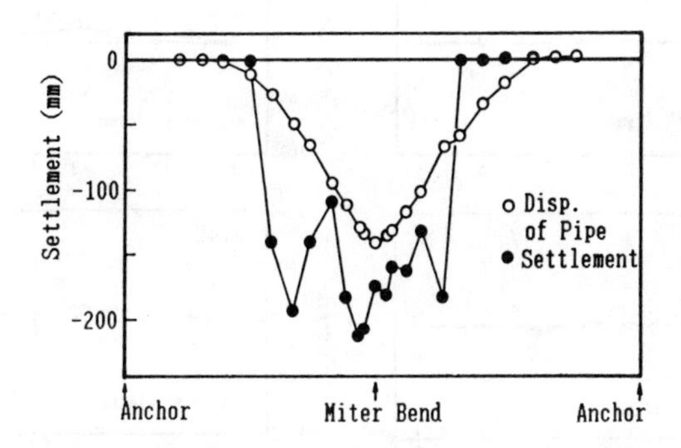

Figure 21. Displacement of pipe and ground settlement

Strains

Figures 22 and 23 compare the measured with the calculated strains at the maximum vertical displacement of the pipe reached 141mm. These data represent a good coincidence each other as shown in the figures.

The longitudinal strains generated at the sides of the pipe are infinitesimally small and the absolute values at the top and the bottom of the pipe are approximately the same value. As far as the strain distribution over the pipe section, it can be recognized that the pipe carried pure bending moment while subjected to the ground settlement. The maximum bending strain of the pipe was generated at a point 2m away from the settled zone, and reached about 75% of the specified yield strain of the pipe material.

The circumferencial strains generated near the concrete anchors were very small, however, the strains induced in the settled zone increased in accordance with the magnitude of the bending moment. These strains have

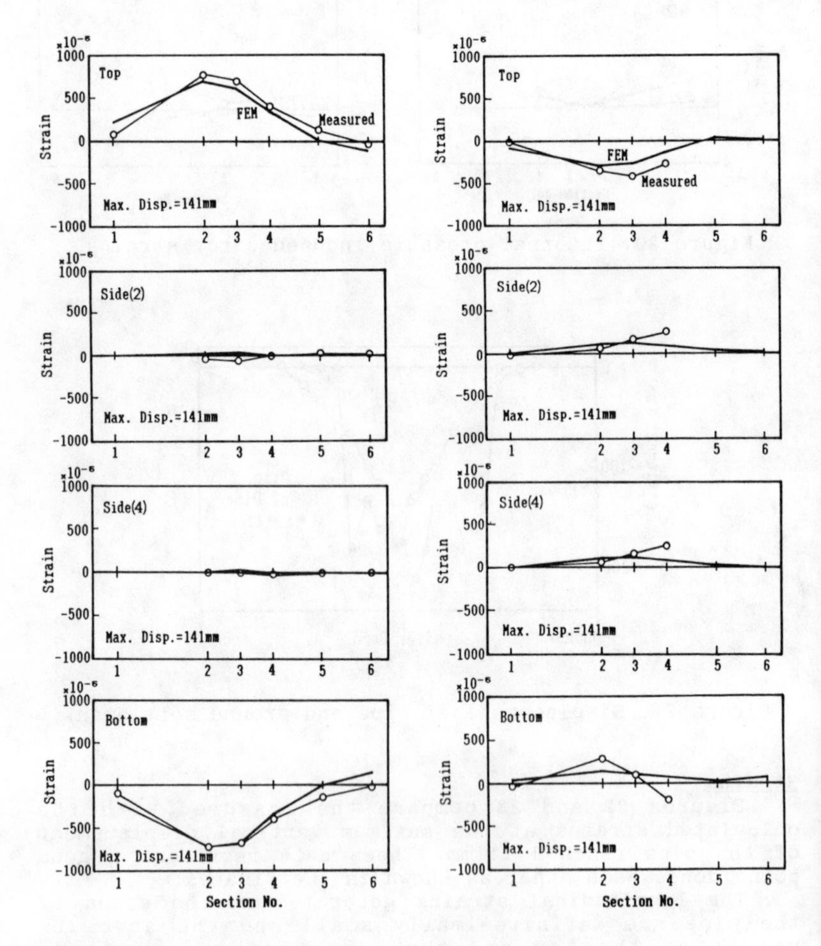

Figure 22. Longitudinal strain by settlement

Figure 23. Circumferential strain by settlement

very close relationship to the sectional deformation of the pipe to be discussed in the subsequent section.

Sectional Deformation of the Pipe

Figure 24 plots the changes in diameters during the ground settlement. Observing the figure, the vertical diameters decrease and the horizontal diameters increase as the increase in the ground settlement and we can recognize larger deformations near the edge of the settled zone.

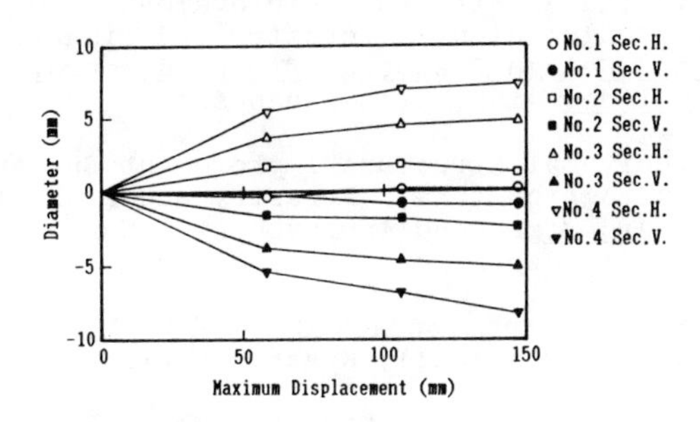

Figure 24. Variation of diameters due to settlement

Conclusions

From the experimental and the analytical results of the hydraulic pressure test and the deformation test due to the ground settlement, we can obtain the following conclusions.

(1) Buried continuous steel pipelines are advantageous to minimize the deformation caused by the internal pressure induced thrust force acting on pipe bends without the massive thrust block due to the continuity of the pipe material.

(2) The elliptical cross section of the pipe generated during backfilling can be rerounded by the internal pressure, which also results in reduction of the bending stress induced in the pipe wall during backfilling.

(3) Welded continuous pipelines are also advantageous to endure ground settlement due to the high ductility of the pipe material and the flexibility of the pipelines.

Acknowledgements

The work presented herein was supported by the Japan Association of Water Steel Pipe in Tokyo. This support is gratefully acknowledged. The authors also acknowledged that the field experiment was conducted in the test field of the National Research Institute of Agricultural Engineering of Ministry of Agriculture, Forestry and Fisheries.

References

[1] Agricultural Structure Improvement Bureau of Ministry of Agriculture, Forestry and Fisheries, : Planning and Design Standard for Land Improvement Project, -Pipeline-, 1977. (in Japanese)

[2] Agricultural Structure Improvement Bureau of Ministry of Agriculture, Forestry and Fisheries, : Design Standard for Land Improvement (Manual of Pipeline), 1983. (in Japanese)

[3] Roark,R.J. and Young,W.C., : Formulas for Stress and Strain, McGraw-Hill, Kogakusha LTD., p.216-238, 1975.

[4] Nakano,R., : 'Structural Design of Pipeline,'Textbook of Japanese Society of Irrigation, Drainage and Reclamation Engineering, p.65-115, 1978. (in Japanese)

SUBJECT INDEX

Page number refers to first page of paper.

AUTHOR INDEX

Page number refers to first page of paper.